# Multiscale Multibody Dynamics

Jielong Wang

# Multiscale Multibody Dynamics

## Motion Formalism Implementation

Springer

Jielong Wang
COMAC Beijing Aircraft Technology
Research Institute
Beijing, China

ISBN 978-981-19-8440-2 ISBN 978-981-19-8441-9 (eBook)
https://doi.org/10.1007/978-981-19-8441-9



This Springer imprint is published by the registered company Springer Nature Singapore Pte Ltd.
The registered company address is: 152 Beach Road, #21-01/04 Gateway East, Singapore 189721, Singapore

*To my wife, Jianqin Yang, and my family*

# Preface

Theory of multiscale multibody dynamics is named in this book at the first time. As a branch of solid mechanics, multibody dynamics studies the motion, vibration and deformation of a multibody system, which is consisted of flexible body, rigid body and mechanical joints. The theory of solid mechanics formulates the strong fundamentals of multibody dynamics, while with the developing of multibody dynamics, few people have paid attentions to the fundamentals of solid mechanics behind the multibody dynamics.

This book starts from the theory of Cosserat continuum to address in detail the new multiscale theory of multibody system. The unified formulas of governing equations of deformation and motion for generalized Cosserat continua, shells and beams are obtained only with the application of geometric dimension reduction. The manipulations of dimension reduction also formulate the fundamentals of the well-known assumption of scale separation which has been widely applied in the multiscale mechanics. Furthermore, if ignored the directors of material particles or ignored the curvature strains equivalently, the theories of general Cauchy continua, membranes and cables can be readily obtained from the unified descriptions of Cosserat theories. Finally, when all the deformations of Cosserat continuum are ignored, the generalized Cosserat continua will be simplified to rigid bodies. In this sense, the general Cosserat continuum theory affords the unified descriptions for three-dimensional Cosserat continuum, shell, beam, rigid body, three-dimensional Cauchy continuum, membrane and cable, all of them.

By considering the deformations of material directors under the assumption of scale separation, original Cosserat continuum can be decoupled into a global macroscopic model and a local microscopic model. Based on the kinematic description of macro–micro decoupling, the multiscale theories of Cosserat continuum stemming from the unified formula mentioned above are then derived, which is the most complicated theoretical part in this book. The multiscale theories make it possible to realize the fine analysis of multibody system in the level of microscale at the speed of global analysis for a macroscopic model.

Another contribution of the book is the motion formalism descriptions of Cosserat continua, shells, beams and rigid bodies, together with the joints. Since the motion of multibody system contains the translation and rotation and spans a six-dimensional space, it is more efficient to describe the motion by using the dual vector or motion tensor directly instead of defining two types of three-dimensional variables, the translation and rotation, separately. Dealing with motion and deformation by using dual vector and motion tensor verifies to be an efficient and elegant manner, which makes the book more readable. For six low pair of joints, rigid connection and rigid rotation, their recursive formulas also benefit from the motion formalism due to the fact that the unified descriptions of relative motion by using motion tensor simplify the complexity of the problem.

Once the analytical models of Cosserat continua were given, the finite element method can be applied to discretize the governing equations of motion into a set of ordinary differential equations. In the third part of the book, the implicit 2- and 3-stage Radau IIA algorithms featuring the third- and fifth-order accuracy have been well designed to predict the time history of dynamic responses for a multibody system by directly integrating the differential algebraic equations of index-3. Note that the ordinary differential equations can be treated as the special case of unconstrained differential algebraic equations. The finite element implementations of Cosserat continua with the dimensions varying from zero to two and the joints have been written into the computational code AeLas to enrich its structural modeling capabilities. The 2- and 3-stage Radau IIA algorithms are also plugged in AeLas as the core solvers. The in-house code AeLas was developed to simulate the aeroelastic phenomena that frequently occurred in the flight of civil aircraft. Now it services as a routinely used tool. The book briefly introduces two specific cases of numerical simulations completed by AeLas, one is the static aeroelastic analysis of aircraft wind tunnel model and the other the prediction of buffeting response, to validate the Cosserat theories developed in the book.

Except the concept and scope of the book briefly introduced above, the purpose of writing the book originates from my work experience. Now looking back, the idea of developing the analytical model of multiscale multibody system and the related computer codes appeared in 2013, at then I had been working in COMAC for nearly two years. The words frequently talked about by the workmates and partners around me, such as "Let's do some real research", "... do some real designs", "... do some real products", showed the enthusiastic attitude to the work. "Firing before aiming" revealed the eagerness of these guys for the achievements which can be quickly reached. However, the two-year cooperation with domestic universities promoted little to the aeroelasticity analysis and design techniques of civil aircraft. The bad cooperation experience pushed me to make the decision developing the aeroelasticity analysis tool for the design of civil aircraft by myself after deeply considering. At that time, providing analysis tools and technical solutions for the design of specific models of civil aircraft fitted the developing strategy of the company. But it was out of my expectation that this job continues till now. Everything went well at the beginning. I quickly accomplished the finite element modeling of beam, rigid body, cable, multipoint constraints, all six lower pair joints with clearance, contact and friction, etc. With the implementation of Radau IIA algorithm, the finite element code AeLas can simulate most parts of structural components in the early design stage of civil aircraft. Unfortunately, I encountered a challenge when developing shell/plate element in 2018. The problem is not the analysis model of the shell and its finite element implementation itself, but the multiple connections of shell elements along the common edges when they are not coplanar. Because the sixth degree of freedom of shell node, called "drilling rotation", is immaterial, usually abandoned to obtain a minimal set of five degrees of freedom. This minimization operation takes trouble when assembling shell elements into the global ones because the normal vectors could have different directions even on the common nodes of different shell elements. The additional manipulations are required to transpose the elemental data into the same local frame. Although the developed shell features good accuracy, it behaves in a low efficiency when modeling the complicated thin-walled structures of aircraft. Therefore, I had to stop structural element modeling here and converted to the numerical simulation of computational fluid dynamics as well as designing the data interface between structure and flow.

When I talked with my workmates at lunch time about possible approaches to deal with the multiple connections of shell elements efficiently, a workmate, who received a Ph.D. degree from UK, and also engaged in finite element analysis, learned about the difficulties I faced. In view of the heavy burden of the work, he tried to stop me and straight forward pointed out writing an inner code in that situation was a dead way even the work sounded very meaningful. I do appreciate his suggestions; however, I must insist on the work because I believe it was worthy to accomplish with no regrets. It was my luck that I quickly found solutions from the theory of Cosserat continuum. First, the analytical model of shell can be described by the unified formula in an efficient manner. Second, the material properties with physical significance for the "drilling rotation" can be provided through weighted integral on the characteristic volume of the micro-Cosserat model, as already presented in the book. For the shell elements with six degrees of freedom at each node, their multiple connections along the common edges can be readily implemented and are not big issues.

The motion formalism in the book benefits from Prof. Bauchau of the University of Maryland, also my graduate adviser. After graduated from Georgia Tech., I had chances to have lunch with my advisor and asked for valuable advices on the technique problems I encountered in my work. In 2015, he invited me to participate in the research project as to motion formalism he mastered. In view of my situations at that time, I had to give up the opportunity that I can corporate with my adviser again. However, I kept my eyes on the rich publications of my adviser and his coworker about the motion formalism. Until 2020, I read one of his latest published papers about the spectral formulation for geometrically exact beams based on a motion interpolation. In the paper, the efficient and elegant way of using motion tensors to deal with the motion of beams made me decide to describe the motion of the Cosserat continuum in this book with motion formalism. In some extents, the implementation of motion formalism in the book can be treated as the extension of motion formalism from beams and shells to three-dimensional solids.

Obviously, the continuous works and accumulations since 2013 nourished the appearance of the book in its prototype. If the contents of the book mentioned above can provide any references to the peers who work in this field or my experience preparing the book can hint the researchers interested in the topics discussed in the book, it could be considered that the original purpose of writing the book has been realized.

Finally, my loving wife, Jianqin, has stood by me all these hard times of this work. Her caring and encouraging enabled me to do my best to complete the writing of the book.

Beijing, China
September 2022

Jielong Wang

# Contents

# Part I
# Preliminary of Motion and Deformation

# Chapter 1
# Vector and Tensor

The purpose of this chapter is to present part of basic properties of vector and tensor operations, which is frequently used in the rest of the book. Usually, vector and tensor are denoted by using the abstract bold-faced symbols, often make the reader confused. For the easy recognition, this chapter denotes vectors using an underline, such as $\underline{v}$, and denotes second-order tensor by the capitals, i.e., $R$ indicates a $3 \times 3$ rotation tensor or $M$ a $n \times n$ mass matrix. A special notation, a single over bar, like $\bar{n}$, presents a frequently used unit vector. In addition to the above denoting rules, the dual vectors and motion tensors within the framework of dual algebra are represented by mathematical calligraphy letters. Instead of following the Einstein summation convention, the first- and second-order tensor operations in this book are depicted by linear algebraic operation symbols of row array, column array and two-dimensional matrix, making these operations easier to understand.

## 1.1 Vector

The quantities which have not only magnitude but also direction are called vectors [1, 2] if they obey a certain law of addition. Given a vector $\underline{v}$, its magnitude will be denoted by either a scalar $v$ or notation $\|\underline{v}\|$ and then

$$\|\underline{v}\| = v \tag{1.1}$$

Specially, a null vector $\underline{v}$ is a vector with zero magnitude, thus $\underline{v} = 0$ if $v = 0$; a unit vector is a vector of unit magnitude, which can be constructed from a vector $\underline{v}$ by

The original version of this chapter has been revised. In this chapter, Fig. 1.10 was replaced. The correction to this chapter can be found at https://doi.org/10.1007/978-981-19-8441-9_7



J. Wang, *Multiscale Multibody Dynamics*,
https://doi.org/10.1007/978-981-19-8441-9_1

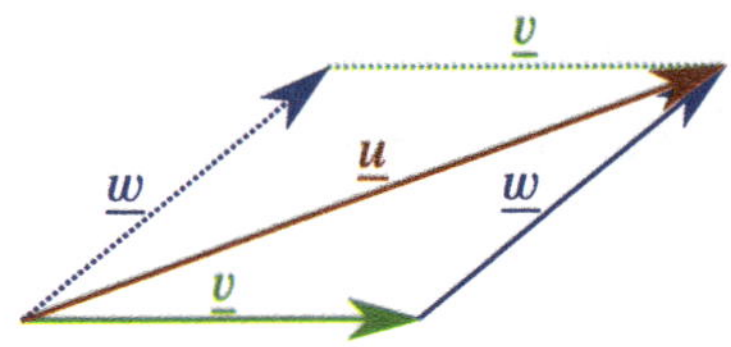


**Fig. 1.1** Law of vector addition $\underline{u} = \underline{v} + \underline{w}$

dividing it by its magnitude

$$\bar{n} = \frac{\underline{v}}{\|\underline{v}\|} \tag{1.2}$$

and hence $n = \|\bar{n}\| = 1$.

### 1.1.1 Law of Vector Addition

As depicted in Fig. 1.1, the certain law of addition that vectors obey is called the law of vector addition. Let $\underline{v}$ and $\underline{w}$ be the two vectors, the vector sum $\underline{u}$ follows the law of addition and is also commutative

$$\underline{u} = \underline{v} + \underline{w} = \underline{w} + \underline{v} \tag{1.3}$$

### 1.1.2 Scalar Multiplication

Figure 1.2 gives a vector $\underline{v}$ with the length of $v$, the multiplication of vector $\underline{v}$ by a scalar $s$ is denoted as

$$\underline{u} = s\,\underline{v} \tag{1.4}$$

By definition, if scalar $s$ is positive, vector $\underline{u}$ points in the same direction as vector $\underline{v}$, and of length $u = s\,v$. If scalar $s$ is negative, vector $\underline{u}$ points in the direction opposite to vector $\underline{v}$, and of length $u = |s|v$.

### 1.1.3 Cartesian Basis

A Cartesian basis [3] is defined in Fig. 1.3 by a set of three rectangular Cartesian coordinates, $x_1$, $x_2$, and $x_3$, or equivalently, by a set of three mutually orthogonal unit vectors, $\bar{\imath}_1, \bar{\imath}_2$, and $\bar{\imath}_3$. It is convenient to introduce the notation, $\mathcal{I}(\bar{\imath}_1, \bar{\imath}_2, \bar{\imath}_3)$, to denote a Cartesian basis. By construction, one of three unit vectors points in the direction of each of the three positive axes. Accordingly, the components of unit vectors, $\bar{\imath}_1$,

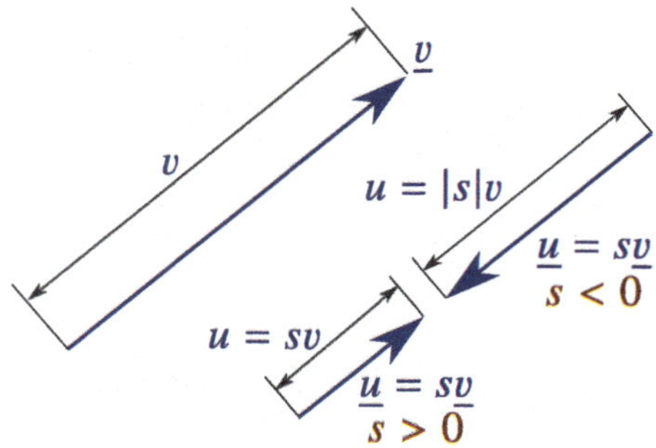


**Fig. 1.2** Multiplication of vector $\underline{v}$ by a scalar $s$

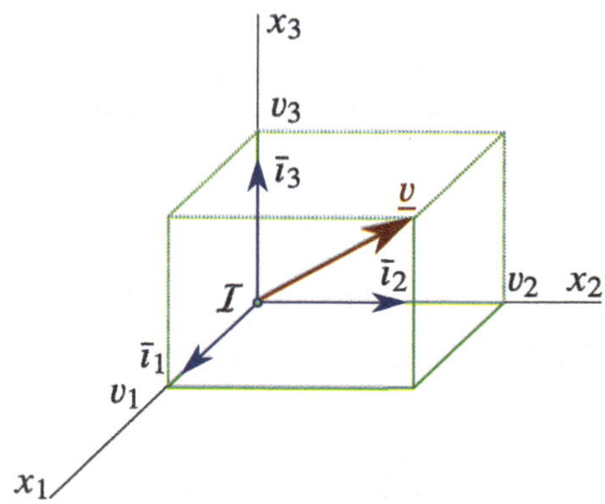


**Fig. 1.3** A Cartesian basis $\mathcal{I}(\bar{\imath}_1, \bar{\imath}_2, \bar{\imath}_3)$

$\bar{\imath}_2$, and $\bar{\imath}_3$, resolved in basis $\mathcal{I}$ are

$$\bar{\imath}_1 = \begin{vmatrix} 1 \\ 0 \\ 0 \end{vmatrix}, \quad \bar{\imath}_2 = \begin{vmatrix} 0 \\ 1 \\ 0 \end{vmatrix}, \quad \bar{\imath}_3 = \begin{vmatrix} 0 \\ 0 \\ 1 \end{vmatrix} \tag{1.5}$$

The identity matrix $I$ of size $3 \times 3$ is readily obtained by assembling the unit vectors, $\bar{\imath}_1$, $\bar{\imath}_2$, and $\bar{\imath}_3$ in order like

$$I = \lfloor \bar{\imath}_1,\ \bar{\imath}_2,\ \bar{\imath}_3 \rfloor = \begin{bmatrix} 1\,0\,0 \\ 0\,1\,0 \\ 0\,0\,1 \end{bmatrix} \tag{1.6}$$

In a convenient manner, the Cartesian basis, a two-dimensional identity matrix of size $3 \times 3$, can be expressed as

$$\mathcal{I}(\bar{\imath}_1, \bar{\imath}_2, \bar{\imath}_3) = I = \lfloor \bar{\imath}_1,\ \bar{\imath}_2,\ \bar{\imath}_3 \rfloor \tag{1.7}$$

As shown in Fig. 1.3, a vector $\underline{v}$ has orthogonal projections in the basis $\mathcal{I}$, which a linear function of the unit vectors

$$\underline{v} = v_1\bar{\imath}_1 + v_2\bar{\imath}_2 + v_3\bar{\imath}_3 \tag{1.8}$$

In view of Eq. (1.5), it yields

$$\underline{v} = v_1 \left|\begin{matrix} 1 \\ 0 \\ 0 \end{matrix}\right| + v_2 \left|\begin{matrix} 0 \\ 1 \\ 0 \end{matrix}\right| + v_3 \left|\begin{matrix} 0 \\ 0 \\ 1 \end{matrix}\right| = \left|\begin{matrix} v_1 \\ v_2 \\ v_3 \end{matrix}\right| \tag{1.9}$$

The transpose of vector $\underline{v}$ is a row array and is denoted with a superscript, $(\cdot)^T$, like

$$\underline{v}^T = \lfloor v_1,\ v_2,\ v_3 \rfloor \tag{1.10}$$

From Fig. 1.3, vector $\underline{v}$ is the diagonal of a rectangular parallelepiped whose edges have lengths $v_1$, $v_2$, and $v_3$. Hence its magnitude $v$ is given by the relation

$$v = \sqrt{v_1^2 + v_2^2 + v_3^2} \tag{1.11}$$

### *1.1.4 The Scalar Product*

Figure 1.4 defines an angle $\phi$ between two vectors $\underline{u}$ and $\underline{v}$, and $\phi$ is constrained to be the smallest nonnegative angle such that $0 \le \phi \le \pi$. The scalar product of vectors $\underline{u}$ and $\underline{v}$ is a scalar

$$\underline{u}^T \underline{v} = u\, v \cos\phi \tag{1.12}$$

and is commutative also

$$\underline{u}^T \underline{v} = \underline{v}^T \underline{u} \tag{1.13}$$

Furthermore, the scalar product is a distributive operation

$$\underline{u}^T (\underline{v} + \underline{w}) = \underline{u}^T \underline{v} + \underline{u}^T \underline{w} \tag{1.14}$$

The scalar product of a vector $\underline{v}$ by itself yields the square of its magnitude

$$\underline{v}^T \underline{v} = v^2 = \|\underline{v}\|^2 \tag{1.15}$$

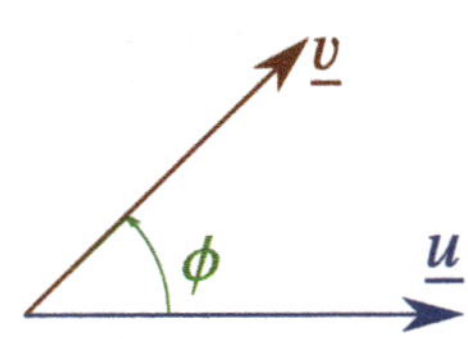


**Fig. 1.4** Angle between two vectors $\underline{u}$ and $\underline{v}$

If vectors $\underline{u}$ and $\underline{v}$ are orthogonal, then $\phi = \pi/2$ and $\underline{u}^T\underline{v} = 0$. However, if statement $\underline{u}^T\underline{v} = 0$ is given, it implies that either $\underline{u} = 0$ or $\underline{v} = 0$, or $\underline{u}$ is orthogonal to $\underline{v}$.

When constructing the Cartesian basis $\mathcal{I}$ in Sect. 1.1.3, three mutually orthogonal unit vectors, $\bar{\imath}_1$, $\bar{\imath}_2$, and $\bar{\imath}_3$, are introduced. Hence, the Cartesian bases are usually called *Orthogonal bases* with the following relations

$$\begin{array}{lll} \bar{\imath}_1^T\bar{\imath}_1 = 1, & \bar{\imath}_1^T\bar{\imath}_2 = 0, & \bar{\imath}_1^T\bar{\imath}_3 = 0 \\ \bar{\imath}_2^T\bar{\imath}_1 = 0, & \bar{\imath}_2^T\bar{\imath}_2 = 1, & \bar{\imath}_2^T\bar{\imath}_3 = 0 \\ \bar{\imath}_3^T\bar{\imath}_1 = 0, & \bar{\imath}_3^T\bar{\imath}_2 = 0, & \bar{\imath}_3^T\bar{\imath}_3 = 1 \end{array} \tag{1.16}$$

Using above properties of a Cartesian basis, Eq. (1.16), the decomposition of a vector, Eq. (1.8), can be represented as

$$\underline{v} = \underline{v}^T\bar{\imath}_1\bar{\imath}_1 + \underline{v}^T\bar{\imath}_2\bar{\imath}_2 + \underline{v}^T\bar{\imath}_3\bar{\imath}_3 \tag{1.17}$$

where the scalar products

$$\underline{v}^T\bar{\imath}_1 = v_1, \quad \underline{v}^T\bar{\imath}_2 = v_2, \quad \underline{v}^T\bar{\imath}_3 = v_3 \tag{1.18}$$

are exactly three components of vector $\underline{v}$, also the projections of vector $\underline{v}$ to each of three mutually orthogonal unit vectors. Furthermore, the scalar product of two vectors $\underline{u}$ and $\underline{v}$ can be calculated from their components

$$\underline{u}^T\underline{v} = u_1v_1 + u_2v_2 + u_3v_3 \tag{1.19}$$

### *1.1.5 The Vector Product*

The vector product of two vectors $\underline{u}$ and $\underline{v}$ is a third vector $\underline{w}$, defined as

$$\underline{w} = \widetilde{u}\,\underline{v} = uv\sin\phi\,\bar{n} \tag{1.20}$$

where $\bar{n}$ is a unit vector perpendicular to both $\underline{u}$ and $\underline{v}$. Its orientation is determined according to the right-hand rule, as shown in Fig. 1.5. The magnitude of vector $\underline{w}$ is $uv\sin\phi$ which is equal to the area of the parallelogram spanned by vectors $\underline{u}$ and $\underline{v}$. According the definition, the following properties could be obtained readily, which stats the vector product is anti-commutative

$$\widetilde{u}\,\underline{v} = -\widetilde{v}\,\underline{u} \tag{1.21}$$

and distributive

$$\widetilde{u}(\underline{v} + \underline{w}) = \widetilde{u}\,\underline{v} + \widetilde{u}\,\underline{w} \tag{1.22}$$

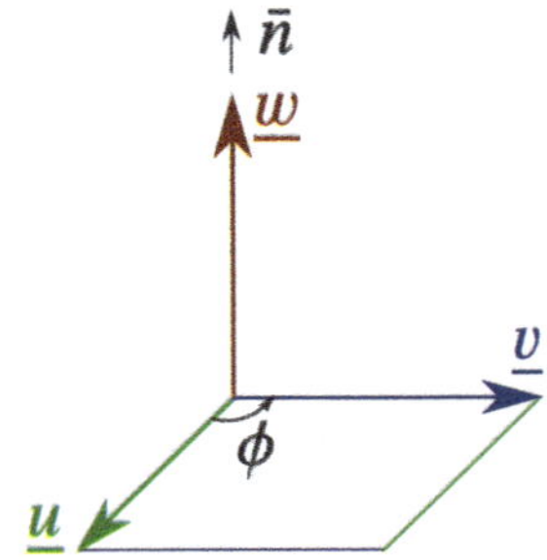


**Fig. 1.5** The vector product of vectors $\underline{u}$ and $\underline{v}$

Furthermore, if vectors $\underline{u}$ and $\underline{v}$ are parallel, then $\phi = 0$ and a null vector appears, $\widetilde{u}\,\underline{v} = 0$. On the contrary, if the vector product $\widetilde{u}\,\underline{v} = 0$ is given, it implies that either $\underline{u} = 0$ or $\underline{v} = 0$, or $\underline{u}$ is parallel to $\underline{v}$.

Considering the vector products of unit vectors, $\bar{\imath}_1$, $\bar{\imath}_2$, and $\bar{\imath}_3$ of a Cartesian basis $\mathcal{I}(\bar{\imath}_1, \bar{\imath}_2, \bar{\imath}_3)$. They are easily obtained from Eq. (1.20) like

$$\begin{array}{lll} \widetilde{\imath}_1\,\bar{\imath}_1 = 0, & \widetilde{\imath}_1\,\bar{\imath}_2 = \bar{\imath}_3, & \widetilde{\imath}_1\,\bar{\imath}_3 = -\bar{\imath}_2 \\ \widetilde{\imath}_2\,\bar{\imath}_1 = -\bar{\imath}_3, & \widetilde{\imath}_2\,\bar{\imath}_2 = 0, & \widetilde{\imath}_2\,\bar{\imath}_3 = \bar{\imath}_1 \\ \widetilde{\imath}_3\,\bar{\imath}_1 = \bar{\imath}_2, & \widetilde{\imath}_3\,\bar{\imath}_2 = -\bar{\imath}_1, & \widetilde{\imath}_3\,\bar{\imath}_3 = 0 \end{array} \tag{1.23}$$

In view of Eq. (1.20), the components of vector product $\underline{w}$ in terms of the components of $\underline{u}$ and $\underline{v}$ can be determined by the conditions that $\underline{w}$ is perpendicular to both $\underline{u}$ and $\underline{v}$, i.e. $\underline{w}^T\,\underline{u} = 0$ and $\underline{w}^T\,\underline{v} = 0$. By Eq. (1.19), these relations can take the form

$$\begin{aligned} w_1u_1 + w_2u_2 + w_3u_3 &= 0 \\ w_1v_1 + w_2v_2 + w_3v_3 &= 0 \end{aligned} \tag{1.24}$$

Solving above equations for $w_1$ and $w_2$ in terms of $w_3$ to find

$$\frac{w_1}{u_2v_3 - u_3v_2} = \frac{w_2}{u_3v_1 - u_1v_3} = \frac{w_3}{u_1v_2 - u_2v_1} \tag{1.25}$$

The common value of these three fractions is denoted by $k$ to preserve symmetry, from which it obtains

$$\begin{aligned} w_1 &= k(u_2v_3 - u_3v_2) \\ w_2 &= k(u_3v_1 - u_1v_3) \\ w_3 &= k(u_1v_2 - u_2v_1) \end{aligned} \tag{1.26}$$

In a special case of the vector product of unit vectors, such as $\widetilde{\imath}_1\,\bar{\imath}_2 = \bar{\imath}_3$, the above relations, Eq. (1.26), should be satisfied also, and then $k$ is readily identified to be $k = 1$. Hence, the components of the vector product become

$$\begin{aligned} \underline{w} &= w_1\bar{\imath}_1 + w_2\bar{\imath}_2 + w_3\bar{\imath}_3 \\ &= (u_2v_3 - u_3v_2)\bar{\imath}_1 + (u_3v_1 - u_1v_3)\bar{\imath}_2 + (u_1v_2 - u_2v_1)\bar{\imath}_3 \end{aligned} \tag{1.27}$$

With the help of Eq. (1.5), the following relationship holds

$$\underline{w} = \begin{vmatrix} w_1 \\ w_2 \\ w_3 \end{vmatrix} = \begin{vmatrix} u_2 v_3 - u_3 v_2 \\ u_3 v_1 - u_1 v_3 \\ u_1 v_2 - u_2 v_1 \end{vmatrix} = \begin{bmatrix} 0 & -u_3 & u_2 \\ u_3 & 0 & -u_1 \\ -u_2 & u_1 & 0 \end{bmatrix} \begin{vmatrix} v_1 \\ v_2 \\ v_3 \end{vmatrix} = \widetilde{u}\,\underline{v} \tag{1.28}$$

It is observed that $\widetilde{u}$ is a second-order, skew-symmetric tensor whose components are

$$\widetilde{u} = \begin{bmatrix} 0 & -u_3 & u_2 \\ u_3 & 0 & -u_1 \\ -u_2 & u_1 & 0 \end{bmatrix} \tag{1.29}$$

This is also a $3 \times 3$ adjoint representation [4] of *Lie algebras so(3)*. Its matrix bases resolved in basis $\mathcal{I}$ are

$$\widetilde{\imath}_1 = \begin{bmatrix} 0 & 0 & 0 \\ 0 & 0 & -1 \\ 0 & 1 & 0 \end{bmatrix}, \quad \widetilde{\imath}_2 = \begin{bmatrix} 0 & 0 & 1 \\ 0 & 0 & 0 \\ -1 & 0 & 0 \end{bmatrix}, \quad \widetilde{\imath}_3 = \begin{bmatrix} 0 & -1 & 0 \\ 1 & 0 & 0 \\ 0 & 0 & 0 \end{bmatrix} \tag{1.30}$$

and

$$\widetilde{u} = u_1 \widetilde{\imath}_1 + u_2 \widetilde{\imath}_2 + u_3 \widetilde{\imath}_3 \tag{1.31}$$

By above construction, the transpose of a skew-symmetric tensor $\widetilde{u}$ is equal to

$$\widetilde{u}^T = -\widetilde{u} \tag{1.32}$$

and hence the anti-commutative attribute of vector product, Eq. (1.21) can be rewritten as

$$\widetilde{u}\,\underline{v} = \widetilde{v}^T \underline{u} \tag{1.33}$$

### 1.1.6 Scalar Triple Product of Vectors

Let $\underline{u}$, $\underline{v}$ and $\underline{w}$ be any three vectors, the scalar $\underline{w}^T \widetilde{u}\,\underline{v}$ is defined as the scalar triple product of these three vectors. Figure 1.6 shows a parallelepiped with vectors $\underline{u}$, $\underline{v}$ and $\underline{w}$ forming adjacent edges. Its volume is given by the absolute value of the scalar triple product

$$V = |\underline{w}^T \widetilde{u}\,\underline{v}| \tag{1.34}$$

Its geometric interpretation is as follows: referring Eq. (1.20), the magnitude of vector product $\widetilde{u}\,\underline{v}$ is equal to the area $A$ of the parallelogram forming the base of the parallelepiped, thus

$$\widetilde{u}\,\underline{v} = A\bar{n} \tag{1.35}$$

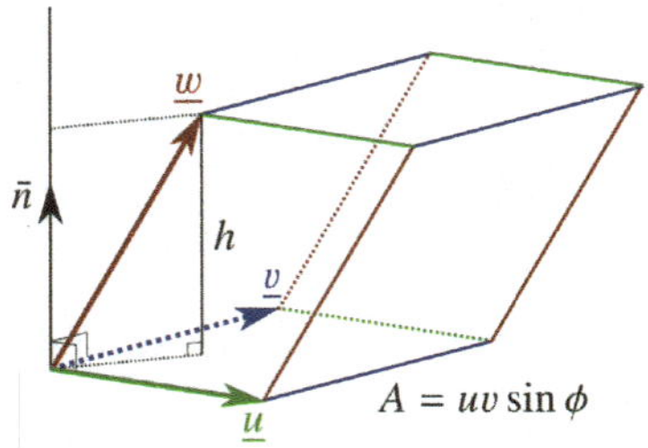


**Fig. 1.6** Scalar triple product of vectors $\underline{u}$, $\underline{v}$ and $\underline{w}$

The absolute value of scalar product $|\underline{w}^T \bar{n}|$ determines the height $h$ of the parallelepiped, $h = |\underline{w}^T \bar{n}|$. It then follows that $V = Ah = |\underline{w}^T \widetilde{u}\, \underline{v}|$. According to the commutative properties of the scalar product, Eq. (1.13), and anti-commutative properties of the vector product, Eq. (1.33), it can be proved

$$\underline{w}^T \widetilde{u}\, \underline{v} = \underline{u}^T \widetilde{v}\, \underline{w} = \underline{v}^T \widetilde{w}\, \underline{u} \tag{1.36}$$

Note that the scalar triple product of vectors can also be rewritten as the determinant of a matrix [5] like

$$\underline{w}^T \widetilde{u}\, \underline{v} = \begin{vmatrix} u_1 & v_1 & w_1 \\ u_2 & v_2 & w_2 \\ u_3 & v_3 & w_3 \end{vmatrix} \tag{1.37}$$

### *1.1.7 Vector Triple Product of Vectors*

Let $\underline{u}$, $\underline{v}$ and $\underline{w}$ be any three vectors, the expression

$$\underline{r} = \widetilde{u}\, \widetilde{v}\, \underline{w} \tag{1.38}$$

creates a vector called a vector triple product of three vectors. Introducing a new symbol $\underline{s}$ to denote

$$\underline{s} = \widetilde{v}\, \underline{w} \tag{1.39}$$

and then vector triple product is simplified to $\underline{r} = \widetilde{u}\, \underline{s}$. By Eq. (1.27), the component $r_1$ of vector $\underline{r}$ becomes

$$\begin{aligned} r_1 &= u_2 s_3 - u_3 s_2 \\ &= u_2(v_1 w_2 - v_2 w_1) - u_3(v_3 w_1 - v_1 w_3) \end{aligned}$$

Rearranging the components on the right hand side of the above equation produces

$$\begin{aligned} r_1 &= v_1(u_2w_2 + u_3w_3) - w_1(u_2v_2 + u_3v_3) \\ &= v_1(\underline{u}^T\underline{w} - u_1w_1) - w_1(\underline{u}^T\underline{v} - u_1v_1) = v_1\underline{u}^T\underline{w} - w_1\underline{u}^T\underline{v} \end{aligned} \tag{1.40}$$

Similarly,

$$r_2 = v_2\underline{u}^T\underline{w} - w_2\underline{u}^T\underline{v} \tag{1.41}$$

and

$$r_3 = v_3\underline{u}^T\underline{w} - w_3\underline{u}^T\underline{v} \tag{1.42}$$

After assembling all these components $r_1$, $r_2$ and $r_3$ into a column array, an identity is obtained to be

$$\underline{r} = \underline{v}\,\underline{u}^T\underline{w} - \underline{w}\,\underline{u}^T\underline{v} \tag{1.43}$$

or alternatively,

$$\tilde{u}\,\tilde{v}\,\underline{w} = \underline{v}\,\underline{u}^T\underline{w} - \underline{w}\,\underline{u}^T\underline{v} \tag{1.44}$$

It answers that a vector triple product produces a vector in the plane of $\underline{v}$ and $\underline{w}$. This is a rather important identity, which will be used frequently. Through simple linear algebra operations, Eq. (1.44) can be rewritten as

$$\left[\tilde{u}\,\tilde{v} - \left(\underline{v}\,\underline{u}^T - \underline{u}^T\underline{v}\,I\right)\right]\underline{w} = 0 \tag{1.45}$$

The necessary condition that $\underline{w}$ is an arbitrary vector will lead to

$$\tilde{u}\,\tilde{v} = \underline{v}\,\underline{u}^T - \underline{u}^T\underline{v}I \tag{1.46}$$

that is another important identity. Applying both commutative properties, Eq. (1.13), and anti-commutative properties, Eq. (1.33), to the identity, Eq. (1.44), results in

$$\widetilde{\tilde{v}\,\underline{w}}^T\underline{u} = \underline{v}\,\underline{w}^T\underline{u} - \underline{w}\,\underline{v}^T\underline{u} \tag{1.47}$$

From which another identity can be obtained with the help of an necessary condition that vector $\underline{u}$ is an arbitrary vector, such as

$$\widetilde{\tilde{v}\,\underline{w}} = \underline{w}\,\underline{v}^T - \underline{v}\,\underline{w}^T \tag{1.48}$$

In view of Eq. (1.46), the tensor products $\underline{w}\,\underline{v}^T$ and $\underline{v}\,\underline{w}^T$ appeared in the right hand side of above equation can be replaced by

$$\widetilde{\tilde{v}\,\underline{w}} = \tilde{v}\,\tilde{w} + \underline{v}^T\underline{w}I - (\tilde{w}\,\tilde{v} + \underline{w}^T\underline{v}I) = \tilde{v}\,\tilde{w} - \tilde{w}\,\tilde{v} \tag{1.49}$$

This identity will remain unchanged for any two vectors $\underline{u}$ and $\underline{v}$, and hence

$$\widetilde{\widetilde{u}\,\underline{v}} = \widetilde{u}\,\widetilde{v} - \widetilde{v}\,\widetilde{u} \tag{1.50}$$

It is necessary to note that the strict proof of identities, Eqs. (1.46), (1.48) and (1.50), can be accomplished by expanding the products with the help of Eq. (1.29).

### 1.1.8 The Tensor Product

Given two vectors $\underline{u}$ and $\underline{v}$, the *tensor product* of these two vectors is a second order tensor defined as

$$T = \underline{u}\,\underline{v}^T \tag{1.51}$$

featuring a fundamental property

$$T\underline{w} = \underline{v}^T\underline{w}\,\underline{u} \tag{1.52}$$

for any arbitrary vector $\underline{w}$. If a unit normal vector $\bar{n}$ is selected to initialize $\underline{u} = \underline{v} = \bar{n}$, the tensor product $T$ describes a projection of vector $\underline{w}$ along the normal direction $\bar{n}$, hence

$$\bar{n}\bar{n}^T\underline{w} = \bar{n}^T\underline{w}\,\bar{n} \tag{1.53}$$

By Eq. (1.44), the projection of vector $\underline{w}$ in normal vector $\bar{n}$ associated null space can be determined to be

$$\widetilde{n}\,\widetilde{n}^T\underline{w} = \underline{w} - \bar{n}\,\bar{n}^T\underline{w} \tag{1.54}$$

From which or by Eq. (1.46) equivalently, the following identity is readily verified

$$\widetilde{n}\,\widetilde{n} = \bar{n}\,\bar{n}^T - I \tag{1.55}$$

By multiplying $\widetilde{n}$ at both sides of above equation, another important identity

$$\widetilde{n}\,\widetilde{n}\,\widetilde{n} = -\widetilde{n} \tag{1.56}$$

is easily obtained also with the application of the vector parallel property, which states that $\widetilde{n}\,\bar{n} = 0$.

The tensor products of unit vectors, $\bar{\imath}_1$, $\bar{\imath}_2$, and $\bar{\imath}_3$, defining Cartesian basis $\mathcal{I}(\bar{\imath}_1, \bar{\imath}_2, \bar{\imath}_3)$, are found to be

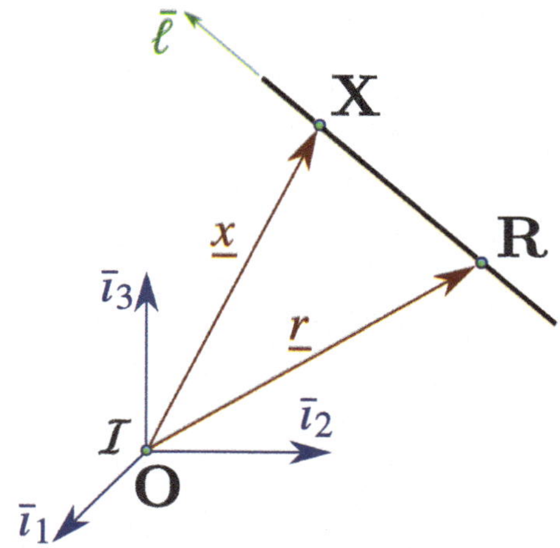


**Fig. 1.7** The definition of a straight line

$$\begin{aligned}
\bar{\imath}_1\,\bar{\imath}_1^T &= \begin{bmatrix} 1\,0\,0 \\ 0\,0\,0 \\ 0\,0\,0 \end{bmatrix}, & \bar{\imath}_1\,\bar{\imath}_2^T &= \begin{bmatrix} 0\,1\,0 \\ 0\,0\,0 \\ 0\,0\,0 \end{bmatrix}, & \bar{\imath}_1\,\bar{\imath}_3^T &= \begin{bmatrix} 0\,0\,1 \\ 0\,0\,0 \\ 0\,0\,0 \end{bmatrix} \\
\bar{\imath}_2\,\bar{\imath}_1^T &= \begin{bmatrix} 0\,0\,0 \\ 1\,0\,0 \\ 0\,0\,0 \end{bmatrix}, & \bar{\imath}_2\,\bar{\imath}_2^T &= \begin{bmatrix} 0\,0\,0 \\ 0\,1\,0 \\ 0\,0\,0 \end{bmatrix}, & \bar{\imath}_2\,\bar{\imath}_3^T &= \begin{bmatrix} 0\,0\,0 \\ 0\,0\,1 \\ 0\,0\,0 \end{bmatrix} \\
\bar{\imath}_3\,\bar{\imath}_1^T &= \begin{bmatrix} 0\,0\,0 \\ 0\,0\,0 \\ 1\,0\,0 \end{bmatrix}, & \bar{\imath}_3\,\bar{\imath}_2^T &= \begin{bmatrix} 0\,0\,0 \\ 0\,0\,0 \\ 0\,1\,0 \end{bmatrix}, & \bar{\imath}_3\,\bar{\imath}_3^T &= \begin{bmatrix} 0\,0\,0 \\ 0\,0\,0 \\ 0\,0\,1 \end{bmatrix}
\end{aligned} \tag{1.57}$$

If $u_i$ and $v_i$ for $i = 1, 2, 3$ are the components of vectors $\underline{u}$ and $\underline{v}$, respectively, the tensor product can be written in detail like

$$\underline{u}\,\underline{v}^T = \begin{vmatrix} u_1 \\ u_2 \\ u_3 \end{vmatrix} \lfloor v_1,\ v_2,\ v_3 \rfloor = \begin{bmatrix} u_1v_1 & u_1v_2 & u_1v_3 \\ u_2v_1 & u_2v_2 & u_2v_3 \\ u_3v_1 & u_3v_2 & u_3v_3 \end{bmatrix} \tag{1.58}$$

### *1.1.9 The Equation of a Line*

In the case of vectors are associated with a specific point in space, they are called *bound vectors*. For example, the position vector $\underline{r}$ in Fig. 1.7, specified the position of point **R** in three-dimensional space with respect to a reference point **O**, is a bound vector.

Note that a Cartesian basis $\mathcal{I}(\bar{\imath}_1, \bar{\imath}_2, \bar{\imath}_3)$ was introduced with the origin of this orthonormal basis undefined in Sect. 1.1.3. Let point **O** be the common origin of the three mutually orthonormal vectors $\bar{\imath}_1$, $\bar{\imath}_2$, and $\bar{\imath}_3$ of the basis. It means the Cartesian

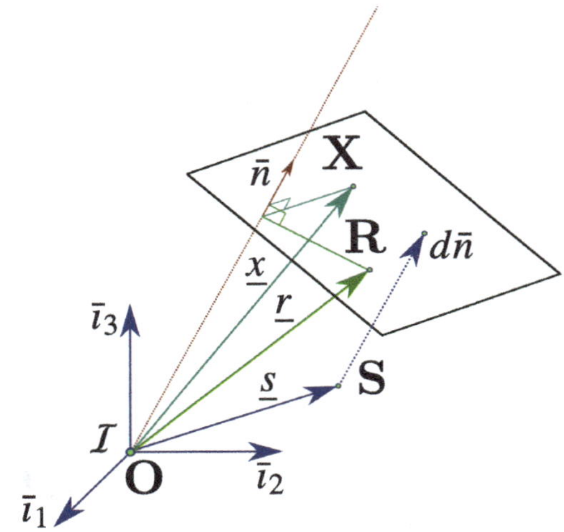


**Fig. 1.8** The definition of a plane

basis is bounded now, and named as a reference frame, $\mathcal{F}_\mathcal{I} = [\mathbf{O}, \mathcal{I}]$, consisting of an orthogonal basis $\mathcal{I}$, with its origin at point **O**.

As depicted in Fig. 1.7, the position vector $\underline{r}$ of an arbitrary point **R** on a straight line determines the position of the line in space. A unit vector $\bar{\ell}$ along the direction of the line shows the orientation of the line. Hence, the point **R**, together with the unit vector $\bar{\ell}$, strictly describes a straight line. An arbitrary point **X** on the line has a position vector, $\underline{x}$, satisfying the parallel condition

$$(\widetilde{x} - \widetilde{r})\bar{\ell} = 0 \tag{1.59}$$

exactly the desired equation of a line.

### *1.1.10 The Equation of a Plane*

Similar to the description of a line in Sect. 1.1.9, a plane is defined by the position vector, $\underline{r}$, of an arbitrary point **R** of the plane, and the unit vector, $\bar{n}$, perpendicular to the plane. As depicted in Fig. 1.8, an arbitrary point **X** of the plane has a position vector, $\underline{x}$, satisfying the perpendicular condition

$$\bar{n}^T(\underline{x} - \underline{r}) = 0 \tag{1.60}$$

also the desired equation of a plane. The vector-perpendicular, $\underline{d} = d\bar{n}$, from a point **S** to the plane can be found from the equation of a plane, Eq. (1.60), like

$$\bar{n}^T\underline{r} = \bar{n}^T\underline{s} + d \tag{1.61}$$

where $\underline{s}$ denotes the position-vector of the point **S**, and the magnitude $d$ of vector $\underline{d}$ is also the perpendicular distance from the point **S** to the plane. When $d$ is determined from

$$d = \bar{n}^T(\underline{r} - \underline{s}) \tag{1.62}$$

the vector-perpendicular $\underline{d} = d\bar{n}$ is readily obtained as

$$\underline{d} = \bar{n}\,\bar{n}^T(\underline{r} - \underline{s}) \tag{1.63}$$

If $d = 0$, it means the point **S** lies on the plane.

## 1.2 Dual Vector

As discussed in Sect. 1.1.9, a straight line is defined by the position vector $\underline{r}$ and the unit vector $\bar{\ell}$, where $\underline{r}$ specifies the position of the line with respect to a reference point **O**, and $\bar{\ell}$ gives the direction of the line. Within the framework of dual algebra, a straight line can be described by a *dual vector* [6, 7]

$$\underline{\mathcal{L}} = \bar{\ell} + \epsilon\,\tilde{r}\,\bar{\ell} = \bar{\ell} + \epsilon\,\underline{\ell}^o \tag{1.64}$$

which is a combination of two vectors $\bar{\ell}$ and $\underline{\ell}^o$. The primary part is unit vector $\bar{\ell}$ parallel to the line and the dual part $\underline{\ell}^o$ is the moment of vector $\bar{\ell}$ about the point **O** and

$$\underline{\ell}^o = \tilde{r}\,\bar{\ell} \tag{1.65}$$

The dual unit, or operator $\epsilon$, follows special rules $\epsilon^n = 0$ for $n \geq 2$.

By Eq. (1.59), it is readily to find

$$\tilde{x}\,\bar{\ell} = \tilde{r}\,\bar{\ell} = \underline{\ell}^o \tag{1.66}$$

where $\underline{x}$ is a position vector of an arbitrary point **X** on the line. If replacing vector $\underline{x}$ by a special position vector $\underline{r}_0$, which is perpendicular to the straight line, the above relations still hold on

$$\tilde{r}_0\,\bar{\ell} = \underline{\ell}^o \tag{1.67}$$

Vector multiplying by $\bar{\ell}$ to both side of the equation generates

$$\tilde{\ell}\,\tilde{r}_0\,\bar{\ell} = \tilde{\ell}\,\underline{\ell}^o \tag{1.68}$$

With the aid of vector triple product identity, Eq. (1.44), and the perpendicular attribute, $\underline{\ell}^T\underline{r}_0 = 0$, position vector $\underline{r}_0$ is determined to be

$$\underline{r}_0 = \tilde{\ell}\,\underline{\ell}^o \tag{1.69}$$

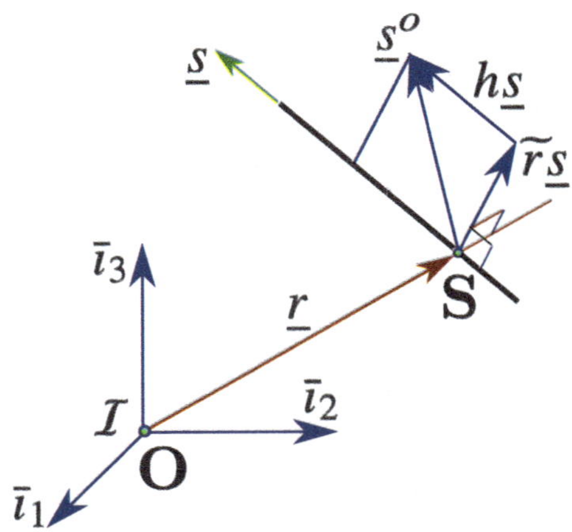


**Fig. 1.9** Geometric description of dual vector $\underline{\mathcal{S}} = \underline{s} + \epsilon \underline{s}^o$

It is noted that the dual vector $\underline{\mathcal{L}}$, Eq. (1.64), defines a straight line with zero pitch

$$h = \bar{\ell}^T \underline{\ell}^o = 0 \tag{1.70}$$

In view of this, $\underline{\mathcal{L}}$ is also named as a *line vector*. More generally, the pitch $h$ is nonzero. As depicted in Fig. 1.9, the dual vector $\underline{\mathcal{S}}$ with finite pitch $h$

$$\underline{\mathcal{S}} = \underline{s} + \epsilon\, \underline{s}^o = \underline{s} + \epsilon\, (\widetilde{r}\, \underline{s} + h\underline{s}) \tag{1.71}$$

gives the dual representation of *screw*, where the primary part $\underline{s}$ defines the screw axis, and the dual part specifies the position of screw axis and also the pitch of the screw. Usually, the *Plücker coordinates* [8] are used to recast the dual vector to the six-dimensional form or the adjoint representation namely

$$\underline{\mathcal{S}} = \begin{vmatrix} \underline{s}^o \\ \underline{s} \end{vmatrix} = \begin{vmatrix} \widetilde{r}\, \underline{s} + h\underline{s} \\ \underline{s} \end{vmatrix}, \quad \widehat{\mathcal{S}} = \begin{bmatrix} \underline{s} & \underline{s}^o \\ \underline{0} & \underline{s} \end{bmatrix} \tag{1.72}$$

where the symbol $\widehat{(\cdot)}$ represents the dual matrix of size $6 \times 2$ constructed by an arbitrary dual vector, and $\underline{0}$ represents a zero vector of size $3 \times 1$. The algebraic operations of dual vector in its Plücker coordinates, Eq. (1.72), will automatically satisfy the algebraic operation rules of row array and matrix [9]. In this work, there is no strict distinction about describing the dual vector $\underline{\mathcal{S}}$ in its dual vector form, Eq. (1.71), or its adjoint representation, Eq. (1.72). In essential, both representations are equivalent each other, however the adjoint representation is convenient for algebraic operations as mentioned above.

The scalar product of primary part and dual part

$$\underline{s}^T \underline{s}^o = \underline{s}^T (\widetilde{r}\, \underline{s} + h\underline{s}) = h\underline{s}^T \underline{s} \tag{1.73}$$

can be used to predict the pitch $h$ of the screw like

$$h = \frac{\underline{s}^T \underline{s}^o}{\underline{s}^T \underline{s}} \tag{1.74}$$

where the identity, $\underline{s}^T \widetilde{r}\, \underline{s} = 0$, shows that $h$ is an invariant independent of position vector $\underline{r}$. Similarly, the position vector $\underline{r}_0$ perpendicular to the screw axis can be obtained through the following vector triple product

$$\widetilde{s}\, \underline{s}^o = \widetilde{s}\,(\widetilde{r}\, \underline{s} + h\underline{s}) = \widetilde{s}\, \widetilde{r}_0\, \underline{s} \tag{1.75}$$

with the attribute, $\widetilde{r}_0\, \underline{s} = \widetilde{r}\, \underline{s}$. By Eq. (1.44), and the perpendicular attribute, $\underline{r}_0^T \underline{s} = 0$, the position vector $\underline{r}_0$ perpendicular to the screw axis is determined to be

$$\underline{r}_0 = \frac{\widetilde{s}\, \underline{s}^o}{\underline{s}^T \underline{s}} \tag{1.76}$$

### *1.2.1 Law of Dual Vector Addition*

Similarly to the law of vector addition, the law of dual vector addition states: Let $\underline{S}_1 = \underline{s}_1 + \epsilon \underline{s}_1^o$ and $\underline{S}_2 = \underline{s}_2 + \epsilon \underline{s}_2^o$ be the two dual vectors, their primary parts and dual parts follow the law of vector addition respectively, and are commutative also. Hence, the vector sum $\underline{S}$ is readily verified to be

$$\underline{S} = \underline{S}_1 + \underline{S}_2 = \underline{s}_1 + \underline{s}_2 + \epsilon(\underline{s}_1^o + \underline{s}_2^o) = \underline{S}_2 + \underline{S}_1 \tag{1.77}$$

Denoting the vector sum as $\underline{S} = \underline{s} + \epsilon \underline{s}^o$, its primary part and dual part can be explicitly expressed as

$$\underline{s} = \underline{s}_1 + \underline{s}_2, \quad \underline{s}^o = \underline{s}_1^o + \underline{s}_2^o \tag{1.78}$$

### *1.2.2 Scalar Multiplication of Dual Vector*

Given a dual number [10] $\phi^d$ with primary part $\phi$ and dual part $\delta$, *i.e.* $\phi^d = \phi + \epsilon\, \delta$, the multiplication of an arbitrary dual vector $\underline{S}$ by the dual number $\phi^d$ is written as

$$\underline{S}_d = \phi^d \underline{S} \tag{1.79}$$

Resolving the dual vector $\underline{S}$ into its primary part $\underline{s}$ and dual part $\underline{s}^o$, the scalar multiplication of dual vector can be written in detail like

$$\underline{S}_d = (\phi + \epsilon\, \delta)(\underline{s} + \epsilon\, \underline{s}^o) = \phi\, \underline{s} + \epsilon(\phi\, \underline{s}^o + \delta\, \underline{s}) \tag{1.80}$$

Here again, the scalar multiplication can be recasted to its adjoint representations

$$\begin{vmatrix} \phi\,\underline{s}^o + \delta\,\underline{s} \\ \phi\,\underline{s} \end{vmatrix} = \begin{bmatrix} \underline{s} & \underline{s}^o \\ \underline{0} & \underline{s} \end{bmatrix} \begin{vmatrix} \delta \\ \phi \end{vmatrix}, \quad \begin{bmatrix} \phi\,\underline{s} & \phi\,\underline{s}^o + \delta\,\underline{s} \\ 0 & \phi\,\underline{s} \end{bmatrix} = \begin{bmatrix} \underline{s} & \underline{s}^o \\ \underline{0} & \underline{s} \end{bmatrix} \begin{bmatrix} \phi & \delta \\ 0 & \phi \end{bmatrix} \tag{1.81}$$

or in compact expressions equivalently

$$\underline{\mathcal{S}}_d = \widehat{\mathcal{S}}\underline{\phi}^d, \quad \widehat{\mathcal{S}}_d = \widehat{\mathcal{S}}\widehat{\phi}^d \tag{1.82}$$

with the aid of the following introduced symbols

$$\underline{\phi}^d = \begin{vmatrix} \delta \\ \phi \end{vmatrix}, \quad \widehat{\phi}^d = \begin{bmatrix} \phi & \delta \\ 0 & \phi \end{bmatrix} \tag{1.83}$$

Similar to the definition of unit vector, Eq. (1.2), a dual vector is called unit dual vector if the magnitude is unit. Denoting an unit dual vector as

$$\underline{\mathcal{N}} = \bar{n} + \epsilon\,\bar{n}^o \tag{1.84}$$

its magnitude is a dual number with zero dual part, $1 + \epsilon\,0$. Any dual vector $\underline{\mathcal{S}}$ can be written as a multiplication of its magnitude, a dual number $\phi^d$, and an unit dual vector $\underline{\mathcal{N}}$, like

$$\underline{\mathcal{S}} = \phi^d \underline{\mathcal{N}} \tag{1.85}$$

or in detail

$$\underline{s} + \epsilon\,\underline{s}^o = (\phi + \epsilon\,\delta)(\bar{n} + \epsilon\,\bar{n}^o) = \phi\,\bar{n} + \epsilon\,(\phi\,\bar{n}^o + \delta\,\bar{n}) \tag{1.86}$$

From which its magnitude $\phi^d$ is determined to be

$$\phi^d = s + \epsilon \frac{\underline{s}^T \underline{s}^o}{s} \tag{1.87}$$

where $s$ represents the magnitude of vector $\underline{s}$ according to the definition of Eq. (1.1), and then the associated unit dual vector $\underline{\mathcal{N}}$ can be determined

$$\underline{\mathcal{N}} = \frac{\underline{s}}{s} + \epsilon(I - \frac{\underline{s}\,\underline{s}^T}{s^2})\frac{\underline{s}^o}{s} \tag{1.88}$$

with the aid of orthogonal property, $\bar{n}^T \bar{n}^o = 0$, which means the unit dual vector $\underline{\mathcal{N}}$ is a line vector by default. In a special case $\underline{s} = 0$, it is readily to get

$$\phi^d = \epsilon\,\|\underline{s}^o\| = \epsilon s^o, \quad \underline{\mathcal{N}} = \frac{\underline{s}^o}{s^o} \tag{1.89}$$

In summary, the magnitude $\phi^d$ and associated unit dual vector $\underline{\mathcal{N}}$ of an arbitrary dual vector $\underline{\mathcal{S}}$ can be computed from [11]

$$\phi^d = \begin{cases} s + \epsilon \dfrac{\underline{s}^T \underline{s}^o}{s}, & s \neq 0 \\ \epsilon s^o, & s = 0 \end{cases} \tag{1.90}$$

and

$$\underline{\mathcal{N}} = \begin{cases} \dfrac{\underline{s}}{s} + \epsilon (I - \dfrac{\underline{s}\,\underline{s}^T}{s^2}) \dfrac{\underline{s}^o}{s}, & s \neq 0 \\ \dfrac{\underline{s}^o}{s^o}, & s = 0 \end{cases} \tag{1.91}$$

If the expression, Eq. (1.85), is considered to be the dual vector form of scalar multiplication, the loose or detailed adjoint representations of the same equation can be found to be

$$\begin{vmatrix} \underline{s}^o \\ \underline{s} \end{vmatrix} = \begin{bmatrix} \bar{n} & \bar{n}^o \\ \underline{0} & \bar{n} \end{bmatrix} \begin{vmatrix} \delta \\ \phi \end{vmatrix}, \quad \begin{bmatrix} \underline{s} & \underline{s}^o \\ \underline{0} & \underline{s} \end{bmatrix} = \begin{bmatrix} \bar{n} & \bar{n}^o \\ \underline{0} & \bar{n} \end{bmatrix} \begin{bmatrix} \phi & \delta \\ 0 & \phi \end{bmatrix} \tag{1.92}$$

or in the compact formulas equivalently

$$\underline{\mathcal{S}} = \widehat{\mathcal{N}}\, \underline{\phi}^d, \quad \widehat{\mathcal{S}} = \widehat{\mathcal{N}}\, \widehat{\phi}^d \tag{1.93}$$

### 1.2.3 The Scalar Product of Dual Vectors

Given two dual vectors $\underline{\mathcal{S}}_1 = \underline{s}_1 + \epsilon\, \underline{s}_1^o$ and $\underline{\mathcal{S}}_2 = \underline{s}_2 + \epsilon\, \underline{s}_2^o$, the scalar product of two dual vectors is a dual number $\phi^s$ obtained by the following dual operations

$$\phi^s = \underline{\mathcal{S}}_1^T \underline{\mathcal{S}}_2 = (\underline{s}_1 + \epsilon\, \underline{s}_1^o)^T (\underline{s}_2 + \epsilon\, \underline{s}_2^o) = \underline{s}_1^T \underline{s}_2 + \epsilon\, (\underline{s}_1^T \underline{s}_2^o + \underline{s}_1^{oT} \underline{s}_2) \tag{1.94}$$

where the dual part, $\underline{s}_1^T \underline{s}_2^o + \underline{s}_1^{o\,T} \underline{s}_2$, is called mutual invariant of screw, also known as reciprocal product. The scalar product can be described by the dual matrix-vector products

$$\begin{vmatrix} \underline{s}_1^T \underline{s}_2^o + \underline{s}_1^{oT} \underline{s}_2 \\ \underline{s}_1^T \underline{s}_2 \end{vmatrix} = \begin{bmatrix} \underline{s}_1^T & \underline{s}_1^{oT} \\ \underline{0}^T & \underline{s}_1^T \end{bmatrix} \begin{vmatrix} \underline{s}_2^o \\ \underline{s}_2 \end{vmatrix} = \begin{bmatrix} \underline{s}_2^T & \underline{s}_2^{oT} \\ \underline{0}^T & \underline{s}_2^T \end{bmatrix} \begin{vmatrix} \underline{s}_1^o \\ \underline{s}_1 \end{vmatrix} \tag{1.95}$$

or the dual matrix-matrix products

$$\begin{bmatrix} \underline{s}_1^T \underline{s}_2 & \underline{s}_1^T \underline{s}_2^o + \underline{s}_1^{oT} \underline{s}_2 \\ \underline{0} & \underline{s}_1^T \underline{s}_2 \end{bmatrix} = \begin{bmatrix} \underline{s}_1^T & \underline{s}_1^{oT} \\ \underline{0}^T & \underline{s}_1^T \end{bmatrix} \begin{bmatrix} \underline{s}_2 & \underline{s}_2^o \\ \underline{0} & \underline{s}_2 \end{bmatrix} = \begin{bmatrix} \underline{s}_2^T & \underline{s}_2^{oT} \\ \underline{0}^T & \underline{s}_2^T \end{bmatrix} \begin{bmatrix} \underline{s}_1 & \underline{s}_1^o \\ \underline{0} & \underline{s}_1 \end{bmatrix} \tag{1.96}$$

These expressions also show the commutative attributes of the scalar product of dual vectors. Equivalently, they can be written in the compact forms as

$$\underline{\phi}^s = \widehat{\underline{S}}_1^T \underline{S}_2 = \widehat{\underline{S}}_2^T \underline{S}_1, \quad \widehat{\underline{\phi}}^s = \widehat{\underline{S}}_1^T \widehat{\underline{S}}_2 = \widehat{\underline{S}}_2^T \widehat{\underline{S}}_1 \tag{1.97}$$

where the symbol $(\cdot)^T$ represents the transpose of dual matrix, which is composed of the transpose of primary part and the transpose of dual part. For example

$$\widehat{\underline{S}}_1^T = \begin{bmatrix} \underline{s}_1^T & \underline{s}_1^{oT} \\ \underline{0}^T & \underline{s}_1^T \end{bmatrix}, \quad \widehat{\underline{S}}_2^T = \begin{bmatrix} \underline{s}_2^T & \underline{s}_2^{oT} \\ \underline{0}^T & \underline{s}_2^T \end{bmatrix} \tag{1.98}$$

Next, a special case is discussed in detail, and then the geometric description of scalar product of dual vectors is presented. The special case considers a scalar product of a given dual vector by itself

$$\underline{S}^T \underline{S} = (\underline{s} + \epsilon\, \underline{s}^o)^T (\underline{s} + \epsilon\, \underline{s}^o) = \underline{s}^T \underline{s} + \epsilon\, 2\underline{s}^T \underline{s}^o \tag{1.99}$$

When rewriting the dual vector $\underline{S}$ as a multiplication of its magnitude $\underline{\phi}^d$ and associated unit dual vector $\underline{\mathcal{N}}$, the scalar product becomes

$$\widehat{\underline{S}}^T \underline{S} = \widehat{\underline{\phi}}^d \widehat{\underline{\mathcal{N}}}^T \widehat{\underline{\mathcal{N}}} \underline{\phi}^d = \widehat{\underline{\phi}}^d \underline{\phi}^d \tag{1.100}$$

where the following identities

$$\widehat{\underline{\mathcal{N}} \underline{\phi}^d} = \widehat{\underline{\mathcal{N}}} \widehat{\underline{\phi}}^d, \quad \widehat{\underline{\mathcal{N}} \underline{\phi}^d}^T = \widehat{\underline{\phi}}^d \widehat{\underline{\mathcal{N}}}^T \tag{1.101}$$

or in detail

$$\begin{bmatrix} \phi\bar{n} & \phi\bar{n}^o + \delta\bar{n} \\ \underline{0} & \phi\bar{n} \end{bmatrix} = \begin{bmatrix} \bar{n} & \bar{n}^o \\ \underline{0} & \bar{n} \end{bmatrix} \begin{bmatrix} \phi & \delta \\ 0 & \phi \end{bmatrix} \tag{1.102}$$

and

$$\begin{bmatrix} \phi\bar{n}^T & \phi\bar{n}^{oT} + \delta\bar{n}^T \\ \underline{0}^T & \phi\bar{n}^T \end{bmatrix} = \begin{bmatrix} \phi & \delta \\ 0 & \phi \end{bmatrix} \begin{bmatrix} \bar{n}^T & \bar{n}^{oT} \\ \underline{0}^T & \bar{n}^T \end{bmatrix} \tag{1.103}$$

are applied. Note the final equalities of Eq. (1.100) follow from the property that dual matrix product of unit dual vector

$$\widehat{\underline{\mathcal{N}}}^T \widehat{\underline{\mathcal{N}}} = \begin{bmatrix} \bar{n}^T & \bar{n}^{oT} \\ \underline{0}^T & \bar{n}^T \end{bmatrix} \begin{bmatrix} \bar{n} & \bar{n}^o \\ \underline{0} & \bar{n} \end{bmatrix} = \begin{bmatrix} 1 & 0 \\ 0 & 1 \end{bmatrix} \tag{1.104}$$

becomes identity matrix. Featuring the same property as Eq. (1.15) states that the scalar product of a vector by itself yields the square of its magnitude, Eq. (1.100) shows that the scalar product of a dual vector by itself is equal to the square of its magnitude. In view of Eq. (1.90), the scalar product of a dual vector by itself can be computed explicitly from

$$\widehat{\underline{S}}^T \underline{S} = \begin{bmatrix} \phi & \delta \\ 0 & \phi \end{bmatrix} \begin{vmatrix} \delta \\ \phi \end{vmatrix} = \begin{vmatrix} 2\phi\,\delta \\ \phi^2 \end{vmatrix} = \begin{vmatrix} 2\underline{s}^T \underline{s}^o \\ s^2 \end{vmatrix} \tag{1.105}$$

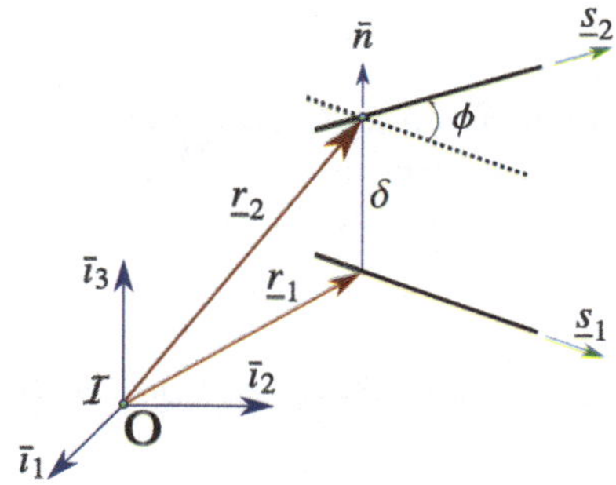


**Fig. 1.10** Angle between two dual vectors $\underline{S}_1$ and $\underline{S}_2$

It is interesting to note that the scalar product of unit dual vector $\underline{\mathcal{N}}$ by itself produces an unit magnitude

$$\widehat{\mathcal{N}}^T\underline{\mathcal{N}} = \begin{bmatrix} \bar{n}^T & \bar{n}^{oT} \\ 0 & \bar{n}^T \end{bmatrix} \begin{vmatrix} \bar{n}^o \\ \bar{n} \end{vmatrix} = \begin{vmatrix} 0 \\ 1 \end{vmatrix} \tag{1.106}$$

or in its dual number form as $1 + \epsilon\, 0$.

The geometric description of scalar product of dual vectors is shown in Fig. 1.10, where two arbitrary screws are given with their screw axes $\underline{s}_1$ and $\underline{s}_2$ determined by the position vectors $\underline{r}_1$ and $\underline{r}_2$, respectively. The smallest nonnegative angle $\phi$ between screw axes $\underline{s}_1$ and $\underline{s}_2$ is defined under the guarantee of $0 \le \phi \le \pi$. Meanwhile, a unit vector $\bar{n}$ is presented to describe the orientation of the common normal of vectors $\underline{s}_1$ and $\underline{s}_2$ by the following definition

$$\bar{n} = \frac{\widetilde{s}_1\, \underline{s}_2}{\|\widetilde{s}_1\, \underline{s}_2\|} \tag{1.107}$$

Once the pitches $h_1$ and $h_2$ of the screws are given, the scalar product $\underline{S}_1^T\underline{S}_2$ can be rewritten in detail like

$$\begin{aligned} \underline{S}_1^T\underline{S}_2 &= \underline{s}_1^T\underline{s}_2 + \epsilon\left[\underline{s}_1^T(\widetilde{r}_2\,\underline{s}_2 + h_2\underline{s}_2) + \underline{s}_2^T(\widetilde{r}_1\,\underline{s}_1 + h_1\underline{s}_1)\right] \\ &= \underline{s}_1^T\underline{s}_2 + \epsilon\left[(h_1 + h_2)\underline{s}_1^T\underline{s}_2 + (\underline{r}_2 - \underline{r}_1)^T\widetilde{s}_1^T\,\underline{s}_2\right] \end{aligned} \tag{1.108}$$

If the perpendicular distance between two screws $\underline{S}_1$ and $\underline{S}_2$ is given as $\delta$, then $\underline{r}_2 - \underline{r}_1 = \delta\,\bar{n}$. In view of the definitions of $\delta$ and vector products, Eqs. (1.12) and (1.20), the associated scalar product of dual vectors, Eq. (1.108), becomes

$$\begin{aligned} \underline{S}_1^T\underline{S}_2 &= \underline{s}_1^T\underline{s}_2 + \epsilon\left[(h_1 + h_2)\underline{s}_1^T\underline{s}_2 + \delta\bar{n}^T\widetilde{s}_1^T\,\underline{s}_2\right] \\ &= s_1 s_2\cos\phi + \epsilon\left[(h_1 + h_2)s_1 s_2\cos\phi - \delta\, s_1 s_2\sin\phi\right] \end{aligned} \tag{1.109}$$

If the pitches are zero, $h_1 = h_2 = 0$, the dual vectors or screws $\underline{S}_1$ and $\underline{S}_2$ reduce to the line vectors. The scalar product is further simplified to

$$\underline{S}_1^T \underline{S}_2 = s_1 s_2 \cos\phi - \epsilon\,\delta\, s_1 s_2 \sin\phi = s_1 s_2 \cos\phi^d \tag{1.110}$$

where the cosinoidal dual function [12] is introduced, $\cos\phi^d = \cos\phi - \epsilon\,\delta\sin\phi$, to make the expression, Eq. (1.110), features consistent description with the definition of vector scalar product, Eq. (1.12).

### 1.2.4 The Vector Product of Dual Vectors

Let $\underline{S}_1$ and $\underline{S}_2$ be two arbitrary dual vectors, the vector product of these two dual vectors is a third dual vector computed by dual operations as

$$\widetilde{\underline{S}}_1 \underline{S}_2 = (\widetilde{s}_1 + \epsilon\widetilde{s}_1^o)(\underline{s}_2 + \epsilon\underline{s}_2^o) = \widetilde{s}_1 \underline{s}_2 + \epsilon(\widetilde{s}_1 \underline{s}_2^o + \widetilde{s}_1^o \underline{s}_2) \tag{1.111}$$

With the aid of Plücker coordinates, the vector product of dual vectors can be recasted to its adjoint representation of Lie algebras (3), containing its 6× 6 Lie brackets

$$\widetilde{\underline{S}}_1 \underline{S}_2 = \begin{bmatrix} \widetilde{s}_1 & \widetilde{s}_1^o \\ 0 & \widetilde{s}_1 \end{bmatrix} \begin{vmatrix} \underline{s}_2^o \\ \underline{s}_2 \end{vmatrix} \tag{1.112}$$

It is easily verified that the vector product of dual vectors is anti-commutative

$$\widetilde{\underline{S}}_1 \underline{S}_2 = \begin{bmatrix} \widetilde{s}_2^T & \widetilde{s}_2^{o\,T} \\ 0 & \widetilde{s}_2^T \end{bmatrix} \begin{vmatrix} \underline{s}_1^o \\ \underline{s}_1 \end{vmatrix} = \widetilde{\underline{S}}_2^T \underline{S}_1 = -\widetilde{\underline{S}}_2 \underline{S}_1 \tag{1.113}$$

where a skew symmetric dual tensor, also the 6 × 6 Lie brackets, and its dual transpose are introduced

$$\widetilde{\underline{S}} = \begin{bmatrix} \widetilde{s} & \widetilde{s}^o \\ 0 & \widetilde{s} \end{bmatrix}, \quad \widetilde{\underline{S}}^T = \begin{bmatrix} \widetilde{s}^T & \widetilde{s}^{o\,T} \\ 0 & \widetilde{s}^T \end{bmatrix} \tag{1.114}$$

It is necessary to emphasize that the dual transpose is totally different from the matrix transpose $\widetilde{\underline{S}}^T \neq \widetilde{S}^T$. While, similar to the anti-commutative attribute, Eq. (1.113), another identity

$$\widetilde{\underline{S}}_1^T \underline{S}_2 = \begin{bmatrix} \widetilde{s}_1^T & 0 \\ \widetilde{s}_1^{o\,T} & \widetilde{s}_1^T \end{bmatrix} \begin{vmatrix} \underline{s}_2^o \\ \underline{s}_2 \end{vmatrix} = \begin{bmatrix} 0 & \widetilde{s}_2^o \\ \widetilde{s}_2^o & \widetilde{s}_2 \end{bmatrix} \begin{vmatrix} \underline{s}_1^o \\ \underline{s}_1 \end{vmatrix} = \widetilde{\underline{S}}_2^{\mathcal{L}^o} \underline{S}_1 \tag{1.115}$$

needs to be introduced, which is widely applied in the rest of the book. Here, the symbol $(\cdot)^{\mathcal{L}^o}$ is introduced to denote the another Lie bracket of dual vector $\underline{S}$ in the form of

$$\widetilde{\underline{S}}^{\mathcal{L}^o} = \begin{bmatrix} 0 & \widetilde{s}^o \\ \widetilde{s}^o & \widetilde{s} \end{bmatrix} \tag{1.116}$$

Alternatively, the vector product of dual vectors can be written in adjoint representations

$$\begin{bmatrix} \widetilde{s}_1 & \widetilde{s}_1^o \\ 0 & \widetilde{s}_1 \end{bmatrix} \begin{bmatrix} \underline{s}_2 & \underline{s}_2^o \\ 0 & \underline{s}_2 \end{bmatrix} = \begin{bmatrix} \widetilde{s}_2^T & \widetilde{s}_2^{oT} \\ 0 & \widetilde{s}_2^T \end{bmatrix} \begin{bmatrix} \underline{s}_1 & \underline{s}_1^o \\ 0 & \underline{s}_1 \end{bmatrix} = - \begin{bmatrix} \widetilde{s}_2 & \widetilde{s}_2^o \\ 0 & \widetilde{s}_2 \end{bmatrix} \begin{bmatrix} \underline{s}_1 & \underline{s}_1^o \\ 0 & \underline{s}_1 \end{bmatrix} \tag{1.117}$$

or in the compact form equivalently

$$\widetilde{\mathcal{S}}_1 \widehat{\mathcal{S}}_2 = \widetilde{\mathcal{S}}_2^T \widehat{\mathcal{S}}_1 = -\widetilde{\mathcal{S}}_2 \widehat{\mathcal{S}}_1 \tag{1.118}$$

For an unit dual vector $\underline{\mathcal{N}}$, the associated vector product by itself is found to be zero, such as

$$\widetilde{\mathcal{N}} \widehat{\mathcal{N}} = 0 \Leftrightarrow \widetilde{\mathcal{N}} \underline{\mathcal{N}} = \underline{0} \tag{1.119}$$

The geometric description of vector product of dual vector can also be referred to Fig. 1.10. As depicted in this Figure, two arbitrary screws are given with their position vectors $\underline{r}_1$ and $\underline{r}_2$, together with the screw axes $\underline{s}_1$ and $\underline{s}_2$, and pitches $h_1$ and $h_2$. The geometric description of dual vector products can be observed from the following formula containing all these parameters like

$$\begin{aligned} \widetilde{\mathcal{S}}_1 \underline{\mathcal{S}}_2 &= \widetilde{s}_1 \underline{s}_2 + \epsilon \left[ \widetilde{s}_1 (\widetilde{r}_2 \, \underline{s}_2 + h_2 \underline{s}_2) + (\widetilde{\widetilde{r}_1 \, \underline{s}_1 + h_1 \underline{s}_1}) \underline{s}_2 \right] \\ &= \widetilde{s}_1 \underline{s}_2 + \epsilon \left[ \widetilde{s}_1 \widetilde{r}_2 \, \underline{s}_2 + h_2 \widetilde{s}_1 \, \underline{s}_2 - \widetilde{s}_2 \widetilde{r}_1 \, \underline{s}_1 + h_1 \widetilde{s}_1 \, \underline{s}_2 \right] \\ &= \widetilde{s}_1 \underline{s}_2 + \epsilon \left[ \underline{s}_1^T \underline{s}_2 (\underline{r}_2 - \underline{r}_1) + (\widetilde{r}_1 + h_2 + h_1) \widetilde{s}_1 \, \underline{s}_2 \right] \end{aligned} \tag{1.120}$$

In view of Eq. (1.107), the above dual vector product is reformulated to

$$\begin{aligned} \widetilde{\mathcal{S}}_1 \underline{\mathcal{S}}_2 &= \widetilde{s}_1 \underline{s}_2 + \epsilon \left[ \underline{s}_1^T \underline{s}_2 \, \delta \, \bar{n} + (\widetilde{r}_1 + h_2 + h_1) \widetilde{s}_1 \, \underline{s}_2 \right] \\ &= s_1 s_2 \{ \sin\phi \, \bar{n} + \epsilon \left[ \cos\phi \, \delta \, \bar{n} + (\widetilde{r}_1 + h_2 I + h_1 I) \sin\phi \, \bar{n} \right] \} \end{aligned} \tag{1.121}$$

where $\phi$ is the smallest nonnegative angle between two vectors $\underline{s}_1$ and $\underline{s}_2$. It is known that if screw pitches are zero, $h_1 = h_2 = 0$, screws $\underline{\mathcal{S}}_1$ and $\underline{\mathcal{S}}_2$ will recover to the line vectors. For this special situation, the dual vector product will be further simplified to

$$\widetilde{\mathcal{S}}_1 \underline{\mathcal{S}}_2 = s_1 s_2 \left[ \sin\phi \, \bar{n} + \epsilon (\cos\phi \, \delta \, \bar{n} + \widetilde{r}_1 \sin\phi \, \bar{n}) \right] = s_1 s_2 \sin\phi^d \, \underline{\mathcal{N}} \tag{1.122}$$

where the sinusoidal dual function [12], $\sin\phi^d = \sin\phi + \epsilon \, \delta \cos\phi$, and a unit dual vector, $\underline{\mathcal{N}} = \bar{n} + \epsilon \, \widetilde{r}_1 \, \bar{n}$, are introduced. The magnitude of vector product of dual vectors is a dual number $s_1 s_2 \sin\phi^d$, and unit dual vector $\underline{\mathcal{N}}$ shows the direction of dual vector product that is perpendicular to both line directions $\underline{s}_1$ and $\underline{s}_2$.

### 1.2.5 Tensor Product of Dual Vectors

Given two dual vectors $\underline{S}_1$ and $\underline{S}_2$, the tensor product of these two dual vectors is a dual tensor denoted as $\mathcal{S}$ and

$$\mathcal{S} = \underline{S}_1 \underline{S}_2^T = (\underline{s}_1 + \epsilon\, \underline{s}_1^o)(\underline{s}_2 + \epsilon\, \underline{s}_2^o)^T = \underline{s}_1 \underline{s}_2^T + \epsilon\, (\underline{s}_1 \underline{s}_2^{oT} + \underline{s}_1^o \underline{s}_2^T) \tag{1.123}$$

With the aid of Plücker coordinates, the above definition can be rewritten in its adjoint representation

$$\begin{bmatrix} \underline{s}_1 \underline{s}_2^T & \underline{s}_1 \underline{s}_2^{oT} + \underline{s}_1^o \underline{s}_2^T \\ 0 & \underline{s}_1 \underline{s}_2^T \end{bmatrix} = \begin{bmatrix} \underline{s}_1 & \underline{s}_1^o \\ 0 & \underline{s}_1 \end{bmatrix} \begin{bmatrix} \underline{s}_2^T & \underline{s}_2^{oT} \\ 0 & \underline{s}_2^T \end{bmatrix} \tag{1.124}$$

or in the compact form equivalently

$$\mathcal{S} = \widehat{\mathcal{S}}_1 \widehat{\mathcal{S}}_2^T \tag{1.125}$$

According to the identity, Eq. (1.46), the tensor product of two vectors $\underline{s}_1$ and $\underline{s}_2$ can be represented by

$$\underline{s}_1 \underline{s}_2^T = \widetilde{s}_2 \widetilde{s}_1 + \underline{s}_1^T \underline{s}_2 I \tag{1.126}$$

and then the dual tensor $\mathcal{S}$ is modified to

$$\begin{aligned} \mathcal{S} &= \begin{bmatrix} \widetilde{s}_2 \widetilde{s}_1 & \widetilde{s}_2 \widetilde{s}_1^o + \widetilde{s}_2^o \widetilde{s}_1 \\ 0 & \widetilde{s}_2 \widetilde{s}_1 \end{bmatrix} + \begin{bmatrix} \underline{s}_1^T \underline{s}_2 I & (\underline{s}_1^T \underline{s}_2^o + \underline{s}_1^{oT} \underline{s}_2) I \\ 0 & \underline{s}_1^T \underline{s}_2 I \end{bmatrix} \\ &= \begin{bmatrix} \widetilde{s}_2 & \widetilde{s}_2^o \\ 0 & \widetilde{s}_2 \end{bmatrix} \begin{bmatrix} \widetilde{s}_1 & \widetilde{s}_1^o \\ 0 & \widetilde{s}_1 \end{bmatrix} + \begin{bmatrix} \underline{s}_1^T \underline{s}_2 I & (\underline{s}_1^T \underline{s}_2^o + \underline{s}_1^{oT} \underline{s}_2) I \\ 0 & \underline{s}_1^T \underline{s}_2 I \end{bmatrix} \end{aligned} \tag{1.127}$$

or rewritten it into the compact form like

$$\widehat{\mathcal{S}}_1 \widehat{\mathcal{S}}_2^T = \widetilde{\mathcal{S}}_2 \widetilde{\mathcal{S}}_1 + \widehat{\mathcal{S}}_1^T \widehat{\mathcal{S}}_2 \mathcal{I} \tag{1.128}$$

where the symbol $\widehat{\mathcal{S}}_1^T \widehat{\mathcal{S}}_2 \mathcal{I}$ denotes the $6 \times 6$ dual diagonal matrix constructed by the dual number $\underline{S}_1^T \underline{S}_2$ and in detail

$$\widehat{\mathcal{S}}_1^T \widehat{\mathcal{S}}_2 \mathcal{I} = \begin{bmatrix} \underline{s}_1^T \underline{s}_2 I & (\underline{s}_1^T \underline{s}_2^o + \underline{s}_1^{oT} \underline{s}_2) I \\ 0 & \underline{s}_1^T \underline{s}_2 I \end{bmatrix} \tag{1.129}$$

In the case of a special situation, such that both of the given dual vectors $\underline{S}_1$ and $\underline{S}_2$ are replaced by an unit dual vector $\underline{\mathcal{N}}$, the dual tensor attribute, Eq. (1.128), will be simplified to

$$\widehat{\mathcal{N}} \widehat{\mathcal{N}}^T = \widetilde{\mathcal{N}} \widetilde{\mathcal{N}} + \mathcal{I} \tag{1.130}$$

where the scalar product of unit dual vector by itself, $\widehat{\mathcal{N}}^T \underline{\mathcal{N}}$, becomes a unit number, and then the diagonal matrix, $\widehat{\mathcal{N}}^T \widehat{\mathcal{N}} \mathcal{I}$, reduces to a $6 \times 6$ identity matrix $\mathcal{I}$. By moving the identity matrix $\mathcal{I}$ from the right hand side of the above equation to the left hand side, the identity can be rearranged to

$$\widetilde{\mathcal{N}}\widetilde{\mathcal{N}} = \widehat{\mathcal{N}}\widehat{\mathcal{N}}^T - \mathcal{I} \tag{1.131}$$

Multiplying both side of the above equation by $\widetilde{\mathcal{N}}$, and in view of the properties of dual vector product, Eq. (1.119), it will produce

$$\widetilde{\mathcal{N}}\widetilde{\mathcal{N}}\widetilde{\mathcal{N}} = -\widetilde{\mathcal{N}} \tag{1.132}$$

### *1.2.6 Scalar Triple Product of Dual Vectors*

Let $\underline{S}_1$, $\underline{S}_2$ and $\underline{S}_3$ be any three dual vectors, the dual number $\underline{S}_1^T\widetilde{\underline{S}}_2\underline{S}_3$ is defined as the scalar triple product of these three dual vectors. With the aid of the vector product of dual vectors, Eq. (1.111), the dual vector product $\widetilde{\underline{S}}_2\underline{S}_3$ becomes

$$\widetilde{\underline{S}}_2\underline{S}_3 = (\widetilde{\underline{s}}_2 + \epsilon\,\widetilde{\underline{s}}_2^o)(\underline{s}_3 + \epsilon\,\underline{s}_3^o) = \widetilde{\underline{s}}_2\underline{s}_3 + \epsilon\,(\widetilde{\underline{s}}_2\underline{s}_3^o + \widetilde{\underline{s}}_2^o\underline{s}_3) \tag{1.133}$$

and then the scalar triple product of dual vectors can be computed from

$$\begin{aligned}\underline{S}_1^T\widetilde{\underline{S}}_2\underline{S}_3 &= (\underline{s}_1^T + \epsilon\,\underline{s}_1^{oT})\left[\widetilde{\underline{s}}_2\underline{s}_3 + \epsilon\,(\widetilde{\underline{s}}_2\underline{s}_3^o + \widetilde{\underline{s}}_2^o\underline{s}_3)\right]\\ &= \underline{s}_1^T\widetilde{\underline{s}}_2\underline{s}_3 + \epsilon\,(\underline{s}_1^{oT}\widetilde{\underline{s}}_2\underline{s}_3 + \underline{s}_1^T\widetilde{\underline{s}}_2\underline{s}_3^o + \underline{s}_1^T\widetilde{\underline{s}}_2^o\underline{s}_3)\end{aligned} \tag{1.134}$$

In view of the anti-commutativity property of the vector products, Eq. (1.36), which states that

$$\underline{s}_1^T\widetilde{\underline{s}}_2\underline{s}_3 = \underline{s}_2^T\widetilde{\underline{s}}_3\underline{s}_1 = \underline{s}_3^T\widetilde{\underline{s}}_1\underline{s}_2 \tag{1.135}$$

the same anti-commutativity property of dual vector products in its dual vector form

$$\underline{S}_1^T\widetilde{\underline{S}}_2\underline{S}_3 = \underline{S}_2^T\widetilde{\underline{S}}_3\underline{S}_1 = \underline{S}_3^T\widetilde{\underline{S}}_1\underline{S}_2 \tag{1.136}$$

or in its dual matrix form

$$\widehat{\underline{S}}_1^T\widetilde{\underline{S}}_2\underline{S}_3 = \widehat{\underline{S}}_2^T\widetilde{\underline{S}}_3\underline{S}_1 = \widehat{\underline{S}}_3^T\widetilde{\underline{S}}_1\underline{S}_2 \tag{1.137}$$

still hold. Here only the proof of one identity of above relations is presented, and the rest of them can be proved following the same approach. Applying the commutative attributes of dual vector products, Eq. (1.96), and then the anti-commutative attributes of dual vector products, Eq. (1.113), to the first equalities of Eq. (1.136) finds

$$\underline{S}_1^T\widetilde{\underline{S}}_2\underline{S}_3 = (\widetilde{\underline{S}}_2\underline{S}_3)^T\underline{S}_1 = \underline{S}_3^T\widetilde{\underline{S}}_2^T\underline{S}_1 = \underline{S}_3^T\widetilde{\underline{S}}_1\underline{S}_2 \tag{1.138}$$

or in its dual matrix form equivalently

$$\widehat{\underline{S}}_1^T \widetilde{\underline{S}}_2 \underline{S}_3 = \widehat{\widetilde{\underline{S}}_2 \underline{S}_3}^T \underline{S}_1 = \widehat{\underline{S}}_3^T \widetilde{\underline{S}}_2^T \underline{S}_1 = \widehat{\underline{S}}_3^T \widetilde{\underline{S}}_1 \underline{S}_2 \tag{1.139}$$

where the identities

$$\widehat{\widetilde{\underline{S}}_2 \underline{S}_3} = \widetilde{\underline{S}}_2 \widehat{\underline{S}}_3, \quad \widehat{\widetilde{\underline{S}}_2 \underline{S}_3}^T = \widehat{\underline{S}}_3^T \widetilde{\underline{S}}_2^T \tag{1.140}$$

has been applied, which can be expanded in detail as follows

$$\begin{bmatrix} \widetilde{s}_2 \underline{s}_3 & \widetilde{s}_2 \underline{s}_3^o + \widetilde{s}_2^o \underline{s}_3 \\ \underline{0} & \widetilde{s}_2 \underline{s}_3 \end{bmatrix} = \begin{bmatrix} \widetilde{s}_2 & \widetilde{s}_2^o \\ \underline{0} & \widetilde{s}_2 \end{bmatrix} \begin{bmatrix} \underline{s}_3 & \underline{s}_3^o \\ \underline{0} & \underline{s}_3 \end{bmatrix} \tag{1.141}$$

and its transpose

$$\begin{bmatrix} \underline{s}_3^T \widetilde{s}_2^T & \underline{s}_3^{oT} \widetilde{s}_2^T + \underline{s}_3^T \widetilde{s}_2^{oT} \\ \underline{0}^T & \underline{s}_3^T \widetilde{s}_2^T \end{bmatrix} = \begin{bmatrix} \underline{s}_3^T & \underline{s}_3^{oT} \\ \underline{0}^T & \underline{s}_3^T \end{bmatrix} \begin{bmatrix} \widetilde{s}_2^T & \widetilde{s}_2^{oT} \\ \underline{0}^T & \widetilde{s}_2^T \end{bmatrix} \tag{1.142}$$

### *1.2.7 Vector Triple Product of Dual Vectors*

Let $\underline{S}_1$, $\underline{S}_2$ and $\underline{S}_3$ be any three dual vectors, the expression $\widetilde{\underline{S}}_1 \widetilde{\underline{S}}_2 \underline{S}_3$ creates a dual vector called a vector triple product of three dual vectors. According to the definition of dual vector product, Eq. (1.111), the product $\widetilde{\underline{S}}_2 \underline{S}_3$ is expanded to

$$\widetilde{\underline{S}}_2 \underline{S}_3 = (\widetilde{s}_2 + \epsilon\, \widetilde{s}_2^o)(\underline{s}_3 + \epsilon\, \underline{s}_3^o) = \widetilde{s}_2 \underline{s}_3 + \epsilon\, (\widetilde{s}_2 \underline{s}_3^o + \widetilde{s}_2^o \underline{s}_3) \tag{1.143}$$

and then repeating the same dual operations, the vector triple product of dual vectors can be computed from

$$\begin{aligned} \widetilde{\underline{S}}_1 \widetilde{\underline{S}}_2 \underline{S}_3 &= (\widetilde{s}_1 + \epsilon\, \widetilde{s}_1^o) \left[\widetilde{s}_2 \underline{s}_3 + \epsilon\, (\widetilde{s}_2 \underline{s}_3^o + \widetilde{s}_2^o \underline{s}_3)\right] \\ &= \widetilde{s}_1 \widetilde{s}_2 \underline{s}_3 + \epsilon \left[\widetilde{s}_1^o \widetilde{s}_2 \underline{s}_3 + \widetilde{s}_1 (\widetilde{s}_2 \underline{s}_3^o + \widetilde{s}_2^o \underline{s}_3)\right] \end{aligned} \tag{1.144}$$

In view of the identity of the vector products, Eq. (1.44), which states that

$$\widetilde{s}_1 \widetilde{s}_2 \underline{s}_3 = \underline{s}_3^T \underline{s}_1 \underline{s}_2 - \underline{s}_2^T \underline{s}_1 \underline{s}_3 \tag{1.145}$$

the vector triple product of dual vectors can be resolved into

$$\begin{aligned} \widetilde{\underline{S}}_1 \widetilde{\underline{S}}_2 \underline{S}_3 &= \underline{s}_3^T \underline{s}_1 \underline{s}_2 + \epsilon\, (\underline{s}_3^T \underline{s}_1^o \underline{s}_2 + \underline{s}_3^{oT} \underline{s}_1 \underline{s}_2 + \underline{s}_3^T \underline{s}_1 \underline{s}_2^o) \\ &\quad - \underline{s}_2^T \underline{s}_1 \underline{s}_3 - \epsilon\, (\underline{s}_2^T \underline{s}_1^o \underline{s}_3 + \underline{s}_2^T \underline{s}_1 \underline{s}_3^o + \underline{s}_2^{oT} \underline{s}_1 \underline{s}_3) \end{aligned} \tag{1.146}$$

After merging the similar terms, the above dual vector triple products are reassembled into

$$
\begin{aligned}
\widetilde{S}_1\widetilde{S}_2\underline{S}_3 &= (\underline{s}_3 + \epsilon\, \underline{s}_3^o)^T \left[\tilde{s}_1\underline{s}_2 + \epsilon\, (\tilde{s}_1^o\underline{s}_2 + \tilde{s}_1\underline{s}_2^o)\right] \\
&\quad -(\underline{s}_2 + \epsilon\, \underline{s}_2^o)^T \left[\tilde{s}_1\underline{s}_3 + \epsilon\, (\tilde{s}_1^o\underline{s}_3 + \tilde{s}_1\underline{s}_3^o)\right] \\
&= (\underline{s}_3 + \epsilon\, \underline{s}_3^o)^T(\underline{s}_1 + \epsilon\, \underline{s}_1^o)(\underline{s}_2 + \epsilon\, \underline{s}_2^o) - (\underline{s}_2 + \epsilon\, \underline{s}_2^o)^T(\underline{s}_1 + \epsilon\, \underline{s}_1^o)(\underline{s}_3 + \epsilon\, \underline{s}_3^o)
\end{aligned}
\tag{1.147}
$$

From which, an important identity is obtained as

$$
\widetilde{S}_1\widetilde{S}_2\underline{S}_3 = \underline{S}_3^T\underline{S}_1\underline{S}_2 - \underline{S}_2^T\underline{S}_1\underline{S}_3 \tag{1.148}
$$

and can be also rewritten in the dual matrix form

$$
\widetilde{S}_1\widetilde{S}_2\underline{S}_3 = \widehat{S}_2\widehat{S}_3^T\underline{S}_1 - \widehat{S}_3\widehat{S}_2^T\underline{S}_1 \tag{1.149}
$$

For easy understanding, the identity in its Plücker coordinates is given in detail

$$
\begin{aligned}
\begin{bmatrix} \tilde{s}_1 & \tilde{s}_1^o \\ 0 & \tilde{s}_1 \end{bmatrix}\begin{bmatrix} \tilde{s}_2 & \tilde{s}_2^o \\ 0 & \tilde{s}_2 \end{bmatrix}\left|\begin{matrix} \underline{s}_3^o \\ \underline{s}_3 \end{matrix}\right| &= \begin{bmatrix} \underline{s}_2 & \underline{s}_2^o \\ 0 & \underline{s}_2 \end{bmatrix}\begin{bmatrix} \underline{s}_3^T & \underline{s}_3^{oT} \\ 0 & \underline{s}_3^T \end{bmatrix}\left|\begin{matrix} \underline{s}_1^o \\ \underline{s}_1 \end{matrix}\right| \\
&\quad - \begin{bmatrix} \underline{s}_3 & \underline{s}_3^o \\ 0 & \underline{s}_3 \end{bmatrix}\begin{bmatrix} \underline{s}_2^T & \underline{s}_2^{oT} \\ 0 & \underline{s}_2^T \end{bmatrix}\left|\begin{matrix} \underline{s}_1^o \\ \underline{s}_1 \end{matrix}\right|
\end{aligned}
\tag{1.150}
$$

In view of anti-commutative attribute of dual vector product, Eq. (1.113), the triple dual vector product can be recasted to

$$
-\widetilde{\widetilde{S}_2\underline{S}_3}\underline{S}_1 = \widehat{S}_2\widehat{S}_3^T\underline{S}_1 - \widehat{S}_3\widehat{S}_2^T\underline{S}_1 \tag{1.151}
$$

From which an homogeneous equation in the form of

$$
\left[\widetilde{\widetilde{S}_2\underline{S}_3} + (\widehat{S}_2\widehat{S}_3^T - \widehat{S}_3\widehat{S}_2^T)\right]\underline{S}_1 = 0 \tag{1.152}
$$

is produced. Assuming that $\underline{S}_1$ is an arbitrary dual vector, it will lead to another important identity like

$$
\widetilde{\widetilde{S}_2\underline{S}_3} = \widehat{S}_3\widehat{S}_2^T - \widehat{S}_2\widehat{S}_3^T \tag{1.153}
$$

Obviously, the above identity, Eq. (1.153), is only a necessary condition that satisfies the homogeneous equations, not the absolute conditions. This straight forward approach is only applied to obtain this identity from the homogeneous equations. It is necessary to note that the strict proof of identity, Eq. (1.153) can be accomplished by expanding the dual products with the help of Eqs. (1.114), (1.72) and (1.98). In view of Eq. (1.128), the dual tensor products appeared in the right hand side of the above equation can be replaced by

$$\widetilde{\widetilde{S}_2\underline{S}_3} = (\widetilde{S}_2\widetilde{S}_3 + \widehat{S}_3^T\widehat{S}_2\mathcal{I}) - (\widetilde{S}_3\widetilde{S}_2 + \widehat{S}_2^T\widehat{S}_3\mathcal{I}) = \widetilde{S}_2\widetilde{S}_3 - \widetilde{S}_3\widetilde{S}_2 \tag{1.154}$$

This identity will remain unchanged for any two dual vectors $\underline{S}_1$ and $\underline{S}_2$, and hence

$$\widetilde{\widetilde{S}_1\underline{S}_2} = \widetilde{S}_1\widetilde{S}_2 - \widetilde{S}_2\widetilde{S}_1 \tag{1.155}$$

## 1.3 Tensor

It is known that tensors are mathematical or physical concepts which follow certain specific laws of transformation when there is a change in the coordinate system. In this section, the fundamental attributes of second-order tensors are introduced at first, and then tensors calculus are defined. Some kinds of second-order tensors, such as rotation tensors, curvature tensors and metric tensors of curve, surface and three-dimensional entity, which are frequently appeared in the rest of the book, will be presented next. Finally, the motion tensor [5], an important physical concept, is addressed in detail for incremental motion with small amplitude.

### *1.3.1 Second-order Tensors*

Mathematically, the second-order tensor is a $3 \times 3$ matrix containing 9 elements. Since the tensor must obey certain laws of transformation, the 9 elements of the second-order tensor are not totally independent in general. For example, the rotation tensor, also a second-order tensor, is precisely defined by 3 independent parameters only even though it is usually depicted by a full rank matrix containing 9 non-zero elements. Note that the second-order tensors automatically satisfy the fundamental algebra operations of the matrix. Such as the determinant of a second-order tensor can be computed from

$$\det(A) = \begin{vmatrix} a_{11} & a_{12} & a_{13} \\ a_{21} & a_{22} & a_{23} \\ a_{31} & a_{32} & a_{33} \end{vmatrix} \tag{1.156}$$

or alternatively

$$\begin{aligned} \det(A) = & \, a_{11}a_{22}a_{33} + a_{12}a_{23}a_{31} + a_{13}a_{32}a_{21} \\ & -a_{23}a_{32}a_{11} - a_{31}a_{13}a_{22} - a_{12}a_{21}a_{33} \end{aligned} \tag{1.157}$$

The trace of a second-order tensor is a scalar defined as

$$\text{tra}(A) = a_{11} + a_{22} + a_{33} \tag{1.158}$$

A second-order tensor is said to be a symmetric tensor if $a_{ij} = a_{ji}$. On the contrary, a second-order tensor is said to be a skew-symmetric tensor if $a_{ij} = -a_{ji}$. This implies that the diagonal terms vanish, $a_{ii} = 0$ for $i = 1, 2, 3$. By the definitions, an arbitrary tensor can always be decomposed into its symmetric part and skew symmetric part

$$A = \frac{A + A^T}{2} + \frac{A - A^T}{2} = \text{sym}(A) + \text{ske}(A) \tag{1.159}$$

where sym$(A)$ represents the symmetric part of the tensor

$$\text{sym}(A) = \frac{1}{2}\begin{bmatrix} 2a_{11} & a_{12} + a_{21} & a_{13} + a_{31} \\ a_{21} + a_{12} & 2a_{22} & a_{23} + a_{32} \\ a_{31} + a_{13} & a_{32} + a_{23} & 2a_{33} \end{bmatrix} \tag{1.160}$$

and ske$(A)$ its skew-symmetric part

$$\text{ske}(A) = \frac{1}{2}\begin{bmatrix} 0 & -(a_{21} - a_{12}) & a_{13} - a_{31} \\ a_{21} - a_{12} & 0 & -(a_{32} - a_{23}) \\ -(a_{13} - a_{31}) & a_{32} - a_{23} & 0 \end{bmatrix} \tag{1.161}$$

The axial vector $\underline{a}$ associated with a second-order tensor $A$ is defined as follows

$$\underline{a} = \text{axi}(A) \Leftrightarrow \tilde{a} = \text{ske}(A) \tag{1.162}$$

From which, it is readily verified that

$$\underline{a} = \begin{vmatrix} a_1 \\ a_2 \\ a_3 \end{vmatrix} = \frac{1}{2}\begin{vmatrix} a_{32} - a_{23} \\ a_{13} - a_{31} \\ a_{21} - a_{12} \end{vmatrix} \tag{1.163}$$

### 1.3.2 Tensors Calculus

The derivative of a scalar function $s(t)$ with respect to time $t$ is defined as

$$\frac{ds}{dt} = \dot{s} = \lim_{\Delta t \to 0} \frac{s(t + \Delta t) - s(t)}{\Delta t} \tag{1.164}$$

The notation $\dot{(\cdot)}$ represents a derivative with respect to time throughout this book. If there is a value of a vector $\underline{u}$ corresponding to each value of the scalar $s$, then $\underline{u}$ is said to be a function of $s$. The derivative of $\underline{u}(s)$ with respect to $s$ is denoted by

$$\frac{d\underline{u}}{ds} = \lim_{\Delta s \to 0} \frac{\underline{u}(s + \Delta s) - \underline{u}(s)}{\Delta s} \tag{1.165}$$

Following the chain rule for differentiation, the time derivative of vector $\underline{u}$ can be computed from

$$\frac{d\underline{u}}{dt} = \frac{d\underline{u}}{ds}\frac{ds}{dt} \tag{1.166}$$

Let $\underline{u}(t)$ and $\underline{v}(t)$ be any two vectors which are time functions, the following formulas hold on

$$\begin{aligned}
\frac{d}{dt}(\underline{u}+\underline{v}) &= \dot{\underline{u}} + \dot{\underline{v}} \\
\frac{d}{dt}(s\underline{u}) &= \dot{s}\underline{u} + s\dot{\underline{u}} \\
\frac{d}{dt}(\underline{u}^T\underline{v}) &= \dot{\underline{u}}^T\underline{v} + \underline{u}^T\dot{\underline{v}} \\
\frac{d}{dt}(\widetilde{u}\,\underline{v}) &= \dot{\widetilde{u}}\,\underline{v} - \widetilde{v}\,\dot{\underline{u}}
\end{aligned} \tag{1.167}$$

Given an unit vector $\bar{n}$ and such that $\bar{n}^T\bar{n} = 1$, its time derivative is orthogonal to the unit vector itself

$$\frac{d}{dt}(\bar{n}^T\bar{n}) = 2\dot{\bar{n}}^T\bar{n} = 0 \tag{1.168}$$

Similarly, considering two mutually orthogonal vectors $\underline{u}$ and $\underline{v}$ with the attribute of $\underline{u}^T\underline{v} = 0$, time derivative of this expression yields

$$\dot{\underline{u}}^T\underline{v} = -\underline{u}^T\dot{\underline{v}} \tag{1.169}$$

If vector $\underline{u}(t)$ is a function of time $t$, it means all components of the vector are time functions and then

$$\dot{\underline{u}} = \dot{u}_1\bar{\imath}_1 + \dot{u}_2\bar{\imath}_2 + \dot{u}_3\bar{\imath}_3 \tag{1.170}$$

### *1.3.3 Rotation Tensor of a Curve*

As depicted in Fig. 1.11, a curve consists of a set of points the position vectors of which satisfy the relationship

$$\underline{p} = \underline{p}(s) \tag{1.171}$$

where $s$ is the curvilinear coordinate that measures length along the curve called the intrinsic parameterization or natural parameterization of the curve. It is noted if the set of points $\underline{p}$ comprising a curve lie in a single plane, the curve is said to be a plane curve. If this set of points does not lie in a single plane, the curve is said to be a skew curve. The square of a differential element of length $ds$ along the curve is computed from

$$ds^2 = d\underline{p}^T d\underline{p} \tag{1.172}$$

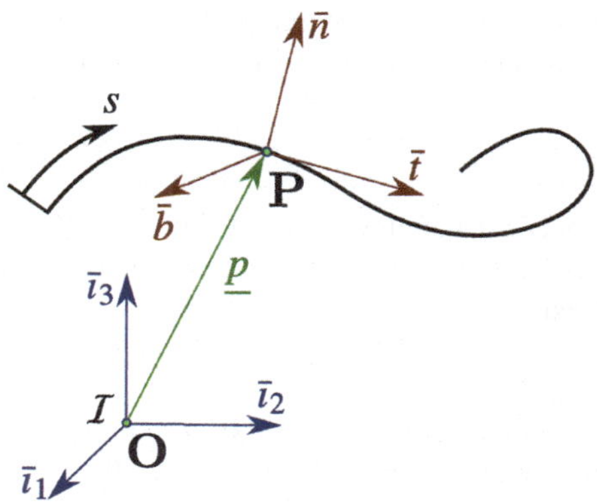


**Fig. 1.11** Configuration of a curve in space

and then it follows

$$\frac{d\underline{p}^T}{ds}\frac{d\underline{p}}{ds} = 1 \tag{1.173}$$

From which, the tangent vector to the curve is found to be

$$\bar{t} = \frac{d\underline{p}}{ds} \tag{1.174}$$

By construction, this is a unit vector because $\bar{t}^T\bar{t} = 1$. Next, according to Eq. (1.168), a derivative of this orthogonal attribute with respect to the curvilinear coordinate $s$ yields

$$\bar{t}^T\frac{d\bar{t}}{ds} = 0 \tag{1.175}$$

From this orthogonal condition, the unit normal vector

$$\bar{n} = \frac{1}{\kappa}\frac{d\bar{t}}{ds} \tag{1.176}$$

perpendicular to the tangent vector is readily defined, together with the curvature of the curve

$$\kappa = \|\frac{d\bar{t}}{ds}\| \tag{1.177}$$

is introduced. An orthonormal basis, also called the principal triad or Serret-Frenet's triad, can be constructed by defining the binormal vector as the cross product of the tangent by the normal vectors

$$\bar{b} = \widetilde{t}\,\bar{n} \tag{1.178}$$

Note that the plane, determined by two unit vectors $\bar{t}$ and $\bar{n}$ and perpendicular to $\bar{b}$, is called the osculating plane of the curve. The plane through point $\underline{p}$ and perpendicular to $\bar{t}$ is called the normal plane of the curve. By assembling the mutually orthogonal unit vectors $\bar{t}$, $\bar{n}$ and $\bar{b}$ into a $3 \times 3$ matrix, an rotation tensor

$$R_F = \lfloor \bar{t}, \bar{n}, \bar{b} \rfloor \tag{1.179}$$

are constructed with the attributes of $R_F^T R_F = I$. This tensor can be interpreted as the space-dependent rotation tensor that bring the reference Cartesian basis $\mathcal{I}$ to Frenet's basis

$$\mathcal{B}_F(\bar{t}, \bar{n}, \bar{b}) = R_F = \lfloor \bar{t}, \bar{n}, \bar{b} \rfloor \tag{1.180}$$

Given an arbitrary point **P** on the curve, and associated Frenet's basis $\mathcal{B}_F$, a body attached reference frame, $\mathcal{F}_F = [\mathbf{P}, \mathcal{B}_F]$, is then defined to describe the orientation of a skew curve.

### 1.3.4 Curvature Tensor of a Curve

The curvature tensor of a curve can be computed from the derivatives of Frenet's basis with respect to the curvilinear coordinate $s$, which can be resolved in Frenet's basis in the form of

$$\begin{aligned}\frac{d\bar{t}}{ds} &= c_{11}\,\bar{t} + c_{12}\,\bar{n} + c_{13}\,\bar{b}\\ \frac{d\bar{n}}{ds} &= c_{21}\,\bar{t} + c_{22}\,\bar{n} + c_{23}\,\bar{b}\\ \frac{d\bar{b}}{ds} &= c_{31}\,\bar{t} + c_{32}\,\bar{n} + c_{33}\,\bar{b}\end{aligned} \tag{1.181}$$

where $c_{ij}$ for $i, j = 1, 2, 3$ are unknown coefficients. Pre-multiplying these equations by unit vectors $\bar{t}^T$, $\bar{n}^T$ and $\bar{b}^T$, respectively, yields

$$\begin{aligned}c_{11} &= \bar{t}^T \frac{d\bar{t}}{ds}, & c_{12} &= \bar{n}^T \frac{d\bar{t}}{ds}, & c_{13} &= \bar{b}^T \frac{d\bar{t}}{ds}\\ c_{21} &= \bar{t}^T \frac{d\bar{n}}{ds}, & c_{22} &= \bar{n}^T \frac{d\bar{n}}{ds}, & c_{23} &= \bar{b}^T \frac{d\bar{n}}{ds}\\ c_{31} &= \bar{t}^T \frac{d\bar{b}}{ds}, & c_{32} &= \bar{n}^T \frac{d\bar{b}}{ds}, & c_{33} &= \bar{b}^T \frac{d\bar{b}}{ds}\end{aligned} \tag{1.182}$$

Similar to Eq. (1.175), the derivatives of orthogonal attributes, $\bar{n}^T\bar{n} = 1$ and $\bar{b}^T\bar{b} = 1$, reveal the coefficients $c_{11}$, $c_{22}$ and $c_{33}$ are all zeros. The rest unknowns can be determined from the specific scalar product of vectors. At first, the spatial derivative of expression, $\bar{t}^T\bar{n} = 0$, leads to

$$\frac{d(\bar{t}^T\bar{n})}{ds} = \bar{n}^T \frac{d\bar{t}}{ds} + \bar{t}^T \frac{d\bar{n}}{ds} = \kappa \bar{n}^T \bar{n} + \bar{t}^T \frac{d\bar{n}}{ds} = \kappa + \bar{t}^T \frac{d\bar{n}}{ds} = 0 \tag{1.183}$$

From which the following result

$$c_{12} = \kappa, \quad c_{21} = \bar{t}^T \frac{d\bar{n}}{ds} = -\kappa \tag{1.184}$$

is obtained. Next, the derivatives of mutually orthogonal attributes, $\bar{b}^T \bar{n} = 0$, with respect to $s$ is applied to get

$$\frac{d(\bar{b}^T \bar{n})}{ds} = \bar{n}^T \frac{d\bar{b}}{ds} + \bar{b}^T \frac{d\bar{n}}{ds} = 0 \tag{1.185}$$

Now the twist $\tau$ of the curve is defined

$$\bar{n}^T \frac{d\bar{b}}{ds} = -\bar{b}^T \frac{d\bar{n}}{ds} = -\tau \tag{1.186}$$

and two unknown coefficients are determined to be $c_{32} = -\tau$ and $c_{23} = \tau$. Finally, the spatial derivatives of mutually orthogonal attributes, $\bar{b}^T \bar{t} = 0$, yields

$$\frac{d(\bar{b}^T \bar{t})}{ds} = \bar{t}^T \frac{d\bar{b}}{ds} + \bar{b}^T \frac{d\bar{t}}{ds} = 0 \tag{1.187}$$

Applying the orthogonal condition $\bar{b}^T \bar{n} = 0$ leads to

$$\bar{t}^T \frac{d\bar{b}}{ds} = -\bar{b}^T \frac{d\bar{t}}{ds} = -\kappa \bar{b}^T \bar{n} = 0 \tag{1.188}$$

It answers the last two unknown coefficients become zero, $c_{31} = c_{13} = 0$. In summary, the derivatives of Frenet's triad with respect to the curvilinear coordinate $s$ are determined to be

$$\begin{aligned} \frac{d\bar{t}}{ds} &= \quad 0\bar{t} + \kappa\bar{n} + 0\bar{b} \\ \frac{d\bar{n}}{ds} &= -\kappa\bar{t} + 0\bar{n} + \tau\bar{b} \\ \frac{d\bar{b}}{ds} &= \quad 0\bar{t} - \tau\bar{n} + 0\bar{b} \end{aligned} \tag{1.189}$$

It can be assembled into matrix form

$$\frac{d}{ds}\lfloor \bar{t}, \bar{n}, \bar{b} \rfloor = \lfloor \bar{t}, \bar{n}, \bar{b} \rfloor \begin{bmatrix} 0 & -\kappa & 0 \\ \kappa & 0 & -\tau \\ 0 & \tau & 0 \end{bmatrix} \tag{1.190}$$

or in compact equivalently

$$\frac{d}{ds}R_F = R_F\widetilde{\kappa} \tag{1.191}$$

where the curvature tensor, a skew-symmetric tensor also, is defined

$$\widetilde{\kappa} = R_F^T R_F' \tag{1.192}$$

with the notation

$$(\cdot)' = \frac{d(\cdot)}{ds} \tag{1.193}$$

is used throughout the book to represent a derivative with respect to curvilinear coordinates $s$ along the curve. The curvature vector

$$\underline{\kappa}^T = \lfloor \tau,\ 0,\ \kappa \rfloor \tag{1.194}$$

contains only two non-vanishing components, the twist and curvature of the curve when this curvature vector resolved in Frenet's basis.

### *1.3.5 The First Metric Tensor of a Surface*

As depicted in Fig. 1.12, a surface is defined as the locus of the points generated by two parameters, $\alpha_1$ and $\alpha_2$. The position vector of such points can be expressed as

$$\underline{p} = \underline{p}(\alpha_1, \alpha_2) \tag{1.195}$$

When one parameter remains unchanged and the other parameter varies, the position vector will define a curve embedded in the surface. Such kinds of two curves are shown in Fig. 1.12 and called $\alpha_1$ curve and $\alpha_2$ curve, respectively. The surface base vectors are defined as follows

$$\underline{a}_1 = \frac{\partial \underline{p}}{\partial \alpha_1}, \quad \underline{a}_2 = \frac{\partial \underline{p}}{\partial \alpha_2} \tag{1.196}$$

Apparently, vectors $\underline{a}_1$ and $\underline{a}_2$ are tangent to the $\alpha_1$ and $\alpha_2$ curves, and lie in the plane tangent to the surface at point **P** consequently. Note that parameters $\alpha_1$ and $\alpha_2$ are two arbitrary parameters, not a natural parameterization, and hence vectors $a_1$ and $a_2$ are not unit tangent vectors, and are not orthogonal to each other as well. The first metric tensor $A$ of a surface is denoted as

$$A = \begin{bmatrix} a_{11} & a_{12} \\ a_{21} & a_{22} \end{bmatrix} = \begin{bmatrix} \underline{a}_1^T \underline{a}_1 & \underline{a}_1^T \underline{a}_2 \\ \underline{a}_2^T \underline{a}_1 & \underline{a}_2^T \underline{a}_2 \end{bmatrix} \tag{1.197}$$

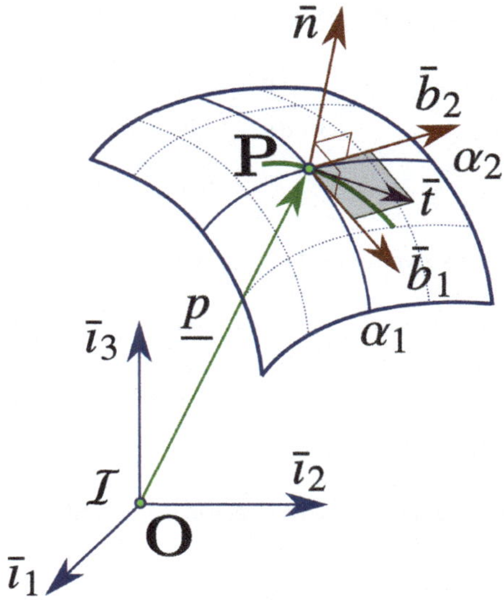


**Fig. 1.12** Configuration of a surface in space

Because the base vectors define the plane tangent to the surface, the unit vector $\bar{n}$ normal to the surface is readily found as

$$\bar{n} = \frac{\widetilde{a}_1 \underline{a}_2}{\|\widetilde{a}_1 \underline{a}_2\|} = \frac{\widetilde{a}_1 \underline{a}_2}{\sqrt{a}} \tag{1.198}$$

The area of a differential element of the surface then becomes

$$dA = \|\widetilde{a}_1 \underline{a}_2 d\alpha_1 d\alpha_2\| = \|\widetilde{a}_1 \underline{a}_2\| d\alpha_1 d\alpha_2 = \sqrt{a} d\alpha_1 d\alpha_2 \tag{1.199}$$

where the determinant of the first metric tensor, $a = \det(A)$, can be computed from

$$a = \|\widetilde{a}_1 \underline{a}_2\|^2 = \underline{a}_1^T \underline{a}_1 \cdot \underline{a}_2^T \underline{a}_2 - (\underline{a}_1^T \underline{a}_2)^2 = a_{11} a_{22} - a_{12} a_{21} \tag{1.200}$$

As shown in Fig. 1.12, a curve is entirely contained within a surface. Let the curve be defined by its natural parameter $s$, the curvilinear coordinate along curve, its unit tangent vector $\bar{t}$ is defined by Eq. (1.174). The scalar product of tangent vector by itself

$$\bar{t}^T \bar{t} = \frac{d\underline{p}^T}{ds} \frac{d\underline{p}}{ds} = 1 \tag{1.201}$$

can be rewritten into the detailed expression like

$$\frac{d\underline{p}^T}{ds}\frac{d\underline{p}}{ds} = (\frac{\partial \underline{p}^T}{\partial \alpha_1}\frac{d\alpha_1}{ds} + \frac{\partial \underline{p}^T}{\partial \alpha_2}\frac{d\alpha_2}{ds})(\frac{\partial \underline{p}}{\partial \alpha_1}\frac{d\alpha_1}{ds} + \frac{\partial \underline{p}}{\partial \alpha_2}\frac{d\alpha_2}{ds})$$
$$= (\underline{a}_1^T\frac{d\alpha_1}{ds} + \underline{a}_2^T\frac{d\alpha_2}{ds})(\underline{a}_1\frac{d\alpha_1}{ds} + \underline{a}_2\frac{d\alpha_2}{ds}) \tag{1.202}$$

or resolved into the matrix form

$$\frac{d\underline{p}^T}{ds}\frac{d\underline{p}}{ds} = \underline{\varphi}^T \begin{bmatrix} \underline{a}_1^T \\ \underline{a}_2^T \end{bmatrix} \lfloor \underline{a}_1, \ \underline{a}_2 \rfloor \underline{\varphi}$$
$$= \underline{\varphi}^T \begin{bmatrix} \underline{a}_1^T\underline{a}_1 & \underline{a}_1^T\underline{a}_2 \\ \underline{a}_2^T\underline{a}_1 & \underline{a}_2^T\underline{a}_2 \end{bmatrix} \underline{\varphi} \tag{1.203}$$

where the following vector is defined

$$\underline{\varphi}^T = \lfloor \frac{d\alpha_1}{ds}, \ \frac{d\alpha_2}{ds} \rfloor \tag{1.204}$$

Finally, the attribute, $\bar{t}^T\bar{t} = 1$, of unit tangent vector is applied to obtain the following identity

$$\underline{\varphi}^T A \underline{\varphi} = 1 \tag{1.205}$$

### *1.3.6 The Second Metric Tensor of a Surface*

As discussed in Sect. 1.3.5, the unit tangent vector $\bar{t}$ on a curve entirely contained within a surface clearly lies in the plane tangent to the surface. Hence, $\bar{t}$ is perpendicular to the unit normal vector, Eq. (1.198), of the surface. However it does not mean the unit normal vector $\bar{n}$ of the surface coincides with the unit normal vector $\bar{n}_c$ of the curve. In general, they are two distinct normal vectors. By Eq. (1.176), the curvature vector $\bar{n}_c$ of a curve

$$\frac{d\bar{t}}{ds} = \kappa\,\bar{n}_c \tag{1.206}$$

have components in and out of the tangent plane of a surface like

$$\frac{d\bar{t}}{ds} = \kappa_n\bar{n} + \kappa_g\bar{t}_g \tag{1.207}$$

where $\bar{t}_g$ lies in this tangent plane, and $\kappa_g$ the associated geodesic curvature. The normal curvature $\kappa_n$ can be computed from

$$\kappa_n = \bar{n}^T \frac{d\bar{t}}{ds} \tag{1.208}$$

with the aid of mutually orthogonal conditions $\bar{n}^T \bar{t}_g = 0$. By Eq. (1.169), the normal curvature $\kappa_n$ is recalculated as

$$\kappa_n = -\bar{t}^T \frac{d\bar{n}}{ds} = -\frac{d\underline{p}^T}{ds}\frac{d\bar{n}}{ds} \tag{1.209}$$

According to the chain rule of derivatives, the normal curvature $\kappa_n$ can be rewritten in detail like

$$\begin{aligned} -\frac{d\underline{p}^T}{ds}\frac{d\bar{n}}{ds} &= -(\frac{\partial \underline{p}^T}{\partial \alpha_1}\frac{d\alpha_1}{ds} + \frac{\partial \underline{p}^T}{\partial \alpha_2}\frac{d\alpha_2}{ds})(\frac{\partial \bar{n}}{\partial \alpha_1}\frac{d\alpha_1}{ds} + \frac{\partial \bar{n}}{\partial \alpha_2}\frac{d\alpha_2}{ds}) \\ &= -(\underline{a}_1^T \frac{d\alpha_1}{ds} + \underline{a}_2^T \frac{d\alpha_2}{ds})(\frac{\partial \bar{n}}{\partial \alpha_1}\frac{d\alpha_1}{ds} + \frac{\partial \bar{n}}{\partial \alpha_2}\frac{d\alpha_2}{ds}) \end{aligned} \tag{1.210}$$

When resolving $\kappa_n$ into matrix form, it can be recasted to

$$\begin{aligned} -\frac{d\underline{p}^T}{ds}\frac{d\bar{n}}{ds} &= -\underline{\varphi}^T \begin{bmatrix} \underline{a}_1^T \\ \underline{a}_2^T \end{bmatrix} \lfloor \frac{\partial \bar{n}}{\partial \alpha_1}, \frac{\partial \bar{n}}{\partial \alpha_2} \rfloor \underline{\varphi} \\ &= -\underline{\varphi}^T \begin{bmatrix} \underline{a}_1^T \frac{\partial \bar{n}}{\partial \alpha_1} & \underline{a}_1^T \frac{\partial \bar{n}}{\partial \alpha_2} \\ \underline{a}_2^T \frac{\partial \bar{n}}{\partial \alpha_1} & \underline{a}_2^T \frac{\partial \bar{n}}{\partial \alpha_2} \end{bmatrix} \underline{\varphi} = -\underline{\varphi}^T \begin{bmatrix} \underline{a}_1^T \bar{n}_{,1} & \underline{a}_1^T \bar{n}_{,2} \\ \underline{a}_2^T \bar{n}_{,1} & \underline{a}_2^T \bar{n}_{,2} \end{bmatrix} \underline{\varphi} \end{aligned} \tag{1.211}$$

where the symbol $(\cdot)_{,i}$ presents the derivatives with respect to the parameters $\alpha_i$ like

$$(\cdot)_{,i} = \frac{\partial(\cdot)}{\partial \alpha_i} \tag{1.212}$$

for $i = 1, 2$. From which, the second metric tensor of the surface is observed to be

$$B = \begin{bmatrix} b_{11} & b_{12} \\ b_{21} & b_{22} \end{bmatrix} = -\begin{bmatrix} \underline{a}_1^T \bar{n}_{,1} & \underline{a}_1^T \bar{n}_{,2} \\ \underline{a}_2^T \bar{n}_{,1} & \underline{a}_2^T \bar{n}_{,2} \end{bmatrix} \tag{1.213}$$

with the determinant is defined as $b = \det(B)$. Finally, the normal curvature can be computed from

$$\kappa_n = \underline{\varphi}^T B \underline{\varphi} \tag{1.214}$$

### 1.3.7 Rotation Tensor of a Surface

Given a point **P** with the position vector $\underline{p}(\alpha_1, \alpha_2)$ on a surface, it can be imagining there exist numerous curves on the surface that pass through the point. Each curve features a distinct normal curvature $\kappa_n$ forming a set of normal curvatures. The maximum value of normal curvatures is called the principal curvature. The principal curvature can be sought from Eq. (1.214) under the normality constraint, Eq. (1.205). With the application of Lagrange multiplier to enforce the constraint, this maximum problem can be expressed mathematically as

$$\begin{aligned} f(\kappa_n^{\max}) &= \max_{\underline{\varphi},\lambda}\left[\kappa_n - \lambda(\underline{\varphi}^T A\underline{\varphi} - 1)\right] \\ &= \max_{\underline{\varphi},\lambda}\left[\underline{\varphi}^T B\underline{\varphi} - \lambda(\underline{\varphi}^T A\underline{\varphi} - 1)\right] \end{aligned} \tag{1.215}$$

Taking the first derivative of maximum function with respect to variables $\underline{\varphi}$ leads to

$$\frac{\partial f(\kappa_n^{\max})}{\partial \underline{\varphi}} = (B - \lambda A)\underline{\varphi} = 0 \tag{1.216}$$

Pre-multiplying above equation by $\underline{\varphi}^T$ yields the physical interpretation of the Lagrange multiplier

$$\underline{\varphi}^T(B - \lambda A)\underline{\varphi} = 0 \tag{1.217}$$

In view of the attribute of first metric tensor, $\underline{\varphi}^T A\underline{\varphi} = 1$, it is readily verified that

$$\lambda = \underline{\varphi}^T B\underline{\varphi} = \kappa_n \tag{1.218}$$

which means the Lagrange's multiplier can be interpreted as the normal curvature itself. Replacing the Lagrange's multiplier $\lambda$ by the normal curvature $\kappa_n$ in Eq. (1.216), the following harmonic equations will be obtained

$$\begin{bmatrix} b_{11} - \kappa_n a_{11} & b_{12} - \kappa_n a_{12} \\ b_{21} - \kappa_n a_{21} & b_{22} - \kappa_n a_{22} \end{bmatrix}\underline{\varphi} = 0 \tag{1.219}$$

When the non-trivial solutions exist, $\underline{\varphi} \neq 0$, it corresponds to the quadratic equation, $\det(B - \kappa_n A) = 0$, has a solution. In fact, this is a parabolic equation

$$(b_{11} - \kappa_n a_{11})(b_{22} - \kappa_n a_{22}) - (b_{12} - \kappa_n a_{12})(b_{21} - \kappa_n a_{21}) = 0 \tag{1.220}$$

and can be simplified to

$$\kappa_n^2 - 2\kappa_0\kappa_n + \frac{b}{a} = 0 \tag{1.221}$$

where

$$\kappa_0 = \frac{a_{11}b_{22} + a_{22}b_{11} - 2a_{12}b_{12}}{2a} \tag{1.222}$$

called mean curvature, and the symbols $a$ and $b$ are the determinants of the first metric tensor $A$ and the second metric tensor $B$, respectively. The solutions of this quadratic equation will be the principal curvatures

$$\kappa_1, \kappa_2 = \kappa_0 \pm \sqrt{\kappa_0^2 - \frac{b}{a}} \tag{1.223}$$

The mean curvature $\kappa_0$ is related to

$$\kappa_0 = \frac{\kappa_1 + \kappa_2}{2} \tag{1.224}$$

and the Gaussian curvature is defined as

$$\kappa_1\kappa_2 = \frac{b}{a} \tag{1.225}$$

After computing the principal curvatures from Eq. (1.223), a line of curvature of a surface is defined as: a curve whose tangent vector always points along the principal curvature directions of the surface. Selecting two special parameters $\alpha_1$ and $\alpha_2$, such that $a_{12} = b_{12} = 0$. By Eq. (1.223),

$$\kappa_1 = \frac{b_{11}}{a_{11}}, \quad \kappa_2 = \frac{b_{22}}{a_{22}} \tag{1.226}$$

On the other hand, according to Eq. (1.205), $a_{12} = b_{12} = 0$, leads to

$$\underline{\varphi}^T = \lfloor \frac{1}{\sqrt{a_{11}}}, 0 \rfloor \quad \text{or} \quad \underline{\varphi}^T = \lfloor 0, \frac{1}{\sqrt{a_{22}}} \rfloor \tag{1.227}$$

then from the definition of normal curvatures, Eq. (1.214), it is readily verified that

$$\kappa_n = \frac{b_{11}}{a_{11}}, \quad \text{or} \quad \kappa_n = \frac{b_{22}}{a_{22}} \tag{1.228}$$

It is now clear when $a_{12} = b_{12} = 0$, the tangent vectors $\underline{a}_1$ and $\underline{a}_2$ are mutually orthogonal equivalently. The associated $\alpha_1$ and $\alpha_2$ curves are indeed the lines of curvatures.

Apparently, the tangent vectors, $\underline{a}_1$ and $\underline{a}_2$, and the normal vector $\bar{n}$ form a set of mutually orthogonal vectors, although the first two are not necessarily unit vectors. A rotation tensor can be constructed by using the set of mutually orthogonal vectors like

$$R_S = \lfloor \bar{b}_1, \bar{b}_2, \bar{b}_3 \rfloor \tag{1.229}$$

where the unit vectors are selected to be

$$\bar{b}_1 = \frac{\underline{a}_1}{\|\underline{a}_1\|}, \quad \bar{b}_2 = \frac{\underline{a}_2}{\|\underline{a}_2\|}, \quad \bar{b}_3 = \bar{n} \tag{1.230}$$

Next, the chain rule for derivatives is used to write

$$\underline{a}_1 = \frac{\partial \underline{p}}{\partial \alpha_1} = \frac{\partial \underline{p}}{\partial s_1}\frac{ds_1}{d\alpha_1} = \frac{ds_1}{d\alpha_1}\bar{b}_1 \tag{1.231}$$

where the parameter $s_1$ is specified to be the arc length measured along the $\alpha_1$ curve. Hence, $\bar{b}_1$ becomes the unit tangent vector to the $\alpha_1$ curve, see Eq. (1.174), it then follows that

$$\begin{aligned} \|\underline{a}_1\| &= h_1 = \frac{ds_1}{d\alpha_1}\|\bar{b}_1\| = \frac{ds_1}{d\alpha_1} \\ \|\underline{a}_2\| &= h_2 = \frac{ds_2}{d\alpha_2}\|\bar{b}_2\| = \frac{ds_2}{d\alpha_2} \end{aligned} \tag{1.232}$$

where $h_i$ for $i = 1, 2$ is scale factors, the ratios of the infinitesimal increment in length, $ds_i$, to the infinitesimal increment in parameter $d\alpha_i$, along the curve. Meanwhile, three mutually orthogonal vectors, $\bar{b}_1$, $\bar{b}_2$, and $\bar{b}_3$, set up a surface basis like

$$\mathcal{B}(\bar{b}_1, \bar{b}_2, \bar{b}_3) = R_S = \lfloor \bar{b}_1, \bar{b}_2, \bar{b}_3 \rfloor \tag{1.233}$$

Given a point **P**, and surface basis $\mathcal{B}$, a body attached reference frame, $\mathcal{F}_\mathcal{B} = [\mathbf{P}, \mathcal{B}]$, is then defined to describe the orientation of point **P** on a spatial surface.

Additionally, the lines of curvatures are frequently used to simplify the geometric description of a shell mid-surface in the modeling of shell element in this book. It is necessary to afford an efficient approach to compute the principal curvatures when the metric tensors have been determined by using two arbitrary parameters $\beta_1$ and $\beta_2$. An efficient approach relates the base vectors defined in two different curvilinear coordinate systems by

$$\begin{aligned} \underline{a}_1 &= \frac{\partial \underline{p}_0}{\partial \alpha_1} = \frac{\partial \underline{p}_0}{\partial \beta_1}\frac{\partial \beta_1}{\partial \alpha_1} + \frac{\partial \underline{p}_0}{\partial \beta_2}\frac{\partial \beta_2}{\partial \alpha_1} = \underline{a}_r\frac{\partial \beta_1}{\partial \alpha_1} + \underline{a}_s\frac{\partial \beta_2}{\partial \alpha_1} \\ \underline{a}_2 &= \frac{\partial \underline{p}_0}{\partial \alpha_2} = \frac{\partial \underline{p}_0}{\partial \beta_1}\frac{\partial \beta_1}{\partial \alpha_2} + \frac{\partial \underline{p}_0}{\partial \beta_2}\frac{\partial \beta_2}{\partial \alpha_2} = \underline{a}_r\frac{\partial \beta_1}{\partial \alpha_2} + \underline{a}_s\frac{\partial \beta_2}{\partial \alpha_2} \end{aligned} \tag{1.234}$$

or equivalently

$$\left|\begin{matrix} \underline{a}_1^T \\ \underline{a}_2^T \end{matrix}\right| = \begin{bmatrix} \frac{\partial \beta_1}{\partial \alpha_1} & \frac{\partial \beta_2}{\partial \alpha_1} \\ \frac{\partial \beta_1}{\partial \alpha_2} & \frac{\partial \beta_2}{\partial \alpha_2} \end{bmatrix} \left|\begin{matrix} \underline{a}_r^T \\ \underline{a}_s^T \end{matrix}\right| \tag{1.235}$$

From which, the following Jacobian matrix

$$\mathcal{J} = \begin{bmatrix} \frac{\partial\beta_1}{\partial\alpha_1} & \frac{\partial\beta_2}{\partial\alpha_1} \\ \frac{\partial\beta_1}{\partial\alpha_2} & \frac{\partial\beta_2}{\partial\alpha_2} \end{bmatrix} \tag{1.236}$$

is introduced to map the first metric tensors in two different coordinate systems

$$\begin{vmatrix} \underline{a}_1^T \\ \underline{a}_2^T \end{vmatrix} \lfloor \underline{a}_1, \ \underline{a}_2 \rfloor = \mathcal{J} \begin{vmatrix} \underline{a}_r^T \\ \underline{a}_s^T \end{vmatrix} \lfloor \underline{a}_r, \ \underline{a}_s \rfloor \mathcal{J}^T \tag{1.237}$$

and in detail

$$\begin{bmatrix} a_{11} & a_{12} \\ a_{21} & a_{22} \end{bmatrix} = \begin{bmatrix} J_{11} & J_{12} \\ J_{21} & J_{22} \end{bmatrix} \begin{bmatrix} a_{rr} & a_{rs} \\ a_{sr} & a_{ss} \end{bmatrix} \begin{bmatrix} J_{11} & J_{21} \\ J_{12} & J_{22} \end{bmatrix} \tag{1.238}$$

Now the problem can be stated clearly: the components $a_{rr}$, $a_{ss}$ and $a_{rs}$ are given and $a_{12} = a_{21} = 0$ with undetermined Jacobian matrix $\mathcal{J}$, the unknowns of principal curvatures, $a_{11}$ and $a_{22}$ need to be solved. The same situations are also suitable for the second metric tensor

$$\begin{bmatrix} b_{11} & b_{12} \\ b_{21} & b_{22} \end{bmatrix} = \begin{bmatrix} J_{11} & J_{12} \\ J_{21} & J_{22} \end{bmatrix} \begin{bmatrix} b_{rr} & b_{rs} \\ b_{sr} & b_{ss} \end{bmatrix} \begin{bmatrix} J_{11} & J_{21} \\ J_{12} & J_{22} \end{bmatrix} \tag{1.239}$$

Even though Eqs. (1.238) and (1.239) will be used to solve the principal curvatures and the components of Jacobians, the number of equations is one less than the number of unknowns. Hence, an additional constraint, $J_{11} = J_{22} = 1$, needs to be provided to get

$$\begin{bmatrix} a_{11} & 0 \\ 0 & a_{22} \end{bmatrix} = \begin{bmatrix} 1 & J_{12} \\ J_{21} & 1 \end{bmatrix} \begin{bmatrix} a_{rr} & a_{rs} \\ a_{sr} & a_{ss} \end{bmatrix} \begin{bmatrix} 1 & J_{21} \\ J_{12} & 1 \end{bmatrix} \tag{1.240}$$

and

$$\begin{bmatrix} b_{11} & 0 \\ 0 & b_{22} \end{bmatrix} = \begin{bmatrix} 1 & J_{12} \\ J_{21} & 1 \end{bmatrix} \begin{bmatrix} b_{rr} & b_{rs} \\ b_{sr} & b_{ss} \end{bmatrix} \begin{bmatrix} 1 & J_{21} \\ J_{12} & 1 \end{bmatrix} \tag{1.241}$$

From which, the unknowns $a_{11}, a_{22}, b_{11}, b_{22}$ and $J_{12}$, $J_{21}$ are solved from the associated parabolic equations as follow

$$\begin{aligned} a_{11} &= a_{rr} + 2J_{12}a_{rs} + J_{12}^2 a_{ss}, \quad & b_{11} &= b_{rr} + 2J_{12}b_{rs} + J_{12}^2 b_{ss} \\ a_{22} &= a_{ss} + 2J_{21}a_{rs} + J_{21}^2 a_{rr}, \quad & b_{22} &= b_{ss} + 2J_{21}b_{rs} + J_{21}^2 b_{rr} \end{aligned} \tag{1.242}$$

and

$$J_{12} = \frac{(a_{rr}b_{ss} - a_{ss}b_{rr}) \pm \sqrt{(a_{ss}b_{rr} - a_{rr}b_{ss})^2 + \Delta}}{2(a_{ss}b_{rs} - a_{rs}b_{ss})}$$
$$J_{21} = \frac{-(a_{rr}b_{ss} - a_{ss}b_{rr}) \pm \sqrt{(a_{rr}b_{ss} - a_{ss}b_{rr})^2 + \Delta}}{2(a_{rr}b_{rs} - a_{rs}b_{rr})} \tag{1.243}$$

where

$$\Delta = 4\,(a_{rs}b_{ss} - a_{ss}b_{rs})(a_{rs}b_{rr} - a_{rr}b_{rs}) \tag{1.244}$$

## 1.3.8 Curvature Tensor of a Surface

The curvature tensors of a surface can be determined by the second-order derivatives of the base vectors $\underline{a}_1$ and $\underline{a}_2$ with respect to the parameters $\alpha_1$ and $\alpha_2$ like

$$\frac{\partial^2 \underline{p}}{\partial\alpha_1 \partial\alpha_2} = \frac{\partial \underline{a}_1}{\partial\alpha_2} = \frac{\partial \underline{a}_2}{\partial\alpha_1} = \frac{\partial(h_1\bar{b}_1)}{\partial\alpha_2} = \frac{\partial(h_2\bar{b}_2)}{\partial\alpha_1} \tag{1.245}$$

Expanding the above second-order derivatives leads to

$$\frac{\partial h_1}{\partial\alpha_2}\bar{b}_1 + h_1\frac{\partial\bar{b}_1}{\partial\alpha_2} = \frac{\partial h_2}{\partial\alpha_1}\bar{b}_2 + h_2\frac{\partial\bar{b}_2}{\partial\alpha_1} \tag{1.246}$$

Pre-multiplying this relationship by $\bar{b}_1^T$ yields

$$\frac{\partial h_1}{\partial\alpha_2}\bar{b}_1^T\bar{b}_1 + h_1\bar{b}_1^T\frac{\partial\bar{b}_1}{\partial\alpha_2} = \frac{\partial h_2}{\partial\alpha_1}\bar{b}_1^T\bar{b}_2 + h_2\bar{b}_1^T\frac{\partial\bar{b}_2}{\partial\alpha_1} \tag{1.247}$$

In view of the mutually orthogonal attributes of base vectors $\bar{b}_1$ and $\bar{b}_2$ like $\bar{b}_1^T\bar{b}_2 = 0$, $\bar{b}_1^T\bar{b}_1 = 1$ and then $\bar{b}_1^T\bar{b}_{1,2} = 0$, the above expression is simplified to

$$\bar{b}_1^T\frac{\partial\bar{b}_2}{\partial\alpha_1} = \frac{1}{h_2}\frac{\partial h_1}{\partial\alpha_2} \tag{1.248}$$

In terms of natural parameterization $s_1$ and $s_2$, the application of chain rule of derivatives yields

$$\bar{b}_1^T\frac{\partial\bar{b}_2}{\partial s_1}\frac{ds_1}{d\alpha_1} = \frac{1}{h_2}\frac{\partial h_1}{\partial s_2}\frac{ds_2}{d\alpha_2} \tag{1.249}$$

From which, the following identity is obtained finally

$$\bar{b}_1^T\frac{\partial\bar{b}_2}{\partial s_1} = \frac{1}{h_1}\frac{\partial h_1}{\partial s_2} \tag{1.250}$$

and then the first twist $\tau_1$ of the surface is introduced to be

$$\bar{b}_1^T \frac{\partial \bar{b}_2}{\partial s_1} = -\bar{b}_2^T \frac{\partial \bar{b}_1}{\partial s_1} = \frac{1}{h_1}\frac{\partial h_1}{\partial s_2} = \tau_1 \tag{1.251}$$

Next, Pre-multiplying Eq. (1.246) by $\bar{b}_2^T$ yields

$$\bar{b}_2^T \frac{\partial \bar{b}_1}{\partial \alpha_2} = \frac{1}{h_1}\frac{\partial h_2}{\partial \alpha_1} \tag{1.252}$$

and repeating the above operations again can obtain the second twist $\tau_2$ of the surface. In terms of natural parameterization $s_1$ and $s_2$, it yields

$$\bar{b}_2^T \frac{\partial \bar{b}_1}{\partial s_2}\frac{ds_2}{d\alpha_2} = \frac{1}{h_1}\frac{\partial h_2}{\partial s_1}\frac{ds_1}{\alpha_1} \tag{1.253}$$

With the aid of the definition of $h_i$, Eq. (1.232), this expression is simplified to

$$\bar{b}_2^T \frac{\partial \bar{b}_1}{\partial s_2} = \frac{1}{h_2}\frac{\partial h_2}{\partial s_1} \tag{1.254}$$

and then the second twist $\tau_2$ of the surface is defined as

$$\bar{b}_2^T \frac{\partial \bar{b}_1}{\partial s_2} = -\bar{b}_1^T \frac{\partial \bar{b}_2}{\partial s_2} = \frac{1}{h_2}\frac{\partial h_2}{\partial s_1} = \tau_2 \tag{1.255}$$

Since the parameterization defines lines of curvatures, $b_{12} = -\underline{a}_1^T \bar{n}_{,2} = 0$, it then yields

$$\underline{a}_1^T \frac{\partial \bar{n}}{\partial \alpha_2} = h_1 \bar{b}_1^T \frac{\partial \bar{n}}{\partial \alpha_2} = h_1 \bar{b}_1^T \frac{\partial \bar{n}}{\partial s_2}\frac{ds_2}{d\alpha_2} = 0 \tag{1.256}$$

From which the following identity is obtained to be

$$\bar{b}_1^T \frac{\partial \bar{n}}{\partial s_2} = \bar{n}^T \frac{\partial \bar{b}_1}{\partial s_2} = 0 \tag{1.257}$$

Similar operations on the attribute, $b_{21} = \underline{a}_2^T \bar{n}_{,1} = 0$, produces

$$h_2 \bar{b}_2^T \frac{\partial \bar{n}}{\partial \alpha_1} = h_2 \bar{b}_2^T \frac{\partial \bar{n}}{\partial s_1}\frac{ds_1}{d\alpha_1} = 0 \tag{1.258}$$

and this expression reduces to

$$\bar{b}_2^T \frac{\partial \bar{n}}{\partial s_1} = \bar{n}^T \frac{\partial \bar{b}_2}{\partial s_1} = 0 \tag{1.259}$$

According to the definitions of the diagonal term, $b_{11}$, of the second metric tensor, Eq. (1.213), and the definitions of principal curvature, Eq. (1.226), the following identity is readily verified to be

$$b_{11} = -\underline{a}_1^T \frac{\partial \bar{n}}{\partial \alpha_1} = -h_1 \bar{b}_1^T \frac{\partial \bar{n}}{\partial s_1} \frac{ds_1}{d\alpha_1} = \kappa_1 a_{11} \tag{1.260}$$

By identity, $a_{11} = \underline{a}_1^T \underline{a}_1 = h_1^2$, it yields

$$\bar{b}_1^T \frac{\partial \bar{n}}{\partial s_1} = -\bar{n}^T \frac{\partial \bar{b}_1}{\partial s_1} = -\kappa_1 \tag{1.261}$$

Operating on the diagonal term $b_{22}$ of the second metric tensor in the same manner produces the identity like

$$\bar{b}_2^T \frac{\partial \bar{n}}{\partial s_2} = -\bar{n}^T \frac{\partial \bar{b}_2}{\partial s_2} = -\kappa_2 \tag{1.262}$$

In summary, the derivatives of the surface base vectors $\bar{b}_1$, $\bar{b}_2$ and $\bar{b}_3$ with respect to curvilinear coordinate $s_1$ can be resolved in the following manner

$$\begin{aligned} \frac{\partial \bar{b}_1}{\partial s_1} &= 0\,\bar{b}_1 - \tau_1 \bar{b}_2 + \kappa_1 \bar{b}_3 \\ \frac{\partial \bar{b}_2}{\partial s_1} &= \tau_1 \bar{b}_1 + 0\,\bar{b}_2 + 0\,\bar{b}_3 \\ \frac{\partial \bar{b}_3}{\partial s_1} &= -\kappa_1 \bar{b}_1 + 0\,\bar{b}_2 + 0\,\bar{b}_3 \end{aligned} \tag{1.263}$$

or can be rewritten into the matrix form

$$\frac{\partial}{\partial s_1} \lfloor \bar{b}_1, \bar{b}_2, \bar{b}_3 \rfloor = \lfloor \bar{b}_1, \bar{b}_2, \bar{b}_3 \rfloor \begin{bmatrix} 0 & \tau_1 & -\kappa_1 \\ -\tau_1 & 0 & 0 \\ \kappa_1 & 0 & 0 \end{bmatrix} \tag{1.264}$$

The derivatives of the surface base vectors with respect to curvilinear coordinate $s_2$ are found in a similar manner

$$\begin{aligned} \frac{\partial \bar{b}_1}{\partial s_2} &= 0\,\bar{b}_1 + \tau_2 \bar{b}_2 + 0\,\bar{b}_3 \\ \frac{\partial \bar{b}_2}{\partial s_2} &= -\tau_2 \bar{b}_1 + 0\,\bar{b}_2 + \kappa_2 \bar{b}_3 \\ \frac{\partial \bar{b}_3}{\partial s_2} &= 0\,\bar{b}_1 - \kappa_2 \bar{b}_2 + 0\,\bar{b}_3 \end{aligned} \tag{1.265}$$

or in matrix form

$$\frac{\partial}{\partial s_2}\lfloor \bar{b}_1, \bar{b}_2, \bar{b}_3 \rfloor = \lfloor \bar{b}_1, \bar{b}_2, \bar{b}_3 \rfloor \begin{bmatrix} 0 & -\tau_2 & 0 \\ \tau_2 & 0 & -\kappa_2 \\ 0 & \kappa_2 & 0 \end{bmatrix} \tag{1.266}$$

Both of them can be rewritten into the compact form equivalently like

$$R_{S,1} = R_S \widetilde{\kappa}_1, \quad R_{S,2} = R_S \widetilde{\kappa}_2 \tag{1.267}$$

where the curvature tensors of a surface, the skew-symmetric tensors also, are defined as

$$\widetilde{\kappa}_1 = R_S^T R_{S,1}, \quad \widetilde{\kappa}_2 = R_S^T R_{S,2} \tag{1.268}$$

The associated curvature vectors

$$\underline{\kappa}_1^T = \lfloor 0, \ -\kappa_1, \ -\tau_1 \rfloor, \quad \underline{\kappa}_2^T = \lfloor \kappa_2, \ 0, \ \tau_2 \rfloor \tag{1.269}$$

contains only two non-vanishing components, the twist and curvature of the surface when the curvature vectors resolved in the body attached basis $\mathcal{B}$.

### *1.3.9 Metric Tensor of a Three-Dimensional Mapping*

The differential geometry of mappings of the three-dimensional space onto itself will be investigated in this section. Some fundamental attributes of a three-dimensional mapping are discussed in detail, such as the Jacobian of transformation, the base vectors and covariant base vectors, metric tensors etc. As depicted in Fig. 1.13, the position vector of an arbitrary point generated by three parameters $\alpha_1$, $\alpha_2$ and $\alpha_3$ is written in the Cartesian basis like

$$\underline{p}(\alpha_1, \alpha_2, \alpha_3) = x_1(\alpha_1, \alpha_2, \alpha_3)\bar{\imath}_1 + x_2(\alpha_1, \alpha_2, \alpha_3)\bar{\imath}_2 + x_3(\alpha_1, \alpha_2, \alpha_3)\bar{\imath}_3 \tag{1.270}$$

Actually, the above equation defines a mapping between the parameters and the Cartesian coordinates, and is rewritten into the components form

$$x_1 = x_1(\alpha_1, \alpha_2, \alpha_3), \quad x_2 = x_2(\alpha_1, \alpha_2, \alpha_3), \quad x_3 = x_3(\alpha_1, \alpha_2, \alpha_3) \tag{1.271}$$

It can be imagined a general curve in three-dimensional space could be generated by varying one of three parameters while holding the other two constant. Next, a general surface in three-dimensional space is generated by varying two of three parameters while holding the other one constant. Apparently, a point in space with parameters $(\alpha_1, \alpha_2, \alpha_3)$ is at the intersection of three $\alpha_1$, $\alpha_2$ and $\alpha_3$ curves, or at the intersection of three $\alpha_1$, $\alpha_2$ and $\alpha_3$ surfaces. Furthermore, a $\alpha_1$ curve forms the intersection of $\alpha_2$ and $\alpha_3$ surfaces, and vice versa.

If the Jacobian of the transformation

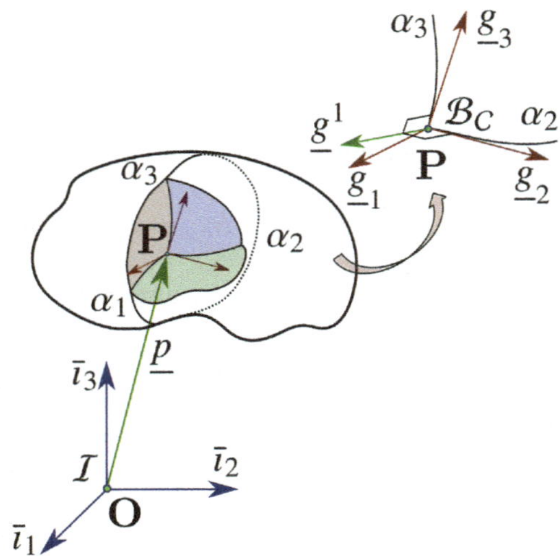


**Fig. 1.13** Configuration of a body in space

$$J = \begin{bmatrix} \frac{\partial x_1}{\partial \alpha_1} & \frac{\partial x_1}{\partial \alpha_2} & \frac{\partial x_1}{\partial \alpha_3} \\ \frac{\partial x_2}{\partial \alpha_1} & \frac{\partial x_2}{\partial \alpha_2} & \frac{\partial x_2}{\partial \alpha_3} \\ \frac{\partial x_3}{\partial \alpha_1} & \frac{\partial x_3}{\partial \alpha_2} & \frac{\partial x_3}{\partial \alpha_3} \end{bmatrix} \tag{1.272}$$

does not vanish anywhere, or has a non-vanishing determinant at all points in space, $\det(J) \neq 0$, then Eq. (1.271) may be solved for $\alpha_1$, $\alpha_2$ and $\alpha_3$ to yield the inverse mapping defines the parameters as functions of the Cartesian coordinates

$$\alpha_1 = \alpha_1(x_1, x_2, x_3), \quad \alpha_2 = \alpha_2(x_1, x_2, x_3), \quad \alpha_3 = \alpha_3(x_1, x_2, x_3) \tag{1.273}$$

Next, the covariant base vectors associated with the parameters are defined as

$$\underline{g}_1 = \frac{\partial \underline{p}}{\partial \alpha_1}, \quad \underline{g}_2 = \frac{\partial \underline{p}}{\partial \alpha_2}, \quad \underline{g}_3 = \frac{\partial \underline{p}}{\partial \alpha_3} \tag{1.274}$$

Note that the covariant vectors represent tangent vectors to the three $\alpha_1$, $\alpha_2$ and $\alpha_3$ curves. These vectors can be resolved into the form of components to get

$$\begin{aligned} \underline{g}_1 &= \frac{\partial x_1}{\partial \alpha_1}\bar{\imath}_1 + \frac{\partial x_2}{\partial \alpha_1}\bar{\imath}_2 + \frac{\partial x_3}{\partial \alpha_1}\bar{\imath}_3 \\ \underline{g}_2 &= \frac{\partial x_1}{\partial \alpha_2}\bar{\imath}_1 + \frac{\partial x_2}{\partial \alpha_2}\bar{\imath}_2 + \frac{\partial x_3}{\partial \alpha_2}\bar{\imath}_3 \\ \underline{g}_3 &= \frac{\partial x_1}{\partial \alpha_3}\bar{\imath}_1 + \frac{\partial x_2}{\partial \alpha_3}\bar{\imath}_2 + \frac{\partial x_3}{\partial \alpha_3}\bar{\imath}_3 \end{aligned} \tag{1.275}$$

By Eq. (1.5), it can be recasted to the matrix from

$$\lfloor \underline{g}_1, \underline{g}_2, \underline{g}_3 \rfloor = \begin{bmatrix} \frac{\partial x_1}{\partial \alpha_1} & \frac{\partial x_1}{\partial \alpha_2} & \frac{\partial x_1}{\partial \alpha_3} \\ \frac{\partial x_2}{\partial \alpha_1} & \frac{\partial x_2}{\partial \alpha_2} & \frac{\partial x_2}{\partial \alpha_3} \\ \frac{\partial x_3}{\partial \alpha_1} & \frac{\partial x_3}{\partial \alpha_2} & \frac{\partial x_3}{\partial \alpha_3} \end{bmatrix} \tag{1.276}$$

Comparing the expressions with the definition of Jacobian matrix, Eq. (1.272), it is readily verified that

$$J = \lfloor \underline{g}_1, \underline{g}_2, \underline{g}_3 \rfloor \tag{1.277}$$

For an arbitrary parameterization, the covariant base vectors will not be unit vectors, nor will they be mutually orthogonal. However, since the parameters $\alpha_1$, $\alpha_2$ and $\alpha_3$ are independent coordinates, and the vectors $\underline{g}_1$, $\underline{g}_2$ and $\underline{g}_3$ are linearly independent vectors and form a right-handed set of vectors. The scalar triple product of based vectors yields

$$\sqrt{g} = \underline{g}_3^T \widetilde{g}_1 \underline{g}_2 = \underline{g}_1^T \widetilde{g}_2 \underline{g}_3 = \underline{g}_2^T \widetilde{g}_3 \underline{g}_1 \tag{1.278}$$

Referring the attributes of scalar triple product, Eq. (1.37), the scalar triple product is nothing but the determinant of Jacobian matrix

$$\sqrt{g} = \det(J) \tag{1.279}$$

With the aid of the definition of base vectors, the differential element of length $ds$ can be found as

$$\begin{aligned} ds^2 &= d\underline{p}^T d\underline{p} \\ &= (\underline{g}_1^T d\alpha_1 + \underline{g}_2^T d\alpha_2 + \underline{g}_3^T d\alpha_3)(\underline{g}_1 d\alpha_1 + \underline{g}_2 d\alpha_2 + \underline{g}_3 d\alpha_3) \end{aligned} \tag{1.280}$$

Reassembling the differential element into the matrix form will produce

$$\begin{aligned} d\underline{p}^T d\underline{p} &= \lfloor d\alpha_1, d\alpha_2, d\alpha_3 \rfloor \begin{bmatrix} \underline{g}_1^T \\ \underline{g}_2^T \\ \underline{g}_3^T \end{bmatrix} \lfloor \underline{g}_1, \underline{g}_2, \underline{g}_3 \rfloor \begin{vmatrix} d\alpha_1 \\ d\alpha_2 \\ d\alpha_3 \end{vmatrix} \\ &= \lfloor d\alpha_1, d\alpha_2, d\alpha_3 \rfloor \begin{bmatrix} \underline{g}_1^T \underline{g}_1 & \underline{g}_1^T \underline{g}_2 & \underline{g}_1^T \underline{g}_3 \\ \underline{g}_2^T \underline{g}_1 & \underline{g}_2^T \underline{g}_2 & \underline{g}_2^T \underline{g}_3 \\ \underline{g}_3^T \underline{g}_1 & \underline{g}_3^T \underline{g}_2 & \underline{g}_3^T \underline{g}_3 \end{bmatrix} \begin{vmatrix} d\alpha_1 \\ d\alpha_2 \\ d\alpha_3 \end{vmatrix} \end{aligned} \tag{1.281}$$

From which the metric tensor is observed to be

$$G = J^T J = \begin{bmatrix} \underline{g}_1^T \underline{g}_1 & \underline{g}_1^T \underline{g}_2 & \underline{g}_1^T \underline{g}_3 \\ \underline{g}_2^T \underline{g}_1 & \underline{g}_2^T \underline{g}_2 & \underline{g}_2^T \underline{g}_3 \\ \underline{g}_3^T \underline{g}_1 & \underline{g}_3^T \underline{g}_2 & \underline{g}_3^T \underline{g}_3 \end{bmatrix} \tag{1.282}$$

Apparently, its determinant is equal to the square of scalar triple product of based vectors like

$$\det(G) = g \tag{1.283}$$

Notice in the above definitions that the vectors $\underline{g}_1$, $\underline{g}_2$ and $\underline{g}_3$ have subscripts for indices. The subscripts for indices mean these vectors are covariant base vectors. These linearly independent vectors can form the so-called covariant bases

$$\mathcal{B}_C = \lfloor \underline{g}_1, \ \underline{g}_2, \ \underline{g}_3 \rfloor \tag{1.284}$$

Next, as shown in Fig. 1.13, the superscripts for indices will be used to denotes the contravariant base vectors, which are defined by the vector products of covariant base vectors such that

$$\underline{g}^1 = \frac{\widetilde{g}_2 \underline{g}_3}{\sqrt{g}}, \quad \underline{g}^2 = \frac{\widetilde{g}_3 \underline{g}_1}{\sqrt{g}}, \quad \underline{g}^3 = \frac{\widetilde{g}_1 \underline{g}_2}{\sqrt{g}} \tag{1.285}$$

Due to the attributes of the vector product, Eq. (1.23), the covariant vectors and the contravariant vectors are biorthogonal sets of vectors satisfying the following relations

$$\begin{matrix} \underline{g}_1^T \underline{g}^1 = 1, & \underline{g}_1^T \underline{g}^2 = 0, & \underline{g}_1^T \underline{g}^3 = 0 \\ \underline{g}_2^T \underline{g}^1 = 0, & \underline{g}_2^T \underline{g}^2 = 1, & \underline{g}_2^T \underline{g}^3 = 0 \\ \underline{g}_3^T \underline{g}^1 = 0, & \underline{g}_3^T \underline{g}^2 = 0, & \underline{g}_3^T \underline{g}^3 = 1 \end{matrix} \tag{1.286}$$

or in the matrix form

$$\begin{bmatrix} \underline{g}_1^T \\ \underline{g}_2^T \\ \underline{g}_3^T \end{bmatrix} \lfloor \underline{g}^1, \ \underline{g}^2, \ \underline{g}^3 \rfloor = I \tag{1.287}$$

When resolving the contravariant base vector $\underline{g}^1$ into the covariant bases

$$\underline{g}^1 = c_1 \underline{g}_1 + c_2 \underline{g}_2 + c_3 \underline{g}_3 \tag{1.288}$$

the contravariant components $c_1$, $c_2$ and $c_3$ of based vector $\underline{g}^1$ is determined in the usual way by taking the scalar product of $\underline{g}^1$ with the appropriate base vectors

$$c_1 = \underline{g}^{1T} \underline{g}^1, \quad c_2 = \underline{g}^{2T} \underline{g}^1, \quad c_3 = \underline{g}^{3T} \underline{g}^1 \tag{1.289}$$

Substituting above expressions into Eq. (1.288) and merging the similar terms yields

$$\underline{g}^1 = \underline{g}_1\underline{g}^{1T}\underline{g}^1 + \underline{g}_2\underline{g}^{2T}\underline{g}^1 + \underline{g}_3\underline{g}^{3T}\underline{g}^1 = (\underline{g}_1\underline{g}^{1T} + \underline{g}_2\underline{g}^{2T} + \underline{g}_3\underline{g}^{3T})\underline{g}^1 \quad (1.290)$$

In view of, $\underline{g}^1 \neq 0$, it is verified the base tensor becomes the second order identity tensor

$$\underline{g}_1\underline{g}^{1T} + \underline{g}_2\underline{g}^{2T} + \underline{g}_3\underline{g}^{3T} = \lfloor \underline{g}_1, \ \underline{g}_2, \ \underline{g}_3 \rfloor \begin{bmatrix} \underline{g}^{1T} \\ \underline{g}^{2T} \\ \underline{g}^{3T} \end{bmatrix} = I \quad (1.291)$$

It is important to note that the book shall consider only the case when the three vectors $\underline{g}_1$, $\underline{g}_2$ and $\underline{g}_3$ are mutually perpendicular. Hence, the unit curvilinear vectors can be computed from

$$\bar{b}_1 = \frac{\underline{g}_1}{\|\underline{g}_1\|}, \quad \bar{b}_2 = \frac{\underline{g}_2}{\|\underline{g}_2\|}, \quad \bar{b}_3 = \frac{\underline{g}_3}{\|\underline{g}_3\|} \quad (1.292)$$

Alternatively, if introducing parameters $s_1, s_2$ and $s_3$ as the arc length measured along the $\alpha_1, \alpha_2$ and $\alpha_3$ curves, the chain rule for derivatives reveals that

$$\begin{aligned} \underline{g}_1 &= \frac{\partial \underline{p}}{\partial \alpha_1} = \frac{\partial \underline{p}}{\partial s_1}\frac{ds_1}{d\alpha_1} = \bar{b}_1\frac{ds_1}{d\alpha_1} \\ \underline{g}_2 &= \frac{\partial \underline{p}}{\partial \alpha_2} = \frac{\partial \underline{p}}{\partial s_2}\frac{ds_2}{d\alpha_2} = \bar{b}_2\frac{ds_2}{d\alpha_2} \\ \underline{g}_3 &= \frac{\partial \underline{p}}{\partial \alpha_3} = \frac{\partial \underline{p}}{\partial s_3}\frac{ds_3}{d\alpha_3} = \bar{b}_3\frac{ds_3}{d\alpha_3} \end{aligned} \quad (1.293)$$

where the definitions of unit vectors $\bar{b}_1, \bar{b}_2$ and $\bar{b}_3$ tangent to the $\alpha_1, \alpha_2$ and $\alpha_3$ curves

$$\bar{b}_1 = \frac{\partial \underline{p}}{\partial s_1}, \quad \bar{b}_2 = \frac{\partial \underline{p}}{\partial s_2}, \quad \bar{b}_3 = \frac{\partial \underline{p}}{\partial s_3} \quad (1.294)$$

are applied to obtain the formulas of Lame's coefficients that are the scale factors, and also the ratios of the infinitesimal increments in length, $ds_1$, $ds_2$ and $ds_3$, to the infinitesimal increments in parameters, $\alpha_1$, $\alpha_2$ and $\alpha_3$, along the curves

$$h_1 = \frac{ds_1}{d\alpha_1} = \|\underline{g}_1\|, \quad h_2 = \frac{ds_2}{d\alpha_2} = \|\underline{g}_2\|, \quad h_3 = \frac{ds_3}{d\alpha_3} = \|\underline{g}_3\| \quad (1.295)$$

Such that the covariant base vectors are recasted to

$$\underline{g}_1 = h_1\bar{b}_1, \quad \underline{g}_2 = h_2\bar{b}_2, \quad \underline{g}_3 = h_3\bar{b}_3 \quad (1.296)$$

and then the determinant of Jacobian matrix becomes

$$\sqrt{g} = h_1 h_2 h_3 \bar{b}_1^T \tilde{\bar{b}}_2 \underline{b}_3 \tag{1.297}$$

When the curvilinear coordinates $s_1$, $s_2$ and $s_3$ are said to be orthogonal, the determinant reduces to

$$\sqrt{g} = h_1 h_2 h_3 \tag{1.298}$$

and the rotation tensor can be constructed by using the set of mutually orthogonal vectors like

$$R_C = \lfloor \bar{b}_1,\ \bar{b}_2,\ \bar{b}_3 \rfloor \tag{1.299}$$

### 1.3.10 Curvature Tensor of a Three-Dimensional Mapping

The above Sect. 1.3.9 investigates the differential geometry of mappings of the three-dimensional space onto itself. The covariant base vectors, $\underline{g}_1$, $\underline{g}_2$ and $\underline{g}_3$, were defined along the base vectors of the mapping. These vectors were shown to form an orthonormal basis, $\mathcal{B}_C$, in the case of orthogonal curvilinear coordinate systems. The rotation tensor $R_C$ that brings the reference Cartesian basis $\mathcal{I}$ to orthonormal basis $\mathcal{B}_C$ is introduced also. The curvature tensors of the orthogonal curvilinear coordinate system are now introduced in this section with the aid of the derivatives of covariant base vectors. Apparently, the derivatives of base vectors are still a vector with components projected to the base vectors. By Eq. (1.168), it is clearly now

$$\bar{b}_1^T \frac{\partial \bar{b}_1}{\partial s_i} = 0, \quad \bar{b}_2^T \frac{\partial \bar{b}_2}{\partial s_i} = 0, \quad \bar{b}_3^T \frac{\partial \bar{b}_3}{\partial s_i} = 0 \tag{1.300}$$

for $i = 1, 2, 3$. Hence, the derivatives of base vectors becomes

$$\begin{aligned}
\frac{\partial \bar{b}_1}{\partial \underline{s}_i} &= \bar{b}_2^T \frac{\partial \bar{b}_1}{\partial \underline{s}_i} \bar{b}_2 + \bar{b}_3^T \frac{\partial \bar{b}_1}{\partial \underline{s}_i} \bar{b}_3 \\
\frac{\partial \bar{b}_2}{\partial \underline{s}_i} &= \bar{b}_1^T \frac{\partial \bar{b}_2}{\partial \underline{s}_i} \bar{b}_1 + \bar{b}_3^T \frac{\partial \bar{b}_2}{\partial \underline{s}_i} \bar{b}_3 \\
\frac{\partial \bar{b}_3}{\partial s_i} &= \bar{b}_1^T \frac{\partial \bar{b}_3}{\partial s_i} \bar{b}_1 + \bar{b}_2^T \frac{\partial \bar{b}_3}{\partial s_i} \bar{b}_2
\end{aligned}$$

The derivative components resolved in the curvilinear orthogonal basis $\mathcal{B}_C$ should be identified one by one. At first, the 6 unknowns will be reduced to 3 of them with the aid of attributes, Eq. (1.169), such that

$$\bar{b}_1^T \frac{\partial \bar{b}_2}{\partial s_i} = -\bar{b}_2^T \frac{\partial \bar{b}_1}{\partial s_i}, \quad \bar{b}_2^T \frac{\partial \bar{b}_3}{\partial s_i} = -\bar{b}_3^T \frac{\partial \bar{b}_2}{\partial s_i}, \quad \bar{b}_3^T \frac{\partial \bar{b}_1}{\partial s_i} = -\bar{b}_1^T \frac{\partial \bar{b}_3}{\partial s_i} \tag{1.301}$$

and the derivatives of base vectors are recasted to

$$
\begin{aligned}
\frac{\partial \bar{b}_1}{\partial \underline{s}_i} &= \bar{b}_2^T \frac{\partial \bar{b}_1}{\partial s_i} \bar{b}_2 + \bar{b}_3^T \frac{\partial \bar{b}_1}{\partial s_i} \bar{b}_3 \\
\frac{\partial \bar{b}_2}{\partial \underline{s}_i} &= -\bar{b}_2^T \frac{\partial \bar{b}_1}{\partial \underline{s}_i} \bar{b}_1 + \bar{b}_3^T \frac{\partial \bar{b}_2}{\partial s_i} \bar{b}_3 \\
\frac{\partial \bar{b}_3}{\partial s_i} &= -\bar{b}_3^T \frac{\partial \bar{b}_1}{\partial s_i} \bar{b}_1 - \bar{b}_3^T \frac{\partial \bar{b}_2}{\partial s_i} \bar{b}_2
\end{aligned} \tag{1.302}
$$

with only 3 unknowns. In view of second derivatives of position vector respect to parameters $\alpha_1$, $\alpha_2$ and $\alpha_3$, the following identities can be found to be

$$
\begin{aligned}
\frac{\partial^2 \underline{p}}{\partial \alpha_1 \partial \alpha_2} &= \frac{\partial \underline{g}_1}{\partial \alpha_2} = \frac{\partial \underline{g}_2}{\partial \alpha_1} \\
\frac{\partial^2 \underline{p}}{\partial \alpha_2 \partial \alpha_3} &= \frac{\partial \underline{g}_2}{\partial \alpha_3} = \frac{\partial \underline{g}_3}{\partial \alpha_2} \\
\frac{\partial^2 \underline{p}}{\partial \alpha_3 \partial \alpha_1} &= \frac{\partial \underline{g}_3}{\partial \alpha_1} = \frac{\partial \underline{g}_1}{\partial \alpha_3}
\end{aligned} \tag{1.303}
$$

With the help of Lame's constants, Eq. (1.296), the derivatives of base vectors are rewritten as

$$
\begin{aligned}
\frac{\partial \underline{g}_1}{\partial \alpha_2} &= \frac{\partial h_1}{\partial \alpha_2} \bar{b}_1 + \frac{\partial \bar{b}_1}{\partial \underline{\alpha}_2} h_1, & \frac{\partial \underline{g}_2}{\partial \alpha_1} &= \frac{\partial h_2}{\partial \alpha_1} \bar{b}_2 + \frac{\partial \bar{b}_2}{\partial \underline{\alpha}_1} h_2 \\
\frac{\partial \underline{g}_2}{\partial \alpha_3} &= \frac{\partial h_2}{\partial \alpha_3} \bar{b}_2 + \frac{\partial \bar{b}_2}{\partial \underline{\alpha}_3} h_2, & \frac{\partial \underline{g}_3}{\partial \alpha_2} &= \frac{\partial h_3}{\partial \alpha_2} \bar{b}_3 + \frac{\partial \bar{b}_3}{\partial \underline{\alpha}_2} h_3 \\
\frac{\partial \underline{g}_3}{\partial \alpha_1} &= \frac{\partial h_3}{\partial \alpha_1} \bar{b}_3 + \frac{\partial \bar{b}_3}{\partial \alpha_1} h_3, & \frac{\partial \underline{g}_1}{\partial \alpha_3} &= \frac{\partial h_1}{\partial \alpha_3} \bar{b}_1 + \frac{\partial \bar{b}_1}{\partial \alpha_3} h_1
\end{aligned} \tag{1.304}
$$

and then the above identities still holds on in terms of Lame's constants like

$$
\begin{aligned}
\frac{\partial h_1}{\partial \alpha_2} \bar{b}_1 + \frac{\partial \bar{b}_1}{\partial \underline{\alpha}_2} h_1 &= \frac{\partial h_2}{\partial \alpha_1} \bar{b}_2 + \frac{\partial \bar{b}_2}{\partial \underline{\alpha}_1} h_2 \\
\frac{\partial h_2}{\partial \alpha_3} \bar{b}_2 + \frac{\partial \bar{b}_2}{\partial \underline{\alpha}_3} h_2 &= \frac{\partial h_3}{\partial \alpha_2} \bar{b}_3 + \frac{\partial \bar{b}_3}{\partial \underline{\alpha}_2} h_3 \\
\frac{\partial h_3}{\partial \alpha_1} \bar{b}_3 + \frac{\partial \bar{b}_3}{\partial \alpha_1} h_3 &= \frac{\partial h_1}{\partial \alpha_3} \bar{b}_1 + \frac{\partial \bar{b}_1}{\partial \alpha_3} h_1
\end{aligned} \tag{1.305}
$$

A set of unit mutually orthogonal vectors $\bar{b}_1$, $\bar{b}_2$ and $\bar{b}_3$ answers that pre-multiplying the above identities by the transpose of these unit vectors will yield the following relationships

$$
\begin{aligned}
\bar{b}_3^T \frac{\partial \bar{b}_1}{\partial \underline{\alpha}_2} h_1 &= \bar{b}_3^T \frac{\partial \bar{b}_2}{\partial \underline{\alpha}_1} h_2 = -\bar{b}_2^T \frac{\partial \bar{b}_3}{\partial \underline{\alpha}_1} h_2 \\
\bar{b}_1^T \frac{\partial \bar{b}_2}{\partial \underline{\alpha}_3} h_2 &= \bar{b}_1^T \frac{\partial \bar{b}_3}{\partial \underline{\alpha}_2} h_3 = -\bar{b}_3^T \frac{\partial \bar{b}_1}{\partial \underline{\alpha}_2} h_3 \\
\bar{b}_2^T \frac{\partial \bar{b}_3}{\partial \alpha_1} h_3 &= \bar{b}_2^T \frac{\partial \bar{b}_1}{\partial \alpha_3} h_1 = -\bar{b}_1^T \frac{\partial \bar{b}_2}{\partial \alpha_3} h_1
\end{aligned} \tag{1.306}
$$

From the first equation of above expressions, the following relations

$$\bar{b}_3^T \frac{\partial \bar{b}_1}{\partial \alpha_2} h_1 = -\bar{b}_2^T \frac{\partial \bar{b}_3}{\partial \alpha_1} h_2 = \frac{h_1 h_2}{h_3} \bar{b}_1^T \frac{\partial \bar{b}_2}{\partial \alpha_3} = -\bar{b}_3^T \frac{\partial \bar{b}_1}{\partial \alpha_2} h_1 \tag{1.307}$$

can be verified which answers that the derivatives projected to the base vector $\bar{b}_3$ becomes zero, such that

$$\bar{b}_3^T \frac{\partial \bar{b}_1}{\partial \alpha_2} = -\bar{b}_3^T \frac{\partial \bar{b}_1}{\partial \alpha_2} \Leftrightarrow \bar{b}_3^T \frac{\partial \bar{b}_1}{\partial \alpha_2} = 0 \tag{1.308}$$

The similar relationships can be observed from the rest two of Eq. (1.306), and hence all the non-diagonal zero projections are determined and summarized as follow

$$\begin{aligned} \bar{b}_3^T \frac{\partial \bar{b}_1}{\partial \alpha_2} &= \bar{b}_3^T \frac{\partial \bar{b}_2}{\partial \alpha_1} = 0 \\ \bar{b}_1^T \frac{\partial \bar{b}_2}{\partial \alpha_3} &= \bar{b}_1^T \frac{\partial \bar{b}_3}{\partial \alpha_2} = 0 \\ \bar{b}_2^T \frac{\partial \bar{b}_3}{\partial \alpha_1} &= \bar{b}_2^T \frac{\partial \bar{b}_1}{\partial \alpha_3} = 0 \end{aligned} \tag{1.309}$$

With the application of chain rules for derivatives

$$\frac{\partial (\cdot)}{\partial \alpha_i} = \frac{\partial (\cdot)}{\partial s_i} \frac{d s_i}{d \alpha_i} = \frac{\partial (\cdot)}{\partial s_i} h_i \tag{1.310}$$

the zero components of base vector derivatives with respect to orthogonal curvilinear coordinates $s_1$, $s_2$ and $s_3$ are found to be

$$\begin{aligned} \bar{b}_3^T \frac{\partial \bar{b}_1}{\partial s_2} &= \bar{b}_3^T \frac{\partial \bar{b}_2}{\partial s_1} = 0 \\ \bar{b}_1^T \frac{\partial \bar{b}_2}{\partial s_3} &= \bar{b}_1^T \frac{\partial \bar{b}_3}{\partial s_2} = 0 \\ \bar{b}_2^T \frac{\partial \bar{b}_3}{\partial s_1} &= \bar{b}_2^T \frac{\partial \bar{b}_1}{\partial s_3} = 0 \end{aligned} \tag{1.311}$$

Meanwhile, the non-zero components are known as the curvatures of the system

$$\begin{aligned} \tau_1 &= \bar{b}_2^T \frac{\partial \bar{b}_1}{\partial s_1}, & \kappa_1 &= -\bar{b}_3^T \frac{\partial \bar{b}_1}{\partial s_1} \\ \tau_2 &= \bar{b}_2^T \frac{\partial \bar{b}_1}{\partial s_2}, & \kappa_2 &= \bar{b}_3^T \frac{\partial \bar{b}_2}{\partial s_2} \\ \tau_3 &= -\bar{b}_3^T \frac{\partial \bar{b}_1}{\partial s_3}, & \kappa_3 &= \bar{b}_3^T \frac{\partial \bar{b}_2}{\partial s_3} \end{aligned} \tag{1.312}$$

Finally, all the derivatives of base vectors are identified to be

$$\begin{aligned}
&\frac{\partial \bar{b}_1}{\partial s_1} = \tau_1 \bar{b}_2 - \kappa_1 \bar{b}_3, && \frac{\partial \bar{b}_1}{\partial s_2} = \tau_2 \bar{b}_2, && \frac{\partial \bar{b}_1}{\partial s_3} = -\tau_3 \bar{b}_3 \\
&\frac{\partial \bar{b}_2}{\partial s_1} = -\tau_1 \bar{b}_1, && \frac{\partial \bar{b}_2}{\partial s_2} = -\tau_2 \bar{b}_1 + \kappa_2 \bar{b}_3, && \frac{\partial \bar{b}_2}{\partial s_3} = \kappa_3 \bar{b}_3 \\
&\frac{\partial \bar{b}_3}{\partial s_1} = \kappa_1 \bar{b}_1, && \frac{\partial \bar{b}_3}{\partial s_2} = -\kappa_2 \bar{b}_2, && \frac{\partial \bar{b}_3}{\partial s_3} = \tau_3 \bar{b}_1 - \kappa_3 \bar{b}_2
\end{aligned} \tag{1.313}$$

They can be rewritten into the compact form like

$$R_{C,1} = R_C \widetilde{\kappa}_1, \quad R_{C,2} = R_C \widetilde{\kappa}_2, \quad R_{C,3} = R_C \widetilde{\kappa}_3 \tag{1.314}$$

where the curvature tensors $\widetilde{\kappa}_1$, $\widetilde{\kappa}_2$ and $\widetilde{\kappa}_3$ of the orthogonal curvilinear coordinate system are defined as follow

$$\widetilde{\kappa}_1 = \begin{bmatrix} 0 & -\tau_1 & \kappa_1 \\ \tau_1 & 0 & 0 \\ -\kappa_1 & 0 & 0 \end{bmatrix}, \quad \widetilde{\kappa}_2 = \begin{bmatrix} 0 & -\tau_2 & 0 \\ \tau_2 & 0 & -\kappa_2 \\ 0 & \kappa_2 & 0 \end{bmatrix} \tag{1.315}$$

together with

$$\widetilde{\kappa}_3 = \begin{bmatrix} 0 & 0 & \tau_3 \\ 0 & 0 & -\kappa_3 \\ -\tau_3 & \kappa_3 & 0 \end{bmatrix} \tag{1.316}$$

The associated curvature vectors are introduced also

$$\underline{\kappa}_1 = \begin{vmatrix} 0 \\ \kappa_1 \\ \tau_1 \end{vmatrix}, \quad \underline{\kappa}_2 = \begin{vmatrix} \kappa_2 \\ 0 \\ \tau_2 \end{vmatrix}, \quad \underline{\kappa}_3 = \begin{vmatrix} \kappa_3 \\ \tau_3 \\ 0 \end{vmatrix} \tag{1.317}$$

Each of them contains only two non-vanishing components when the curvature vectors are resolved in the body attached basis $\mathcal{B}_C$.

## 1.4 Motion Tensor

Referring Eq. (1.114), a skew-symmetric dual tensor has been defined which is the first type of dual tensors we touched in the book. In this section, the concept of motion tensor, other important dual tensor, will be introduced with the aid of line vectors. Now the problems state: given a material line vector

$$\underline{\mathcal{L}}_1 = \bar{\ell}_1 + \epsilon \widetilde{r}_1 \bar{\ell}_1 \tag{1.318}$$

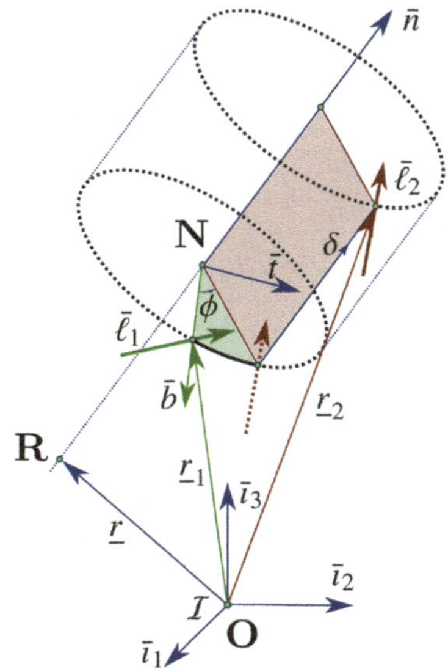


**Fig. 1.14** Geometric description of small motion

it rotates an infinitesimal angle $\phi$ around a unit normal vector $\bar{n}$, and then translates an infinitesimal distance $\delta$ along the same normal vector to the new position of

$$\underline{\mathcal{L}}_2 = \bar{\ell}_2 + \epsilon\,\widetilde{r}_2\bar{\ell}_2 \tag{1.319}$$

Could this motion be accurately described by using the so-called motion tensor?

The question has been answered by the Chasles' theorem [13] states that the general motion of a rigid body can be represented by a screw motion, which consists of a translation along the axis of the screw followed by a rotation about the same axis. This axis, also known as the Mozzi-Chasles axis, is usually appeared in its Plücker coordinates. When an arbitrary material line vector subjected to a screw motion, it's Plücker coordinates, readily evaluated [14], are shown to transform by the action of the motion tensor. In the case of infinitesimal motion, the Pücker coordinates also transform in the same manner.

The geometric description of infinitesimal motion are shown in Fig. 1.14. The magnitude of small motion is defined by a dual number

$$\phi^d = \phi + \epsilon\,\delta \tag{1.320}$$

where both the rotation angle $\phi$ and amplitude of translation $\delta$ are small, $\phi \ll 1$ and $\delta \ll 1$, or the magnitude of motion is small equivalently. The direction of rotating and translating is represented by the Chasles' line of Plücker coordinates, a unit line vector

$$\underline{\mathcal{N}} = \bar{n} + \epsilon\,\widetilde{r}\bar{n} \tag{1.321}$$

Hence, the infinitesimal motion is characterized by its Chasles' line of Plücker coordinates $\underline{\mathcal{N}}$ and the magnitude of the motion, $\phi^d$. The scalar multiplication of line vector $\underline{\mathcal{N}}$ by the dual number $\phi^d$ produces the dual representation of screw like

$$\phi^d \underline{\mathcal{N}} = (\phi + \epsilon\, \delta)(\bar{n} + \epsilon\, \tilde{r}\bar{n}) = \phi\, \bar{n} + \epsilon(\delta\, \bar{n} + \tilde{r}\underline{\phi}) = \underline{\phi} + \epsilon\, \underline{u}_o \tag{1.322}$$

where the infinitesimal rotating vector, $\underline{\phi} = \phi\, \bar{n}$, and the displacement vector, $\underline{u}_o = \delta\, \bar{n} + \tilde{r}\underline{\phi}$, are defined. Assuming a rigid body with infinite volume, $\underline{u}_o$ represents the displacement of a material point in the rigid body that instantaneously coincide with the origin of reference frame $\mathcal{F}_{\mathcal{I}}$. With the aid of definition, Eq. (1.114), the following skew-symmetric dual tensor and associated Plücker coordinates are constructed

$$\phi^d \widetilde{\mathcal{N}} = \begin{bmatrix} \widetilde{\phi} & \widetilde{u}_o \\ 0 & \widetilde{\phi} \end{bmatrix}, \quad \phi^d \underline{\mathcal{N}} = \begin{vmatrix} \underline{u}_o \\ \underline{\phi} \end{vmatrix} \tag{1.323}$$

The detailed expression of motion tensor can be obtained through the following geometric operations. At first, the tangent vector and binormal vector, both perpendicular to the normal vector $\bar{n}$, are determined

$$\bar{b} = \tilde{\ell}_1 \bar{n}, \quad \bar{t} = \tilde{n}\bar{b} \tag{1.324}$$

to form a right-hand set, $\lfloor \bar{t}, \bar{n}, \bar{b} \rfloor$, and then the unit orientation vector $\bar{\ell}_1$ is resolved into this basis only with two non-zero components like

$$\bar{\ell}_1 = \underline{\ell}_1^n + \underline{\ell}_1^t \tag{1.325}$$

where $\underline{\ell}_1^n$ and $\underline{\ell}_1^t$ are the two non-zero components of $\bar{\ell}_1$ projected onto the direction of $\bar{n}$ and $\bar{t}$, respectively. Now the rotation of $\bar{\ell}_1$ at a small angle $\phi$ to $\bar{\ell}_2$ will only cause the same rotation of $\underline{\ell}_1^t$, while keeping $\underline{\ell}_1^n$ unchanged. From this simple geometric relation, the new orientation vector $\bar{\ell}_2$ is readily verified to be

$$\bar{\ell}_2 = \underline{\ell}_1^n + \underline{\ell}_1^t - \phi\, \ell_1^t \bar{b} = (I + \phi\tilde{n})(\underline{\ell}_1^n + \underline{\ell}_1^t) = R\, \bar{\ell}_1 \tag{1.326}$$

where the identities, $\tilde{n}\underline{\ell}_1^n = 0$, and, $-\ell_1^t \bar{b} = \tilde{n}\underline{\ell}_1^t$, are applied and the rotation tensor $R = I + \phi\, \tilde{n} = I + \underset{\sim}{\phi}$ which is applicable to the infinitesimal rotation is introduced.

Second, from the definition of line vector, it is known that the vector $\underline{r}_1$ describes the spatial position of the line vector $\mathcal{L}_1$. After the motion, new position of the vector $\underline{r}_1$ is determined in the same manner as the director $\bar{\ell}_2$ to be

$$\underline{r}_2 = \underline{r}_1 + \phi\, \tilde{n}(\underline{r}_1 - \underline{r}) + \delta\bar{n} = (I + \widetilde{\phi})\, \underline{r}_1 - \widetilde{\phi}\, \underline{r} + \delta\, \bar{n} = R\, \underline{r}_1 + \underline{u}_o \tag{1.327}$$

Finally, the expressions of line vector $\underline{\mathcal{L}}_2$ in terms of $\bar{\ell}_1$ and $\underline{r}_1$ are found to be

$$\underline{\mathcal{L}}_2 = R\, \bar{\ell}_1 + \epsilon \left(\widetilde{R\underline{r}_1} + \tilde{u}_o\right) R\bar{\ell}_1 = R\bar{\ell}_1 + \epsilon\, (R\tilde{r}_1 + \tilde{u}_o R)\, \bar{\ell}_1 \tag{1.328}$$

where the identity, $\widetilde{R\underline{r}_1 R} = R\widetilde{r}_1$, can be proved by algebraic operations neglecting the higher order terms $O(\phi^2)$. With the aid of the Plučker coordinates, the line vector $\underline{\mathcal{L}}_2$ can be recasted to

$$\underline{\mathcal{L}}_2 = \begin{vmatrix} \widetilde{r}_2\bar{\ell}_2 \\ \bar{\ell}_2 \end{vmatrix} = \begin{bmatrix} R & \widetilde{u}_o R \\ 0 & R \end{bmatrix} \begin{vmatrix} \widetilde{r}_1\bar{\ell}_1 \\ \bar{\ell}_1 \end{vmatrix} \tag{1.329}$$

From which, the *motion tensor* is observed to be

$$\mathcal{R} = \begin{bmatrix} R & \widetilde{u}_o R \\ 0 & R \end{bmatrix} = \begin{bmatrix} I + \widetilde{\phi} & \widetilde{u}_o(I + \widetilde{\phi}) \\ 0 & I + \widetilde{\phi} \end{bmatrix} \tag{1.330}$$

Since above motion tensor is only suitable for small motion, it can be simplified with the aid of Eq. (1.323) to get

$$\mathcal{R} \approx \mathcal{I} + \phi^d \underline{\widetilde{\mathcal{N}}} \tag{1.331}$$

where $\mathcal{I}$ is a $6 \times 6$ identity matrix, and the higher order term $\widetilde{u}_o\,\widetilde{\phi}$ has been neglected. From above approximations, the dual vector with infinitesimal magnitude is determined to be the *axial vector* of motion tensors like

$$\phi^d \underline{\mathcal{N}} = \mathrm{axi}(\mathcal{R}) \tag{1.332}$$

Note that the motion tensors that are applicable to finite motions will be discussed in the next chapter.

## References


1. Hay, G.E.: Vector and Tensor Analysis. Dover, New York (1953)
2. Brand, L.: Vector and Tensor Analysis. Wiley, New York (1947)
3. Rudnicki, J.W.: Fundamentals of Continuum Mechanics. Wiley, New York (2014)
4. Dai, J.S.: Geometrical Foundations and Screw Algebra for Mechanisms and Robotics. Higher Education Press, Beijing (2014)
5. Bauchau, O.A.: Flexible Multibody Dynamics. Springer, Dordrecht (2011)
6. Clifford, W.K.: Preliminary sketch of biquaternions. Proc. London Math. Soc. **4**, 381–395 (1873)
7. Pennestrí, E., Stefanelli, R.: Linear algebra and numerical algorithms using dual numbers. Mult. Syst. Dyn. **18**, 323–344 (2007)
8. Bottema, O., Roth, B.: Theoretical Kinematics. Dover, New York (1979)
9. Macdonald, A.: Linear and Geometric Algebra. CreateSpace, North Charleston (2011)
10. Study, E.: Geometrie der dynamen. Verlag Teubner, Leipzig (1903)
11. Condurache, D., Burlacu, A.: Dual lie algebra representations of the rigid body motion. In: AIAA/AAS Astrodynamics Specialist Conference (2014). https://doi.org/10.2514/6.2014-4347
12. Pennestrí, E., Valentini, P.P.: Linear dual algebra algorithms and their application to kinematics. In: Multibody Dynamics (2009). https://doi.org/10.1007/978-1-4020-8829-2_11

13. Chasles, M.: Note sur les propriétés géné rales du systéme de deux corps semblables entre eux et placés d'une manière quelconque dans l'espace; et sur le déplacement fini, ou infiniment petit d'un corps solide libre. Bulletin des Sciences Mathématiques de Férussac **14**, 321–326 (1830)
14. Angeles, J.: Fundamentals of Robotic Mechanical Systems: Theory, Methods, and Algorithms. Springer, New York (1997)

# Chapter 2
# Motion and Deformation

When observing the world where the people are living, it is found that the motion is an essential characteristic of the world. From the movement of celestial bodies in the universe to the swimming of a Paramecium in a drop of water, motion becomes a common phenomenon in reality. In classical mechanics, the Newton's law reveals that it is the force determines the behavior of a moving body. Even though the force is invisible, it causes the changing of a body in its position, orientation and shape, i.e. the changing of both motion and deformation. Here, the motion of a body is defined as the changing of its position and orientation altogether, and the deformation of a body is defined as the changing of its shape before and after the position and orientation changing. When a body changes it position, orientation and shape under the external or internal force, it can be concluded that both motion and deformation occur simultaneously. From a physical point of view, they are closely related each other. However, the motion and deformation can be mathematically decomposed into rigid body motion and deformation. This chapter presents the strict motion definitions of the Cosserat [1] solid bodies in its motion formalism [2], and then the mathematical description of deformations with the aid of the introduction of Green-Lagrange strain tensors.

## 2.1 Material Coordinate

As shown in Fig. 2.1, a body is considered to be a collection of material particles. Assuming each material particle in this collection features infinitesimal mass and infinitesimal volume, all the material particles will occupy a spatial region, which is called the *reference configuration* [3] of the body. If the region occupied by all the material particles was a smooth regular region bounded by a closed surface, the body will be a continuous medium without broken. When the body moves from its

 
J. Wang, *Multiscale Multibody Dynamics*,
https://doi.org/10.1007/978-981-19-8441-9_2

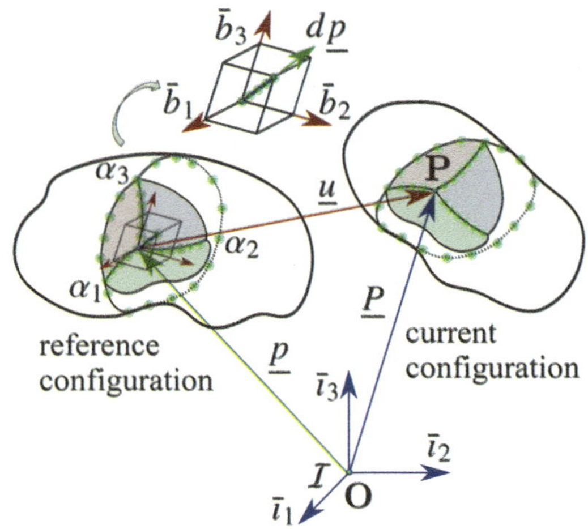


**Fig. 2.1** Configuration of a body

reference configuration to a new region under the external force, its position, attitude and shape will usually change. In other words, the body is occurred displacement accompanied by deformation. The new region occupied by the body is often referred to as the *current configuration* [3]. Figure 2.1 illustrates the changes of a body configuration in a Cartesian reference frame. For easy of description, each material particle of the body is identified by a label to distinguish one particle from another. The label consists of a triplet of real numbers $(\alpha_1, \alpha_2, \alpha_3)$, that remains attached to the material particle during motion and deformation. In general, the label is defined as the *material coordinates* of material particle. As discussed in Sect. 1.3.9, it is readily to confirm that the material coordinates is nothing but a type of curvilinear coordinates.

Given the position vector $\underline{p}(\alpha_1, \alpha_2, \alpha_3)$ of a material particle **P**, a local curvilinear basis at **P** could be constructed to describe the local properties of the body around this material particle. The orientation of the local curvilinear basis is described by three parametric curves that intersect at **P** and span into infinite in space. According to the definition of curvilinear basis in Sect. 1.3.9, the parametric curve of $\alpha_1$ is constructed by varying the material coordinate $\alpha_1$, while keeping $\alpha_2$ and $\alpha_3$ constant. Similarly, the parametric curves of $\alpha_2$ and $\alpha_3$ can be obtained through the same operation. All these parametric curves can also be imagined as an ensemble of material particles forming these parametric curves. As depicted in Fig. 2.1, a *material line* that exactly coincide with the infinitesimal vector $d\underline{p}$ can be produced by the material particles standing along $d\underline{p}$. It is straightforward that changing of position, orientation and length of a material line directly reflects the change of the motion and deformation of a body. If the direction of the material line is defined by using a unit vector $\bar{\ell}$, and the spatial locations of the material line is determined by a vector $\underline{r}$, then the material line can be presented by a line vector

$$\underline{\mathcal{L}} = \bar{\ell} + \epsilon \tilde{r} \bar{\ell} \tag{2.1}$$

or a screw

$$\underline{S} = \bar{\ell} + \epsilon\,(\widetilde{r}\,\bar{\ell} + h\bar{\ell}) \tag{2.2}$$

in a more general sense. The motion and deformation of the body can be conveniently described by using the material line.

## 2.2 Motion

As mentioned before, an arbitrary displacement of a body is usually accompanied by deformation. Therefore, it is not appropriate to consider only motion or deformation separately when the configuration of a body is changed. Mathematically, the configuration changing can be decomposed into rigid body motion and deformation. This section will describe the detail of rigid body motion which is consisted of translation and rotation. Translation describes the spatial position changing from the initial position to the current position, and rotation describes the variation of a body attitude meanwhile.

### *2.2.1 Kinematics of a Material Particle*

The position vector $\underline{p}$ of a material particle **P** described by its material coordinates $\alpha_1$, $\alpha_2$ and $\alpha_3$ is given in Sect. 2.1. Meanwhile, the position vector can also be represented by its Cartesian coordinates $x_1$, $x_2$, and $x_3$ in the Cartesian reference frame $\mathcal{F}_{\mathcal{I}} = [\mathbf{O}, \mathcal{I}]$, illustrated in Fig. 2.1, like

$$\underline{p} = x_1\bar{\imath}_1 + x_2\bar{\imath}_2 + x_3\bar{\imath}_3 \tag{2.3}$$

As discussed in Sect. 1.3.9, there exists a one-to-one mapping between material coordinates and Cartesian coordinates if the Jacobian of transformation, Eq. (1.272), is non-zero. This is an important property of continuum body, which becomes the foundation for defining strain tensor to measure the body deformation.

The position vectors of all material particles uniquely determine the time-independent reference configuration of a body. When all the material particles translate a spatial distance, measured by a time-dependent displacement vector $\underline{u}(t)$, a new region called current configuration will be occupied by the particles. The new position vector of particle **P** can be expressed as

$$\underline{P}(t) = \underline{p} + \underline{u}(t) = [x_1 + u_1(t)]\,\bar{\imath}_1 + [x_2 + u_2(t)]\,\bar{\imath}_2 + [x_3 + u_3(t)]\,\bar{\imath}_3 \tag{2.4}$$

The components of the velocity vector of particle **P** are readily obtained by differentiating the expression for the position vector, Eq. (2.4), to find

$$\underline{v}(t) = \underline{\dot{P}}(t) = \underline{\dot{u}} = \dot{u}_1(t)\bar{\imath}_1 + \dot{u}_2(t)\bar{\imath}_2 + \dot{u}_3(t)\bar{\imath}_3 \tag{2.5}$$

or rewritten in detail

$$v_1 = \dot{u}_1(t), \; v_2 = \dot{u}_2(t), \; v_3 = \dot{u}_3(t) \tag{2.6}$$

after introducing the Cartesian components of velocity vector, $v_1$, $v_2$ and $v_3$. Similarly, the acceleration vector of particle **P** is obtained by taking a time derivative of the velocity vector, like

$$\underline{a}(t) = \underline{\dot{v}} = \underline{\ddot{P}}(t) = \underline{\ddot{u}} = \ddot{u}_1(t)\bar{\imath}_1 + \ddot{u}_2(t)\bar{\imath}_2 + \ddot{u}_3(t)\bar{\imath}_3 \tag{2.7}$$

or in components

$$a_1 = \ddot{u}_1(t), \; a_2 = \ddot{u}_2(t), \; a_3 = \ddot{u}_3(t) \tag{2.8}$$

### 2.2.2 Geometric Description of Rotation

Euler's rotation theorem [4, 5] states that an arbitrary motion of a rigid body, that leaves one of its points fixed, is equivalent to a single rotation of magnitude $\phi$ about a unit vector $\bar{n}$ passing through the fixed point. According to this theorem, the geometric description of single rotation is presented in this section. It determines the explicit expression of rotation tensor $R$ as a function of $\phi$ and $\bar{n}$, and shows that the rotation tensor $R$ possesses a unit eigenvalue, $\lambda = 1$, associated with eigenvector $\bar{n}$. As depicted in Fig. 2.2, an arbitrary material line of a body in its line vector formula

$$\underline{\mathcal{S}}_1 = \bar{\ell}_1 + \epsilon \, \tilde{s}_1 \bar{\ell}_1 \tag{2.9}$$

rotates a finite angle of magnitude $\phi$ about the normal vector $\bar{n}$ to its current configuration

$$\underline{\mathcal{S}}_2 = \bar{\ell}_2 + \epsilon \, \tilde{s}_2 \bar{\ell}_2 \tag{2.10}$$

Due to rigid rotation, the vector $\underline{s}_1$ translates to a new position $\underline{s}_2$, and the direction $\bar{\ell}_1$ of the material line has been rotated to $\bar{\ell}_2$. The rotation axis is also a straight line

$$\underline{\mathcal{N}} = \bar{n} + \epsilon \, \tilde{r} \, \bar{n} \tag{2.11}$$

where vector $\underline{r}$ defines the position of an arbitrary point **R** on the rotation axis. By the law of vector addition, the position vectors $\underline{s}_1$ and $\underline{s}_s$ on the material lines decompose into

$$\underline{s}_1 = \underline{r} + \underline{r}_1, \; \underline{s}_2 = \underline{r} + \underline{r}_2 \tag{2.12}$$

For the easy description of finite rotations, the common normal vector of material line $\underline{\mathcal{S}}_1$ and the rotation axis $\underline{\mathcal{N}}$ is determined through

$$\bar{b} = \tilde{\ell}_1 \bar{n} \tag{2.13}$$

and the distance between these two lines is denoted by $h$. It is also noted that this unit vector $\bar{b}$ intersects lines $\underline{\mathcal{S}}_1$ and $\underline{\mathcal{N}}$ at points **N** and $\mathbf{S}_1$, respectively. In view of the tensor product properties, Eq. (1.54), the projection of vector $\underline{r}_1$ in normal vector $\bar{n}$ associated null space can be determined from

$$h\bar{b} = \underline{r}_1 - \bar{n}\,\bar{n}^T \underline{r}_1 = \widetilde{n}\widetilde{n}^T \underline{r}_1 \tag{2.14}$$

Furthermore, a tangent vector

$$\bar{t} = \tilde{n}\,\bar{b} \tag{2.15}$$

is readily produced to formulate a local orthogonal basis $\lfloor \bar{t},\ \bar{n},\ \bar{b} \rfloor$. If position vectors $\underline{r}$ and $\underline{r}_1$ were given, the position vector $\underline{r}_2$, resolved in the local orthogonal basis, can be represented by

$$\begin{aligned} \underline{r}_2 &= \bar{n}^T \underline{r}_1 \bar{n} + h \cos\phi\, \bar{b} + h \sin\phi\, \bar{t} \\ &= \bar{n}^T \underline{r}_1 \bar{n} + h \cos\phi\, \bar{b} + h \sin\phi\, \tilde{n}\, \bar{b} \end{aligned} \tag{2.16}$$

By Eq. (2.14), the vector $\underline{r}_2$ reformulates to

$$\begin{aligned} \underline{r}_2 &= (\underline{r}_1 + \widetilde{n}\widetilde{n}\underline{r}_1) + \cos\phi\, \widetilde{n}\widetilde{n}^T \underline{r}_1 + \sin\phi\, \widetilde{n}\widetilde{n}\widetilde{n}^T \underline{r}_1 \\ &= \underline{r}_1 + \sin\phi\, \tilde{n}\underline{r}_1 + (1 - \cos\phi)\widetilde{n}\widetilde{n}\underline{r}_1 \end{aligned} \tag{2.17}$$

where the property of identity, Eq. (1.56), is applied. From above formulation, the *rotation tensor* $R$ is found to be

$$R = I + \sin\phi\, \tilde{n} + (1 - \cos\phi)\, \tilde{n}\,\tilde{n} \tag{2.18}$$

This equation is also called Euler-Rodrigues formula [6]. Note this formula fails whenever $\phi = 0$, since $\bar{n}$ is then undetermined. The relation between vectors $\underline{r}_1$ and $\underline{r}_2$ before and after finite rotation is simplified to

$$\underline{r}_2 = R\, \underline{r}_1 \tag{2.19}$$

By Eq. (2.12), the position vectors $\underline{s}_1$ and $\underline{s}_2$ are related with

$$\underline{s}_2 = \underline{r} + R\, \underline{r}_1 = \underline{r} + R\, (\underline{s}_1 - \underline{r}) \tag{2.20}$$

and then

$$\underline{s}_2 = R\, \underline{s}_1 + (I - R)\, \underline{r} \tag{2.21}$$

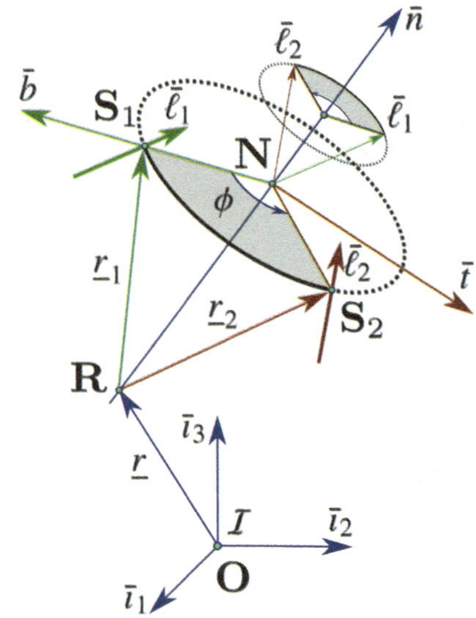


**Fig. 2.2** Geometric description of rotation

The above formulation shows the vector $\underline{s}_2$ that determines the current position of the material line undergoes a rotation $R$ and a translation $(I - R)\underline{r}$ due to finite rotation of the material line.

Unlike the position vector, $\underline{r}$, which is a bound vector, the unit vectors $\bar{\ell}_1$ and $\bar{\ell}_2$ are free vectors that can be moved to any spatial positions under the guarantee of remaining the directions unchanged. As depicted in Fig. 2.2, the unit vectors $\bar{\ell}_1$ and $\bar{\ell}_2$ have translated to point **N**. It can be observed that $\bar{\ell}_1$ rotates to $\bar{\ell}_2$ when the vector $\underline{r}_1$ rotates a finite angle of magnitude $\phi$ about the normal vector $\bar{n}$ to the vector $\underline{r}_2$. This observation shows that

$$\bar{\ell}_2 = R\,\bar{\ell}_1 \tag{2.22}$$

By vector product, $\tilde{n}\,\bar{n} = 0$, the following relation

$$\bar{n} = R\,\bar{n} \tag{2.23}$$

is easy to obtain from Eq. (2.18), and features clear physical meaning, that is, the normal vector keeps its orientation unchanged when a material line rotates a finite rotation about the normal vector. The first fundamental property of the rotation tensor arising from this relation states that it possesses a unit eigenvalue, $\lambda = 1$, associated with eigenvector $\bar{n}$. Furthermore, by Eq. (2.18), it is readily verified that

$$R^T\,R = R\,R^T = I \tag{2.24}$$

which answers $R$ is an orthogonal tensor, the second fundamental property of the rotation tensor.

### 2.2.3 Composition of Rotation Tensor

As discussed in Sect. 1.3.9, the curvilinear orthogonal basis $\mathcal{B}_i = \lfloor \bar{b}_1,\ \bar{b}_2,\ \bar{b}_3 \rfloor$ can be constructed after given the position vector of a material point in the body. Now a rotation tensor $R_i$ brings the Cartesian basis $\mathcal{I} = \lfloor \bar{\imath}_1,\ \bar{\imath}_2,\ \bar{\imath}_3 \rfloor$ to the body attached basis $\mathcal{B}_i$. With the aid of Eq. (2.22), resolving this tensor relationship in basis $\mathcal{I}$ then yields

$$\lfloor \bar{b}_1,\ \bar{b}_2,\ \bar{b}_3 \rfloor = R_i \lfloor \bar{\imath}_1,\ \bar{\imath}_2,\ \bar{\imath}_3 \rfloor \tag{2.25}$$

Note that $\mathcal{I} = \lfloor \bar{\imath}_1,\ \bar{\imath}_2,\ \bar{\imath}_3 \rfloor = I_{3\times 3}$, it shows

$$\mathcal{B}_i = R_i = \lfloor \bar{b}_1,\ \bar{b}_2,\ \bar{b}_3 \rfloor \tag{2.26}$$

which means the rotation tensor $R_i$ is identical with the body attached basis $\mathcal{B}_i$. When a rotation tensor $R$ brings a body with reference curvilinear basis $\mathcal{B}_i$ to its current basis $\mathcal{B}_f = \lfloor \bar{B}_1,\ \bar{B}_2,\ \bar{B}_3 \rfloor$, the relations

$$\mathcal{B}_f = R_f = \lfloor \bar{B}_1,\ \bar{B}_2,\ \bar{B}_3 \rfloor \tag{2.27}$$

and

$$\lfloor \bar{B}_1,\ \bar{B}_2,\ \bar{B}_3 \rfloor = R \lfloor \bar{b}_1,\ \bar{b}_2,\ \bar{b}_3 \rfloor \tag{2.28}$$

still hold, where the rotation tensor $R_f$ is identical with the body attached basis $\mathcal{B}_f$. In compact formula,

$$R_f = R\, R_i \tag{2.29}$$

is called the *composition of rotations*. This operation combines two rotations, that from the Cartesian basis $\mathcal{I}$ to the body attached basis $\mathcal{B}_i$, and that from $\mathcal{B}_i$ to $\mathcal{B}_f$, into a single rotation $R_f$ from basis $\mathcal{I}$ to $\mathcal{B}_f$.

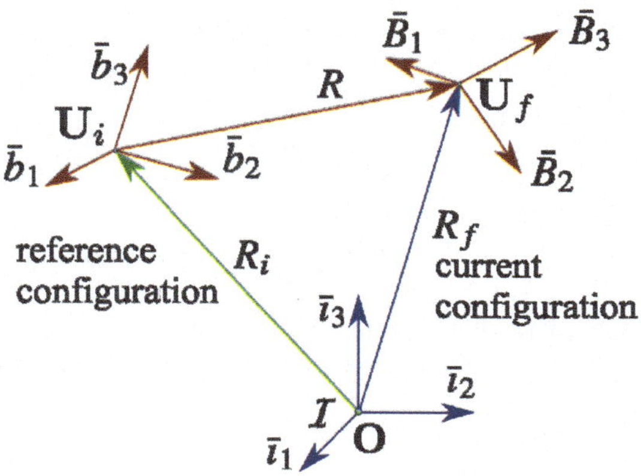


**Fig. 2.3** Composition of rotations

It should be kept in mind that all of the rotation tensors, $R_f$, $R_i$ and $R$ are resolved in the Cartesian basis $\mathcal{I}$ as depicted in Fig. 2.3. Actually, the rotation tensors can be resolved in any basis. By Eq. (1.17), the rotation tensor $R_f$ or the body attached basis $\mathcal{B}_f$, equivalently, can be resolved in the local basis $\mathcal{B}_i$ by projecting the mutually orthogonal unit vectors $\bar{B}_1$, $\bar{B}_2$ and $\bar{B}_3$ into $\mathcal{B}_i$ like

$$\bar{j}_1 = \begin{vmatrix} \bar{b}_1^T \bar{B}_1 \\ \bar{b}_2^T \bar{B}_1 \\ \bar{b}_3^T \bar{B}_1 \end{vmatrix}, \ \bar{j}_2 = \begin{vmatrix} \bar{b}_1^T \bar{B}_2 \\ \bar{b}_2^T \bar{B}_2 \\ \bar{b}_3^T \bar{B}_2 \end{vmatrix}, \ \bar{j}_3 = \begin{vmatrix} \bar{b}_1^T \bar{B}_3 \\ \bar{b}_2^T \bar{B}_3 \\ \bar{b}_3^T \bar{B}_3 \end{vmatrix} \tag{2.30}$$

or in assembly,

$$\lfloor \bar{j}_1, \ \bar{j}_2, \ \bar{j}_3 \rfloor = \begin{bmatrix} \bar{b}_1^T \\ \bar{b}_2^T \\ \bar{b}_3^T \end{bmatrix} \lfloor \bar{B}_1, \ \bar{B}_2, \ \bar{B}_3 \rfloor = R_i^T R_f \tag{2.31}$$

where the unit vectors $\bar{j}_1$, $\bar{j}_2$ and $\bar{j}_3$ are verified to be mutually orthogonal

$$\begin{aligned} \bar{j}_1^T \bar{j}_1 &= \bar{B}_1^T \bar{B}_1 = 1, \ \bar{j}_1^T \bar{j}_2 = \bar{B}_1^T \bar{B}_2 = 0, \ \bar{j}_1^T \bar{j}_3 = \bar{B}_1^T \bar{B}_3 = 0 \\ \bar{j}_2^T \bar{j}_1 &= \bar{B}_2^T \bar{B}_1 = 0, \ \bar{j}_2^T \bar{j}_2 = \bar{B}_2^T \bar{B}_2 = 1, \ \bar{j}_2^T \bar{j}_3 = \bar{B}_2^T \bar{B}_3 = 0 \\ \bar{j}_3^T \bar{j}_1 &= \bar{B}_3^T \bar{B}_1 = 0, \ \bar{j}_3^T \bar{j}_2 = \bar{B}_3^T \bar{B}_2 = 0, \ \bar{j}_3^T \bar{j}_3 = \bar{B}_3^T \bar{B}_3 = 1 \end{aligned} \tag{2.32}$$

Introducing new symbols

$$\mathcal{B}^* = R^* = \lfloor \bar{j}_1, \ \bar{j}_2, \ \bar{j}_3 \rfloor \tag{2.33}$$

where the symbol $(\cdot)^*$ represents the quantity measured in the body attached basis $\mathcal{B}_i$. The composition of rotation is recasted to

$$R_f = R_i R^* \tag{2.34}$$

which means the rotation tensor $R^*$, a local curvilinear orthogonal basis $\mathcal{B}^*$ equivalently, describes the configuration of basis $\mathcal{B}_f$ in basis $\mathcal{B}_i$. Comparing the above expression, Eq. (2.34), with the composition of rotations, Eq. (2.29), readily find

$$R = R_i R^* R_i^T, \ R^* = R_i^T R R_i \tag{2.35}$$

or equivalently,

$$R R_i = R_i R^* \tag{2.36}$$

It indicates that the rotation tensor $R$ resolved in the Cartesian basis $\mathcal{I}$ and $R^*$ resolved in the body attached basis $\mathcal{B}_i$, are related by the above transformation law.

### 2.2.4 Change of Basis Operations

Similar to this transformation of rotation tensor, Eq. (2.35), the vector components in various orthonormal bases are transformed in a similar approach. Given an arbitrary vector

$$\underline{v} = v_1 \bar{\imath}_1 + v_2 \bar{\imath}_2 + v_3 \bar{\imath}_3 \tag{2.37}$$

by Eq. (1.8), its components $v_1$, $v_2$ and $v_3$ are resolved in the Cartesian basis $\mathcal{I}$. When a rotation tensor $R_i$ brings the Cartesian basis $\mathcal{I}$ to the body attached basis $\mathcal{B}_i$, rotating vector $\underline{v}$ generates a new vector $\underline{v}_i$. Pure rotation guarantees the components of vector $\underline{v}_i$ measured in the body attached frame $\mathcal{B}_i$ are identical with $v_1$, $v_2$ and $v_3$. Resolving vector $\underline{v}_i$ in basis $\mathcal{B}_i$ yields

$$\underline{v}_i^* = \begin{vmatrix} v_1 \\ v_2 \\ v_3 \end{vmatrix} = \underline{v} \tag{2.38}$$

Now, projecting the vector $\underline{v}_i$ in the basis $\mathcal{B}_i$ obtains

$$\underline{v}_i = v_1 \bar{b}_1 + v_2 \bar{b}_2 + v_3 \bar{b}_3 = \lfloor \bar{b}_1,\ \bar{b}_2,\ \bar{b}_3 \rfloor \underline{v}_i^* \tag{2.39}$$

Alternatively, the following transformations

$$\underline{v}_i = R_i\, \underline{v}_i^*,\ \underline{v}_i^* = R_i^T\, \underline{v}_i \tag{2.40}$$

are equivalent to the rotation of a bound vector, Eq. (2.19).

Similarly, the vector tensor products, $T_i = \underline{u}_i\, \underline{v}_i^T$ and $T_i^* = \underline{u}_i^*\, \underline{v}_i^{*T}$, resolved in the Cartesian basis $\mathcal{I}$ and the body attached basis $\mathcal{B}_i$, respectively, are related by

$$T_i = \underline{u}_i\, \underline{v}_i^T = R_i\, \underline{u}_i^* \underline{v}_i^{*T} R_i^T = R_i\, T_i^* R_i^T \tag{2.41}$$

Once again with the aid of the property of orthogonal tensors, Eq. (2.24), the transformation laws for the components of second-order tensors are found to be

$$T_i = R_i\, T_i^* R_i^T,\ T_i^* = R_i^T\, T_i\, R_i \tag{2.42}$$

Referring the definition of tensors in Sect. 1.3, it has mentioned the specific laws of transformation must be followed by tensors. Now the rigorous definitions for first-order tensors and second-order tensors are clarified: first-order tensors whose components resolved in two bases are related by Eq. (2.40); while the transformation of second-order tensors follows Eq. (2.42). The transformation of rotation tensor, Eq. (2.35), apparently satisfies the law of transformation for second-order tensors. Another example is the skew-symmetric, second order tensor $\widetilde{v}$ formulated by an arbitrary vector $\underline{v}$. Its components resolved in two bases must be related by Eq. (2.42). Hence, the tensorial nature of $\widetilde{v}$ will be established if and only if

$$\widetilde{v} = R_i\, \widetilde{v}^*\, R_i^T \Leftrightarrow \underline{v} = R_i \underline{v}^* \tag{2.43}$$

This statement can be proved through simple algebraic manipulations, taking into account the fact that $R_i$ is an orthogonal tensor.

### 2.2.5 *Derivatives of Rotation Operations*

As depicted in Fig. 2.3, the rotation tensor $R$ brings reference basis $\mathcal{B}_i$ to current basis $\mathcal{B}_f$; the components of this tensor resolved in basis $\mathcal{B}_i$ are $R^* = \lfloor \bar{j}_1,\ \bar{j}_2,\ \bar{j}_3 \rfloor$, Eq. (2.33). If rotation tensor $R^*$ is a function of time, $R^* = R^*(t)$, and it then follows that

$$R^{*T}\dot{R}^* = \begin{bmatrix} 0 & \bar{j}_1^T \dot{\bar{j}}_2 & \bar{j}_1^T \dot{\bar{j}}_3 \\ \bar{j}_2^T \dot{\bar{j}}_1 & 0 & \bar{j}_2^T \dot{\bar{j}}_3 \\ \bar{j}_3^T \dot{\bar{j}}_1 & \bar{j}_3^T \dot{\bar{j}}_2 & 0 \end{bmatrix} = \begin{bmatrix} 0 & -\bar{j}_2^T \dot{\bar{j}}_1 & \bar{j}_1^T \dot{\bar{j}}_3 \\ \bar{j}_2^T \dot{\bar{j}}_1 & 0 & -\bar{j}_3^T \dot{\bar{j}}_2 \\ -\bar{j}_1^T \dot{\bar{j}}_3 & \bar{j}_3^T \dot{\bar{j}}_2 & 0 \end{bmatrix} \tag{2.44}$$

where the time derivatives of orthogonal attributes, $\bar{j}_i^T\,\bar{j}_i = 1$ for $i = 1, 2, 3$, are found to be zero, $\bar{j}_i^T\,\dot{\bar{j}}_i = 0$. Apparently the above time derivatives are a skew-symmetric, second order tensor formulated by the vector $\underline{\omega}^*$ called the *angular velocity vector*. In compact formula,

$$\widetilde{\omega}^* = R^{*T}\dot{R}^* \tag{2.45}$$

In view of $R_f(t) = R_i\, R^*(t)$, it can be verified

$$\widetilde{\omega}^* = R_f^T \dot{R}_f \tag{2.46}$$

Even though the angular velocity $\underline{\omega}^*$ is introduced by using the time derivatives of rotation tensor $R^*$ resolved in body attached basis $\mathcal{B}_i$, it does not mean $\underline{\omega}^*$ is resolved in the same basis as well. Actually, the components of angular velocity $\underline{\omega}^*$ are measured in basis $\mathcal{B}_f$. The angular velocity $\underline{\omega}^*$ can be transformed into the Cartesian basis $\mathcal{I}$ with the aid of the law of transformation for second order tensor, Eq. (2.43), like

$$\underline{\omega} = R_f\, \underline{\omega}^* \Leftrightarrow \widetilde{\omega} = R_f\, \widetilde{\omega}^* R_f^T \Leftrightarrow \widetilde{\omega} = \dot{R}_f R_f^T \tag{2.47}$$

With the aid of $R_f(t) = R(t)R_i$, it is readily to find

$$\widetilde{\omega} = \dot{R}\, R^T \tag{2.48}$$

If $R$ is defined by Eq. (2.18), the angular velocity vector $\underline{\omega}$ can be expressed explicitly as a function of $\phi$ and $\bar{n}$ together with the time derivatives

$$\underline{\omega} = \dot{\phi}\bar{n} + \sin\phi\dot{\bar{n}} + (1 - \cos\phi)\widetilde{\bar{n}}\dot{\bar{n}} \tag{2.49}$$

In view of the composition of rotations, $R_f = R_i R^* = R\, R_i$, the angular velocity resolved in different bases can be related by

$$\underline{\omega} = R_i R^* \underline{\omega}^* = R\, R_i \underline{\omega}^* \Leftrightarrow R_i^T \underline{\omega} = R^* \underline{\omega}^* \Leftrightarrow R^T \underline{\omega} = R_i \underline{\omega}^* \tag{2.50}$$

The angular velocity vector $\underline{\omega}$ enables the evaluation of time derivative of an orthonormal basis, $\mathcal{B}_f$, as

$$\begin{aligned} \lfloor \dot{\bar{B}}_1,\ \dot{\bar{B}}_2,\ \dot{\bar{B}}_3 \rfloor &= \dot{R} \lfloor \bar{b}_1,\ \bar{b}_2,\ \bar{b}_3 \rfloor = \dot{R}_f \lfloor \bar{\imath}_1,\ \bar{\imath}_2,\ \bar{\imath}_3 \rfloor \\ &= \dot{R}_f R_f^T R_f \lfloor \bar{\imath}_1,\ \bar{\imath}_2,\ \bar{\imath}_3 \rfloor = \widetilde{\omega}\, R_f \lfloor \bar{\imath}_1,\ \bar{\imath}_2,\ \bar{\imath}_3 \rfloor \end{aligned} \tag{2.51}$$

alternatively,

$$\lfloor \dot{\bar{B}}_1,\ \dot{\bar{B}}_2,\ \dot{\bar{B}}_3 \rfloor = \widetilde{\omega} \lfloor \bar{B}_1,\ \bar{B}_2,\ \bar{B}_3 \rfloor \Leftrightarrow \dot{R}_f = \widetilde{\omega}\, R_f \tag{2.52}$$

Spatial derivatives of rotation operations are treated in a similar manner to its time derivatives. Consider now $\mathcal{B}_f$ is a space-dependent orthonormal basis, $R_f = R_f(s)$, a spatial derivative of this expression yields

$$\widetilde{\kappa}^* = R_f^T R_f' \Leftrightarrow \widetilde{\kappa} = R_f' R_f^T \Leftrightarrow \underline{\kappa} = R_f\, \underline{\kappa}^* \tag{2.53}$$

It is also readily to find

$$\lfloor \bar{B}_1',\ \bar{B}_2',\ \bar{B}_3' \rfloor = \widetilde{\kappa} \lfloor \bar{B}_1,\ \bar{B}_2,\ \bar{B}_3 \rfloor \Leftrightarrow R_f' = \widetilde{\kappa}\, R_f \tag{2.54}$$

Note that the twist and curvature of the curve, defined in Eq. (1.194), are the two non-vanishing components of the curvature vector resolved in Frenet's basis. The curvature tensors of the surface have defined in Sect. 1.3.8, and the curvature tensors of the orthogonal curvilinear coordinate system can be found in Sect. 1.3.10.

### 2.2.6 *Vectorial Parameterization of Rotation Tensor*

Mathematically, the rotation tensor $R$ also known as the second-order orthogonal tensor, is a $3 \times 3$ matrix with nine parameters. The first fundamental property of rotation tensor constrains three parameters among of them. Meanwhile, the orthogonality of rotation tensor constrains additional three parameters. As a result, the rotation tensor $R$ can be strictly defined by a minimum set of only three independent parameters. This minimal presentation of rotation tensor is implemented through a process of vectorial parameterization [6]. In general, the formula of vectorial parameterization of rotation is defined as

$$\underline{p} = p(\phi)\bar{n} \tag{2.55}$$

where the arbitrary function $p(\phi)$ is called the generating function. Taking a time derivative of the rotation parameter vector $\underline{p}$ yields

$$\underline{\dot{p}} = p'\dot{\phi}\bar{n} + p\dot{\bar{n}} \tag{2.56}$$

where $p' = dp/d\phi$. Pre-multiplying $\widetilde{n}$ at both side of this equation leads to, $\widetilde{n}\underline{\dot{p}} = p\widetilde{n}\dot{\bar{n}}$, due to the attribute of $\widetilde{n}\bar{n} = 0$. By Eq. (1.55) and the identity, $\bar{n}^T\dot{\bar{n}} = 0$, it can be verified quickly

$$\widetilde{n}\widetilde{n}\underline{\dot{p}} = p(\bar{n}\,\bar{n}^T - I)\dot{\bar{n}} = -p\dot{\bar{n}} \tag{2.57}$$

Substituting this equality into Eq. (2.56), another key identity is sought

$$\dot{\phi}\bar{n} = \frac{1}{p'}(I + \widetilde{n}\widetilde{n})\underline{\dot{p}} \tag{2.58}$$

Next, the explicit formulation of the rotation tensor in terms of $\underline{p}$ is easily obtained from Rodrigues' rotation formula, Eq. (2.18), like

$$R = I + \frac{\sin\phi}{p}\widetilde{p} + \frac{1-\cos\phi}{p^2}\widetilde{p}\widetilde{p} \tag{2.59}$$

and then the relationship between the angular velocity vector $\underline{\omega}$ and the vectorial parameterization vector $\underline{p}$ and its time derivative is related by

$$\underline{\omega} = H(\underline{p})\underline{\dot{p}} \tag{2.60}$$

where the tangent operator $H(\underline{p})$ is identified from Eq. (2.49) with the aid of above formulas, Eqs. (2.57) and (2.58). The explicit expression of tangent operator is found to be

$$H(\underline{p}) = \frac{1}{p'}I + \frac{1-\cos\phi}{p^2}\widetilde{p} + \frac{1}{p^2}(\frac{1}{p'} - \frac{\sin\phi}{p})\widetilde{p}\widetilde{p} \tag{2.61}$$

Taking matrix inverse of tangent operator, $H(\underline{p})$, the time derivative of the vectorial parameterization vector is related to the angular velocity vector by

$$\underline{\dot{p}} = H^{-1}(\underline{p})\underline{\omega} \tag{2.62}$$

where the inverse of tangent operator is given as

$$H^{-1}(\underline{p}) = p'I - \frac{1}{2}\widetilde{p} + \frac{1}{p^2}(p' - \frac{p}{2\tan\phi/2})\widetilde{p}\widetilde{p} \tag{2.63}$$

It is important to note that the tangent operator $H$ features remarkable properties as follow

$$R = H\,H^{-T} = H^{-T} H \tag{2.64}$$

From which, the components of the angular velocity vector resolved in local frame, $R^T \underline{\omega}$, can be written as the function of $\underline{p}$ and $\dot{\underline{p}}$ like

$$R^T \underline{\omega} = H^T(\underline{p})\dot{\underline{p}} \tag{2.65}$$

Even though the relations of rotation tensor $R$ and angular velocity $\underline{\omega}$ in terms of $\underline{p}$ and its time derivatives $\dot{\underline{p}}$ are discussed above in detail, the generating function $p(\phi)$ is still arbitrary. Now two types of specific vectorial parameterizations, the widely used approaches in the rest of the book, are presented in this section; one is known as the *Cartesian rotation vector*, and another one the *Wiener-Milenković parameters*.

For the first type of parameterization, the simplest choice of generating function, $p(\phi) = \phi$, leads to

$$\underline{p}(\phi) = \phi\bar{n} = \underline{\phi} \tag{2.66}$$

Clearly, $p' = 1$. By inserting $p(\phi) = \phi$ into Eq. (2.59), the expression for the rotation tensor is readily found

$$R = I + \frac{\sin\phi}{\phi}\widetilde{\phi} + \frac{1-\cos\phi}{\phi^2}\widetilde{\phi}\widetilde{\phi} \tag{2.67}$$

According to Taylor expansion of trigonometric functions

$$\begin{aligned}
\sin\phi &= \phi - \frac{1}{3!}\phi^3 + \frac{1}{5!}\phi^5 - \cdots \\
\cos\phi &= 1 - \frac{1}{2!}\phi^2 + \frac{1}{4!}\phi^4 - \frac{1}{6!}\phi^6 + \cdots
\end{aligned} \tag{2.68}$$

the rotation tensor $R$ can be rewritten as a polynomial function of $\underline{\phi} = \phi\,\bar{n}$ or $\widetilde{\phi} = \phi\,\widetilde{n}$ equivalently, and in detail

$$\begin{aligned}
R &= I + (\phi - \frac{1}{3!}\phi^3 + \frac{1}{5!}\phi^5 + \cdots)\widetilde{n} + (\frac{1}{2!}\phi^2 - \frac{1}{4!}\phi^4 + \frac{1}{6!}\phi^6 - \cdots)\widetilde{n}\,\widetilde{n} \\
&= I + \phi\,\widetilde{n} + \frac{1}{2!}(\phi\,\widetilde{n})^2 + \frac{1}{3!}(\phi\,\widetilde{n})^3 + \frac{1}{4!}(\phi\,\widetilde{n})^4 + \frac{1}{5!}(\phi\,\widetilde{n})^5 + \cdots \\
&= I + \widetilde{\phi} + \frac{1}{2!}\widetilde{\phi}^2 + \frac{1}{3!}\widetilde{\phi}^3 + \frac{1}{4!}\widetilde{\phi}^4 + \cdots
\end{aligned} \tag{2.69}$$

where the identity, $\widetilde{n}\widetilde{n}\widetilde{n} = -\widetilde{n}$, is applied. In view of exponential mapping between Lie algebra, $so(3)$, and Lie group, $SO(3)$, *i.e.*

$$\exp(\widetilde{\phi}) = \sum_{k=0}^{\infty}\frac{\widetilde{\phi}^k}{k!} = I + \widetilde{\phi} + \frac{1}{2!}\widetilde{\phi}^2 + \frac{1}{3!}\widetilde{\phi}^3 + \frac{1}{4!}\widetilde{\phi}^4 + \cdots \tag{2.70}$$

the *exponential map* of rotation is obtained as

$$R = \exp(\phi \widetilde{n}) = \exp(\widetilde{\phi}) \tag{2.71}$$

For very small rotations, only the first-order term of the exponential map is kept, to yield

$$R = I + \widetilde{\phi} = \begin{bmatrix} 1 & -\phi_3 & \phi_2 \\ \phi_3 & 1 & -\phi_1 \\ -\phi_2 & \phi_1 & 1 \end{bmatrix} \tag{2.72}$$

where $\phi_1$, $\phi_2$, and $\phi_3$ are three components of a rotation parameter vector $\underline{\phi}^T = \lfloor \phi_1, \phi_2, \phi_3 \rfloor$. Substituting generating function $p(\phi) = \phi$ into Eqs. (2.61) and (2.63), the expressions for operator $H$ and its inverse $H^{-1}$ are identified

$$H = I + \frac{1 - \cos\phi}{\phi^2}\widetilde{\phi} + \frac{1}{\phi^2}(1 - \frac{\sin\phi}{\phi})\widetilde{\phi}\widetilde{\phi} \tag{2.73}$$

and

$$H^{-1} = I - \frac{1}{2}\widetilde{\phi} + \frac{1}{\phi^2}(1 - \frac{\phi}{2\tan\phi/2})\widetilde{\phi}\widetilde{\phi} \tag{2.74}$$

For the second type of vectorial parameterization, the selection of generating function, $p(\phi) = 4\tan(\phi/4)$, yields Wiener-Milenković parameters

$$\underline{c} = 4\tan\frac{\phi}{4}\,\bar{n} \tag{2.75}$$

This representation is limited to rotation angles $|\phi| < 2\pi$ since it presents a singularity $\underline{c} \to \infty$ when $|\phi| \to 2\pi$. The rotation tensor expressed in terms of Wiener-Milenković parameters is readily found from Eq. (2.59) as

$$R(\underline{c}) = I + \frac{\nu_c^2}{2}c_0\widetilde{c} + \frac{\nu_c^2}{2}\widetilde{c}\widetilde{c} \tag{2.76}$$

where $c_0 = 2(1 - \tan^2\phi/4) = 2 - \frac{1}{8}\underline{c}^T\underline{c}$, and $\nu_c = 2/(4 - c_0) = \cos^2\phi/4$. The similar operations applied to Eqs. (2.61) and (2.63) yields the explicit expressions of tangent operator $H(\underline{c})$ and its inverse $H^{-1}(\underline{c})$ in the form of

$$H(\underline{c}) = \nu_c(I + \frac{\nu_c}{2}\widetilde{c} + \frac{\nu_c}{8}\widetilde{c}\widetilde{c}) \tag{2.77}$$

and

$$H^{-1}(\underline{c}) = \frac{1}{\nu_c}I - \frac{1}{2}\widetilde{c} + \frac{1}{8}\widetilde{c}\widetilde{c} \tag{2.78}$$

Wiener-Milenković parameters always require the rescaling operation to avoid the singularity when rotation angle $\phi$ close to $2\pi$, possible occurs in the successive composition of rotations. Let $\underline{c}_i$ and $\underline{c}$ be the Wiener-Milenković parameters of two successive rotations, and $\underline{c}_f$ the conformal rotation parameters of the composed rotation, such that $R_f(\underline{c}_f) = R(\underline{c})R_i(\underline{c}_i)$. The composition of rotation tensor by the Wiener-Milenković parameters is combined the rescaling of rotation tensor into a single operation [2]

$$\underline{c}_f = \begin{cases} 4(c_0\underline{c}_i + c_0^i\underline{c} + \widetilde{c}\underline{c}_i)/(\Delta_1 + \Delta_2) & \text{if } \Delta_2 \geq 0 \\ -4(c_0\underline{c}_i + c_0^i\underline{c} + \widetilde{c}\underline{c}_i)/(\Delta_1 - \Delta_2) & \text{if } \Delta_2 < 0 \end{cases} \tag{2.79}$$

where $\Delta_1 = (4 - c_0)(4 - c_0^i)$ and $\Delta_2 = c_0c_0^i - \underline{c}^T\underline{c}_i$. The rescaling condition automatically selects the largest denominator, also guaranteeing the most accurate numerical evaluation of the composed rotation.

### 2.2.7 Geometric Description of Motion

Chasles' theorem [7] states that the most general motion of a rigid body consists of a rotation about a line followed by a translation along the same line. Hence, a general motion is characterized by the magnitudes $\phi$ of the rotation and intrinsic displacement $\delta$ about its Chasles' line

$$\underline{\mathcal{N}} = \bar{n} + \epsilon\,\widetilde{r}\bar{n} \tag{2.80}$$

where $\underline{r}$ gives the position of an arbitrary point **R** on the Chasles' line with unit normal vector $\bar{n}$ specifying the direction of rotation and translation. As depicted in Fig. 2.4, an arbitrary material line of a body undergoes spatial motion and changes its configuration from

$$\underline{\mathcal{S}}_1 = \bar{\ell}_1 + \epsilon\,\widetilde{s}_1\bar{\ell}_1 \tag{2.81}$$

to

$$\underline{\mathcal{S}}_2 = \bar{\ell}_2 + \epsilon\,\widetilde{s}_2\bar{\ell}_2 \tag{2.82}$$

where the line vectors $\underline{\mathcal{S}}_1$ and $\underline{\mathcal{S}}_2$ describe the reference configuration and current configuration of the material line, respectively. An arbitrary position vector $\underline{s}_1$ specifies the position of the material line with respect to a reference point **O**, and $\bar{\ell}_1$ gives the direction of the material line. The expressions of position vectors $\underline{s}_1$ and $\underline{s}_2$ in terms of $\underline{r}$, $\underline{r}_1$ and $\underline{r}_2$ are given by Eq. (2.12). According to Chasles' theorem, the motion of material line from $\underline{\mathcal{S}}_1$ to $\underline{\mathcal{S}}_2$ can be decomposed into a finite rotation of magnitude $\phi$ about the Chasles' line $\underline{\mathcal{N}}$ and then followed by a translation $\delta$ along the same line.

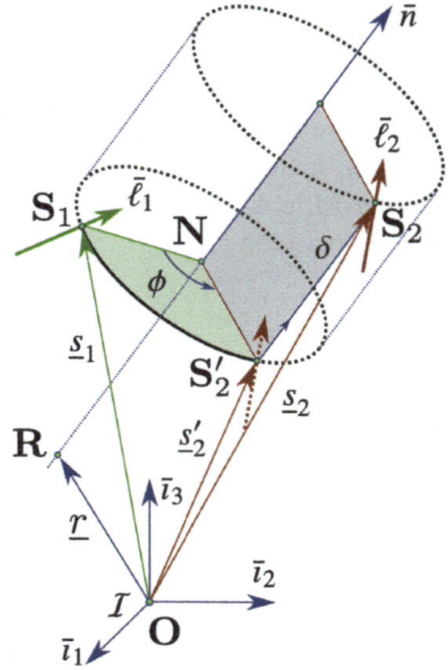


**Fig. 2.4** Material line of a body before and after motion

As discussed in Sect. 2.2.2, pure rotation of material line $\underline{\mathcal{S}}_1$ about the Chasles' line $\underline{\mathcal{N}}$ from point $\mathbf{S}_1$ to $\mathbf{S}_2'$ is specified by a rotation tensor $R$, Eq. (2.18), and the current direction $\bar{\ell}_2$ of material line is determined by

$$\bar{\ell}_2 = R\bar{\ell}_1 \tag{2.83}$$

Referring the expressions, Eq. (2.21), the updated position vector is readily obtained

$$\underline{s}_2' = R\,\underline{s}_1 + (I - R)\,\underline{r} \tag{2.84}$$

Next, a translation $\delta$ along the Chasles' line $\underline{\mathcal{N}}$ brings point $\mathbf{S}_2'$ to $\mathbf{S}_2$. The position vector of material line at current configuration is updated to

$$\underline{s}_2 = R\,\underline{s}_1 + (I - R)\,\underline{r} + \delta\bar{n} = R\,\underline{s}_1 + \underline{u}_o \tag{2.85}$$

where the displacement vector

$$\underline{u}_o = (I - R)\,\underline{r} + \delta\bar{n} \tag{2.86}$$

describes the motion of a particle in the body coincident with the reference point **O**. This expression shows the displacement $\underline{u}_o$ of the reference point **O** consisted two parts, one contributes from the rotation, $(I - R)\,\underline{r}$, and the other dues to the translation $\delta\,\bar{n}$. It is very important to note that the vector $\underline{u}_o$ is different from the displacement

$$\underline{u} = \underline{s}_2 - \underline{s}_1 \tag{2.87}$$

of a material particle $\mathbf{S}_1$ to which the material line $\underline{\mathcal{S}}_1$ is attached. As defined in Sect. 2.2.1, the vector $\underline{u}$ is a very straightforward description of the displacement of a material particle moved from its reference position $\mathbf{S}_1$ to current position $\mathbf{S}_2$. These two definitions are related by

$$\underline{u} = (R - I)\,\underline{s}_1 + \underline{u}_o \tag{2.88}$$

Further simplification leads to

$$\underline{u} = (R - I)\,\underline{r}_1 + \delta\,\bar{n} \tag{2.89}$$

the first part $(R - I)\,\underline{r}_1$ of displacement $\underline{u}$ is exactly the contribution of rotation of material line $\underline{\mathcal{S}}_1$ about the Chasles' line stated in Chasles' theorem. In view of above, the material lines $\underline{\mathcal{S}}_1$ and $\underline{\mathcal{S}}_2$ in the Plücker coordinates can be related by the so called *motion tensor* like

$$\begin{vmatrix} \tilde{s}_2\bar{\ell}_2 \\ \bar{\ell}_2 \end{vmatrix} = \begin{bmatrix} R & \tilde{u}_o R \\ 0 & R \end{bmatrix} \begin{vmatrix} \tilde{s}_1\bar{\ell}_1 \\ \bar{\ell}_1 \end{vmatrix} \tag{2.90}$$

or in compact form

$$\underline{\mathcal{S}}_2 = \mathcal{R}\,\underline{\mathcal{S}}_1 \tag{2.91}$$

where the motion tensor are introduced

$$\mathcal{R} = \begin{bmatrix} R & \tilde{u}_o R \\ 0 & R \end{bmatrix} \tag{2.92}$$

Clearly, the motion tensor is the operator that transforms the Plücker coordinates of material line $\underline{\mathcal{S}}_1$ in the reference configuration to its counterpart $\underline{\mathcal{S}}_2$ in the current configuration [8–10]. The factorized form of motion tensor

$$\mathcal{R} = \begin{bmatrix} R & \tilde{u}_o R \\ 0 & R \end{bmatrix} = \begin{bmatrix} I & \tilde{u}_o \\ 0 & I \end{bmatrix} \begin{bmatrix} R & 0 \\ 0 & R \end{bmatrix} \tag{2.93}$$

exactly matches the motion decomposition of a material particle $\mathbf{O}$ in Chasles' theorem, first a pure rotation then followed by a translation. Similar to rotation tensor, motion tensor possesses two fundamental properties. The first one

$$\underline{\mathcal{N}} = \mathcal{R}\,\underline{\mathcal{N}} \tag{2.94}$$

states that motion tensor has a unit eigenvalue, $\lambda = 1$, associated with eigenvector $\underline{\mathcal{N}}$. In detail,

$$\begin{vmatrix} \tilde{r}\,\bar{n} \\ \bar{n} \end{vmatrix} = \begin{bmatrix} R & \tilde{u}_o R \\ 0 & R \end{bmatrix} \begin{vmatrix} \tilde{r}\,\bar{n} \\ \bar{n} \end{vmatrix} = \begin{vmatrix} R\,\tilde{r}\,\bar{n} + \tilde{u}_o\,R\,\bar{n} \\ R\,\bar{n} \end{vmatrix} \tag{2.95}$$

The second equation, $\bar{n} = R\,\bar{n}$, is nothing but the first fundamental property of rotation tensor, Eq. (2.23). With the aid of this property and the definition of displacement vector, Eq. (2.86), the first equation is readily verified to be

$$\begin{aligned} R\tilde{r}\,\bar{n} + \tilde{u}_o\,R\,\bar{n} &= R\tilde{r}\,\bar{n} + [\widetilde{(I-R)\underline{r}} + \delta\,\bar{n}]\,\bar{n} \\ &= R\tilde{r}\,\bar{n} + \tilde{r}\,\bar{n} - \widetilde{R\underline{r}}\,\bar{n} = \tilde{r}\,\bar{n} \end{aligned} \tag{2.96}$$

where the attributes, $\widetilde{R\underline{r}}\,\bar{n} = R\,\tilde{r}\,R^T\bar{n} = R\,\tilde{r}\,\bar{n}$, and, $\tilde{n}\bar{n} = 0$, are applied to eliminate the dummy terms. The second fundamental property of motion tensor

$$\mathcal{R}^T\mathcal{R} = \mathcal{R}\mathcal{R}^T = \mathcal{I} = I_{6\times 6} \tag{2.97}$$

states that motion tensor is orthogonal tensor, where $\mathcal{R}^T$ represents the transpose of motion tensor defined as

$$\mathcal{R}^T = \begin{bmatrix} R^T & (\tilde{u}_o R)^T \\ 0 & R^T \end{bmatrix} \tag{2.98}$$

Note this motion transpose is defined in the same manner as the dual transpose of skew symmetric dual tensor, Eq. (1.114). Hence, it will not distinct anymore in the rest of the book about the matrix operation called dual transpose or motion transpose. They are considered to be the same operations. Substituting Eq. (2.98) into Eq. (2.97), the orthogonality of motion tensor can be proved as follow

$$\begin{aligned} \mathcal{R}^T\mathcal{R} &= \begin{bmatrix} R^T & (\tilde{u}_o R)^T \\ 0 & R^T \end{bmatrix}\begin{bmatrix} R & \tilde{u}_o R \\ 0 & R \end{bmatrix} = \begin{bmatrix} R^T R & R^T\tilde{u}_o R + (\tilde{u}_o R)^T R \\ 0 & R^T R \end{bmatrix} \\ &= \begin{bmatrix} R^T R & R^T(\tilde{u}_o - \tilde{u}_o^T)R \\ 0 & R^T R \end{bmatrix} = \begin{bmatrix} I_{3\times 3} & 0 \\ 0 & I_{3\times 3} \end{bmatrix} = I_{6\times 6} \end{aligned} \tag{2.99}$$

and

$$\begin{aligned} \mathcal{R}\mathcal{R}^T &= \begin{bmatrix} R & \tilde{u}_o R \\ 0 & R \end{bmatrix}\begin{bmatrix} R^T & (\tilde{u}_o R)^T \\ 0 & R^T \end{bmatrix} = \begin{bmatrix} R\,R^T & R(\tilde{u}_o R)^T + \tilde{u}_o R\,R^T \\ 0 & R\,R^T \end{bmatrix} \\ &= \begin{bmatrix} R\,R^T & R\,R^T\tilde{u}_o^T + \tilde{u}_o R\,R^T \\ 0 & R\,R^T \end{bmatrix} = \begin{bmatrix} I_{3\times 3} & 0 \\ 0 & I_{3\times 3} \end{bmatrix} = I_{6\times 6} \end{aligned} \tag{2.100}$$

### *2.2.8 Composition of Motion Tensors*

As defined in Sect. 1.1.9, an inertial frame $\mathcal{F}_\mathcal{I} = [\mathbf{O}, \mathcal{I}]$ is constructed by attaching a Cartesian basis $\mathcal{I}(\bar{\imath}_1, \bar{\imath}_2, \bar{\imath}_3)$ to a given point $\mathbf{O}$. If assuming the point $\mathbf{O}$ labels a material particle in a body that can move from the origin to the point $\mathbf{P}_i$ with the position vector $\underline{u}_i$, a body attached reference frame $\mathcal{F}_i = [\mathbf{P}_i, \mathcal{B}_i]$ can be regarded as a new configuration of the original frame experienced a general motion. Referring

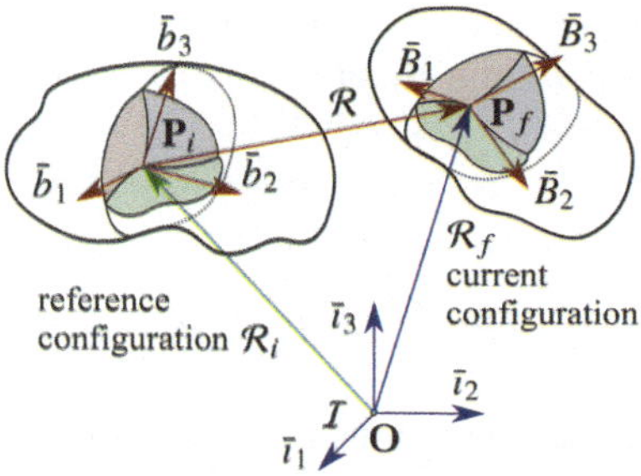


**Fig. 2.5** Composition of motion tensors

the definition of a material line, Eq. (1.64), the point $\mathbf{P}_i$ together with three mutually orthogonal unit vectors, $\bar{b}_1$, $\bar{b}_2$ and $\bar{b}_3$, determine three mutually orthogonal material lines (Fig. 2.5)

$$\underline{\mathcal{B}}^i_1 = \bar{b}_1 + \epsilon\, \widetilde{u}_i\, \bar{b}_1,\ \ \underline{\mathcal{B}}^i_2 = \bar{b}_2 + \epsilon\, \widetilde{u}_i\, \bar{b}_2,\ \ \underline{\mathcal{B}}^i_3 = \bar{b}_3 + \epsilon\, \widetilde{u}_i\, \bar{b}_3 \tag{2.101}$$

In view of Eq. (2.25), the following matrix formula

$$\begin{bmatrix} \lfloor \bar{b}_1, \bar{b}_2, \bar{b}_3 \rfloor & \widetilde{u}_i \lfloor \bar{b}_1, \bar{b}_2, \bar{b}_3 \rfloor \\ 0 & \lfloor \bar{b}_1, \bar{b}_2, \bar{b}_3 \rfloor \end{bmatrix} = \begin{bmatrix} R_i & \widetilde{u}_i R_i \\ 0 & R_i \end{bmatrix} \begin{bmatrix} \lfloor \bar{\imath}_1, \bar{\imath}_2, \bar{\imath}_3 \rfloor & 0 \\ 0 & \lfloor \bar{\imath}_1, \bar{\imath}_2, \bar{\imath}_3 \rfloor \end{bmatrix} \tag{2.102}$$

or in compact formula

$$\begin{bmatrix} \mathcal{B}_i & \widetilde{u}_i \mathcal{B}_i \\ 0 & \mathcal{B}_i \end{bmatrix} = \begin{bmatrix} R_i & \widetilde{u}_i\, R_i \\ 0 & R_i \end{bmatrix} = \mathcal{R}_i \tag{2.103}$$

apparently reveals that the motion tensor $\mathcal{R}_i$ brings the inertial frame $\mathcal{F}_\mathcal{I}$ to the reference frame $\mathcal{F}_i = [\mathbf{P}_i, \mathcal{B}_i]$. Similarly, the motion tensor

$$\mathcal{R}_f = \begin{bmatrix} \mathcal{B}_f & \widetilde{u}_f \mathcal{B}_f \\ 0 & \mathcal{B}_f \end{bmatrix} \tag{2.104}$$

brings the inertial frame $\mathcal{F}_\mathcal{I}$ to the current frame $\mathcal{F}_f = [\mathbf{P}_f, \mathcal{B}_f]$, where a basis $\mathcal{B}_f(\bar{B}_1, \bar{B}_2, \bar{B}_3) = R_f$ is attached to the point $\mathbf{P}_f$ described by the position vector $\underline{u}_f$. If decomposing the rotation tensor $R_f$ into

$$R_f = R\, R_i \tag{2.105}$$

and giving the displacement vectors

$$\underline{u} = \underline{u}_f - \underline{u}_i,\ \ \underline{u}_o = (I - R)\underline{u}_i + \underline{u} = \underline{u}_f - R\underline{u}_i \tag{2.106}$$

three mutually orthogonal material lines defined by the point $\mathbf{P}_f$ and the directors, $\bar{B}_1$, $\bar{B}_2$ and $\bar{B}_3$, are then expressed in terms of

$$\begin{aligned}
\bar{B}_1 + \epsilon\,\widetilde{u}_f\,\bar{B}_1 &= \bar{B}_1 + \epsilon\,(\widetilde{u}_o + \widetilde{R\,\underline{u}_i})\,\bar{B}_1 = R\,\bar{b}_1 + \epsilon\,(\widetilde{u}_o R + R\,\widetilde{u}_i)\,\bar{b}_1 \\
\bar{B}_2 + \epsilon\,\widetilde{u}_f\,\bar{B}_2 &= \bar{B}_2 + \epsilon\,(\widetilde{u}_o + \widetilde{R\,\underline{u}_i})\,\bar{B}_2 = R\,\bar{b}_2 + \epsilon\,(\widetilde{u}_o R + R\,\widetilde{u}_i)\,\bar{b}_2 \\
\bar{B}_3 + \epsilon\,\widetilde{u}_f\,\bar{B}_3 &= \bar{B}_3 + \epsilon\,(\widetilde{u}_o + \widetilde{R\,\underline{u}_i})\,\bar{B}_3 = R\,\bar{b}_3 + \epsilon\,(\widetilde{u}_o R + R\,\widetilde{u}_i)\,\bar{b}_3
\end{aligned} \tag{2.107}$$

these expressions can be assembled into the Plučker coordinates form

$$\begin{bmatrix} \lfloor \bar{B}_1, \bar{B}_2, \bar{B}_3 \rfloor & \widetilde{u}_f \lfloor \bar{B}_1, \bar{B}_2, \bar{B}_3 \rfloor \\ 0 & \lfloor \bar{B}_1, \bar{B}_2, \bar{B}_3 \rfloor \end{bmatrix} = \begin{bmatrix} R & \widetilde{u}_o R \\ 0 & R \end{bmatrix} \begin{bmatrix} \lfloor \bar{b}_1, \bar{b}_2, \bar{b}_3 \rfloor & \widetilde{u}_i \lfloor \bar{b}_1, \bar{b}_2, \bar{b}_3 \rfloor \\ 0 & \lfloor \bar{b}_1, \bar{b}_2, \bar{b}_3 \rfloor \end{bmatrix}$$

In view of the definition of the rotation tensors $R_f$ and $R_i$, the above formula is recasted to

$$\begin{bmatrix} R_f & \widetilde{u}_f R_f \\ 0 & R_f \end{bmatrix} = \begin{bmatrix} R & \widetilde{u}_o R \\ 0 & R \end{bmatrix} \begin{bmatrix} R_i & \widetilde{u}_i R_i \\ 0 & R_i \end{bmatrix} \tag{2.108}$$

or in compact form

$$\mathcal{R}_f = \mathcal{R}\,\mathcal{R}_i \tag{2.109}$$

Alternatively, the current frame $\mathcal{F}_f$ can be described by using the motion tensor $\mathcal{R}^*$ resolved in reference frame $\mathcal{F}_i$. In view of $R_f = R_i R^*$, three mutually orthogonal material lines forming the current frame $\mathcal{F}_f$ are represented by

$$\begin{aligned}
\bar{B}_1 + \epsilon\,\widetilde{u}_f\,\bar{B}_1 &= \bar{B}_1 + \epsilon\,(\widetilde{u}_i + \widetilde{u})\,\bar{B}_1 = R_i \bar{\jmath}_1 + \epsilon\,(\widetilde{u}_i\,R_i + R_i \widetilde{R_i^T \underline{u}})\,\bar{\jmath}_1 \\
\bar{B}_2 + \epsilon\,\widetilde{u}_f\,\bar{B}_2 &= \bar{B}_2 + \epsilon\,(\widetilde{u}_i + \widetilde{u})\,\bar{B}_2 = R_i \bar{\jmath}_2 + \epsilon\,(\widetilde{u}_i\,R_i + R_i \widetilde{R_i^T \underline{u}})\,\bar{\jmath}_2 \\
\bar{B}_3 + \epsilon\,\widetilde{u}_f\,\bar{B}_3 &= \bar{B}_3 + \epsilon\,(\widetilde{u}_i + \widetilde{u})\,\bar{B}_3 = R_i \bar{\jmath}_3 + \epsilon\,(\widetilde{u}_i\,R_i + R_i \widetilde{R_i^T \underline{u}})\,\bar{\jmath}_3
\end{aligned} \tag{2.110}$$

where three mutually orthogonal unit vectors $\bar{\jmath}_1$, $\bar{\jmath}_2$ and $\bar{\jmath}_3$ are defined by Eq. (2.30). Their Plücker coordinates can be readily assembled into

$$\begin{bmatrix} \lfloor \bar{B}_1, \bar{B}_2, \bar{B}_3 \rfloor & \widetilde{u}_f \lfloor \bar{B}_1, \bar{B}_2, \bar{B}_3 \rfloor \\ 0 & \lfloor \bar{B}_1, \bar{B}_2, \bar{B}_3 \rfloor \end{bmatrix} = \begin{bmatrix} R_i & \widetilde{u}_i R_i \\ 0 & R_i \end{bmatrix} \begin{bmatrix} \lfloor \bar{\jmath}_1, \bar{\jmath}_2, \bar{\jmath}_3 \rfloor & \widetilde{R_i^T \underline{u}} \lfloor \bar{\jmath}_1, \bar{\jmath}_2, \bar{\jmath}_3 \rfloor \\ 0 & \lfloor \bar{\jmath}_1, \bar{\jmath}_2, \bar{\jmath}_3 \rfloor \end{bmatrix}$$

equivalently, by Eq. (2.33), the above adjoint mappings are rewritten as

$$\begin{bmatrix} R_f & \widetilde{u}_f R_f \\ 0 & R_f \end{bmatrix} = \begin{bmatrix} R_i & \widetilde{u}_i R_i \\ 0 & R_i \end{bmatrix} \begin{bmatrix} R^* & \widetilde{R_i^T \underline{u}} R^* \\ 0 & R^* \end{bmatrix} \tag{2.111}$$

or in dual matrix form

$$\mathcal{R}_f = \mathcal{R}_i \mathcal{R}^* \tag{2.112}$$

in which the motion tensor $\mathcal{R}^*$, also the local frame $\mathcal{F}^*$,

$$\mathcal{R}^* = \begin{bmatrix} R^* & \widetilde{R_i^T \underline{u} R^*} \\ 0 & R^* \end{bmatrix} = \begin{bmatrix} \lfloor \bar{j}_1, \bar{j}_2, \bar{j}_3 \rfloor & \widetilde{R_i^T \underline{u}} \lfloor \bar{j}_1, \bar{j}_2, \bar{j}_3 \rfloor \\ 0 & \lfloor \bar{j}_1, \bar{j}_2, \bar{j}_3 \rfloor \end{bmatrix} = \mathcal{F}^* \tag{2.113}$$

describes the configuration of current frame $\mathcal{B}_f$ resolved in the associated reference frame $\mathcal{B}_i$. Comparing the above expressions, Eqs. (2.109) and (2.112), it readily finds

$$\mathcal{R}\mathcal{R}_i = \mathcal{R}_i\mathcal{R}^* \tag{2.114}$$

or

$$\mathcal{R} = \mathcal{R}_i\mathcal{R}^*\mathcal{R}_i^T, \ \mathcal{R}^* = \mathcal{R}_i^T\mathcal{R}\mathcal{R}_i \tag{2.115}$$

It indicates the motion tensor $\mathcal{R}$ resolved in the Cartesian frame $\mathcal{F}_\mathcal{I}$ and $\mathcal{R}^*$ resolved in the body attached frame $\mathcal{F}_i$, are related by the above transformation law.

### 2.2.9 Derivatives of Motion Operations

The time derivatives of motion tensor $\mathcal{R}_f$ is considered at first, which can be readily identified by taking time derivatives of all the components like

$$\dot{\mathcal{R}}_f = \begin{bmatrix} \dot{R}_f & \dot{\tilde{u}}_f R_f + \tilde{u}_f \dot{R}_f \\ 0 & \dot{R}_f \end{bmatrix} \tag{2.116}$$

Right multiplying the motion transpose of $\mathcal{R}_f$ will produce the velocities in the skew-symmetric motion tensor form like

$$\begin{aligned} \dot{\mathcal{R}}_f\mathcal{R}_f^T &= \begin{bmatrix} \dot{R}_f & \dot{\tilde{u}}_f R_f + \tilde{u}_f \dot{R}_f \\ 0 & \dot{R}_f \end{bmatrix} \begin{bmatrix} R_f^T & (\tilde{u}_f R_f)^T \\ 0 & R_f^T \end{bmatrix} \\ &= \begin{bmatrix} \tilde{\omega} & \tilde{\omega}\tilde{u}_f^T + \dot{\tilde{u}}_f + \tilde{u}_f\tilde{\omega} \\ 0 & \tilde{\omega} \end{bmatrix} = \begin{bmatrix} \tilde{\omega} & \dot{\tilde{u}}_f + \widetilde{\tilde{u}_f \underline{\omega}} \\ 0 & \tilde{\omega} \end{bmatrix} = \tilde{\mathcal{V}} \end{aligned} \tag{2.117}$$

The above Eq. (2.117) are nothing but the definitions of velocity measured in the initial Cartesian frame. The compact formula and its motion transpose can be written as

$$\tilde{\mathcal{V}} = \dot{\mathcal{R}}_f\mathcal{R}_f^T, \ \tilde{\mathcal{V}}^T = \mathcal{R}_f\dot{\mathcal{R}}_f^T \tag{2.118}$$

or equivalently,

$$\tilde{\mathcal{V}}\mathcal{R}_f = \dot{\mathcal{R}}_f, \ \mathcal{R}_f^T\tilde{\mathcal{V}}^T = \dot{\mathcal{R}}_f^T \tag{2.119}$$

In detail, the adjoint mapping of velocity is composed of linear and rotation velocity like

$$\underline{\mathcal{V}} = \begin{vmatrix} \underline{\dot{u}}_f + \widetilde{u}_f \underline{\omega} \\ \underline{\omega} \end{vmatrix} \tag{2.120}$$

Left multiplying the motion transpose $\mathcal{R}_f^T$ to $\dot{\mathcal{R}}_f$ produces the expression of velocity defined in the body attached frame instead of the initial frame, like

$$\begin{aligned} \mathcal{R}_f^T \dot{\mathcal{R}}_f &= \begin{bmatrix} R_f^T & (\widetilde{u}_f R_f)^T \\ 0 & R_f^T \end{bmatrix} \begin{bmatrix} \dot{R}_f & \widetilde{u}_f \dot{R}_f + \widetilde{\dot{u}}_f R_f \\ 0 & \dot{R}_f \end{bmatrix} \\ &= \begin{bmatrix} \widetilde{\omega}^* & \widetilde{R_f^T \underline{\dot{u}}_f} + R_f^T \widetilde{u}_f \dot{R}_f + (\widetilde{u}_f R_f)^T \dot{R}_f \\ 0 & \widetilde{\omega}^* \end{bmatrix} \\ &= \begin{bmatrix} \widetilde{\omega}^* & \widetilde{R_f^T \underline{\dot{u}}_f} \\ 0 & \widetilde{\omega}^* \end{bmatrix} = \widetilde{\mathcal{V}}^* \end{aligned} \tag{2.121}$$

According to the above definitions, the skew-symmetric dual tensor of velocity measures and its motion transpose can be readily found as the following motion tensor products

$$\widetilde{\mathcal{V}}^* = \mathcal{R}_f^T \dot{\mathcal{R}}_f, \ \widetilde{\mathcal{V}}^{*T} = \dot{\mathcal{R}}_f^T \mathcal{R}_f \tag{2.122}$$

or

$$\mathcal{R}_f \widetilde{\mathcal{V}}^* = \dot{\mathcal{R}}_f, \ \widetilde{\mathcal{V}}^{*T} \mathcal{R}_f^T = \dot{\mathcal{R}}_f^T \tag{2.123}$$

from which the adjoint representation of velocity measured in the body attached frame $\mathcal{F}_f$ is explicitly expressed to be

$$\underline{\mathcal{V}}^* = \begin{vmatrix} R_f^T \underline{\dot{u}}_f \\ \underline{\omega}^* \end{vmatrix} \tag{2.124}$$

Following the laws of motion tensor transformation, it can be proved that

$$\underline{\mathcal{V}} = \mathcal{R}_f \underline{\mathcal{V}}^*, \ \widetilde{\mathcal{V}} = \mathcal{R}_f \widetilde{\mathcal{V}}^* \mathcal{R}_f^T \tag{2.125}$$

In view of only $\mathcal{R}(t)$ and $\mathcal{R}^*(t)$ are the functions of time, such that $\mathcal{R}_f(t) = \mathcal{R}(t)\mathcal{R}_i$ and $\mathcal{R}_f(t) = \mathcal{R}_i \mathcal{R}^*(t)$, the velocities in the skew-symmetric dual tensor form can be computed from

$$\widetilde{\mathcal{V}} = \dot{\mathcal{R}}_f \mathcal{R}_f^T = \dot{\mathcal{R}} \mathcal{R}^T \tag{2.126}$$

and vice versa

$$\widetilde{\mathcal{V}}^* = \mathcal{R}_f^T \dot{\mathcal{R}}_f = \mathcal{R}^{*T} \dot{\mathcal{R}}^* \tag{2.127}$$

By taking derivatives of all the components of motion tensor $\mathcal{R}$, defined by Eq. (2.108), with respected to time, another formulation of velocity measured in the initial Cartesian frame is identified from Eq. (2.126) to be

$$\begin{aligned}\dot{\mathcal{R}}\mathcal{R}^T &= \begin{bmatrix} \dot{R} & \dot{\widetilde{u}}_o R + \widetilde{u}_o \dot{R} \\ 0 & \dot{R} \end{bmatrix} \begin{bmatrix} R^T & (\widetilde{u}_o R)^T \\ 0 & R^T \end{bmatrix} \\ &= \begin{bmatrix} \widetilde{\omega} & \widetilde{\omega}\widetilde{u}_o^T + \dot{\widetilde{u}}_o + \widetilde{u}_o \widetilde{\omega} \\ 0 & \widetilde{\omega} \end{bmatrix} = \begin{bmatrix} \widetilde{\omega} & \dot{\widetilde{u}}_o + \widetilde{\widetilde{u}_o \underline{\omega}} \\ 0 & \widetilde{\omega} \end{bmatrix} = \widetilde{\mathcal{V}} \end{aligned} \tag{2.128}$$

The same operations applied to the motion tensor $\mathcal{R}^*$ and its motion transform yield

$$\begin{aligned}\mathcal{R}^{*T}\dot{\mathcal{R}}^* &= \begin{bmatrix} R^{*T} & (\widetilde{R_i^T \underline{u}} R^*)^T \\ 0 & R^{*T} \end{bmatrix} \begin{bmatrix} \dot{R}^* & \widetilde{R_i^T \dot{\underline{u}}} R^* + \widetilde{R_i^T \underline{u}} \dot{R}^* \\ 0 & \dot{R}^* \end{bmatrix} \\ &= \begin{bmatrix} \widetilde{\omega}^* & R^{*T} \widetilde{R_i^T \dot{\underline{u}}} R^* + R^{*T} \widetilde{R_i^T \underline{u}} \dot{R}^* + (\widetilde{R_i^T \underline{u}} R^*)^T \dot{R}^* \\ 0 & \widetilde{\omega}^* \end{bmatrix} \\ &= \begin{bmatrix} \widetilde{\omega}^* & \widetilde{R_f^T \dot{\underline{u}}} \\ 0 & \widetilde{\omega}^* \end{bmatrix} = \widetilde{\mathcal{V}}^* \end{aligned} \tag{2.129}$$

With the aid of the definitions of displacement vectors, $\underline{u} = \underline{u}_f - \underline{u}_i$, $\underline{u}_o = \underline{u}_f - R\underline{u}_i$, it is not difficult to verify that two expressions of velocities in its Plucker coordinates are identical each other when all the quantities are measured in the initial frame

$$\underline{\mathcal{V}} = \begin{vmatrix} \dot{\underline{u}}_f + \widetilde{u}_f \underline{\omega} \\ \underline{\omega} \end{vmatrix} = \begin{vmatrix} \dot{\underline{u}}_o + \widetilde{u}_o \underline{\omega} \\ \underline{\omega} \end{vmatrix} \tag{2.130}$$

and the same properties still hold when all the quantities are measured in the body attached frame

$$\underline{\mathcal{V}}^* = \begin{vmatrix} R_f^T \dot{\underline{u}}_f \\ \underline{\omega}^* \end{vmatrix} = \begin{vmatrix} R_f^T \dot{\underline{u}} \\ \underline{\omega}^* \end{vmatrix} \tag{2.131}$$

In view of the composition of motions, $\mathcal{R}_f = \mathcal{R}_i \mathcal{R}^* = \mathcal{R}\mathcal{R}_i$, the general velocity resolved in different bases can be related by

$$\underline{\mathcal{V}} = \mathcal{R}_i \mathcal{R}^* \underline{\mathcal{V}}^* = \mathcal{R}\mathcal{R}_i \underline{\mathcal{V}}^* \Leftrightarrow \mathcal{R}_i^T \underline{\mathcal{V}} = \mathcal{R}^* \underline{\mathcal{V}}^* \Leftrightarrow \mathcal{R}^T \underline{\mathcal{V}} = \mathcal{R}_i \underline{\mathcal{V}}^* \tag{2.132}$$

Next, the derivatives of motion tensor respective to the spatial coordinates performs in the same manner to obtain the explicit expression of curvatures. If defining three parameters, $s_1$, $s_2$ and $s_3$, as the arc length coordinates, then the curvature tensor is determined from the following spatial derivatives

$$\mathcal{R}_{f,i} = \begin{bmatrix} R_{f,i} & \widetilde{u}_{f,i} R_f + \widetilde{u}_f R_{f,i} \\ 0 & R_{f,i} \end{bmatrix} \tag{2.133}$$

by right multiplying the motion transpose of motion tensor

$$
\begin{aligned}
\mathcal{R}_{f,i}\mathcal{R}_f^T &= \begin{bmatrix} R_{f,i} & \widetilde{u}_{f,i}R_f + \widetilde{u}_f R_{f,i} \\ 0 & R_{f,i} \end{bmatrix} \begin{bmatrix} R_f^T & (\widetilde{u}_f R_f)^T \\ 0 & R_f^T \end{bmatrix} \\
&= \begin{bmatrix} R_{f,i}R_f^T & R_{f,i}(\widetilde{u}_f R_f)^T + \widetilde{u}_{f,i} + \widetilde{u}_f R_{f,i} R_f^T \\ 0 & R_{f,i}R_f^T \end{bmatrix} \\
&= \begin{bmatrix} \widetilde{\kappa}_i & \widetilde{u}_{f,i} + \widetilde{\widetilde{u}_f \underline{\kappa}_i} \\ 0 & \widetilde{\kappa}_i \end{bmatrix} = \widetilde{\mathcal{K}}_i
\end{aligned} \tag{2.134}
$$

where the curvature in its skew-symmetric tensor form

$$
\widetilde{\kappa}_i = R_{f,i} R_f^T \tag{2.135}
$$

for $i = 1, 2, 3$ is defined to simplify above Lie bracket. The curvatures in their Plücker coordinates measured in Cartesian frame are extracted as the axial vector of the Lie bracket $\widetilde{\mathcal{K}}_i$ and in detail

$$
\underline{\mathcal{K}}_1 = \begin{vmatrix} \underline{u}_{f,1} + \widetilde{u}_f \underline{\kappa}_1 \\ \underline{\kappa}_1 \end{vmatrix}, \ \underline{\mathcal{K}}_2 = \begin{vmatrix} \underline{u}_{f,2} + \widetilde{u}_f \underline{\kappa}_2 \\ \underline{\kappa}_2 \end{vmatrix}, \ \underline{\mathcal{K}}_3 = \begin{vmatrix} \underline{u}_{f,3} + \widetilde{u}_f \underline{\kappa}_3 \\ \underline{\kappa}_3 \end{vmatrix} \tag{2.136}
$$

Left multiplying the transpose of motion tensor will produce

$$
\begin{aligned}
\mathcal{R}_f^T \mathcal{R}_{f,i} &= \begin{bmatrix} R_f^T & (\widetilde{u}_f R_f)^T \\ 0 & R_f^T \end{bmatrix} \begin{bmatrix} R_{f,i} & \widetilde{u}_{f,i}R_f + \widetilde{u}_f R_{f,i} \\ 0 & R_{f,i} \end{bmatrix} \\
&= \begin{bmatrix} \widetilde{\kappa}_i^* & R_f^T \widetilde{u}_{f,i} R_f + R_f^T \widetilde{u}_f R_{f,i} + (\widetilde{u}_f R_f)^T R_{f,i} \\ 0 & \widetilde{\kappa}_i^* \end{bmatrix} \\
&= \begin{bmatrix} \widetilde{\kappa}_i^* & \widetilde{R_f^T \underline{u}_{f,i}} \\ 0 & \widetilde{\kappa}_i^* \end{bmatrix} = \widetilde{\mathcal{K}}_i^*
\end{aligned} \tag{2.137}
$$

from which the curvatures measured in body attached frame $\mathcal{F}_f$ are identified to be

$$
\underline{\mathcal{K}}_1^* = \begin{vmatrix} R_f^T \underline{u}_{f,1} \\ \underline{\kappa}_1^* \end{vmatrix}, \ \underline{\mathcal{K}}_2^* = \begin{vmatrix} R_f^T \underline{u}_{f,2} \\ \underline{\kappa}_2^* \end{vmatrix}, \ \underline{\mathcal{K}}_3^* = \begin{vmatrix} R_f^T \underline{u}_{f,3} \\ \underline{\kappa}_3^* \end{vmatrix} \tag{2.138}
$$

The curvature measures can be recasted to the compact form of skew-symmetric dual tensor in inertial frame to get

$$
\widetilde{\mathcal{K}}_1 = \mathcal{R}_{f,1}\mathcal{R}_f^T, \ \widetilde{\mathcal{K}}_2 = \mathcal{R}_{f,2}\mathcal{R}_f^T, \ \widetilde{\mathcal{K}}_3 = \mathcal{R}_{f,3}\mathcal{R}_f^T \tag{2.139}
$$

and then transform to the local frame like

$$
\widetilde{\mathcal{K}}_1^* = \mathcal{R}_f^T \mathcal{R}_{f,1}, \ \widetilde{\mathcal{K}}_2^* = \mathcal{R}_f^T \mathcal{R}_{f,2}, \ \widetilde{\mathcal{K}}_3^* = \mathcal{R}_f^T \mathcal{R}_{f,3} \tag{2.140}
$$

The following expressions are then readily obtained in a straight forward approach to be

$$\mathcal{R}_{f,1} = \widetilde{\mathcal{K}}_1 \mathcal{R}_f,\ \mathcal{R}_{f,2} = \widetilde{\mathcal{K}}_2 \mathcal{R}_f,\ \mathcal{R}_{f,3} = \widetilde{\mathcal{K}}_3 \mathcal{R}_f \tag{2.141}$$

and

$$\mathcal{R}_{f,1} = \mathcal{R}_f \widetilde{\mathcal{K}}_1^*,\ \mathcal{R}_{f,2} = \mathcal{R}_f \widetilde{\mathcal{K}}_2^*,\ \mathcal{R}_{f,3} = \mathcal{R}_f \widetilde{\mathcal{K}}_3^* \tag{2.142}$$

Furthermore, it can be verified that the curvatures measured in different frames can be related by the law of dual tensor transform as

$$\widetilde{\mathcal{K}}_1 = \mathcal{R}_f \widetilde{\mathcal{K}}_1^* \mathcal{R}_f^T,\ \widetilde{\mathcal{K}}_2 = \mathcal{R}_f \widetilde{\mathcal{K}}_2^* \mathcal{R}_f^T,\ \widetilde{\mathcal{K}}_3 = \mathcal{R}_f \widetilde{\mathcal{K}}_3^* \mathcal{R}_f^T \tag{2.143}$$

or alternatively

$$\widetilde{\mathcal{K}}_1^* = \mathcal{R}_f^T \widetilde{\mathcal{K}}_1 \mathcal{R}_f,\ \widetilde{\mathcal{K}}_2^* = \mathcal{R}_f^T \widetilde{\mathcal{K}}_2 \mathcal{R}_f,\ \widetilde{\mathcal{K}}_3^* = \mathcal{R}_f^T \widetilde{\mathcal{K}}_3 \mathcal{R}_f \tag{2.144}$$

The initial curvatures of the reference configurations are identified in the same manner. The components of inertial curvatures could be measured in the inertial frame or the body attached reference frame, alternatively. For initial curvatures defined in the inertial frame, the definitions are given as

$$\widetilde{\mathcal{K}}_1^i = \mathcal{R}_{i,1} \mathcal{R}_i^T,\ \widetilde{\mathcal{K}}_2^i = \mathcal{R}_{i,2} \mathcal{R}_i^T,\ \widetilde{\mathcal{K}}_3^i = \mathcal{R}_{i,3} \mathcal{R}_i^T \tag{2.145}$$

Meanwhile, the definitions in the body attached reference frame are expressed to be

$$\widetilde{\mathcal{K}}_1^{i*} = \mathcal{R}_i^T \mathcal{R}_{i,1},\ \widetilde{\mathcal{K}}_2^{i*} = \mathcal{R}_i^T \mathcal{R}_{i,2},\ \widetilde{\mathcal{K}}_3^{i*} = \mathcal{R}_i^T \mathcal{R}_{i,3} \tag{2.146}$$

The above relations can be directly rewritten as

$$\mathcal{R}_{i,1} = \widetilde{\mathcal{K}}_1^i \mathcal{R}_i,\ \mathcal{R}_{i,2} = \widetilde{\mathcal{K}}_2^i \mathcal{R}_i,\ \mathcal{R}_{i,3} = \widetilde{\mathcal{K}}_3^i \mathcal{R}_i \tag{2.147}$$

and

$$\mathcal{R}_{i,1} = \mathcal{R}_i \widetilde{\mathcal{K}}_1^{i*},\ \mathcal{R}_{i,2} = \mathcal{R}_i \widetilde{\mathcal{K}}_2^{i*},\ \mathcal{R}_{i,3} = \mathcal{R}_i \widetilde{\mathcal{K}}_3^{i*} \tag{2.148}$$

It is readily verified that both formulas are related to

$$\widetilde{\mathcal{K}}_1^i = \mathcal{R}_i \widetilde{\mathcal{K}}_1^{i*} \mathcal{R}_i^T,\ \widetilde{\mathcal{K}}_2^i = \mathcal{R}_i \widetilde{\mathcal{K}}_2^{i*} \mathcal{R}_i^T,\ \widetilde{\mathcal{K}}_3^i = \mathcal{R}_i \widetilde{\mathcal{K}}_3^{i*} \mathcal{R}_i^T \tag{2.149}$$

and vice versa

$$\widetilde{\mathcal{K}}_1^{i*} = \mathcal{R}_i^T \widetilde{\mathcal{K}}_1 \mathcal{R}_i,\ \widetilde{\mathcal{K}}_2^{i*} = \mathcal{R}_i^T \widetilde{\mathcal{K}}_2 \mathcal{R}_i,\ \widetilde{\mathcal{K}}_3^{i*} = \mathcal{R}_i^T \widetilde{\mathcal{K}}_3 \mathcal{R}_i \tag{2.150}$$

Specially, the initial curvatures measured in body attached reference frame are found to be in detail like

$$\underline{\mathcal{K}}_1^{i*} = \begin{vmatrix} R_i^T \underline{u}_{i,1} \\ \underline{\kappa}_1^{i*} \end{vmatrix}, \ \underline{\mathcal{K}}_2^{i*} = \begin{vmatrix} R_i^T \underline{u}_{i,2} \\ \underline{\kappa}_2^{i*} \end{vmatrix}, \ \underline{\mathcal{K}}_3^{i*} = \begin{vmatrix} R_i^T \underline{u}_{i,3} \\ \underline{\kappa}_3^{i*} \end{vmatrix} \tag{2.151}$$

where $\underline{\kappa}_j^{i*}$ represents the initial curvature tensor and $\underline{\kappa}_j^{i*} = \text{axi}(R_{i,j} R_i^T)$ for $j = 1, 2, 3$. Now, the Lagrangian strain measures are defined as the difference between the curvature dual vectors of the deformed and reference configurations

$$\underline{\epsilon}_1 = \underline{\mathcal{K}}_1^* - \underline{\mathcal{K}}_1^{i*}, \ \underline{\epsilon}_2 = \underline{\mathcal{K}}_2^* - \underline{\mathcal{K}}_2^{i*}, \ \underline{\epsilon}_3 = \underline{\mathcal{K}}_3^* - \underline{\mathcal{K}}_3^{i*} \tag{2.152}$$

Finally, the definitions of virtual motion vectors are found to be similar to the velocity definition in its adjoint representation like

$$\begin{aligned} \mathcal{R}_f^T \delta \mathcal{R}_f &= \begin{bmatrix} R_f^T & (\widetilde{u}_f R_f)^T \\ 0 & R_f^T \end{bmatrix} \begin{bmatrix} \delta R_f & \delta \widetilde{u}_f R_f + \widetilde{u}_f \delta R_f \\ 0 & \delta R_f \end{bmatrix} \\ &= \begin{bmatrix} R_f^T \delta R_f & R_f^T \delta \widetilde{u}_f R_f + R_f^T \widetilde{u}_f \delta R_f + (\widetilde{u}_f R_f)^T \delta R_f \\ 0 & R_f^T \delta R_f \end{bmatrix} \\ &= \begin{bmatrix} \delta\widetilde{\psi}^* & \widetilde{R_f^T \delta \underline{u}_f} \\ 0 & \delta\widetilde{\psi}^* \end{bmatrix} = \delta\widetilde{\mathcal{U}}^* \end{aligned} \tag{2.153}$$

Since the motion tensor $\mathcal{R}_f$ can be expressed as the composition of motion tensors, $\mathcal{R}_f = \mathcal{R}_i \mathcal{R}^*$, the virtual motions are then recasted to

$$\delta\widetilde{\mathcal{U}}^* = \mathcal{R}_f^T \delta \mathcal{R}_f = \mathcal{R}^{*T} \delta \mathcal{R}^* \tag{2.154}$$

Multiplying the motion tensors at both side of above equations produces

$$\delta \mathcal{R}_f = \mathcal{R}_f \delta\widetilde{\mathcal{U}}^*, \ \delta \mathcal{R}^* = \mathcal{R}^* \delta\widetilde{\mathcal{U}}^* \tag{2.155}$$

It is also necessary to give the explicit expressions of virtual motion vector and

$$\underline{\delta \mathcal{U}^*} = \begin{vmatrix} R_f^T \delta \underline{u}_f \\ \underline{\delta \psi^*} \end{vmatrix} = \begin{vmatrix} R_f^T \delta \underline{u} \\ \underline{\delta \psi^*} \end{vmatrix} \tag{2.156}$$

where $\underline{\delta \psi^*}$ is the virtual rotation vector defined in the body attached frame $\mathcal{F}_f$, and its skew-symmetric tensor is determined from $\delta\widetilde{\psi}^* = R_f^T \delta R_f$.

The *compatibility equations*, resulting from the continuity of the motion tensor with respect to spatial, temporal, or virtual variables, will be discussed at last. In brief, two types of compatibility equations are presented. The compatibility equation that are referred to the transitivity equations in analytical dynamics [11] is introduces at first. The virtual dual velocity vector is readily obtained by taking variation of velocity directly

$$\delta\widetilde{\mathcal{V}}^* = \delta \mathcal{R}_f^T \dot{\mathcal{R}}_f + \mathcal{R}_f^T \delta \dot{\mathcal{R}}_f \tag{2.157}$$

and a time derivative of the virtual motion vectors leads to

$$\dot{\widetilde{\delta \mathcal{U}}}^* = \dot{\mathcal{R}}_f^T \delta \mathcal{R}_f + \mathcal{R}_f^T \delta \dot{\mathcal{R}}_f \tag{2.158}$$

Subtracting these two equations and using the identities of Eqs. (2.127) and (2.154) will yield

$$\widetilde{\delta \mathcal{V}}^* - \dot{\widetilde{\delta \mathcal{U}}}^* = \widetilde{\mathcal{V}}^* \widetilde{\delta \mathcal{U}}^* - \widetilde{\delta \mathcal{U}}^* \mathcal{V}^* \tag{2.159}$$

Applying another important identity of dual vector, Eq. (1.155), leads to the frequently used compatibility equation

$$\delta \underline{\mathcal{V}}^* = \underline{\delta \dot{\mathcal{U}}}^* + \widetilde{\mathcal{V}}^* \underline{\delta \mathcal{U}}^* \tag{2.160}$$

Similarly, the second type of compatibility equations is given next. The variations of Lagrangian strain measures, also variations of curvatures, are related to the virtual motion vector and its spatial derivatives in the similar manner. Here, the second type of compatibility equations about variations of Lagrangian strains are given directly

$$\begin{aligned} \delta \underline{\epsilon}_1 &= \delta \underline{\mathcal{K}}_1^* = \underline{\delta \mathcal{U}}_{,1}^* + \widetilde{\mathcal{K}}_1^* \underline{\delta \mathcal{U}}^* \\ \delta \underline{\epsilon}_2 &= \delta \underline{\mathcal{K}}_2^* = \underline{\delta \mathcal{U}}_{,2}^* + \widetilde{\mathcal{K}}_2^* \underline{\delta \mathcal{U}}^* \\ \delta \underline{\epsilon}_3 &= \delta \underline{\mathcal{K}}_3^* = \underline{\delta \mathcal{U}}_{,3}^* + \widetilde{\mathcal{K}}_3^* \underline{\delta \mathcal{U}}^* \end{aligned} \tag{2.161}$$

### 2.2.10 Vectorial Parameterization of Motion Tensor

As mentioned in Sect. 2.2.6, the rotation tensor requires only a minimal set of three degrees of freedom to describe the rotation exactly, even though the rotation tensor contains nine dependent components. Similarly, the motion tensor can be parameterized by a minimal set of six degrees of freedom to describe the motion measured by the magnitudes of rotation and intrinsic displacement along the Chasles' line. In essential, a vectorial parameterizing of motion tensor is the process that generalizes the vector parameterization of rotation tensor. Before going into details of vectorial parameterization of motion tensor, the *dual function* [12, 13] is introduced at first as follow. A function $\mathcal{P}$ of a dual variable $\phi^d = \phi + \epsilon \delta$ can be represented in the form of [14]

$$\mathcal{P}(\phi^d) = p(\phi, \delta) + \epsilon p^o(\phi, \delta) \tag{2.162}$$

where primary part $p$ and dual part $p^o$ are real functions of real variables $\phi$ and $\delta$. When the dual function $\mathcal{P}(\phi^d)$ is assumed to be analytic, the necessary and sufficient conditions [13]

$$\frac{\partial p}{\partial \delta} = 0, \ \frac{\partial p}{\partial \phi} = \frac{\partial p^o}{\partial \delta} \tag{2.163}$$

must be satisfied, which leads to the equality in a more general sense like

$$\mathcal{P}(\phi^d) = p(\phi) + \epsilon \delta p' \tag{2.164}$$

A bunch of equalities of dual functions can be found in Table 1 of Ref. [13]. The equalities of dual trigonometric functions frequently used in this book are given as

$$\sin \phi^d = \sin \phi + \epsilon \delta \cos \phi, \ \cos \phi^d = \cos \phi - \epsilon \delta \sin \phi \tag{2.165}$$

and

$$\tan \phi^d = \tan \phi + \epsilon \frac{\delta}{\cos^2 \phi} \tag{2.166}$$

with the application of the rule of division by a dual number

$$\frac{\eta^d}{\phi^d} = \frac{\eta + \epsilon \theta}{\phi + \epsilon \delta} = \frac{\eta}{\phi} + \epsilon \frac{\theta \phi - \eta \delta}{\phi^2} \tag{2.167}$$

In Sect. 2.2.7, the generalized motion described by Chasles' theorem is discussed in detail. The Chasles' line is defined as a line vector, $\underline{\mathcal{N}}$, by Eq. (2.80). Now the magnitude $\phi$ of rotation around the Chasles' line and translation $\delta$ along the same line are combined to a dual variable, $\phi^d = \phi + \epsilon \delta$, that measures the magnitude of same motion in a more efficient manner. Through tedious algebraic operations, the motion tensor, Eq. (2.92), can be expressed in terms of dual variable $\phi^d$ and line vector $\underline{\mathcal{N}}$ like

$$\mathcal{R} = \mathcal{I} + \sin \phi^d \widetilde{\mathcal{N}} + (1 - \cos \phi^d) \widetilde{\mathcal{N}} \widetilde{\mathcal{N}} \tag{2.168}$$

or in the form of adjoint representations

$$\begin{aligned} \begin{bmatrix} R & \widetilde{u}_o R \\ 0 & R \end{bmatrix} = \begin{bmatrix} I & \\ & I \end{bmatrix} + \begin{bmatrix} \sin \phi I & \delta \cos \phi I \\ 0 & \sin \phi I \end{bmatrix} \begin{bmatrix} \widetilde{n} & \widetilde{r \widetilde{n}} \\ 0 & \widetilde{n} \end{bmatrix} \\ + \begin{bmatrix} (1 - \cos \phi) I & \delta \sin \phi I \\ 0 & (1 - \cos \phi) I \end{bmatrix} \begin{bmatrix} \widetilde{n} & \widetilde{r \widetilde{n}} \\ 0 & \widetilde{n} \end{bmatrix} \begin{bmatrix} \widetilde{n} & \widetilde{r \widetilde{n}} \\ 0 & \widetilde{n} \end{bmatrix} \end{aligned} \tag{2.169}$$

This expression clearly shows that the primary part of motion tensor $\mathcal{R}$ is exactly the rotation tensor $R$ defined by Eq. (2.18), and the dual part of motion tensor $\mathcal{R}$ is found to be $\widetilde{u}_o R$ with the application of identity

$$\widetilde{n} \widetilde{r \widetilde{n}} \widetilde{n} = 0 \tag{2.170}$$

proved by using Eq. (1.46). Here, an alternatively expression of vector $\underline{u}_o$ that is identity to Eq. (2.86) is dug out

$$\underline{u}_o = [\sin\phi I + (1-\cos\phi)\tilde{n}]\tilde{r}\bar{n} + \delta\bar{n} \tag{2.171}$$

Similarly, substituting the unit line vector $\underline{\mathcal{N}}$ and motion magnitude $\phi^d$ in Eq. (2.128), implies the velocities become the function of dual variable and line vector and their time derivatives

$$\underline{\mathcal{V}} = \dot{\phi}^d\underline{\mathcal{N}} + \sin\phi^d\dot{\underline{\mathcal{N}}} + (1-\cos\phi^d)\tilde{\mathcal{N}}\dot{\underline{\mathcal{N}}} \tag{2.172}$$

The dual representation can be readily formulated

$$\begin{aligned}\begin{vmatrix}\dot{\underline{u}}_o + \tilde{u}_o\underline{\omega}\\ \underline{\omega}\end{vmatrix} &= \begin{bmatrix}\dot{\phi} & \dot{\delta}\\ 0 & \dot{\phi}\end{bmatrix}\begin{vmatrix}\tilde{r}\bar{n}\\ \bar{n}\end{vmatrix} + \begin{bmatrix}\sin\phi I & \delta\cos\phi I\\ 0 & \sin\phi I\end{bmatrix}\begin{vmatrix}\dot{\widetilde{r}\bar{n}}\\ \dot{\bar{n}}\end{vmatrix}\\ &+ \begin{bmatrix}(1-\cos\phi)I & \delta\sin\phi I\\ 0 & (1-\cos\phi)I\end{bmatrix}\begin{bmatrix}\tilde{n} & \widetilde{\tilde{r}\bar{n}}\\ 0 & \tilde{n}\end{bmatrix}\begin{vmatrix}\dot{\widetilde{r}\bar{n}}\\ \dot{\bar{n}}\end{vmatrix}\end{aligned} \tag{2.173}$$

Note that the primary part of above expression is exactly the definition of angular velocity, Eq. (2.49), in terms of unit vector $\bar{n}$ and the magnitude $\phi$ of rotation. Through algebraic operations, the dual part of above expression, an equality, can be proved with the aid of identity

$$\tilde{n}\dot{\tilde{n}}\tilde{n} = 0 \tag{2.174}$$

With above preliminary definitions of dual function, and expressions of motion tensor and velocity in terms of line vector $\underline{\mathcal{N}}$ and dual variable $\phi^d$, the vectorial parameterizing procedure of motion tensor can be addressed with clarity. In general, the vectorial parameterization of motion, only involving six degrees of freedom, is defined as

$$\underline{\mathcal{P}} = \mathcal{P}(\phi^d)\underline{\mathcal{N}} \tag{2.175}$$

where $\mathcal{P}(\phi^d)$, called *generating function*, is an analytical dual function of dual variable $\phi^d$, defined by Eq. (2.164). Obviously, dual vector $\underline{\mathcal{P}}$ is not unit by definition. However, Eq. (2.168) implies $\mathcal{R}\underline{\mathcal{P}} = \underline{\mathcal{P}}$. It means dual vector $\underline{\mathcal{P}}$ is an eigenvector of the motion tensor $\mathcal{R}$ associated with its unit eigenvalue. The explicit expression for the motion tensor $\mathcal{R}$ in terms of motion parameters $\underline{\mathcal{P}}$ is obtained from Eq. (2.168) like

$$\mathcal{R} = \mathcal{I} + \frac{\sin\phi^d}{\mathcal{P}}\tilde{\mathcal{P}} + \frac{(1-\cos\phi^d)}{\mathcal{P}^2}\tilde{\mathcal{P}}\tilde{\mathcal{P}} \tag{2.176}$$

Before identifying the relations of velocities $\underline{\mathcal{V}}$ as a function of motion parameters $\underline{\mathcal{P}}$, the time derivative of motion parameter vector needs to be performed at first

$$\dot{\underline{\mathcal{P}}} = \dot{\mathcal{P}}\underline{\mathcal{N}} + \mathcal{P}\dot{\underline{\mathcal{N}}} = \mathcal{P}'\dot{\phi}^d\underline{\mathcal{N}} + \mathcal{P}\dot{\underline{\mathcal{N}}} \tag{2.177}$$

where the time derivatives of generating function reveal

$$\dot{\mathcal{P}} = \dot{p} + \epsilon\dot{\delta p'} = \mathcal{P}'\dot{\phi}^d \tag{2.178}$$

By identity, Eq. (1.131), and the attribute of $\underline{\mathcal{N}}$ to be a unit dual vector, the relation

$$\tilde{\mathcal{N}}\tilde{\mathcal{N}}\dot{\underline{\mathcal{P}}} = -\mathcal{P}\dot{\underline{\mathcal{N}}} \tag{2.179}$$

are readily found to express time derivative $\dot{\phi}^d\underline{\mathcal{N}}$ in terms of

$$\dot{\phi}^d\underline{\mathcal{N}} = \frac{1}{\mathcal{P}'}(\mathcal{I} + \tilde{\mathcal{N}}\tilde{\mathcal{N}})\dot{\underline{\mathcal{P}}} \tag{2.180}$$

Introducing these results into Eq. (2.172) then leads to

$$\underline{\mathcal{V}} = \mathcal{H}(\underline{\mathcal{P}})\dot{\underline{\mathcal{P}}} \tag{2.181}$$

where the dual matrix $\mathcal{H}(\underline{\mathcal{P}})$, referred as tangent tensor, is identified from Eq. (2.172) like

$$\mathcal{H}(\underline{\mathcal{P}}) = \frac{1}{\mathcal{P}'}\mathcal{I} + \frac{1-\cos\phi^d}{\mathcal{P}^2}\tilde{\mathcal{P}} + \frac{1}{\mathcal{P}^2}(\frac{1}{\mathcal{P}'} - \frac{\sin\phi^d}{\mathcal{P}})\tilde{\mathcal{P}}\tilde{\mathcal{P}} \tag{2.182}$$

By taking dual matrix inverse of tangent tensor, $\mathcal{H}(\underline{\mathcal{P}})$, the time derivative of the vectorial parameters is related to the velocities by

$$\dot{\underline{\mathcal{P}}} = \mathcal{H}^{-1}(\underline{\mathcal{P}})\underline{\mathcal{V}} \tag{2.183}$$

where the inverse of tangent tensor is evaluated from

$$\mathcal{H}^{-1}(\underline{\mathcal{P}}) = \mathcal{P}'I - \frac{1}{2}\tilde{\mathcal{P}} + \frac{1}{\mathcal{P}^2}(\mathcal{P}' - \frac{\mathcal{P}}{2\tan\phi^d/2})\tilde{\mathcal{P}}\tilde{\mathcal{P}} \tag{2.184}$$

Note that the tangent tensor $\mathcal{H}$ features the following remarkable attribute

$$\mathcal{R} = \mathcal{H}\mathcal{H}^{-\mathcal{T}} = \mathcal{H}^{-\mathcal{T}}\mathcal{H} \tag{2.185}$$

By this attribute, the components of velocities resolved in local frame, $\mathcal{R}^{\mathcal{T}}\mathcal{V}$, can be readily expressed in terms of $\underline{\mathcal{P}}$ and $\dot{\underline{\mathcal{P}}}$ like

$$\mathcal{R}^{\mathcal{T}}\underline{\mathcal{V}} = \mathcal{H}^{\mathcal{T}}(\underline{\mathcal{P}})\dot{\underline{\mathcal{P}}} \tag{2.186}$$

Before giving two specific parameterizations, it is very necessary to address the relation of displacement vector $\underline{u}_o$ and the dual part $\underline{p}^o$ of vectorial parameterization. In doing so, the detail of vectorial parameterization of motion is presented

$$\underline{\mathcal{P}} = \underline{p} + \epsilon\underline{p}^o = p\bar{n} + \epsilon(\delta p'\bar{n} + p\tilde{r}\bar{n}) \tag{2.187}$$

Both intrinsic displacement $\delta$ and the moment of vector $\tilde{r}\bar{n}$ can be expressed as a function of $\underline{p}^o$ like

$$\delta = \frac{1}{p'}\bar{n}^T \underline{p}^o,\ \widetilde{r}\bar{n} = \frac{1}{p}(\underline{p}^o - \delta p'\bar{n}) = \frac{1}{p}(I - \bar{n}\bar{n}^T)\underline{p}^o \tag{2.188}$$

Substituting these results in Eq. (2.171), the displacement vector writes

$$\underline{u}_o = [\sin\phi I + (1-\cos\phi)\widetilde{n}]\frac{1}{p}(I - \bar{n}\bar{n}^T)\underline{p}^o + \frac{1}{p'}\bar{n}\bar{n}^T \underline{p}^o \tag{2.189}$$

By identities, Eqs. (1.55) and (1.56), and merging the similar terms, the displacement vector becomes a function of $\underline{p}^o$ in the form of

$$\underline{u}_o = \left[\frac{1}{p'}I + \frac{(1-\cos\phi)}{p^2}\widetilde{p} + \frac{1}{p^2}(\frac{1}{p'} - \frac{\sin\phi}{p})\widetilde{p}\widetilde{p}\right]\underline{p}^o = H(\underline{p})\underline{p}^o \tag{2.190}$$

where $H(\underline{p})$ is the tangent operator, defined by Eq. (2.61), also the primary part of tangent tensor, $\mathcal{H}(\underline{\mathcal{P}})$, as mentioned above.

Now two types of specific vectorial parameterizations, *Cartesian parameterization* and *Wiener-Milenković parameterization*, are presented. Cartesian parameters lead to the following parameterization of motion

$$\underline{\mathcal{P}}(\phi^d) = \phi^d \underline{\mathcal{N}} \tag{2.191}$$

by selecting generating function as $\mathcal{P} = \phi^d$. Clearly, $\mathcal{P}' = 1 + \epsilon\, 0 = 1$. By inserting Cartesian parameters into Eq. (2.176), the motion tensor is readily found to be

$$\mathcal{R} = \mathcal{I} + \frac{\sin\phi^d}{\phi^d}\widetilde{\mathcal{P}} + \frac{1-\cos\phi^d}{\phi^{d\,2}}\widetilde{\mathcal{P}}\widetilde{\mathcal{P}} = \exp(\widetilde{\mathcal{P}}) \tag{2.192}$$

The second equality shows the exponential map of the motion tensor. For small motion, $\underline{\mathcal{P}} \ll 1$, only the first-order term of the exponential map is kept, to yield

$$\mathcal{R} = \mathcal{I} + \widetilde{\mathcal{P}} = \begin{bmatrix} I + \widetilde{p} & \widetilde{p}^o \\ 0 & I + \widetilde{p} \end{bmatrix} \tag{2.193}$$

In a similar approach, the tangent tensor and its inverse are obtained from Eqs. (2.182) and (2.184) like

$$\mathcal{H} = \mathcal{I} + \frac{1-\cos\phi^d}{\phi^{d\,2}}\widetilde{\mathcal{P}} + \frac{1}{\phi^{d\,2}}(1 - \frac{\sin\phi^d}{\phi^d})\widetilde{\mathcal{P}}\widetilde{\mathcal{P}} \tag{2.194}$$

and

$$\mathcal{H}^{-1} = I - \frac{1}{2}\widetilde{\mathcal{P}} + \frac{1}{\phi^{d\,2}}(1 - \frac{\phi^d}{2\tan\phi^d/2})\widetilde{\mathcal{P}}\widetilde{\mathcal{P}} \tag{2.195}$$

For the Wiener-Milenković parameterization [15], the selection of generating function, $\mathcal{P}(\phi^d) = 4\tan(\phi^d/4)$, yields Wiener-Milenković parameters

$$\underline{\mathcal{P}} = 4\tan\frac{\phi^d}{4}\underline{\mathcal{N}} \tag{2.196}$$

Its Plučker coordinates

$$\underline{\mathcal{P}} = \begin{vmatrix} \underline{p}^o \\ \underline{p} \end{vmatrix} = \begin{vmatrix} 4\tan\frac{\phi}{4}\widetilde{r}\bar{n} + \delta\bar{n}/\cos^2\frac{\phi}{4} \\ 4\tan\frac{\phi}{4}\bar{n} \end{vmatrix} \tag{2.197}$$

clearly show that a factor of four enforces the condition $\lim_{\phi\to 0}\underline{\mathcal{P}} = \phi^d\underline{\mathcal{N}}$, such that the infinitesimal motion vector, Eq. (1.322), is recovered for small motions. This representation is limited to rotation angles $|\phi| < 2\pi$ since it presents singularity for $\mathcal{P} \to \infty$ when $|\phi| \to 2\pi$. By the symbols introduced as follow

$$\wp_0 = 2(1 - \tan^2\frac{\phi^d}{4}) = 2 - \frac{1}{8}\underline{\mathcal{P}}^T\underline{\mathcal{P}},\ \wp = \frac{2}{4 - \wp_0} = \cos^2\frac{\phi^d}{4} \tag{2.198}$$

and their Plučker coordinates

$$\wp_0 = \begin{vmatrix} \delta_0 \\ c_0 \end{vmatrix} = \begin{vmatrix} -\frac{1}{4}\underline{p}^T\underline{p}^o \\ 2 - \frac{1}{8}\underline{p}^T\underline{p} \end{vmatrix},\ \wp = \begin{vmatrix} \mu \\ \nu_c \end{vmatrix} = \begin{vmatrix} \frac{2\delta_0}{(4-c_0)^2} \\ \frac{2}{4-c_0} \end{vmatrix} \tag{2.199}$$

the motion tensor follows from Eq. (2.176) as

$$\mathcal{R}(\underline{\mathcal{P}}) = \mathcal{I} + \frac{\wp^2}{2}\wp_0\widetilde{\mathcal{P}} + \frac{\wp^2}{2}\widetilde{\mathcal{P}}\widetilde{\mathcal{P}} \tag{2.200}$$

and associated adjoint map

$$\begin{aligned} \begin{bmatrix} R & \widetilde{u}_o R \\ 0 & R \end{bmatrix} = \begin{bmatrix} I & \\ & I \end{bmatrix} + \begin{bmatrix} \nu_c^2 c_0/2\, I & (\nu_c^2\delta_0/2 + \mu\nu_c c_0)I \\ 0 & \nu_c^2 c_0/2\, I \end{bmatrix}\begin{bmatrix} \widetilde{p} & \widetilde{p}^o \\ 0 & \widetilde{p} \end{bmatrix} \\ + \begin{bmatrix} \nu_c^2/2\, I & \mu\nu_c\, I \\ 0 & \nu_c^2/2\, I \end{bmatrix}\begin{bmatrix} \widetilde{p} & \widetilde{p}^o \\ 0 & \widetilde{p} \end{bmatrix}\begin{bmatrix} \widetilde{p} & \widetilde{p}^o \\ 0 & \widetilde{p} \end{bmatrix} \end{aligned} \tag{2.201}$$

Obviously, the primary part of motion tensor is the rotation tensor, Eq. (2.76), in terms of Wiener-Milenković parameters $\underline{c}$. The dual part of motion tensor, an equality, can be readily proved through dual operations. The similar operations applied to Eqs. (2.182) and (2.184) yields the explicit expressions of tangent operator $\mathcal{H}(\mathcal{P})$ and its inverse $\mathcal{H}^{-1}(\mathcal{P})$ in the form of

$$\mathcal{H}(\underline{\mathcal{P}}) = \wp(\mathcal{I} + \frac{\wp}{2}\widetilde{\mathcal{P}} + \frac{\wp}{8}\widetilde{\mathcal{P}}\widetilde{\mathcal{P}}) \tag{2.202}$$

and

$$\mathcal{H}^{-1}(\underline{\mathcal{P}}) = \frac{1}{\wp}\mathcal{I} - \frac{1}{2}\widetilde{\mathcal{P}} + \frac{1}{8}\widetilde{\mathcal{P}}\widetilde{\mathcal{P}} \tag{2.203}$$

As mentioned before, Wiener-Milenković parameters always require the rescaling operation to avoid the singularity when rotation angle $\phi$ close to $2\pi$, possible occurs in the successive composition of motions.

Let $\underline{\mathcal{P}}_i$, $\underline{\mathcal{P}}$ and $\underline{\mathcal{P}}_f$ be the Wiener-Milenković parameters of three motion tensors, such that $\mathcal{R}_f(\underline{\mathcal{P}}_f) = \mathcal{R}(\underline{\mathcal{P}})\mathcal{R}_i(\underline{\mathcal{P}}_i)$. The composition of motion tensor is then combined the rescaling [2] of motion tensor into a single operation

$$\underline{\mathcal{P}}_f = \begin{cases} 4(\wp_0\underline{\mathcal{P}}_i + \wp_0^i\underline{\mathcal{P}} + \widetilde{\mathcal{P}}\underline{\mathcal{P}}_i)/(\mathcal{D}_1 + \mathcal{D}_2) & \text{if } \mathrm{Re}(\mathcal{D}_2) \geq 0 \\ -4(\wp_0\underline{\mathcal{P}}_i + \wp_0^i\underline{\mathcal{P}} + \widetilde{\mathcal{P}}\underline{\mathcal{P}}_i)/(\mathcal{D}_1 - \mathcal{D}_2) & \text{if } \mathrm{Re}(\mathcal{D}_2) < 0 \end{cases} \tag{2.204}$$

where $\mathcal{D}_1 = (4 - \wp_0)(4 - \wp_0^i)$ and $\mathcal{D}_2 = \wp_0\wp_0^i - \underline{\mathcal{P}}^T\underline{\mathcal{P}}_i$, $\mathrm{Re}(\mathcal{D}_2)$ denotes the primary part of dual variable $\mathcal{D}_2$. The rescaling condition automatically selects the largest denominator, also guaranteeing the most accurate numerical evaluation of the composed motion.

## 2.3 Deformation

This section presents a brief discussion of the state of deformation in the neighborhood of a material particle in a flexible body. As depicted in Fig. 2.6, two configurations of this body will be defined: a reference configuration, and a deformed current configuration. Let point **P** be a material particle of the reference configuration, and the position vectors of this material point are denoted as

$$\begin{aligned} \underline{p}(\alpha_1, \alpha_2, \alpha_3) = x_1(\alpha_1, \alpha_2, \alpha_3)\,\bar{\imath}_1 \\ + x_2(\alpha_1, \alpha_2, \alpha_3)\,\bar{\imath}_2 + x_3(\alpha_1, \alpha_2, \alpha_3)\,\bar{\imath}_3 \end{aligned} \tag{2.205}$$

and

$$\begin{aligned} \underline{P}(\alpha_1, \alpha_2, \alpha_3) = X_1(\alpha_1, \alpha_2, \alpha_3)\,\bar{\imath}_1 \\ + X_2(\alpha_1, \alpha_2, \alpha_3)\,\bar{\imath}_2 + X_3(\alpha_1, \alpha_2, \alpha_3)\,\bar{\imath}_3 \end{aligned} \tag{2.206}$$

in the reference and current configuration, respectively. As defined in Sect. 2.1, each material particle of the body will be identified by a label that remains attached to the material particle throughout the deformation process. This label is called the material coordinates of material particle **P**, and is denoted $(\alpha_1, \alpha_2, \alpha_3)$.

Increments in position vector are denoted $d\underline{p}$ and $d\underline{P}$ in the reference and deformed configurations, respectively, and are expressed as

$$d\underline{p} = \frac{\partial \underline{p}}{\partial \alpha_1} d\alpha_1 + \frac{\partial \underline{p}}{\partial \alpha_2} d\alpha_2 + \frac{\partial \underline{p}}{\partial \alpha_3} d\alpha_3 \tag{2.207}$$

and

$$d\underline{P} = \frac{\partial \underline{P}}{\partial \alpha_1} d\alpha_1 + \frac{\partial \underline{P}}{\partial \alpha_2} d\alpha_2 + \frac{\partial \underline{P}}{\partial \alpha_3} d\alpha_3 \tag{2.208}$$

Fig. 2.6 Deformation of a flexible body

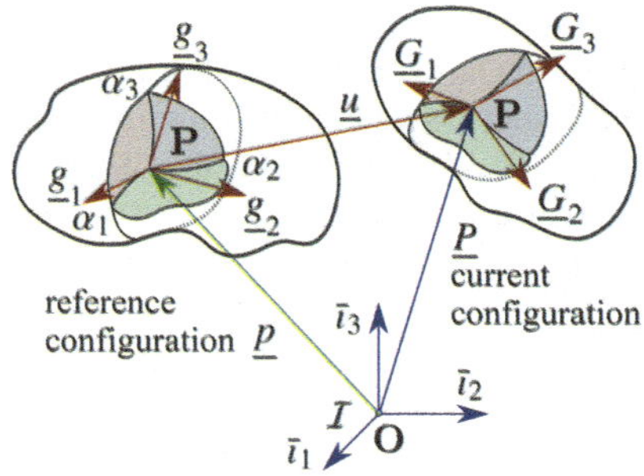


In view of the definitions of covariant base vectors, Eq. (1.274), the base vectors attached to the material particle **P** can be extracted from above formulas to be

$$\underline{g}_1 = \frac{\partial \underline{p}}{\partial \alpha_1},\ \underline{g}_2 = \frac{\partial \underline{p}}{\partial \alpha_2},\ \underline{g}_3 = \frac{\partial \underline{p}}{\partial \alpha_3} \tag{2.209}$$

in the reference configuration and

$$\underline{G}_1 = \frac{\partial \underline{P}}{\partial \alpha_1},\ \underline{G}_2 = \frac{\partial \underline{P}}{\partial \alpha_2},\ \underline{G}_3 = \frac{\partial \underline{P}}{\partial \alpha_3} \tag{2.210}$$

in the deformed configuration, respectively. These base vectors are not mutually orthogonal, nor are they unit vectors, especially after deformation. With the aid of Eq. (1.277), the following relation can be readily verified to be

$$d\underline{p} = \lfloor \underline{g}_1,\ \underline{g}_2,\ \underline{g}_3 \rfloor\, d\underline{\alpha} = J\, d\underline{\alpha} \tag{2.211}$$

where the Jacobian $J$ of transformation is defined by Eq. (1.272), and

$$d\underline{P} = \lfloor \underline{G}_1,\ \underline{G}_2,\ \underline{G}_3 \rfloor\, d\underline{\alpha} \tag{2.212}$$

where the symbol $d\underline{\alpha}$ represents the increments of three curvilinear coordinates, $d\underline{\alpha}^T = \lfloor d\alpha_1,\ d\alpha_2,\ d\alpha_3 \rfloor$.

The incremental lengths, denoted $ds$ and $dS$ in the reference and deformed configurations, respectively, are readily evaluated from

$$\begin{aligned} ds^2 &= d\underline{p}^T d\underline{p} \\ &= (\underline{g}_1^T d\alpha_1 + \underline{g}_2^T d\alpha_2 + \underline{g}_3^T d\alpha_3)(\underline{g}_1 d\alpha_1 + \underline{g}_2 d\alpha_2 + \underline{g}_3 d\alpha_3) \end{aligned} \tag{2.213}$$

and

$$\begin{aligned} dS^2 &= d\underline{P}^T d\underline{P} \\ &= (\underline{G}_1^T d\alpha_1 + \underline{G}_2^T d\alpha_2 + \underline{G}_3^T d\alpha_3)(\underline{G}_1 d\alpha_1 + \underline{G}_2 d\alpha_2 + \underline{G}_3 d\alpha_3) \end{aligned} \tag{2.214}$$

Reassembling these differential elements into the matrix form will produce

$$\begin{aligned} d\underline{p}^T d\underline{p} &= \lfloor d\alpha_1,\ d\alpha_2,\ d\alpha_3 \rfloor \begin{bmatrix} \underline{g}_1^T \\ \underline{g}_2^T \\ \underline{g}_3^T \end{bmatrix} \lfloor \underline{g}_1,\ \underline{g}_2,\ \underline{g}_3 \rfloor \begin{vmatrix} d\alpha_1 \\ d\alpha_2 \\ d\alpha_3 \end{vmatrix} \\ &= \lfloor d\alpha_1,\ d\alpha_2,\ d\alpha_3 \rfloor \begin{bmatrix} \underline{g}_1^T \underline{g}_1 & \underline{g}_1^T \underline{g}_2 & \underline{g}_1^T \underline{g}_3 \\ \underline{g}_2^T \underline{g}_1 & \underline{g}_2^T \underline{g}_2 & \underline{g}_2^T \underline{g}_3 \\ \underline{g}_3^T \underline{g}_1 & \underline{g}_3^T \underline{g}_2 & \underline{g}_3^T \underline{g}_3 \end{bmatrix} \begin{vmatrix} d\alpha_1 \\ d\alpha_2 \\ d\alpha_3 \end{vmatrix} \end{aligned} \tag{2.215}$$

and

$$\begin{aligned} d\underline{P}^T d\underline{P} &= \lfloor d\alpha_1,\ d\alpha_2,\ d\alpha_3 \rfloor \begin{bmatrix} \underline{G}_1^T \\ \underline{G}_2^T \\ \underline{G}_3^T \end{bmatrix} \lfloor \underline{G}_1,\ \underline{G}_2,\ \underline{G}_3 \rfloor \begin{vmatrix} d\alpha_1 \\ d\alpha_2 \\ d\alpha_3 \end{vmatrix} \\ &= \lfloor d\alpha_1,\ d\alpha_2,\ d\alpha_3 \rfloor \begin{bmatrix} \underline{G}_1^T \underline{G}_1 & \underline{G}_1^T \underline{G}_2 & \underline{G}_1^T \underline{G}_3 \\ \underline{G}_2^T \underline{G}_1 & \underline{G}_2^T \underline{G}_2 & \underline{G}_2^T \underline{G}_3 \\ \underline{G}_3^T \underline{G}_1 & \underline{G}_3^T \underline{G}_2 & \underline{G}_3^T \underline{G}_3 \end{bmatrix} \begin{vmatrix} d\alpha_1 \\ d\alpha_2 \\ d\alpha_3 \end{vmatrix} \end{aligned} \tag{2.216}$$

These relations define the components of metric tensors in the reference and deformed configurations

$$g = \begin{bmatrix} \underline{g}_1^T \underline{g}_1 & \underline{g}_1^T \underline{g}_2 & \underline{g}_1^T \underline{g}_3 \\ \underline{g}_2^T \underline{g}_1 & \underline{g}_2^T \underline{g}_2 & \underline{g}_2^T \underline{g}_3 \\ \underline{g}_3^T \underline{g}_1 & \underline{g}_3^T \underline{g}_2 & \underline{g}_3^T \underline{g}_3 \end{bmatrix} \tag{2.217}$$

and

$$G = \begin{bmatrix} \underline{G}_1^T \underline{G}_1 & \underline{G}_1^T \underline{G}_2 & \underline{G}_1^T \underline{G}_3 \\ \underline{G}_2^T \underline{G}_1 & \underline{G}_2^T \underline{G}_2 & \underline{G}_2^T \underline{G}_3 \\ \underline{G}_3^T \underline{G}_1 & \underline{G}_3^T \underline{G}_2 & \underline{G}_3^T \underline{G}_3 \end{bmatrix} \tag{2.218}$$

The symmetry of both metric tensors is apparent from these expressions. With the aid of above expressions, the lengths of these increments can be recasted to the compact formula like

$$d\underline{p}^T d\underline{p} = d\underline{\alpha}^T\, g\, d\underline{\alpha},\ d\underline{P}^T d\underline{P} = d\underline{\alpha}^T\, G\, d\underline{\alpha} \tag{2.219}$$

Meanwhile, the increments in position vector $d\underline{p}$ and $d\underline{P}$, can be expressed in their components form alternatively

$$d\underline{p} = dx_1\, \bar{\imath}_1 + dx_2\, \bar{\imath}_2 + dx_3\, \bar{\imath}_3 = \begin{vmatrix} dx_1 \\ dx_2 \\ dx_3 \end{vmatrix} \tag{2.220}$$

and

$$d\underline{P} = \frac{\partial \underline{P}}{\partial x_1}dx_1 + \frac{\partial \underline{P}}{\partial x_2}dx_2 + \frac{\partial \underline{P}}{\partial x_3}dx_3 = \lfloor \frac{\partial \underline{P}}{\partial x_1}, \frac{\partial \underline{P}}{\partial x_2}, \frac{\partial \underline{P}}{\partial x_3} \rfloor \begin{vmatrix} dx_1 \\ dx_2 \\ dx_3 \end{vmatrix} \tag{2.221}$$

From above formulas, the *deformation gradient tensor*, a widely used strain measure, is observed to be

$$F = \lfloor \frac{\partial \underline{P}}{\partial x_1}, \frac{\partial \underline{P}}{\partial x_2}, \frac{\partial \underline{P}}{\partial x_3} \rfloor = \begin{bmatrix} \frac{\partial X_1}{\partial x_1} & \frac{\partial X_1}{\partial x_2} & \frac{\partial X_1}{\partial x_3} \\ \frac{\partial X_2}{\partial x_1} & \frac{\partial X_2}{\partial x_2} & \frac{\partial X_2}{\partial x_3} \\ \frac{\partial X_3}{\partial x_1} & \frac{\partial X_3}{\partial x_2} & \frac{\partial X_3}{\partial x_3} \end{bmatrix} \tag{2.222}$$

and then the position increments $d\underline{P}$ can be rewritten in compact form like

$$d\underline{P} = F\, d\underline{p} \tag{2.223}$$

In view of Eq. (2.211), the increment $d\underline{P}$ of position vector in deformed configuration is related by

$$d\underline{P} = F\, d\underline{p} = F\, J\, d\underline{\alpha} \tag{2.224}$$

Referring the definitions of covariant base vectors, Eq. (2.212), it is not difficult to find the deformation gradient tensor can be regarded as a linear mapping matrix that transforms the covariant bases from initial reference configuration to the deformed configuration like

$$\lfloor \underline{G}_1, \underline{G}_2, \underline{G}_3 \rfloor = F \lfloor \underline{g}_1, \underline{g}_2, \underline{g}_3 \rfloor \tag{2.225}$$

As discussed in Sect. 1.1.6, the absolute value of the scalar triple product $\underline{g}_3^T \widetilde{g}_1\, \underline{g}_2$ are verified to be the volume of a parallelepiped with base vectors $\underline{g}_1$, $\underline{g}_2$ and $\underline{g}_3$ forming adjacent edges, and

$$V_r = |\underline{g}_3^T \widetilde{g}_1\, \underline{g}_2| \tag{2.226}$$

When the parallelepiped is transformed by the deformation gradient tensor $F$ into the new deformed configurations, its volume are determined by

$$V_d = |\underline{G}_3^T \widetilde{G}_1\, \underline{G}_2| = |(F\underline{g}_3)^T \widetilde{F\underline{g}_1}\, F\underline{g}_2| \tag{2.227}$$

The scalar triple product can also be rewritten as the determinant of a matrix, from which the volume of parallelepiped before and after deformation can be related to each other like

$$V_d = |\underline{G}_3^T \widetilde{G}_1\, \underline{G}_2| = \left| F\underline{g}_1, F\underline{g}_2, F\underline{g}_3 \right| = \det(F) V_r \tag{2.228}$$

Another important identity can be obtained also from above relation through the following expansions

$$|\underline{G}_3^T \widetilde{G}_1 \, \underline{G}_2| = \det(F)|\underline{g}_3^T \widetilde{g}_1 \, \underline{g}_2| \tag{2.229}$$

With the application of transformation, $\underline{G}_3 = F\underline{g}_3$, the volume mapping is reformulated to

$$|\underline{g}_3^T F^T \widetilde{G}_1 \, \underline{G}_2| = \det(F)|\underline{g}_3^T \widetilde{g}_1 \, \underline{g}_2| \tag{2.230}$$

and equivalently

$$\underline{g}_3^T (|F^T \widetilde{G}_1 \, \underline{G}_2| - \det(F)|\widetilde{g}_1 \, \underline{g}_2|) = 0 \tag{2.231}$$

In general, the base vector $\underline{g}_3$ is an arbitrary non-zero vector, such that

$$|F^T \widetilde{G}_1 \, \underline{G}_2| = \det(F)|\widetilde{g}_1 \, \underline{g}_2| \tag{2.232}$$

This is the well-known *Manson's formula*, which is applied to map the vector product or the so-called area vectors before and after deformation.

Except the mappings of volume and area are discussed in detail, the length mapping will be introduced next, from which two types of strain tensors are defined. With the aid of deformation gradient tensor, the length difference of the position increments $d\underline{P}$ and $d\underline{p}$ in the reference and deformed configurations, respectively, is recasted to

$$\begin{aligned} dS^2 - ds^2 &= d\underline{P}^T d\underline{P} - d\underline{p}^T d\underline{p} \\ &= d\underline{p}^T \, (F^T F - I) \, d\underline{p} = d\underline{\alpha}^T \, J^T (F^T F - I) J \, d\underline{\alpha} \\ &= d\underline{\alpha}^T (G - g) \, d\underline{\alpha} \end{aligned} \tag{2.233}$$

From which, the *Green-Lagrange strain tensor*, a widely used strain measure, is defined in the Cartesian basis as

$$\varepsilon = \frac{1}{2}(F^T F - I), \tag{2.234}$$

and it is also resolved into the curvilinear basis like

$$\widehat{\varepsilon} = J^T \varepsilon \, J = \frac{1}{2}(G - g) \tag{2.235}$$

From these definitions, the metric tensor in the deformed configuration is found to be closely related to the deformation gradient tensor like

$$G = J^T F^T F \, J \tag{2.236}$$

and the definition of metric tensor in reference configuration

$$g = J^T J \tag{2.237}$$

is observed also. On the other hand, the undeformed position increments $d\underline{p}$ could be transformed into the deformed configuration by means of the deformation gradient tensor $F$ like

$$d\underline{p} = F^{-1} d\underline{P} \tag{2.238}$$

Thus, the length difference between the position increments $d\underline{P}$ and $d\underline{p}$ measured in the deformed configurations becomes

$$dS^2 - ds^2 = d\underline{P}^T d\underline{P} - d\underline{p}^T d\underline{p} = d\underline{P}^T (I - F^{-T} F^{-1}) d\underline{P} \tag{2.239}$$

where the Almansi strain tensor $\varepsilon^*$ are introduced

$$\varepsilon^* = \frac{1}{2}(I - F^{-T} F^{-1}) \tag{2.240}$$

and related to the Green-Lagrange strain tensor by the law of tensor transformation

$$\varepsilon = F^T \varepsilon^* F \tag{2.241}$$

Note that a convenient choice for the material coordinates consists of the components of the position vector resolved in basis $\mathcal{I}$, which states

$$\underline{p}(\alpha_1, \alpha_2, \alpha_3) = \alpha_1 \bar{\imath}_1 + \alpha_2 \bar{\imath}_2 + \alpha_3 \bar{\imath}_3 \tag{2.242}$$

This particular choice of the material coordinates is called the *Lagrangian representation*. Apparently, the Jacobian of transformation $J$ will reduce to identity matrix $I$. Under this special situation, the Green-Lagrange strain tensor $\varepsilon$ and its transformation $\widehat{\varepsilon}$ becomes identical to each other

$$\varepsilon = \widehat{\varepsilon} = \frac{1}{2}(F^T F - I) = \frac{1}{2}(G - g) \tag{2.243}$$

and $G = F^T F$ together with $g = I$.

## References


1. Rubin, M.B.: Cosserat Theories: Shells, Rods and Points. Springer, Dordrecht (2000)
2. Bauchau, O.A.: Flexible Multibody Dynamics. Springer, Dordrecht (2011)
3. Rudnicki, J.W.: Fundamentals of Continuum Mechanics. Wiley, Verlag (2014)
4. Bauchau, O.A., Choi, J.: The vector parameterization of motion. Nonlinear Dyn. **33**, 165–188 (2003)
5. Bauchau, O.A., Li, L.: Tensorial parameterization of rotation and motion. J. Comput. Nonlinear Dyn. (2011). https://doi.org/10.1115/1.4003176

6. Bauchau, O.A., Trainelli, L.: The vectorial parameterization of rotation. Nonlinear Dyn. (2003). https://doi.org/10.1023/A:1024265401576
7. Chasles, M.: Note sur les propriétés géné rales du systéme de deux corps semblables entre eux et placés d'une manière quelconque dans l'espace; et sur le déplacement fini, ou infiniment petit d'un corps solide libre. Bull. Sci. Math. Férussac **14**, 321–326 (1830)
8. Bottema, O., Roth, B.: Theoretical Kinematics. Dover, New York (1979)
9. Angeles, J.: The application of dual algebra to kinematic analysis. In: Angeles, J., Zakhariev, E. (eds.) Comput. Methods Mech. Syst., pp. 3–31. Springer, Heidelberg (1998)
10. Pradeep, A.K., Yoder, P.J., Mukundan, R.: On the use of dual matrix exponentials in robot kinematics. Int. J. Robot. Res. **8**, 57–66 (1989)
11. Papastavridis, J.G.: On the transitivity equations of rigid-body dynamics. J. Appl. Mech. **59**, 955–962 (1992)
12. Pennestrí, E., Stefanelli, R.: Linear algebra and numerical algorithms using dual numbers. Multibody Syst. Dyn. **18**, 323–344 (2007)
13. Pennestrí, E., Valentini, P.P.: Linear dual algebra algorithms and their application to kinematics. Multibody Dyn. (2009). https://doi.org/10.1007/978-1-4020-8829-2_11
14. Condurache, D., Burlacu, A.: Dual tensors based solutions for rigid body motion parameterization. Mech. Mach. Theory **74**, 390–412 (2014)
15. Han, S., Bauchau, O.A.: Manipulation of motion via dual entities. Nonlinear Dyn. (2016). https://doi.org/10.1007/s11071-016-2703-7

# Part II
# Unified Theory of Cosserat Continuum

# Chapter 3
# Cosserat Continuum

The Cosserat continuum is a notion of generalized or higher-order continuum proposed by the Cosserat brothers [1]. Different from the Cauchy continuum, a collection of material particles with only position vectors, the Cosserat continuum includes a director vector [2] at each material particle in addition to the position vector. The purpose of introducing the director vectors in a Cosserat continuum is to characterize the material fibers. As discussed in Chap. 2, the director of material fiber passing through a particle **P** can be strictly defined by using a local curvilinear basis at the particle. When the material fiber experiences deformations, such as the extension, contraction, twist or bending, the deformed configuration of the fiber can be readily described by using the director or the curvilinear basis, equivalently. In this sense, the theory based on Cosserat continuum allows for more general deformation than the classical theories based on Cauchy continuum. Up to now, after continuously developed more than a century, the theoretical framework of Cosserat continuum has been gradually completed and many summarized literatures [3–6] appeared. The Cosserat theory successfully applied to simulate various mechanical phenomena in many fields, such as granular media [7], composite materials [8–10], polycrystalline solids [11], cellular solids [12], biomaterials [13], liquid crystals [14], foams [15], nanomaterials [16], plasticity [17] and cracks [18]. The extensively application expands the scope of Cosserat theory and accelerates the theory to be matured.

With the aids of the algebraic operations of vector and tensor together with the descriptions of motion and deformation, introduced in Chaps. 1 and 2, respectively, this chapter presents a generalized formula of a three-dimensional Cosserat continuum within the context of the theory of elasticity.



J. Wang, *Multiscale Multibody Dynamics*,
https://doi.org/10.1007/978-981-19-8441-9_3

## 3.1 General Cosserat Continuum Theory

From the point of differential geometry, the Cosserat continuum is apparently different from the Cauchy continuum in the way of describing the configuration of a material particle mathematically. Each material particle in a Cosserat continuum occupies an infinitesimal region, its configuration before and after deformations has to be described by using the position vector together with the associated director. While the description of a material particle in the Cauchy continuum is relatively simple, and only the position vector is defined to measure the spatial position of material particle. In other words, a material particle in a Cosserat continuum behaves like a rigid body [19] with infinitesimal volume. In this section, the generalized Cosserat continuum theory starts with the mathematical description of a material particle configuration. The spatial location of the material particle is determined by its position vector, and the director of the particle is computed from the derivatives of position vector with respected to the curvilinear coordinates, from which a local orthogonal curvilinear basis at the particle is constructed. Unlike the classical mechanics of Cosserat within the framework of small perturbation [2, 19], the general Cosserat theory implemented in this section allows arbitrary micro-motion. Instead of vectorial parameterizing finite micro-rotation [20] solely, the current implementation utilizes the Wiener-Milenković parameterization of motion, described in Sect. 2.2.10, which leads to the nonlinear Cosserat theory. The metric tensors in reference configuration and deformed configuration are then defined, respectively. Through the components of metric tensor components, the rotation-free Green-Lagrange strain tensor could be correlated with Lagrangian stretch tensor and the Lagrangian curvature stain tensor explicitly. In current implementations, the Cosserat theory is limited to orthotropic materials, and the constitutive laws for Cosserat continuum are explicitly derived, especially the part associated with the curvature strain components. With the applications of the principle of virtual work and also the Hamilton's principle, the governing equations for three-dimensional continuum are obtained, and then recasted into its motion formalism [21].

### 3.1.1 Kinematics of Cosserat Continuum

As shown in Fig. 3.1, a material particle with infinitesimal volume in a Cosserat continuum is depicted by its position and director vectors in its reference configuration and current deformed configuration, respectively. In the original reference configuration, the position vector $\underline{p}(\alpha_i)$ of the material particle relative to a point $\mathbf{O}$, the original of reference frame, $\mathcal{F}_{\mathcal{I}} = [\mathbf{O}, \mathcal{I}]$, can be written as

$$\underline{p}(\alpha_1, \alpha_2, \alpha_3) = \underline{x}_0(\alpha_1, \alpha_2, \alpha_3) + R_0(\alpha_1, \alpha_2, \alpha_3)\underline{s}_0(\alpha_1, \alpha_2, \alpha_3) \tag{3.1}$$

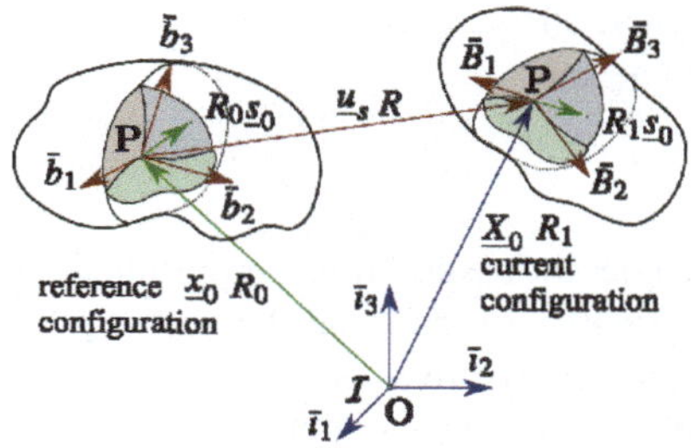


**Fig. 3.1** Kinematics of a material particle with infinitesimal volume in a Cosserat continuum

where the general curvilinear coordinates, $(\alpha_1, \alpha_2, \alpha_3)$, are introduced to label the material particle in the original continuum, and $R_0$ the initial rotation tensor measures the orientation of the reference basis $\mathcal{B}_0$. It has been verified in Sect. 2.2.3 the rotation tensor $R_0$ is identical to the body attached basis $\mathcal{B}_0$, and in detail

$$R_0 = \lfloor \bar{b}_1, \bar{b}_2, \bar{b}_3 \rfloor = \mathcal{B}_0 \tag{3.2}$$

Note that the initial basis $\mathcal{B}_0$ is bounded by the material particle $\mathbf{P}$ to produce the reference frame $\mathcal{F}_0 = [\mathbf{P}, \mathcal{B}_0]$. The local vector $\underline{s}_0$, defined in this reference frame $\mathcal{F}_0$, describes the director of a material particle. In an orthogonal curvilinear coordinate system, the director can be given directly

$$\underline{s}_0 = s_0 \bar{n}(\alpha_1, \alpha_2, \alpha_3) \tag{3.3}$$

where $\bar{n}$ is the unit vector along the direction of $\underline{s}_0$. It is very important to emphasize that the magnitude $s_0$ determines the characteristic length in a Cosserat continuum associated with torsional and bending deformations. It will be initialized according to the material properties of a specific problem. In finite element analysis, the edge size of three-dimensional solid element could be selected as the characteristic length. Obviously, Eq. (3.1) decomposes the position vector $\underline{p}$ of a material particle into a reference position $\underline{x}_0$ and a local vector $\underline{s}_0$. The reference particle with position $\underline{x}_0$ can translate freely, while the rigid director $\underline{s}_0$ attached to the reference particle can only rotate without deformations.

For the description of Cosserat continuum deformations, the covariant base vectors,

$$\underline{g}_i = \frac{\partial \underline{p}}{\partial \alpha_i} \tag{3.4}$$

in the reference configuration are required, and the specific expressions can be obtained by taking derivatives of Eq. (3.1) with respected to the curvilinear coordinates $\alpha_i$ like

$$\begin{aligned} \underline{g}_1 &= \underline{x}_{0,1} + R_{0,1}\underline{s}_0 + R_0\underline{s}_{0,1} \\ \underline{g}_2 &= \underline{x}_{0,2} + R_{0,2}\underline{s}_0 + R_0\underline{s}_{0,2} \\ \underline{g}_3 &= \underline{x}_{0,3} + R_{0,3}\underline{s}_0 + R_0\underline{s}_{0,3} \end{aligned} \tag{3.5}$$

where the symbol $(\cdot)_{,i}$ for $i = 1, 2, 3$ denotes the spatial derivatives with respected to $\alpha_i$ and

$$(\cdot)_{,i} = \frac{\partial(\cdot)}{\partial \alpha_i} \tag{3.6}$$

Under the assumption of constant directors in a Cosserat continuum [2], it is straightforward to verify that the vector $\underline{s}_0$ will not be a functions of $\alpha_i$, which means the spatial derivatives of $\underline{s}_0$ will become zero, $\underline{s}_{0,i} = 0$. This special situation is applicable to the isotropic materials. In current implementation, the nonzero items $\underline{s}_{0,i}$ are retained in the formulations of base vectors. Further operations are taken to resolve the base vectors into the local reference frame $\mathcal{F}_0$ to get

$$\begin{aligned} \underline{g}_1^* &= R_0^T \underline{x}_{0,1} + R_0^T R_{0,1} \underline{s}_0 + \underline{s}_{0,1} \\ \underline{g}_2^* &= R_0^T \underline{x}_{0,2} + R_0^T R_{0,2} \underline{s}_0 + \underline{s}_{0,2} \\ \underline{g}_3^* &= R_0^T \underline{x}_{0,3} + R_0^T R_{0,3} \underline{s}_0 + \underline{s}_{0,3} \end{aligned} \tag{3.7}$$

According to the definitions of initial curvature tensors, Eq. (1.314), the base vectors can be represented by the following formulas like

$$\begin{aligned} \underline{g}_1^* &= R_0^T \underline{x}_{0,1} + \widetilde{\kappa}_1^0 \underline{s}_0 + \underline{s}_{0,1} \\ \underline{g}_2^* &= R_0^T \underline{x}_{0,2} + \widetilde{\kappa}_2^0 \underline{s}_0 + \underline{s}_{0,2} \\ \underline{g}_3^* &= R_0^T \underline{x}_{0,3} + \widetilde{\kappa}_3^0 \underline{s}_0 + \underline{s}_{0,3} \end{aligned} \tag{3.8}$$

where the symbols $\widetilde{\kappa}_i^0 = R_0^T R_{0,i}$ for $i = 1, 2, 3$ denote the initial curvature tensors. All these base vectors form the covariant bases $\mathcal{B}_g$ like

$$\mathcal{B}_g = \lfloor \underline{g}_1^*, \ \underline{g}_2^*, \ \underline{g}_3^* \rfloor \tag{3.9}$$

and then the metric tensor measured in the reference frame $\mathcal{F}_0$ is computed from

$$g = \mathcal{B}_g^T \mathcal{B}_g \tag{3.10}$$

For the easy description of global and local motions of the material particle, the covariant bases $\mathcal{B}_g$ are decomposed into two parts

$$\mathcal{B}_g = \mathcal{B}_g^s + \mathcal{B}_g^\alpha \tag{3.11}$$

with the aid of introduced symbols as

$$\mathcal{B}_g^s = \lfloor \underline{g}_1^s, \ \underline{g}_2^s, \ \underline{g}_3^s \rfloor, \ \mathcal{B}_g^\alpha = \lfloor \underline{g}_1^\alpha, \ \underline{g}_2^\alpha, \ \underline{g}_3^\alpha \rfloor \tag{3.12}$$

where the base vectors are given in detail for global part

$$\begin{aligned} \underline{g}_1^s &= R_0^T \underline{x}_{0,1} + \widetilde{\kappa}_1^0 \underline{s}_0 \\ \underline{g}_2^s &= R_0^T \underline{x}_{0,2} + \widetilde{\kappa}_2^0 \underline{s}_0 \\ \underline{g}_3^s &= R_0^T \underline{x}_{0,3} + \widetilde{\kappa}_3^0 \underline{s}_0 \end{aligned} \tag{3.13}$$

and for local part

$$\underline{g}_1^\alpha = \underline{s}_{0,1}, \; \underline{g}_2^\alpha = \underline{s}_{0,2}, \; \underline{g}_3^\alpha = \underline{s}_{0,3} \tag{3.14}$$

In view of this, the metric tensor $g$ will be consisted of

$$g = g^{ss} + g^{s\alpha} + g^{\alpha s} + g^{\alpha\alpha} \tag{3.15}$$

and in detail

$$g^{ss} = \mathcal{B}_g^{sT} \mathcal{B}_g^s, \; g^{s\alpha} = \mathcal{B}_g^{sT} \mathcal{B}_g^\alpha, \; g^{\alpha s} = \mathcal{B}_g^{\alpha T} \mathcal{B}_g^s, \; g^{\alpha\alpha} = \mathcal{B}_g^{\alpha T} \mathcal{B}_g^\alpha \tag{3.16}$$

when the Cosserat continuum deforms with spatial motions, the material particle that had position vector $\underline{p}$ in the undeformed reference configuration now has position vector $\underline{P}$ in the deformed configuration, such that

$$\underline{P}(\alpha_1, \alpha_2, \alpha_3) = \underline{X}_0(\alpha_1, \alpha_2, \alpha_3) + R_1(\alpha_1, \alpha_2, \alpha_3)\underline{s}_0(\alpha_1, \alpha_2, \alpha_3) \tag{3.17}$$

where $R_1$ is the total rotation tensor that describes the orientation of basis $\mathcal{B}_1$ attached to the deformed configuration and

$$R_1 = \lfloor \bar{B}_1, \bar{B}_2, \bar{B}_3 \rfloor = \mathcal{B}_1 \tag{3.18}$$

The associated frame $\mathcal{F}_1 = [\mathbf{P}, \mathcal{B}_1]$ will be constructed when bounding the current deformed basis $\mathcal{B}_1$ by the material particle $\mathbf{P}$. The position vector $\underline{X}_0$ of reference point in current deformed configuration is recasted to

$$\underline{X}_0(\alpha_1, \alpha_2, \alpha_3) = \underline{x}_0(\alpha_1, \alpha_2, \alpha_3) + \underline{u}_s(\alpha_1, \alpha_2, \alpha_3) \tag{3.19}$$

where the displacement vector $\underline{u}_s$ describes arbitrary translation of reference point. Alternatively, the position vector $\underline{P}$ of the material particle in its deformed configuration can be represented by

$$\underline{P}(\alpha_1, \alpha_2, \alpha_3) = \underline{p}(\alpha_1, \alpha_2, \alpha_3) + \underline{u}(\alpha_1, \alpha_2, \alpha_3) \tag{3.20}$$

where $\underline{u}$ is the displacement of particle in space. It is readily to find that the displacement $\underline{u}$ are related to

$$\underline{u} = \underline{u}_s + (R_1 - R_0)\underline{s}_0 \tag{3.21}$$

with a very clear physical meaning: the displacement $\underline{u}$ is composed of the translation $\underline{u}_s$ of reference particle and the contribution of pure rotation $(R_1 - R_0)\underline{s}_0$ of local director $\underline{s}_0$. According to the definitions of covariant base vectors in deformed configuration

$$\underline{G}_i = \frac{\partial \underline{P}}{\partial \alpha_i} \tag{3.22}$$

the specific formulas can be obtained by taking spatial derivatives of Eq. (3.17) with respected to the curvilinear coordinates $\alpha_i$ for $i = 1, 2, 3$ like

$$\begin{aligned} \underline{G}_1 &= \underline{X}_{0,1} + R_{1,1}\underline{s}_0 + R_1\underline{s}_{0,1} \\ \underline{G}_2 &= \underline{X}_{0,2} + R_{1,2}\underline{s}_0 + R_1\underline{s}_{0,2} \\ \underline{G}_3 &= \underline{X}_{0,3} + R_{1,3}\underline{s}_0 + R_1\underline{s}_{0,3} \end{aligned} \tag{3.23}$$

Once again, the base vectors are resolved into the current local frame $\mathcal{F}_1$ to obtain

$$\begin{aligned} \underline{G}_1^* &= R_1^T\underline{X}_{0,1} + R_1^T R_{1,1}\underline{s}_0 + R_1^T R_1\underline{s}_{0,1} \\ \underline{G}_2^* &= R_1^T\underline{X}_{0,2} + R_1^T R_{1,2}\underline{s}_0 + R_1^T R_1\underline{s}_{0,2} \\ \underline{G}_3^* &= R_1^T\underline{X}_{0,3} + R_1^T R_{1,3}\underline{s}_0 + R_1^T R_1\underline{s}_{0,3} \end{aligned} \tag{3.24}$$

In view of the definitions of deformed curvature tensors $\widetilde{\kappa}_i^1 = R_1^T R_{1,i}$ for $i = 1, 2, 3$, and the orthogonal attributes of rotation tensor, $R_1^T R_1 = I$, the covariant base vectors are simplified to

$$\begin{aligned} \underline{G}_1^* &= R_1^T\underline{X}_{0,1} + \widetilde{\kappa}_1^1\underline{s}_0 + \underline{s}_{0,1} \\ \underline{G}_2^* &= R_1^T\underline{X}_{0,2} + \widetilde{\kappa}_2^1\underline{s}_0 + \underline{s}_{0,2} \\ \underline{G}_3^* &= R_1^T\underline{X}_{0,3} + \widetilde{\kappa}_3^1\underline{s}_0 + \underline{s}_{0,3} \end{aligned} \tag{3.25}$$

All these base vectors could be assembled into the covariant bases $\mathcal{B}_G$ like

$$\mathcal{B}_G = \lfloor \underline{G}_1^*,\ \underline{G}_2^*,\ \underline{G}_3^* \rfloor \tag{3.26}$$

which is decomposed into two parts

$$\mathcal{B}_G = \mathcal{B}_G^s + \mathcal{B}_G^\alpha \tag{3.27}$$

with the help of new symbols defined in detail as

$$\mathcal{B}_G^s = \lfloor \underline{G}_1^s,\ \underline{G}_2^s,\ \underline{G}_3^s \rfloor,\ \mathcal{B}_G^\alpha = \lfloor \underline{G}_1^\alpha,\ \underline{G}_2^\alpha,\ \underline{G}_3^\alpha \rfloor \tag{3.28}$$

where the global and local base vectors are introduced to be

$$\begin{aligned} \underline{G}_1^s &= R_1^T\underline{X}_{0,1} + \widetilde{\kappa}_1^1\underline{s}_0 \\ \underline{G}_2^s &= R_1^T\underline{X}_{0,2} + \widetilde{\kappa}_2^1\underline{s}_0 \\ \underline{G}_3^s &= R_1^T\underline{X}_{0,3} + \widetilde{\kappa}_3^1\underline{s}_0 \end{aligned} \tag{3.29}$$

and

$$\underline{G}_1^\alpha = \underline{s}_{0,1}, \ \underline{G}_2^\alpha = \underline{s}_{0,2}, \ \underline{G}_3^\alpha = \underline{s}_{0,3} \tag{3.30}$$

In view of above expressions of covariant bases, Eq. (3.27), measured in the deformed basis $\mathcal{F}_1$, the associated metric tensor

$$G = \mathcal{B}_G^T \mathcal{B}_G \tag{3.31}$$

can be found to be composed of

$$G = G^{ss} + G^{s\alpha} + G^{\alpha s} + G^{\alpha\alpha} \tag{3.32}$$

with the following definitions

$$G^{ss} = \mathcal{B}_G^{sT} \mathcal{B}_G^s, \ G^{s\alpha} = \mathcal{B}_G^{sT} \mathcal{B}_G^\alpha, \ G^{\alpha s} = \mathcal{B}_G^{\alpha T} \mathcal{B}_G^s, \ G^{\alpha\alpha} = \mathcal{B}_G^{\alpha T} \mathcal{B}_G^\alpha \tag{3.33}$$

### 3.1.2 Strain Tensor of Cosserat Continuum

The rotation-free Green-Lagrange strain tensor [22] is applied to measure the deformation of Cosserat continuum. In doing so, the Lagrangian stretch tensor and the Lagrangian curvature strain tensor [23, 24] in their vector-dyadic form

$$\underline{e}_i = R_1^T \underline{X}_{0,i} - R_0^T \underline{x}_{0,i}, \ \widetilde{\kappa}_i = \widetilde{\kappa}_i^1 - \widetilde{\kappa}_i^0 = R_1^T R_{1,i} - R_0^T R_{0,i} \tag{3.34}$$

for $i = 1, 2, 3$ are required to reformulate the deformed covariant base vectors in terms of

$$\begin{aligned} \underline{G}_1^s &= \underline{e}_1 + R_0^T \underline{x}_{0,1} + \widetilde{\kappa}_1^1 \underline{s}_0 \\ \underline{G}_2^s &= \underline{e}_2 + R_0^T \underline{x}_{0,2} + \widetilde{\kappa}_2^1 \underline{s}_0 \\ \underline{G}_3^s &= \underline{e}_3 + R_0^T \underline{x}_{0,3} + \widetilde{\kappa}_3^1 \underline{s}_0 \end{aligned} \tag{3.35}$$

Note that the Lagrangian stretch $\underline{e}_i$ and curvatures $\underline{\kappa}_i$ are nothing but the components of Lagrangian strain measures in their Plucker coordinates, Eq. (2.152), defined as the difference of curvature dual vectors, such that

$$\underline{\epsilon}_i = \underline{\mathcal{K}}_i^1 - \underline{\mathcal{K}}_i^0 = \begin{vmatrix} R_1^T \underline{X}_{0,i} \\ \underline{\kappa}_i^1 \end{vmatrix} - \begin{vmatrix} R_0^T \underline{x}_{0,i} \\ \underline{\kappa}_i^0 \end{vmatrix} = \begin{vmatrix} \underline{e}_i \\ \underline{\kappa}_i \end{vmatrix} \tag{3.36}$$

for $i = 1, 2, 3$. Referring the decompositions of metric tensors in reference configuration and deformed configuration, Eqs. (3.15) and (3.32), respectively, the rotation-free Green-Lagrangian strain tensor measured in the curvilinear basis

$$\widehat{\varepsilon} = \frac{1}{2}(G - g) \tag{3.37}$$

can be regarded as a combination of two contributions, $\widehat{\varepsilon} = \varepsilon^s + \varepsilon^c$, one from the global deformation

$$\varepsilon^s = \frac{1}{2}(G^{ss} - g^{ss}) \tag{3.38}$$

and the other from the global-local coupling deformation

$$\varepsilon^c = \frac{1}{2}(G^{\alpha s} + G^{s\alpha} - g^{\alpha s} - g^{s\alpha}) \tag{3.39}$$

The contribution of local deformation is zero in classical Cosserat theory

$$\varepsilon^\alpha = \frac{1}{2}(G^{\alpha\alpha} - g^{\alpha\alpha}) = 0 \tag{3.40}$$

Substituting the specific formulations of metric tensors in terms of covariant basis vectors, Eq. (3.35), into the definitions of strain tensor, Eqs. (3.38) and (3.39), the components of rotation-free Green-Lagrange strain tensor, Eq. (3.37), are found to be related to the components of Lagrangian stretch strains and curvature strains. In the vector-dyadic form, the relations are written in detail like

$$\begin{aligned}
\varepsilon_{11} &= (\underline{x}_{0,1}^T R_0 + \underline{\kappa}_1^{0T} \widetilde{s}_0 + \underline{s}_{0,1}^T)(\underline{e}_1 + \widetilde{s}_0^T \underline{\kappa}_1) \\
2\varepsilon_{12} &= (\underline{x}_{0,1}^T R_0 + \underline{\kappa}_1^{0T} \widetilde{s}_0 + \underline{s}_{0,1}^T)(\underline{e}_2 + \widetilde{s}_0^T \underline{\kappa}_2) + (\underline{x}_{0,2}^T R_0 + \underline{\kappa}_2^{0T} \widetilde{s}_0 + \underline{s}_{0,2}^T)(\underline{e}_1 + \widetilde{s}_0^T \underline{\kappa}_1) \\
2\varepsilon_{13} &= (\underline{x}_{0,1}^T R_0 + \underline{\kappa}_1^{0T} \widetilde{s}_0 + \underline{s}_{0,1}^T)(\underline{e}_3 + \widetilde{s}_0^T \underline{\kappa}_3) + (\underline{x}_{0,3}^T R_0 + \underline{\kappa}_3^{0T} \widetilde{s}_0 + \underline{s}_{0,3}^T)(\underline{e}_1 + \widetilde{s}_0^T \underline{\kappa}_1) \\
\varepsilon_{22} &= (\underline{x}_{0,2}^T R_0 + \underline{\kappa}_2^{0T} \widetilde{s}_0 + \underline{s}_{0,2}^T)(\underline{e}_2 + \widetilde{s}_0^T \underline{\kappa}_2) \\
2\varepsilon_{23} &= (\underline{x}_{0,2}^T R_0 + \underline{\kappa}_2^{0T} \widetilde{s}_0 + \underline{s}_{0,2}^T)(\underline{e}_3 + \widetilde{s}_0^T \underline{\kappa}_3) + (\underline{x}_{0,3}^T R_0 + \underline{\kappa}_3^{0T} \widetilde{s}_0 + \underline{s}_{0,3}^T)(\underline{e}_2 + \widetilde{s}_0^T \underline{\kappa}_2) \\
\varepsilon_{33} &= (\underline{x}_{0,3}^T R_0 + \underline{\kappa}_3^{0T} \widetilde{s}_0 + \underline{s}_{0,3}^T)(\underline{e}_3 + \widetilde{s}_0^T \underline{\kappa}_3)
\end{aligned} \tag{3.41}$$

Note that the assumptions of small stretch strain $\underline{e}_i \ll 0$ and small curvature strain $\underline{\kappa}_i \ll 0$ for $i = 1, 2, 3$ are applied such that the higher order terms as to $\underline{e}_i$ and $\underline{\kappa}_i$ for $i = 1, 2, 3$ are all neglected.

Alternatively, a straight forward approach can be applied to obtain the same expressions as Eq. (3.41). From the definitions of covariant base vectors, Eq. (3.8) in reference configuration and Eq. (3.25) in deformed configuration, respectively, the differences of base vectors can be found to be

$$\begin{aligned}
\Delta\underline{G}_1^* &= \underline{G}_1^* - \underline{g}_1^* = R_1^T \underline{X}_{0,1} - R_0^T \underline{x}_{0,1} + (\widetilde{\kappa}_1^1 - \widetilde{\kappa}_1^0)\underline{s}_0 \\
\Delta\underline{G}_2^* &= \underline{G}_2^* - \underline{g}_2^* = R_1^T \underline{X}_{0,2} - R_0^T \underline{x}_{0,2} + (\widetilde{\kappa}_2^1 - \widetilde{\kappa}_2^0)\underline{s}_0 \\
\Delta\underline{G}_3^* &= \underline{G}_3^* - \underline{g}_3^* = R_1^T \underline{X}_{0,3} - R_0^T \underline{x}_{0,3} + (\widetilde{\kappa}_3^1 - \widetilde{\kappa}_3^0)\underline{s}_0
\end{aligned} \tag{3.42}$$

Referring the definitions of Lagrangian stretch and Lagrange curvature measures, Eq. (3.34), the above definitions can be recasted to the compact form like

$$\Delta\underline{G}_1^* = \underline{e}_1 + \widetilde{s}_0^T \underline{\kappa}_1, \quad \Delta\underline{G}_2^* = \underline{e}_2 + \widetilde{s}_0^T \underline{\kappa}_2, \quad \Delta\underline{G}_3^* = \underline{e}_3 + \widetilde{s}_0^T \underline{\kappa}_3 \tag{3.43}$$

The components of Green-Lagrange strain tensor are approximated to be

$$\begin{aligned}
\varepsilon_{11} &= \frac{1}{2}[(\underline{g}_1^* + \Delta\underline{G}_1^*)^T(\underline{g}_1^* + \Delta\underline{G}_1^*) - \underline{g}_1^{*T}\underline{g}_1^*] \approx \underline{g}_1^{*T}\Delta\underline{G}_1^* \\
2\varepsilon_{12} &= (\underline{g}_1^* + \Delta\underline{G}_1^*)^T(\underline{g}_2^* + \Delta\underline{G}_2^*) - \underline{g}_1^{*T}\underline{g}_2^* \approx \underline{g}_1^{*T}\Delta\underline{G}_2^* + \underline{g}_2^{*T}\Delta\underline{G}_1^* \\
2\varepsilon_{13} &= (\underline{g}_1^* + \Delta\underline{G}_1^*)^T(\underline{g}_3^* + \Delta\underline{G}_3^*) - \underline{g}_1^{*T}\underline{g}_3^* \approx \underline{g}_1^{*T}\Delta\underline{G}_3^* + \underline{g}_3^{*T}\Delta\underline{G}_1^* \\
\varepsilon_{22} &= \frac{1}{2}[(\underline{g}_2^* + \Delta\underline{G}_2^*)^T(\underline{g}_2^* + \Delta\underline{G}_2^*) - \underline{g}_2^{*T}\underline{g}_2^*] \approx \underline{g}_2^{*T}\Delta\underline{G}_2^* \\
2\varepsilon_{23} &= (\underline{g}_2^* + \Delta\underline{G}_2^*)^T(\underline{g}_3^* + \Delta\underline{G}_3^*) - \underline{g}_2^{*T}\underline{g}_3^* \approx \underline{g}_2^{*T}\Delta\underline{G}_3^* + \underline{g}_3^{*T}\Delta\underline{G}_2^* \\
\varepsilon_{33} &= \frac{1}{2}[(\underline{g}_3^* + \Delta\underline{G}_3^*)^T(\underline{g}_3^* + \Delta\underline{G}_3^*) - \underline{g}_3^{*T}\underline{g}_3^*] \approx \underline{g}_3^{*T}\Delta\underline{G}_3^*
\end{aligned} \tag{3.44}$$

where the higher order terms $\Delta\underline{G}_i^{*T}\Delta\underline{G}_j^*$ for $i, j = 1, 2, 3$ are all truncated under the assumption of small strains. Once again, the linearized strain measures are obtained to be

$$\begin{aligned}
\varepsilon_{11} &= \underline{g}_1^{*T}(\underline{e}_1 + \widetilde{s}_0^T\underline{\kappa}_1) \\
2\varepsilon_{12} &= \underline{g}_1^{*T}(\underline{e}_2 + \widetilde{s}_0^T\underline{\kappa}_2) + \underline{g}_2^{*T}(\underline{e}_1 + \widetilde{s}_0^T\underline{\kappa}_1) \\
2\varepsilon_{13} &= \underline{g}_1^{*T}(\underline{e}_3 + \widetilde{s}_0^T\underline{\kappa}_3) + \underline{g}_3^{*T}(\underline{e}_1 + \widetilde{s}_0^T\underline{\kappa}_1) \\
\varepsilon_{22} &= \underline{g}_2^{*T}(\underline{e}_2 + \widetilde{s}_0^T\underline{\kappa}_2) \\
2\varepsilon_{23} &= \underline{g}_2^{*T}(\underline{e}_3 + \widetilde{s}_0^T\underline{\kappa}_3) + \underline{g}_3^{*T}(\underline{e}_2 + \widetilde{s}_0^T\underline{\kappa}_2) \\
\varepsilon_{33} &= \underline{g}_3^{*T}(\underline{e}_3 + \widetilde{s}_0^T\underline{\kappa}_3)
\end{aligned} \tag{3.45}$$

which are exactly the same as Eq. (3.41).

### 3.1.3 Stress Tensor of Cosserat Continuum

In the definition of the Green-Lagrange strain tensor, Eq. (2.234), all components are defined with respected to the initial coordinates $(x_1, x_2, x_3)$. If the energy density defined as the product of stress and strain measures is also related to the undeformed volume, the second Piola-Kirchhoff stress tensor $\sigma$ must be selected as the work-conjugate with the Green-Lagrange strain tensor. The stress tensor $\sigma$ is defined as the tensor producing a traction vector $\underline{t}_\sigma$ applied to the infinitesimal surface area of the magnitude $A_0$ in the undeformed reference configuration of a body

$$\underline{t}_\sigma = A_0 \sigma \bar{n} \tag{3.46}$$

where $\bar{n}$ is the normal vector of the surface. In deformed configuration, the Cauchy stress tensor $\tau$ produces a traction vector $\underline{T}_\tau$ applied to the infinitesimal surface area of the magnitude $A_1$ with unit normal vector $\bar{N}$ for a generic body like

$$\underline{T}_\tau = A_1 \tau \bar{N} \tag{3.47}$$

It can be verified that the traction vectors $\underline{t}_\sigma$ and $\underline{T}_\tau$ are related by the deformation gradient tensor $F$ like

$$\underline{T}_\tau = F\,\underline{t}_\sigma \tag{3.48}$$

and in detail

$$A_1 \tau \bar{N} = F\,A_0 \sigma \bar{n} \tag{3.49}$$

The above expression is nothing but the transformation of traction vector $\underline{t}_\sigma$ from initial reference configuration to the deformed configuration. With the application of Manson's formula, Eq. (2.232), the deformed surface area can be determined from

$$F^T A_1 \bar{N} = \det(F) A_0 \bar{n} \tag{3.50}$$

Substituting above equations into Eq. (3.49), it can be found that the following relations hold

$$\det(F) \tau F^{-T} A_0 \bar{n} = F\,A_0 \sigma \bar{n} \tag{3.51}$$

Rearranging the variables in the above relations will yield

$$(\det(F) F^{-1} \tau F^{-T} - \sigma) A_0 \bar{n} = 0 \tag{3.52}$$

Since this relations must hold for any surface area with a non-zero magnitude $A_0$ and arbitrary normal vector $\bar{n}$, then second Piola-Kirchhoff stress tensor $\sigma$ could be associated with the Cauchy stress tensor $\tau$ in the form of

$$\sigma = \det(F) F^{-1} \tau F^{-T} \tag{3.53}$$

and vice versa

$$\tau = \det(F)^{-1} F\,\sigma\,F^T \tag{3.54}$$

Given the volume of Cosserat continuum with the magnitudes of $V_0$ and $V_1$ in reference configuration and deformed configuration, respectively, the strain energy can be measured in its reference configuration as

$$2\mathcal{U}_e = \int\limits_{V_0} \underline{\varepsilon}^T \underline{\sigma} d\,V_0 \tag{3.55}$$

or in its deformed configuration, equivalently

$$2\mathcal{U}_e = \int\limits_{V_1} \underline{\varepsilon}^{*T} \underline{\tau} d\,V_1 \tag{3.56}$$

The above two formulas are mathematically identical each other, but have different physical meanings. The second Piola-Kirchhoff stress tensor $\sigma$ is a nominal stress tensor with little physical meaning, while the Cauchy stress tensor $\tau$ is usually called real stress tensor. The identity of strain energies measured in the initial reference configuration and current deformed configuration, respectively, can be proved by the transformations, Eqs. (2.241) and (3.54). At first, the transformations corresponding to the second-order tensors should be recasted into the vector-dyadic forms

$$\underline{\varepsilon} = F_d^T \underline{\varepsilon}^*, \quad \underline{\tau} = \det(F)^{-1} F_d \underline{\sigma} \tag{3.57}$$

where $F_d$, a $6 \times 6$ matrix, is constructed by the contributions of deformation gradient tensor $F$ and its transpose $F^T$, in above tensor transformations. Starting from the definitions of strain energy related to its reference configuration, it is readily to verify

$$2\mathcal{U}_e = \int\limits_{V_0} \underline{\varepsilon}^T \underline{\sigma}\, dV_0 = \int\limits_{V_0} \underline{\varepsilon}^{*T} F_d F_d^{-1} \underline{\tau} \det(F)\, dV_0 = \int\limits_{V_1} \underline{\varepsilon}^{*T} \underline{\tau}\, dV_1 \tag{3.58}$$

where the increments $dV_0$ and $dV_1$ of characteristic volume are correlated by $dV_1 = \det(F) dV_0$. Finally, it is clarified the Green-Lagrange strain tensor and its work-conjugate, the second Piola-Kirchhoff stress tensor, will be applied in the book, to integral the strain energy in the initial reference configuration.

### *3.1.4 Constitutive Laws for Cosserat Continuum*

For general anisotropic nonlinear elastic materials, that is considered to be ideal materials, the components of second Piola-Kirchhoff stress tensor are correlated to the components of Green-Lagrange strain tensor by the following constitutive equations in its vector-dyadic form like

$$\underline{\sigma} = \mathcal{D} \underline{\varepsilon} \tag{3.59}$$

where $\mathcal{D}$ is the constitutive matrix and usually a full matrix

$$\mathcal{D} = \begin{bmatrix} d_{11}, d_{12}, d_{13}, d_{14}, d_{15}, d_{16} \\ d_{21}, d_{22}, d_{23}, d_{24}, d_{25}, d_{26} \\ d_{31}, d_{32}, d_{33}, d_{34}, d_{35}, d_{36} \\ d_{41}, d_{42}, d_{43}, d_{44}, d_{45}, d_{46} \\ d_{51}, d_{52}, d_{53}, d_{54}, d_{55}, d_{56} \\ d_{61}, d_{62}, d_{63}, d_{64}, d_{65}, d_{66} \end{bmatrix} \tag{3.60}$$

When the Green-Lagrange strain tensor is measured in the curvilinear basis, the associated tensor transformation, $\widehat{\varepsilon} = J^T \varepsilon J$, can be recasted to its vector-dyadic form

$$\underline{\widehat{\varepsilon}} = J_d^T \underline{\varepsilon}, \quad \underline{\varepsilon} = J_d^{-T} \underline{\widehat{\varepsilon}} \tag{3.61}$$

where the Jacobian, $J_d$, a $6 \times 6$ matrix, is constructed by the contributions of Jacobian matrix $J$ and its transpose $J^T$, in above tensor transformation. Given the volume of a Cosserat continuum with the magnitude of $V_0$, the strain energy $\mathcal{U}_e$ can be measured from the integral of stain and stress tensor product in to a closed-form relation like

$$2\mathcal{U}_e = \int_{V_0} \widehat{\underline{\varepsilon}}^T \widehat{\mathcal{D}} \widehat{\underline{\varepsilon}} \sqrt{g} d\alpha_1 d\alpha_2 d\alpha_3 \tag{3.62}$$

where the constitutive matrix $\mathcal{D}$ is transformed into the curvilinear coordinate system

$$\widehat{\mathcal{D}} = J_d^{-1} \mathcal{D} J_d^{-T} \tag{3.63}$$

If selected the Lagrangian representation of the material coordinates, there will be no difference between $\widehat{\mathcal{D}}$ and $\mathcal{D}$. The rest of the book will not distinguish between $\widehat{\mathcal{D}}$ and $\mathcal{D}$ to simplify the description of Cosserat continuum theory.

For orthotropic materials [25, 26], the material properties are symmetric about all three coordinate planes, and the explicit expression of the constitutive matrix $\mathcal{D}$ can be given as follows

$$\mathcal{D} = \begin{bmatrix} (1-\nu_{23}\nu_{32})\lambda_1, & 0, & 0, & (\nu_{12}+\nu_{13}\nu_{32})\lambda_2, & 0, & (\nu_{13}+\nu_{12}\nu_{23})\lambda_3 \\ 0, & G_{12}, & 0, & 0, & 0, & 0 \\ 0, & 0, & G_{13}, & 0, & 0, & 0 \\ (\nu_{12}+\nu_{13}\nu_{32})\lambda_2, & 0, & 0, & (1-\nu_{31}\nu_{13})\lambda_2, & 0, & (\nu_{23}+\nu_{21}\nu_{13})\lambda_3 \\ 0, & 0, & 0, & 0, & G_{23}, & 0 \\ (\nu_{13}+\nu_{12}\nu_{23})\lambda_3, & 0, & 0, & (\nu_{23}+\nu_{21}\nu_{13})\lambda_3, & 0, & (1-\nu_{21}\nu_{12})\lambda_3 \end{bmatrix} \tag{3.64}$$

where $\nu_{12}, \nu_{13}, \nu_{23}$ and $\nu_{21}, \nu_{31}, \nu_{32}$ are two set of Poisson's ratios. Usually only $\nu_{12}, \nu_{13}, \nu_{23}$ are given and the other Poisson's ratios are calculated as

$$\nu_{21} = \nu_{12}\frac{E_2}{E_1}, \quad \nu_{31} = \nu_{13}\frac{E_3}{E_1}, \quad \nu_{32} = \nu_{23}\frac{E_3}{E_2} \tag{3.65}$$

and $E_1, E_2, E_3$ are Young's moduli in three directions, meanwhile $G_{12}, G_{13}, G_{23}$ are shear moduli in three directions. The parameters $\lambda_1, \lambda_2, \lambda_3$ are computed from

$$\lambda_1 = E_1\lambda, \quad \lambda_2 = E_2\lambda, \quad \lambda_3 = E_3\lambda \tag{3.66}$$

with

$$\lambda = 1/(1 - \nu_{12}\nu_{21} - \nu_{23}\nu_{32} - \nu_{31}\nu_{13} - 2\nu_{21}\nu_{13}\nu_{32}) \tag{3.67}$$

For isotropic materials, Young's moduli $E_1, E_2, E_3$ are simplified to only one effective Young's modulus $E$, and shear moduli $G_{12}, G_{13}, G_{23}$ in three directions are replaced by an effective shear modulus $G$. The effective Poisson's ratio $\nu$ is introduced to relate Young's modulus $E$ and shear modulus $G$ as

$$G = \frac{E}{2(1+\nu)} \tag{3.68}$$

The constitutive relations can be expressed in terms of

$$\mathcal{D} = \begin{bmatrix} \frac{E(1-\nu)}{(1+\nu)(1-2\nu)}, & 0, & 0, & \frac{E\nu}{(1+\nu)(1-2\nu)}, & 0, & \frac{E\nu}{(1+\nu)(1-2\nu)} \\ 0, & G, & 0, & 0, & 0, & 0 \\ 0, & 0, & G, & 0, & 0, & 0 \\ \frac{E\nu}{(1+\nu)(1-2\nu)}, & 0, & 0, & \frac{E(1-\nu)}{(1+\nu)(1-2\nu)}, & 0, & \frac{E\nu}{(1+\nu)(1-2\nu)} \\ 0, & 0, & 0, & 0, & G, & 0 \\ \frac{E\nu}{(1+\nu)(1-2\nu)}, & 0, & 0, & \frac{E\nu}{(1+\nu)(1-2\nu)}, & 0, & \frac{E(1-\nu)}{(1+\nu)(1-2\nu)} \end{bmatrix} \tag{3.69}$$

It is necessary to emphasis that the constitutive relation, also the generalized Hooke's law, Eq. (3.59), is usually determined by the experiments. Although it is always used as the constitutive law of Cauchy continuum, we can prove that constitutive matrix $\mathcal{D}$ is also the original source of the constitutive relations of Cosserat continuum. For the easy description of Cosserat constitutive relations, the Green-Lagrange strain tensor in its vector-dyadic form needs to be decomposed into

$$\widehat{\underline{\varepsilon}} = \begin{vmatrix} \varepsilon_{11} \\ 2\varepsilon_{12} \\ 2\varepsilon_{13} \\ \varepsilon_{22} \\ 2\varepsilon_{23} \\ \varepsilon_{33} \end{vmatrix} = \begin{vmatrix} \varepsilon_{11} \\ \varepsilon_{21} \\ \varepsilon_{31} \\ 0 \\ 0 \\ 0 \end{vmatrix} + \begin{vmatrix} 0 \\ \varepsilon_{12} \\ 0 \\ \varepsilon_{22} \\ \varepsilon_{32} \\ 0 \end{vmatrix} + \begin{vmatrix} 0 \\ 0 \\ \varepsilon_{13} \\ 0 \\ \varepsilon_{23} \\ \varepsilon_{33} \end{vmatrix} = \widehat{\underline{\varepsilon}}_1 + \widehat{\underline{\varepsilon}}_2 + \widehat{\underline{\varepsilon}}_3 \tag{3.70}$$

It is noted that the symbols $\varepsilon_{12}$, $\varepsilon_{21}$, $\varepsilon_{23}$, $\varepsilon_{32}$, and $\varepsilon_{31}$, $\varepsilon_{13}$ do not represent the components of Green-Lagrange strain tensor, but are defined in the following relations

$$2\varepsilon_{12} = \varepsilon_{21} + \varepsilon_{12}, \quad 2\varepsilon_{13} = \varepsilon_{31} + \varepsilon_{13}, \quad 2\varepsilon_{23} = \varepsilon_{32} + \varepsilon_{23} \tag{3.71}$$

After filtered out the zero items, the specific vectorial expressions are reduced to

$$\begin{aligned} \underline{\varepsilon}_1 &= \begin{vmatrix} \varepsilon_{11} \\ \varepsilon_{21} \\ \varepsilon_{31} \end{vmatrix} = \begin{vmatrix} (\underline{x}_{0,1}^T R_0 + \underline{\kappa}_1^{0T} \tilde{s}_0 + \underline{s}_{0,1}^T)(\underline{e}_1 + \tilde{s}_0^T \underline{\kappa}_1) \\ (\underline{x}_{0,2}^T R_0 + \underline{\kappa}_2^{0T} \tilde{s}_0 + \underline{s}_{0,2}^T)(\underline{e}_1 + \tilde{s}_0^T \underline{\kappa}_1) \\ (\underline{x}_{0,3}^T R_0 + \underline{\kappa}_3^{0T} \tilde{s}_0 + \underline{s}_{0,3}^T)(\underline{e}_1 + \tilde{s}_0^T \underline{\kappa}_1) \end{vmatrix} \\ \underline{\varepsilon}_2 &= \begin{vmatrix} \varepsilon_{12} \\ \varepsilon_{22} \\ \varepsilon_{32} \end{vmatrix} = \begin{vmatrix} (\underline{x}_{0,1}^T R_0 + \underline{\kappa}_1^{0T} \tilde{s}_0 + \underline{s}_{0,1}^T)(\underline{e}_2 + \tilde{s}_0^T \underline{\kappa}_2) \\ (\underline{x}_{0,2}^T R_0 + \underline{\kappa}_2^{0T} \tilde{s}_0 + \underline{s}_{0,2}^T)(\underline{e}_2 + \tilde{s}_0^T \underline{\kappa}_2) \\ (\underline{x}_{0,3}^T R_0 + \underline{\kappa}_3^{0T} \tilde{s}_0 + \underline{s}_{0,3}^T)(\underline{e}_2 + \tilde{s}_0^T \underline{\kappa}_2) \end{vmatrix} \\ \underline{\varepsilon}_3 &= \begin{vmatrix} \varepsilon_{13} \\ \varepsilon_{23} \\ \varepsilon_{33} \end{vmatrix} = \begin{vmatrix} (\underline{x}_{0,1}^T R_0 + \underline{\kappa}_1^{0T} \tilde{s}_0 + \underline{s}_{0,1}^T)(\underline{e}_3 + \tilde{s}_0^T \underline{\kappa}_3) \\ (\underline{x}_{0,2}^T R_0 + \underline{\kappa}_2^{0T} \tilde{s}_0 + \underline{s}_{0,2}^T)(\underline{e}_3 + \tilde{s}_0^T \underline{\kappa}_3) \\ (\underline{x}_{0,3}^T R_0 + \underline{\kappa}_3^{0T} \tilde{s}_0 + \underline{s}_{0,3}^T)(\underline{e}_3 + \tilde{s}_0^T \underline{\kappa}_3) \end{vmatrix} \end{aligned} \tag{3.72}$$

The above formulas could be assembled into

$$\begin{bmatrix} \varepsilon_{11}, \varepsilon_{12}, \varepsilon_{13} \\ \varepsilon_{21}, \varepsilon_{22}, \varepsilon_{23} \\ \varepsilon_{31}, \varepsilon_{32}, \varepsilon_{33} \end{bmatrix} = \begin{bmatrix} \underline{x}_{0,1}^T R_0 + \underline{\kappa}_1^{0T} \widetilde{s}_0 + \underline{s}_{0,1}^T \\ \underline{x}_{0,2}^T R_0 + \underline{\kappa}_2^{0T} \widetilde{s}_0 + \underline{s}_{0,2}^T \\ \underline{x}_{0,3}^T R_0 + \underline{\kappa}_3^{0T} \widetilde{s}_0 + \underline{s}_{0,3}^T \end{bmatrix} [\, I, \ \widetilde{s}_0^T \,] \begin{bmatrix} \underline{e}_1, \ \underline{e}_2, \ \underline{e}_3 \\ \underline{\kappa}_1, \ \underline{\kappa}_2, \ \underline{\kappa}_3 \end{bmatrix} \tag{3.73}$$

and the associated compact form is found to be

$$\lfloor \underline{\varepsilon}_1, \ \underline{\varepsilon}_2, \ \underline{\varepsilon}_3 \rfloor = \mathcal{S}_0^T \lfloor \underline{\epsilon}_1, \ \underline{\epsilon}_2, \ \underline{\epsilon}_3 \rfloor \tag{3.74}$$

with the introduced symbol

$$\mathcal{S}_0^T = \mathcal{B}_g^T \, [\, I, \ \widetilde{s}_0^T \,] \tag{3.75}$$

The strain energy $\mathcal{U}_e$ can be represented by using the Green-Lagrange strain vectors $\widehat{\underline{\varepsilon}}_1$, $\widehat{\underline{\varepsilon}}_2$, and $\widehat{\underline{\varepsilon}}_3$ like

$$\begin{aligned} 2\mathcal{U}_e &= \int_{V_0} (\widehat{\underline{\varepsilon}}_1^T + \widehat{\underline{\varepsilon}}_2^T + \widehat{\underline{\varepsilon}}_3^T) \mathcal{D} (\widehat{\underline{\varepsilon}}_1 + \widehat{\underline{\varepsilon}}_2 + \widehat{\underline{\varepsilon}}_3) dV_0 \\ &= \int_{V_0} \left[ \widehat{\underline{\varepsilon}}_1^T \mathcal{D} \widehat{\underline{\varepsilon}}_1 + \widehat{\underline{\varepsilon}}_2^T \mathcal{D} \widehat{\underline{\varepsilon}}_2 + \widehat{\underline{\varepsilon}}_3^T \mathcal{D} \widehat{\underline{\varepsilon}}_3 + 2 (\widehat{\underline{\varepsilon}}_2^T \mathcal{D} \widehat{\underline{\varepsilon}}_1 + \widehat{\underline{\varepsilon}}_3^T \mathcal{D} \widehat{\underline{\varepsilon}}_1 + \widehat{\underline{\varepsilon}}_3^T \mathcal{D} \widehat{\underline{\varepsilon}}_2) \right] dV_0 \end{aligned} \tag{3.76}$$

when the zero components of Green-Lagrange strain vectors $\widehat{\underline{\varepsilon}}_1$, $\widehat{\underline{\varepsilon}}_2$, and $\widehat{\underline{\varepsilon}}_3$ are filtered out from above formula, the strain energy will reduce to

$$\begin{aligned} 2\mathcal{U}_e = \int_{V_0} & [\underline{\varepsilon}_1^T \mathcal{D}_{11} \underline{\varepsilon}_1 + \underline{\varepsilon}_2^T \mathcal{D}_{22} \underline{\varepsilon}_2 + \underline{\varepsilon}_3^T \mathcal{D}_{33} \underline{\varepsilon}_3 \\ & + 2 (\underline{\varepsilon}_2^T \mathcal{D}_{21} \underline{\varepsilon}_1 + \underline{\varepsilon}_3^T \mathcal{D}_{31} \underline{\varepsilon}_1 + \underline{\varepsilon}_3^T \mathcal{D}_{32} \underline{\varepsilon}_2)] \, dV_0 \end{aligned} \tag{3.77}$$

where the following constitutive matrix blocks are introduced

$$\mathcal{D}_{11} = \begin{bmatrix} d_{11}, d_{12}, d_{13} \\ d_{21}, d_{22}, d_{23} \\ d_{31}, d_{32}, d_{33} \end{bmatrix}, \ \mathcal{D}_{22} = \begin{bmatrix} d_{22}, d_{24}, d_{25} \\ d_{42}, d_{44}, d_{45} \\ d_{52}, d_{54}, d_{55} \end{bmatrix}, \ \mathcal{D}_{33} = \begin{bmatrix} d_{33}, d_{35}, d_{36} \\ d_{53}, d_{55}, d_{56} \\ d_{63}, d_{65}, d_{66} \end{bmatrix} \tag{3.78}$$

$$\mathcal{D}_{21} = \begin{bmatrix} d_{21}, d_{22}, d_{23} \\ d_{41}, d_{42}, d_{43} \\ d_{51}, d_{52}, d_{53} \end{bmatrix}, \ \mathcal{D}_{31} = \begin{bmatrix} d_{31}, d_{32}, d_{33} \\ d_{51}, d_{52}, d_{53} \\ d_{61}, d_{62}, d_{63} \end{bmatrix}, \ \mathcal{D}_{32} = \begin{bmatrix} d_{32}, d_{34}, d_{35} \\ d_{52}, d_{54}, d_{55} \\ d_{62}, d_{64}, d_{65} \end{bmatrix} \tag{3.79}$$

$$\mathcal{D}_{12} = \begin{bmatrix} d_{12}, d_{14}, d_{15} \\ d_{22}, d_{24}, d_{25} \\ d_{32}, d_{34}, d_{35} \end{bmatrix}, \ \mathcal{D}_{13} = \begin{bmatrix} d_{13}, d_{15}, d_{16} \\ d_{23}, d_{25}, d_{26} \\ d_{33}, d_{35}, d_{36} \end{bmatrix}, \ \mathcal{D}_{23} = \begin{bmatrix} d_{23}, d_{25}, d_{26} \\ d_{43}, d_{45}, d_{46} \\ d_{53}, d_{55}, d_{56} \end{bmatrix} \tag{3.80}$$

Since the constitutive matrix is a symmetric matrix, it is readily to verify that $\mathcal{D}_{12} = \mathcal{D}_{21}^T$, $\mathcal{D}_{13} = \mathcal{D}_{31}^T$ and $\mathcal{D}_{23} = \mathcal{D}_{32}^T$. Referring the relations of Green-Lagrange strain and Lagrangian strain measures, Eq. (3.74), the above strain energy is recasted to

$$2\mathcal{U}_e = \int_{V_0} [\underline{\epsilon}_1^T \mathcal{C}_{11} \underline{\epsilon}_1 + \underline{\epsilon}_2^T \mathcal{C}_{22} \underline{\epsilon}_2 + \underline{\epsilon}_3^T \mathcal{C}_{33} \underline{\epsilon}_3 + 2(\underline{\epsilon}_2^T \mathcal{C}_{21} \underline{\epsilon}_1 + \underline{\epsilon}_3^T \mathcal{C}_{31} \underline{\epsilon}_1 + \underline{\epsilon}_3^T \mathcal{C}_{32} \underline{\epsilon}_2)] \, dV_0 \tag{3.81}$$

where the constitutive matrices of Cosserat continuum are defined

$$\begin{aligned} \mathcal{C}_{11} &= \mathcal{S}_0 \mathcal{D}_{11} \mathcal{S}_0^T, \quad \mathcal{C}_{22} = \mathcal{S}_0 \mathcal{D}_{22} \mathcal{S}_0^T, \quad \mathcal{C}_{33} = \mathcal{S}_0 \mathcal{D}_{33} \mathcal{S}_0^T \\ \mathcal{C}_{21} &= \mathcal{S}_0 \mathcal{D}_{21} \mathcal{S}_0^T, \quad \mathcal{C}_{31} = \mathcal{S}_0 \mathcal{D}_{31} \mathcal{S}_0^T, \quad \mathcal{C}_{32} = \mathcal{S}_0 \mathcal{D}_{32} \mathcal{S}_0^T \end{aligned} \tag{3.82}$$

and the symmetric properties of constitutive matrices are still hold, such that $\mathcal{C}_{21} = \mathcal{C}_{12}, \mathcal{C}_{31} = \mathcal{C}_{13}$ and $\mathcal{C}_{32} = \mathcal{C}_{23}$. Apparently, the constitutive matrices of Cosserat continuum are determined by the characteristic length $s_0$ contained in the matrix $\mathcal{S}_0$ defined by Eq. (3.75). As mentioned before, the characteristic length $s_0$ is required to be initialized by user for a Cosserat continuum. An alternative approach to predict the constitutive matrices is by using the averaged quantities over the characteristic volume $V_c$, such that

$$\begin{aligned} \bar{\mathcal{C}}_{11} &= \frac{\int_{V_c} \mathcal{C}_{11} dV_c}{V_c}, \quad \bar{\mathcal{C}}_{22} = \frac{\int_{V_c} \mathcal{C}_{22} dV_c}{V_c}, \quad \bar{\mathcal{C}}_{33} = \frac{\int_{V_c} \mathcal{C}_{33} dV_c}{V_c} \\ \bar{\mathcal{C}}_{21} &= \frac{\int_{V_c} \mathcal{C}_{21} dV_c}{V_c}, \quad \bar{\mathcal{C}}_{31} = \frac{\int_{V_c} \mathcal{C}_{31} dV_c}{V_c}, \quad \bar{\mathcal{C}}_{32} = \frac{\int_{V_c} \mathcal{C}_{32} dV_c}{V_c} \end{aligned} \tag{3.83}$$

Widely investigation shows that the decomposition described above is seldom found in the published articles. The reasons why such kind of decomposition performed are addressed as follows. For a linear isotropic Cosserat elastic solid, the classical approaches [19, 26] must introduce four additional elasticity moduli in addition to the Lamé constants, and define the associated characteristic lengths. These moduli combined with the characteristic lengths afford a complete description of the constitutive tensor [17], which leads to a complicated expressions. The current implementation decomposes the constitutive matrix $\mathcal{D}$ for orthogonal material into sub-matrices. The product of the decomposed sub-matrices with the skew-symmetric tensor $\widetilde{s}_0$ of characteristic vector produces the explicit expressions of the constitutive tensors, Eq. (3.82). The new approach simplifies the complexity of the original problem. Especially, when reducing the original three-dimensional Cosserat problems to two-dimensional or one-dimensional problems, the physical significance of explicit expressions becomes clear.

### 3.1.5 Variation of Strain Energy of Cosserat Continuum

The second Piola-Kirchhoff stress and Green-Lagrange strain measures are selected so that their product gives an accurate energy density that is related to the undeformed volume of reference configuration. Referring the expressions of strain energy, Eq. (3.81), the strain energy density for a Cosserat continuum is given as

$$\mathcal{U}_\rho = \frac{1}{2}(\underline{\epsilon}_1^T C_{11}\underline{\epsilon}_1 + \underline{\epsilon}_2^T C_{22}\underline{\epsilon}_2 + \underline{\epsilon}_3^T C_{33}\underline{\epsilon}_3) + \underline{\epsilon}_2^T C_{21}\underline{\epsilon}_1 + \underline{\epsilon}_3^T C_{31}\underline{\epsilon}_1 + \underline{\epsilon}_3^T C_{32}\underline{\epsilon}_2 \quad (3.84)$$

The variation of $\mathcal{U}_\rho$ is readily verified to be

$$\begin{aligned}\delta\mathcal{U}_\rho = &\delta\underline{\epsilon}_1^T C_{11}\underline{\epsilon}_1 + \delta\underline{\epsilon}_2^T C_{22}\underline{\epsilon}_2 + \delta\underline{\epsilon}_3^T C_{33}\underline{\epsilon}_3 \\ &+\delta\underline{\epsilon}_2^T C_{21}\underline{\epsilon}_1 + \delta\underline{\epsilon}_3^T C_{31}\underline{\epsilon}_1 + \delta\underline{\epsilon}_3^T C_{32}\underline{\epsilon}_2 \\ &+\delta\underline{\epsilon}_1^T C_{12}\underline{\epsilon}_2 + \delta\underline{\epsilon}_1^T C_{13}\underline{\epsilon}_3 + \delta\underline{\epsilon}_2^T C_{23}\underline{\epsilon}_3\end{aligned} \quad (3.85)$$

After merging the similar terms at the right-hand-side of above equations, the variation of strain energy density can be rearranged into the compact form

$$\delta\mathcal{U}_\rho = \delta\underline{\epsilon}_1^T \underline{F}_{\epsilon\epsilon}^{11} + \delta\underline{\epsilon}_2^T \underline{F}_{\epsilon\epsilon}^{22} + \delta\underline{\epsilon}_3^T \underline{F}_{\epsilon\epsilon}^{33} \quad (3.86)$$

where the stress measures of a Cosserat continuum are defined in the vectorial formulas like

$$\begin{aligned}\underline{F}_{\epsilon\epsilon}^{11} &= C_{11}\underline{\epsilon}_1 + C_{12}\underline{\epsilon}_2 + C_{13}\underline{\epsilon}_3 \\ \underline{F}_{\epsilon\epsilon}^{22} &= C_{21}\underline{\epsilon}_1 + C_{22}\underline{\epsilon}_2 + C_{23}\underline{\epsilon}_3 \\ \underline{F}_{\epsilon\epsilon}^{33} &= C_{31}\underline{\epsilon}_1 + C_{32}\underline{\epsilon}_2 + C_{33}\underline{\epsilon}_3\end{aligned} \quad (3.87)$$

Finally, the variation of strain energy per characteristic volume can be represented by

$$\delta\mathcal{U}_e = \int_{V_0} (\delta\underline{\epsilon}_1^T \underline{F}_{\epsilon\epsilon}^{11} + \delta\underline{\epsilon}_2^T \underline{F}_{\epsilon\epsilon}^{22} + \delta\underline{\epsilon}_3^T \underline{F}_{\epsilon\epsilon}^{33}) dV_0 \quad (3.88)$$

### 3.1.6 Virtual Work of External Forces for Cosserat Continuum

The principle of virtual work is applied to express the Cosserat continuum mechanics problem in current implementation. The virtual work done by a corresponding virtual displacement operating on the external forces is evaluated in this section. According to Eq. (3.17), the virtual displacement of an arbitrary particle in deformed configuration is obtained by

$$\delta\underline{P} = \delta(\underline{X}_0 + R_1\underline{s}_0) = \delta(\underline{x}_0 + \underline{u}_s + R_1\underline{s}_0) \quad (3.89)$$

where $\underline{u}_s$ is the displacement of reference particle **P** and $R_1$ the total rotation tensor that describes the orientation of local vector $\underline{s}_0$ in current deformed configuration. It will be readily verified the virtual displacement $\delta \underline{P}$, also the position variation, is determined to be

$$\delta \underline{P} = \delta \underline{u}_s + \delta R_1 \underline{s}_0 \tag{3.90}$$

After resolved the above equations into the body attached frame $\mathcal{F}_1$, the variation of position vector will relate to the virtual displacements of reference particle like

$$R_1^T \delta \underline{P} = R_1^T \delta \underline{u}_s + \tilde{s}_0^T \delta \underline{\psi} \tag{3.91}$$

where the virtual rotations are defined, $\widetilde{\delta \psi} = R_1^T \delta R_1$. In current implementation, two types of external forces are considered, one is body force $\underline{g}$, the other surface pressure $\underline{p}_s$, both measured in body attached frame $\mathcal{F}_1$. Then, the virtual work of the external forces can be computed from

$$\delta \mathcal{W}_e = \int_{V_0} \delta \underline{P}^T R_1 \underline{g}\, dV_0 + \int_{S_0} \delta \underline{P}^T R_1 \underline{p}_s\, dS_0 \tag{3.92}$$

where $S_0$ represents the area magnitude of a Cosserat continuum where the surface pressure is applied to. Referring the formula of position variation, Eq. (3.91), the virtual work can be expressed in the following detailed formula as

$$\begin{aligned} \delta \mathcal{W}_e &= \int_{V_0} (R_1^T \delta \underline{u}_s + \tilde{s}_0^T \delta \underline{\psi})^T \underline{g}\, dV_0 + \int_{S_0} (R_1^T \delta \underline{u}_s + \tilde{s}_0^T \delta \underline{\psi})^T \underline{p}_s\, dS_0 \\ &= \int_{V_0} \delta \underline{u}_s^T R_1 \underline{g}\, dV_0 + \int_{S_0} \delta \underline{u}_s^T R_1 \underline{p}_s\, dS_0 + \int_{V_0} \delta \underline{\psi}^T \tilde{s}_0 \underline{g}\, dV_0 + \int_{S_0} \delta \underline{\psi}^T \tilde{s}_0 \underline{p}_s\, dS_0 \end{aligned} \tag{3.93}$$

For the easy description of virtual work, the external force and torque vectors are averaged at first by the characteristic volume $V_c$ and the characteristic area $S_c$ like

$$\underline{F} = \frac{\int_{V_c} \underline{g}\, dV_c + \int_{S_c} \underline{p}_s\, dS_c}{V_c}, \quad \underline{T} = \frac{\int_{V_c} \tilde{s}_0 \underline{g}\, dV_c + \int_{S_c} \tilde{s}_0 \underline{p}_s\, dS_c}{V_c} \tag{3.94}$$

then the virtual work can be approximated to be

$$\delta \mathcal{W}_e = \int_{V_0} (\delta \underline{u}_s^T R_1 \underline{F} + \delta \underline{\psi}^T \underline{T})\, dV_0 \tag{3.95}$$

or in the motion formalism like

$$\delta \mathcal{W}_e = \int_{V_0} \underline{\delta \mathcal{U}}^T \underline{F}_{\epsilon\alpha} dV_0 \tag{3.96}$$

where the following symbols are introduced

$$\underline{F}_{\epsilon\alpha} = \begin{vmatrix} \underline{F} \\ \underline{T} \end{vmatrix}, \quad \underline{\delta \mathcal{U}} = \begin{vmatrix} R_1^T \delta \underline{u}_s \\ \delta \underline{\psi} \end{vmatrix} \tag{3.97}$$

### 3.1.7 Governing Equations in Motion Formalism

The principle of virtual work states

$$\delta \mathcal{U}_e - \delta \mathcal{W}_e = 0 \tag{3.98}$$

which answers the virtual work done by the infinitesimal strain operating on the stress measures equals to the virtual work done by a corresponding virtual displacement operating on the external forces. Introducing Eqs. (3.88) and (3.95) leads to

$$\int_{V_0} (\delta \underline{\epsilon}_1^T \underline{F}_{\epsilon\epsilon}^{11} + \delta \underline{\epsilon}_2^T \underline{F}_{\epsilon\epsilon}^{22} + \delta \underline{\epsilon}_3^T \underline{F}_{\epsilon\epsilon}^{33} - \underline{\delta \mathcal{U}}^T \underline{F}_{\epsilon\alpha}) dV_0 = 0 \tag{3.99}$$

In view of Eq. (2.161), the variations of Lagrangian strains can be expressed in terms of

$$\delta \underline{\epsilon}_1 = \underline{\delta \mathcal{U}}_{,1} + \widetilde{\mathcal{K}}_1^1 \underline{\delta \mathcal{U}}, \quad \delta \underline{\epsilon}_2 = \underline{\delta \mathcal{U}}_{,2} + \widetilde{\mathcal{K}}_2^1 \underline{\delta \mathcal{U}}, \quad \delta \underline{\epsilon}_3 = \underline{\delta \mathcal{U}}_{,3} + \widetilde{\mathcal{K}}_3^1 \underline{\delta \mathcal{U}} \tag{3.100}$$

and then the principle of virtual work is recasted to

$$\begin{aligned} \int_{V_0} [(\underline{\delta \mathcal{U}}_{,1} + \widetilde{\mathcal{K}}_1^1 \underline{\delta \mathcal{U}})^T \underline{F}_{\epsilon\epsilon}^{11} + (\underline{\delta \mathcal{U}}_{,2} + \widetilde{\mathcal{K}}_2^1 \underline{\delta \mathcal{U}})^T \underline{F}_{\epsilon\epsilon}^{22} \\ + (\underline{\delta \mathcal{U}}_{,3} + \widetilde{\mathcal{K}}_3^1 \underline{\delta \mathcal{U}})^T \underline{F}_{\epsilon\epsilon}^{33} - \underline{\delta \mathcal{U}}^T \underline{F}_{\epsilon\alpha}] dV_0 = 0 \end{aligned} \tag{3.101}$$

With the application of finite element discretization, the variations of motion vectors and its derivatives can be interpolated directly by using the shape functions $\mathcal{H}_\epsilon$ to get

$$\underline{\delta \mathcal{U}} = \mathcal{H}_\epsilon \underline{\delta \mathcal{U}}_n, \ \underline{\delta \mathcal{U}}_{,i} = \mathcal{H}_{\epsilon,i} \underline{\delta \mathcal{U}}_n \tag{3.102}$$

With the aid of the isoparameteric interpolations, the weak formula becomes

$$\begin{aligned} \underline{\delta \mathcal{U}}_n^T \int_{V_0} [(\mathcal{H}_{\epsilon,1}^T + \mathcal{H}_\epsilon^T \widetilde{\mathcal{K}}_1^{1T}) \underline{F}_{\epsilon\epsilon}^{11} + (\mathcal{H}_{\epsilon,2}^T + \mathcal{H}_\epsilon^T \widetilde{\mathcal{K}}_2^{1T}) \underline{F}_{\epsilon\epsilon}^{22} \\ + (\mathcal{H}_{\epsilon,3}^T + \mathcal{H}_\epsilon^T \widetilde{\mathcal{K}}_3^{1T}) \underline{F}_{\epsilon\epsilon}^{33} - \mathcal{H}_\epsilon^T \underline{F}_{\epsilon\alpha}] dV_0 = 0 \end{aligned} \tag{3.103}$$

From which, the governing equations are obtained to be

$$\int_{V_0} [(\mathcal{H}_{\epsilon,1}^T + \mathcal{H}_\epsilon^T \widetilde{\mathcal{K}}_1^{1T})\underline{F}_{\epsilon\epsilon}^{11} + (\mathcal{H}_{\epsilon,2}^T + \mathcal{H}_\epsilon^T \widetilde{\mathcal{K}}_2^{1T})\underline{F}_{\epsilon\epsilon}^{22}$$
$$+(\mathcal{H}_{\epsilon,3}^T + \mathcal{H}_\epsilon^T \widetilde{\mathcal{K}}_3^{1T})\underline{F}_{\epsilon\epsilon}^{33} - \mathcal{H}_\epsilon^T \underline{F}_{\epsilon\alpha}]dV_0 = 0 \tag{3.104}$$

After introduced the elastic force vector

$$\underline{\mathcal{F}}_e = \int_{V_0} [(\mathcal{H}_{\epsilon,1}^T + \mathcal{H}_\epsilon^T \widetilde{\mathcal{K}}_1^{1T})\underline{F}_{\epsilon\epsilon}^{11} + (\mathcal{H}_{\epsilon,2}^T + \mathcal{H}_\epsilon^T \widetilde{\mathcal{K}}_2^{1T})\underline{F}_{\epsilon\epsilon}^{22}$$
$$+(\mathcal{H}_{\epsilon,3}^T + \mathcal{H}_\epsilon^T \widetilde{\mathcal{K}}_3^{1T})\underline{F}_{\epsilon\epsilon}^{33}]dV_0 \tag{3.105}$$

and the external force vector

$$\underline{\mathcal{F}}_a = \int_{V_0} \mathcal{H}_\epsilon^T \underline{F}_{\epsilon\alpha} dV_0 \tag{3.106}$$

the governing equations can be described by the following compact formula

$$\underline{\mathcal{F}}_e - \underline{\mathcal{F}}_a = 0 \tag{3.107}$$

### 3.1.8 Kinematic Energy of Cosserat Continuum

Similar to the definitions of position variation $\delta \underline{P}$, the velocity of an arbitrary material particle in a Cosserat continuum can be evaluated by taking time derivatives of position vector $\underline{P}$, defined by Eq. (3.17), to get

$$\dot{\underline{P}} = (\underline{x}_0 + \underline{u}_s + R_1 \underline{s}_0)^{\cdot} = \dot{\underline{u}}_s + \dot{R}_1 \underline{s}_0 \tag{3.108}$$

After resolved the time derivatives into the body attached frame $\mathcal{F}_1$, the particle velocity will relate to the linear and angular velocities of reference particle **P** like

$$R_1^T \dot{\underline{P}} = R_1^T \dot{\underline{u}}_s + \widetilde{s}_0^T \underline{\omega} \tag{3.109}$$

where the angular velocity is defined as $\widetilde{\omega} = R_1^T \dot{R}_1$. If denoting the material density as $\rho$, the kinematic energy of a Cosserat continuum can be computed from the integral related to the initial volume like

$$\mathcal{K}_e = \int_{V_0} \frac{1}{2}\rho (R_1^T \dot{\underline{P}})^T R_1^T \dot{\underline{P}} dV_0$$
$$= \int_{V_0} \frac{1}{2} \lfloor \dot{\underline{u}}_s^T R_1, \ \underline{\omega}^T \rfloor \begin{bmatrix} \rho I, & \rho \widetilde{s}_0^T \\ \rho \widetilde{s}_0, & \rho \widetilde{s}_0 \widetilde{s}_0^T \end{bmatrix} \left| \begin{matrix} R_1^T \dot{\underline{u}}_s \\ \underline{\omega} \end{matrix} \right| dV_0 \tag{3.110}$$

The motion formalism of kinematic energy is readily obtained to be

$$\mathcal{K}_e = \int_{V_0} \frac{1}{2} \underline{\mathcal{V}}^T \mathcal{M}_{\epsilon\epsilon} \underline{\mathcal{V}} dV_0 \tag{3.111}$$

with the aid of following definitions

$$\mathcal{M}_{\epsilon\epsilon} = \begin{bmatrix} \rho I, & \rho \tilde{s}_0^T \\ \rho \tilde{s}_0, & \rho \tilde{s}_0 \tilde{s}_0^T \end{bmatrix}, \quad \underline{\mathcal{V}} = \begin{vmatrix} R_1^T \underline{\dot{u}}_s \\ \underline{\omega} \end{vmatrix} \tag{3.112}$$

Once again, the mass matrix $\mathcal{M}_{\epsilon\epsilon}$ could be averaged by using the characteristic volume $V_c$ like

$$\bar{\mathcal{M}}_{\epsilon\epsilon} = \frac{\int_{V_c} \mathcal{M}_{\epsilon\epsilon} dV_c}{V_c} \tag{3.113}$$

and the averaged quantities $\bar{\mathcal{M}}_{\epsilon\epsilon}$ will replace the mass matrix when predicting the kinematic energy of a Cosserat continuum. Referring Eq. (3.111), the variation of kinematic energy is readily obtained to be

$$\delta\mathcal{K}_e = \int_{V_0} \delta\underline{\mathcal{V}}^T \mathcal{M}_{\epsilon\epsilon} \underline{\mathcal{V}} dV_0 \tag{3.114}$$

In view of Eq. (2.160), the virtual dual velocity is related to motion vector $\mathcal{U}$ and its time derivative like

$$\delta\underline{\mathcal{V}} = \underline{\delta\dot{\mathcal{U}}} + \tilde{\mathcal{V}} \underline{\delta\mathcal{U}} \tag{3.115}$$

and the relations are applied to obtain the weak form like

$$\delta\mathcal{K}_e = \int_{V_0} (\underline{\delta\dot{\mathcal{U}}}^T \mathcal{M}_{\epsilon\epsilon} \underline{\mathcal{V}} + \underline{\delta\mathcal{U}}^T \tilde{\mathcal{V}}^T \mathcal{M}_{\epsilon\epsilon} \underline{\mathcal{V}})\, dV_0 \tag{3.116}$$

Integrating the variation of kinematic energy by part in time domain will produce

$$\int_t \delta\mathcal{K}_e dt = \int_t \int_{V_0} (-\underline{\delta\mathcal{U}}^T \mathcal{M}_{\epsilon\epsilon} \underline{\dot{\mathcal{V}}} + \underline{\delta\mathcal{U}}^T \tilde{\mathcal{V}}^T \mathcal{M}_{\epsilon\epsilon} \underline{\mathcal{V}})\, dV_0 dt \tag{3.117}$$

From which the variation of kinematic energy is approximated by

$$\delta\mathcal{K}_e = \int_{V_0} (-\underline{\delta\mathcal{U}}^T \mathcal{M}_{\epsilon\epsilon} \underline{\dot{\mathcal{V}}} + \underline{\delta\mathcal{U}}^T \tilde{\mathcal{V}}^T \mathcal{M}_{\epsilon\epsilon} \underline{\mathcal{V}})\, dV_0 \tag{3.118}$$

With the application of finite element discretization, the velocity and acceleration of reference particle are approximated by the following polynomials of nodal velocities $\underline{\mathcal{V}}_n$ and nodal accelerations $\dot{\underline{\mathcal{V}}}_n$, respectively

$$\underline{\mathcal{V}} = \mathcal{H}_\epsilon \underline{\mathcal{V}}_n, \quad \dot{\underline{\mathcal{V}}} = \mathcal{H}_\epsilon \dot{\underline{\mathcal{V}}}_n \tag{3.119}$$

and the weak form of kinematic energy variation is found to be

$$\delta \mathcal{K}_e = \delta \underline{\mathcal{U}}_n^T \left( - \int_{V_0} \mathcal{H}_\epsilon^T \mathcal{M}_{\epsilon\epsilon} \mathcal{H}_\epsilon d V_0 \dot{\underline{\mathcal{V}}}_n + \int_{V_0} \mathcal{H}_\epsilon^T \widetilde{\mathcal{V}}^T \mathcal{M}_{\epsilon\epsilon} \underline{\mathcal{V}} d V_0 \right) \tag{3.120}$$

After defined the mass matrix and the Coriolis force vector

$$\mathcal{M}_n = \int_{V_0} \mathcal{H}_\epsilon^T \mathcal{M}_{\epsilon\epsilon} \mathcal{H}_\epsilon d V_0, \quad \underline{\mathcal{F}}_g = \int_{V_0} \mathcal{H}_\epsilon^T \widetilde{\mathcal{V}}^T \mathcal{M}_{\epsilon\epsilon} \underline{\mathcal{V}} d V_0 \tag{3.121}$$

the variation of original kinematic energy can be rewritten into the compact formula

$$\delta \mathcal{K}_e = \delta \underline{\mathcal{U}}_n^T (-\mathcal{M}_n \dot{\underline{\mathcal{V}}}_n + \underline{\mathcal{F}}_g) \tag{3.122}$$

### *3.1.9 Extended to Dynamic Problem*

When considering the inertial and the Coriolis forces, the governing equations of motion for a Cosserat continuum can be obtained from the Hamilton's principle. Equivalently, the inertial force and the associated Coriolis force are required to be added into the Eq. (3.107) to get

$$\mathcal{M}_n \dot{\underline{\mathcal{V}}}_n - \underline{\mathcal{F}}_g + \underline{\mathcal{F}}_e = \underline{\mathcal{F}}_a \tag{3.123}$$

## 3.2 General Shell-Like Theory of 5 DOFS

The shell theory, an old topic, has been studied over many years. Up to now, various shell elements are developed to solve engineering problems of shell-like structures. When tracing back 1970s, Ahmad et al. [27] directly reduced the basic equations of three-dimensional Cauchy continuum mechanics to obtain the governing equations of shell. The formulas of these reduced equations [22, 28–30] are found to be simple, and its finite element implementations are also straightforward. The representative works belong to Bathe and Dvorkin [31, 32], Hughes and his colleagues [33–35], and others [36, 37]. Since the large deformations frequently occur in shell-like structures, the accurate prediction of large deformations has been paid special attentions. Typically, Sussman and Bathe [38] develop a 3D shell element with four nodes for a large strain structure where two control vectors at each node describe the large deformations. Meanwhile, Hughes et al. [39] propose the concept of isogeometric analysis for direct design-to-analysis simulations. With this concept, many isogeometric shell elements [40–42] implemented. The efficient ones of these elements are the large deformation, rotation-free elements [40, 42] with three translational degrees of freedom that are valid only for thin shell. Although the rotation-free elements apply the Kirchhoff-Love theory to simplify the description of fiber motion, the Reissner-Mindlin shell elements [43] are still necessary when measuring shear strains through the thickness. Unlike the Kirchhoff-Love theory assuming the orientation of the fiber with respect to the reference surface is fixed, the Reissner-Mindlin shear deformable shell theory defines rotational degrees of freedom to describe the motion of the fiber, resulting in an shell element with five or six degrees of freedom per node. These rotational degrees of freedom are frequently the source of convergence difficulties in implicit structural analysis [40]. Pimenta et al. [44] have applied the Rodrigues rotation vector to vectorial parameterize the fiber rotations. It is also pointed out by the author that the rotation needs to be restricted to be less than $\pi$ to avoid the singularity. Even though, this restriction does not affect the formulation of an updated description, but it does affect the formulation of the total description [22].

In view of the geometric dimensions of shell-like structures, it is apparently that a dimension along the thickness of the shell is usually much smaller than the other two geometric dimensions. This observation motivates the development of two-dimensional shell theories [45–47], where the thin shell is treated by a reduced set of equations governing the middle surface and specifying the displacement field both on the surface and in its normal director. Compared with the reduced approach based on three-dimensional Cauchy continuum mechanics, the two-dimensional shell theories provide superior numerical conditioning and convergence. Many powerful finite element models [48–55] have been formulated under these theories. Among them, Simo et al. [49] proposed a stress-resultant-based geometrically exact shell element, which is formulated entirely in terms of stress resultants. When modeling the shell-like structures by using the shell element, it requires this type of element has the ability to decouple the deformation from arbitrary motion in an efficient manner. The two-dimensional shell theory has this decoupling advantage. It directly decouples the kinematics of an arbitrary material point in the element into the translation of reference point and rotation of the normal director.

The current implementation treats the shell-like structure as a two-dimensional Cosserat continuum, and its characteristic length is naturally selected as the thickness of the shell. The theory of two-dimensional Cosserat continuum is obtained by reducing the governing equations of three-dimensional Cosserat continuum mechanics discussed in previous Sect. 3. The curvilinear coordinate $\alpha_3$ is selected as the "thin" dimension, such that the derivatives with respected to $\alpha_3$ are all neglected. Under certain conditions, it is possible to model a shell-like structure with governing equations that depend on only two spatial coordinates $\alpha_1$ and $\alpha_2$. For the easy understanding, the theory details of two-dimensional Cosserat continuum are introduced in current section, and called general shell-like theory.

### 3.2.1 Kinematics of Shell

As shown in Fig. 3.2, a material particle with infinitesimal volume in a shell-like continuum is depicted by its position and director vectors in its reference configuration and current deformed configuration, respectively. In the original reference configuration, the position vector $\underline{p}(\alpha_i)$ of the material particle relative to a point **O**, the original of reference frame, $\mathcal{F}_{\mathcal{I}} = [\mathbf{O}, \mathcal{I}]$, can be written as

$$\underline{p}(\alpha_1, \alpha_2, \alpha_3) = \underline{x}_0(\alpha_1, \alpha_2) + R_0(\alpha_1, \alpha_2)\underline{s}_0(\alpha_3) \tag{3.124}$$

where the general curvilinear coordinates, $(\alpha_1, \alpha_2, \alpha_3)$, are introduced to label the material particle in the original continuum, and $R_0$ the initial rotation tensor measures the orientation of the reference basis $\mathcal{B}_0$. It has been verified in Sect. 2.2.3 the rotation tensor $R_0$ is identical to the body attached basis $\mathcal{B}_0$, and in detail

$$R_0 = \lfloor \bar{b}_1, \bar{b}_2, \bar{b}_3 \rfloor = \mathcal{B}_0 \tag{3.125}$$

The construction of rotation tensor $R_0$ can be referred to Eq. (1.229). Note that the initial basis $\mathcal{B}_0$ is bounded by the material particle **P** to produce the reference frame $\mathcal{F}_0 = [\mathbf{P}, \mathcal{B}_0]$. The local vector $\underline{s}_0$, defined in this reference frame $\mathcal{F}_0$, describes the director of a material particle. In the orthogonal curvilinear coordinate systems $\mathcal{B}_0$ and $\mathcal{I}$, the director can be given directly

$$\underline{s}_0 = \alpha_3 \bar{\imath}_3, \quad R_0 \underline{s}_0 = \alpha_3 \bar{b}_3 \tag{3.126}$$

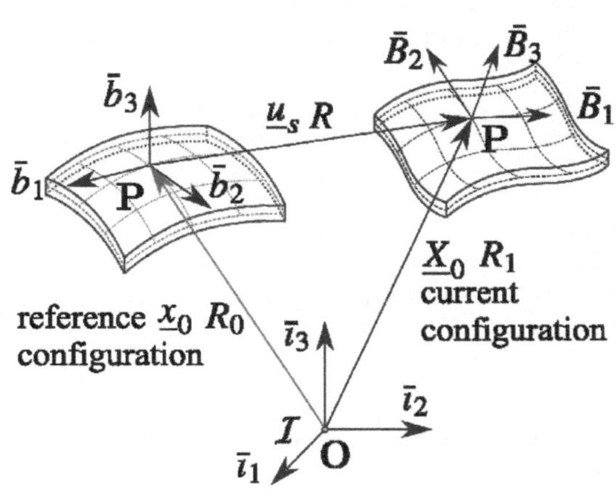


**Fig. 3.2** Kinematics of a material particle with infinitesimal volume in a shell-like continuum

Apparently, the director $\underline{s}_0$ is selected as the normal vector of shell surface. According to above definitions, it is readily verified

$$\underline{s}_{0,1} = \underline{s}_{0,2} = 0, \quad \underline{s}_{0,3} = \bar{\imath}_3 \tag{3.127}$$

It is noted that the thickness $h$ of the shell is exactly the characteristic length in a two-dimensional Cosserat continuum associated with torsional and bending deformations. Obviously, Eq. (3.124) decomposes the position vector $\underline{p}$ of a material particle into a reference position $\underline{x}_0$ and a local vector $\underline{s}_0$. The reference particle with position $\underline{x}_0$ can translate freely, while the rigid director $\underline{s}_0$ attached to the reference particle can only rotate without deformations.

For the description of shell deformations, the covariant base vectors,

$$\underline{g}_i = \frac{\partial \underline{p}}{\partial \alpha_i} \tag{3.128}$$

in the reference configuration are required, and the specific expressions can be obtained by taking derivatives of Eq. (3.124) with respected to the curvilinear coordinates $\alpha_i$ for $i = 1, 2, 3$ like

$$\begin{aligned} \underline{g}_1 &= \underline{x}_{0,1} + R_{0,1}\underline{s}_0 \\ \underline{g}_2 &= \underline{x}_{0,2} + R_{0,2}\underline{s}_0 \\ \underline{g}_3 &= R_0\bar{\imath}_3 \end{aligned} \tag{3.129}$$

Further operations are taken to resolve the base vectors into the local reference frame $\mathcal{F}_0$ to get

$$\begin{aligned} \underline{g}_1^* &= R_0^T\underline{x}_{0,1} + R_0^T R_{0,1}\underline{s}_0 \\ \underline{g}_2^* &= R_0^T\underline{x}_{0,2} + R_0^T R_{0,2}\underline{s}_0 \\ \underline{g}_3^* &= \bar{\imath}_3 \end{aligned} \tag{3.130}$$

According to the definitions of initial curvature tensors, Eq. (1.314), the base vectors can be represented by the following formulas

$$\begin{aligned} \underline{g}_1^* &= R_0^T\underline{x}_{0,1} + \widetilde{\kappa}_1^0\underline{s}_0 \\ \underline{g}_2^* &= R_0^T\underline{x}_{0,2} + \widetilde{\kappa}_2^0\underline{s}_0 \\ \underline{g}_3^* &= \bar{\imath}_3 \end{aligned} \tag{3.131}$$

where the symbols $\widetilde{\kappa}_i^0 = R_0^T R_{0,i}$ for $i = 1, 2$ denote the initial curvature tensors. All these base vectors form the covariant bases $\mathcal{B}_g$ like

$$\mathcal{B}_g = \lfloor \underline{g}_1^*,\ \underline{g}_2^*,\ \underline{g}_3^* \rfloor \tag{3.132}$$

and then the metric tensor in the reference configuration is computed from

$$g = \mathcal{B}_g^T \mathcal{B}_g \tag{3.133}$$

For the easy description of global and local motions of the material particle, the covariant bases $\mathcal{B}_g$ are decomposed into two parts

$$\mathcal{B}_g = \mathcal{B}_g^s + \mathcal{B}_g^\alpha \tag{3.134}$$

with the aid of introduced symbols as

$$\mathcal{B}_g^s = \lfloor \underline{g}_1^s,\ \underline{g}_2^s,\ \underline{g}_3^s \rfloor,\quad \mathcal{B}_g^\alpha = \lfloor \underline{g}_1^\alpha,\ \underline{g}_2^\alpha,\ \underline{g}_3^\alpha \rfloor \tag{3.135}$$

where the base vectors are given in detail for global part like

$$\begin{aligned} \underline{g}_1^s &= R_0^T \underline{x}_{0,1} + \tilde{\kappa}_1^0 \underline{s}_0 \\ \underline{g}_2^s &= R_0^T \underline{x}_{0,2} + \tilde{\kappa}_2^0 \underline{s}_0 \\ \underline{g}_3^s &= \underline{0} \end{aligned} \tag{3.136}$$

and for local part

$$\underline{g}_1^\alpha = \underline{0},\ \underline{g}_2^\alpha = \underline{0},\ \underline{g}_3^\alpha = \bar{\imath}_3 \tag{3.137}$$

In view of this, the metric tensor $g$ will be consisted of

$$g = g^{ss} + g^{s\alpha} + g^{\alpha s} + g^{\alpha\alpha} \tag{3.138}$$

and in detail

$$g^{ss} = \mathcal{B}_g^{sT} \mathcal{B}_g^s,\ g^{s\alpha} = \mathcal{B}_g^{sT} \mathcal{B}_g^\alpha,\ g^{\alpha s} = \mathcal{B}_g^{\alpha T} \mathcal{B}_g^s,\ g^{\alpha\alpha} = \mathcal{B}_g^{\alpha T} \mathcal{B}_g^\alpha \tag{3.139}$$

When the shell changes its configuration, the material particle that had position vector $\underline{p}$ in the undeformed reference configuration now has position vector $\underline{P}$ in the deformed configuration, such that

$$\underline{P}(\alpha_1, \alpha_2, \alpha_3) = \underline{X}_0(\alpha_1, \alpha_2) + R_1(\alpha_1, \alpha_2)\underline{s}_0(\alpha_3) \tag{3.140}$$

where $R_1$ is the total rotation tensor that describes the orientation of basis $\mathcal{B}_1$ attached to the deformed configuration and

$$R_1 = \lfloor \bar{B}_1, \bar{B}_2, \bar{B}_3 \rfloor = \mathcal{B}_1 \tag{3.141}$$

The associated frame $\mathcal{F}_1 = [\mathbf{P}, \mathcal{B}_1]$ will be constructed when bounding the current deformed basis $\mathcal{B}_1$ by the material particle $\mathbf{P}$. The position vector $\underline{X}_0$ of reference point in current deformed configuration can be recasted to

$$\underline{X}_0(\alpha_1, \alpha_2) = \underline{x}_0(\alpha_1, \alpha_2) + \underline{u}_s(\alpha_1, \alpha_2) \tag{3.142}$$

where the displacement vector $\underline{u}_s$ describes arbitrary translation of reference point **P**. Alternatively, the position vector $\underline{P}$ of the material particle in its deformed configuration can be represented by

$$\underline{P}(\alpha_1, \alpha_2, \alpha_3) = \underline{p}(\alpha_1, \alpha_2, \alpha_3) + \underline{u}(\alpha_1, \alpha_2, \alpha_3) \tag{3.143}$$

where $\underline{u}$ is the displacement of particle in space. It is readily to find that the displacement $\underline{u}$ are related to

$$\underline{u} = \underline{u}_s + (R_1 - R_0)\underline{s}_0 \tag{3.144}$$

with a very clear physical meaning: the displacement $\underline{u}$ is composed of the translation $\underline{u}_s$ of reference particle and the pure rotation $(R_1 - R_0)\underline{s}_0$ of local director. According to the definitions of covariant base vectors in deformed configuration

$$\underline{G}_i = \frac{\partial \underline{P}}{\partial \alpha_i} \tag{3.145}$$

the specific formulas can be obtained by taking spatial derivatives of Eq. (3.140) with respected to the curvilinear coordinates $\alpha_i$ for $i = 1, 2, 3$ like

$$\begin{aligned} \underline{G}_1 &= \underline{X}_{0,1} + R_{1,1}\underline{s}_0 \\ \underline{G}_2 &= \underline{X}_{0,2} + R_{1,2}\underline{s}_0 \\ \underline{G}_3 &= R_1\bar{\imath}_3 \end{aligned} \tag{3.146}$$

Here again, the base vectors are resolved into the current local frame $\mathcal{F}_1$ to find

$$\begin{aligned} \underline{G}_1^* &= R_1^T\underline{X}_{0,1} + R_1^T R_{1,1}\underline{s}_0 \\ \underline{G}_2^* &= R_1^T\underline{X}_{0,2} + R_1^T R_{1,2}\underline{s}_0 \\ \underline{G}_3^* &= \bar{\imath}_3 \end{aligned} \tag{3.147}$$

In view of the definitions of final curvature tensors $\widetilde{\kappa}_i^1 = R_1^T R_{1,i}$ for $i = 1, 2$, the covariant base vectors are simplified to

$$\begin{aligned} \underline{G}_1^* &= R_1^T\underline{X}_{0,1} + \widetilde{\kappa}_1^1\underline{s}_0 \\ \underline{G}_2^* &= R_1^T\underline{X}_{0,2} + \widetilde{\kappa}_2^1\underline{s}_0 \\ \underline{G}_3^* &= \bar{\imath}_3 \end{aligned} \tag{3.148}$$

All these base vectors could be assembled into the covariant bases $\mathcal{B}_G$ like

$$\mathcal{B}_G = \lfloor \underline{G}_1^*,\ \underline{G}_2^*,\ \underline{G}_3^* \rfloor \tag{3.149}$$

which is decomposed into two parts

$$\mathcal{B}_G = \mathcal{B}_G^s + \mathcal{B}_G^\alpha \tag{3.150}$$

with the help of new symbols defined in detail as

$$\mathcal{B}_G^s = \lfloor \underline{G}_1^s,\ \underline{G}_2^s,\ \underline{G}_3^s \rfloor,\quad \mathcal{B}_G^\alpha = \lfloor \underline{G}_1^\alpha,\ \underline{G}_2^\alpha,\ \underline{G}_3^\alpha \rfloor \tag{3.151}$$

where the global and local base vectors are introduced to be

$$\begin{aligned} \underline{G}_1^s &= R_1^T \underline{X}_{0,1} + \widetilde{\kappa}_1^1 \underline{s}_0 \\ \underline{G}_2^s &= R_1^T \underline{X}_{0,2} + \widetilde{\kappa}_2^1 \underline{s}_0 \\ \underline{G}_3^s &= \underline{0} \end{aligned} \tag{3.152}$$

and

$$\underline{G}_1^\alpha = \underline{0},\ \underline{G}_2^\alpha = \underline{0},\ \underline{G}_3^\alpha = \bar{\imath}_3 \tag{3.153}$$

In view of above expressions of covariant basis vectors in the deformed configuration, the associated metric tensor

$$G = \mathcal{B}_G^T \mathcal{B}_G \tag{3.154}$$

can be found to be composed of

$$G = G^{ss} + G^{s\alpha} + G^{\alpha s} + G^{\alpha\alpha} \tag{3.155}$$

with the following definitions

$$G^{ss} = \mathcal{B}_G^{sT} \mathcal{B}_G^s,\ G^{s\alpha} = \mathcal{B}_G^{sT} \mathcal{B}_G^\alpha,\ G^{\alpha s} = \mathcal{B}_G^{\alpha T} \mathcal{B}_G^s,\ G^{\alpha\alpha} = \mathcal{B}_G^{\alpha T} \mathcal{B}_G^\alpha \tag{3.156}$$

### 3.2.2 Strain Tensor of Shell

The rotation-free Green-Lagrange strain tensor [22] will be applied to measure the shell deformation. In doing so, the Lagrangian stretch tensor and the Lagrangian curvature strain tensor [23, 24]

$$\underline{e}_i = R_1^T \underline{X}_{0,i} - R_0^T \underline{x}_{0,i},\ \widetilde{\kappa}_i = \widetilde{\kappa}_i^1 - \widetilde{\kappa}_i^0 = R_1^T R_{1,i} - R_0^T R_{0,i} \tag{3.157}$$

for $i = 1, 2$ are required to rewritten the deformed covariant base vectors in terms of

$$\begin{aligned} \underline{G}_1^s &= \underline{e}_1 + R_0^T \underline{x}_{0,1} + \widetilde{\kappa}_1^1 \underline{s}_0 \\ \underline{G}_2^s &= \underline{e}_2 + R_0^T \underline{x}_{0,2} + \widetilde{\kappa}_2^1 \underline{s}_0 \\ \underline{G}_3^s &= \underline{0} \end{aligned} \tag{3.158}$$

Referring the decompositions of metric tensors in reference configuration and deformed configuration, Eqs. (3.138) and (3.155), respectively, the rotation-free Green-Lagrangian strain tensor measured in the curvilinear basis

$$\widehat{\varepsilon} = \frac{1}{2}(G - g) \tag{3.159}$$

can be regarded as a combination of two contributions, $\widehat{\varepsilon} = \varepsilon^s + \varepsilon^c$, one from the global deformation

$$\varepsilon^s = \frac{1}{2}(G^{ss} - g^{ss}) \tag{3.160}$$

and the other from the global-local coupling deformation

$$\varepsilon^c = \frac{1}{2}(G^{\alpha s} + G^{s\alpha} - g^{\alpha s} - g^{s\alpha}) \tag{3.161}$$

The contribution of local deformation is zero in classical Cosserat theory

$$\varepsilon^\alpha = \frac{1}{2}(G^{\alpha\alpha} - g^{\alpha\alpha}) = 0 \tag{3.162}$$

Substituting the specific formulations of metric tensors in terms of covariant basis vectors, Eq. (3.158), into the definitions of strain tensor, Eqs. (3.160) and (3.161), the components of rotation-free Green-Lagrange strain tensor are found to be related to the components of Lagrangian stretch strains and curvature stains. In the vector-dyadic form, the relations are written in detail like

$$\begin{aligned}
\varepsilon_{11} &= (\underline{x}_{0,1}^T R_0 + \underline{\kappa}_1^{0T} \widetilde{s}_0)(\underline{e}_1 + \widetilde{s}_0^T \underline{\kappa}_1) \\
2\varepsilon_{12} &= (\underline{x}_{0,1}^T R_0 + \underline{\kappa}_1^{0T} \widetilde{s}_0)(\underline{e}_2 + \widetilde{s}_0^T \underline{\kappa}_2) + (\underline{x}_{0,2}^T R_0 + \underline{\kappa}_2^{0T} \widetilde{s}_0)(\underline{e}_1 + \widetilde{s}_0^T \underline{\kappa}_1) \\
2\varepsilon_{13} &= \bar{\imath}_3^T (\underline{e}_1 + \widetilde{s}_0^T \underline{\kappa}_1) \\
\varepsilon_{22} &= (\underline{x}_{0,2}^T R_0 + \underline{\kappa}_2^{0T} \widetilde{s}_0)(\underline{e}_2 + \widetilde{s}_0^T \underline{\kappa}_2) \\
2\varepsilon_{23} &= \bar{\imath}_3^T (\underline{e}_2 + \widetilde{s}_0^T \underline{\kappa}_2) \\
\varepsilon_{33} &= 0
\end{aligned} \tag{3.163}$$

Under the assumptions of small stretch strains $\underline{e}_i \ll 0$ and small curvature strains $\underline{\kappa}_i \ll 0$ for $i = 1, 2, 3$, the higher order terms as to $\underline{e}_i$ and $\underline{\kappa}_i$ for $i = 1, 2$ are all neglected. Note that except the assumptions of above small strains, there will be no further assumptions are introduced. The ensuing derivations of shell governing equations are then strictly kept within the framework of these assumptions, which reproduces the concept of *geometrically exact* shell model.

Here again, a straight forward approach is applied to obtain the same expressions as Eq. (3.163). From the definitions of covariant base vectors, Eq. (3.131) in reference configuration and Eq. (3.148) in deformed configuration, respectively, the differences of base vectors can be found to be

$$\begin{aligned}
\Delta\underline{G}_1^* &= \underline{G}_1^* - \underline{g}_1^* = R_1^T \underline{X}_{0,1} - R_0^T \underline{x}_{0,1} + (\widetilde{\kappa}_1^1 - \widetilde{\kappa}_1^0)\underline{s}_0 \\
\Delta\underline{G}_2^* &= \underline{G}_2^* - \underline{g}_2^* = R_2^T \underline{X}_{0,2} - R_0^T \underline{x}_{0,2} + (\widetilde{\kappa}_2^1 - \widetilde{\kappa}_2^0)\underline{s}_0 \\
\Delta\underline{G}_3^* &= \underline{G}_3^* - \underline{g}_3^* = \underline{0}
\end{aligned} \tag{3.164}$$

Referring the definitions of Lagrangian stretch and Lagrange curvature measures, Eq. (2.152), the above definitions can be recasted to the compact form like

$$\Delta \underline{G}_1^* = \underline{e}_1 + \widetilde{s}_0^T \underline{\kappa}_1, \quad \Delta \underline{G}_2^* = \underline{e}_2 + \widetilde{s}_0^T \underline{\kappa}_2, \quad \Delta \underline{G}_3^* = \underline{0} \tag{3.165}$$

The components of Green-Lagrange strain tensor are approximated to be

$$\begin{aligned}
\varepsilon_{11} &= \frac{1}{2}[(\underline{g}_1^* + \Delta \underline{G}_1^*)^T (\underline{g}_1^* + \Delta \underline{G}_1^*) - \underline{g}_1^{*T} \underline{g}_1^*] \approx \underline{g}_1^{*T} \Delta \underline{G}_1^* \\
2\varepsilon_{12} &= (\underline{g}_1^* + \Delta \underline{G}_1^*)^T (\underline{g}_2^* + \Delta \underline{G}_2^*) - \underline{g}_1^{*T} \underline{g}_2^* \approx \underline{g}_1^{*T} \Delta \underline{G}_2^* + \underline{g}_2^{*T} \Delta \underline{G}_1^* \\
2\varepsilon_{13} &= (\underline{g}_1^* + \Delta \underline{G}_1^*)^T (\underline{g}_3^* + \Delta \underline{G}_3^*) - \underline{g}_1^{*T} \underline{g}_3^* \approx \underline{g}_3^{*T} \Delta \underline{G}_1^* \\
\varepsilon_{22} &= \frac{1}{2}[(\underline{g}_2^* + \Delta \underline{G}_2^*)^T (\underline{g}_2^* + \Delta \underline{G}_2^*) - \underline{g}_2^{*T} \underline{g}_2^*] \approx \underline{g}_2^{*T} \Delta \underline{G}_2^* \\
2\varepsilon_{23} &= (\underline{g}_2^* + \Delta \underline{G}_2^*)^T (\underline{g}_3^* + \Delta \underline{G}_3^*) - \underline{g}_2^{*T} \underline{g}_3^* \approx \underline{g}_3^{*T} \Delta \underline{G}_2^* \\
\varepsilon_{33} &= \frac{1}{2}[(\underline{g}_3^* + \Delta \underline{G}_3^*)^T (\underline{g}_3^* + \Delta \underline{G}_3^*) - \underline{g}_3^{*T} \underline{g}_3^*] = 0
\end{aligned} \tag{3.166}$$

where the higher order terms $\Delta \underline{G}_i^{*T} \Delta \underline{G}_j^*$ for $i, j = 1, 2$ are all truncated under the same assumption of small strains. Once again, the linearized strain measures are obtained to be

$$\begin{aligned}
\varepsilon_{11} &= \underline{g}_1^{*T} (\underline{e}_1 + \widetilde{s}_0^T \underline{\kappa}_1) \\
2\varepsilon_{12} &= \underline{g}_1^{*T} (\underline{e}_2 + \widetilde{s}_0^T \underline{\kappa}_2) + \underline{g}_2^{*T} (\underline{e}_1 + \widetilde{s}_0^T \underline{\kappa}_1) \\
2\varepsilon_{13} &= \underline{g}_3^{*T} (\underline{e}_1 + \widetilde{s}_0^T \underline{\kappa}_1) \\
\varepsilon_{22} &= \underline{g}_2^{*T} (\underline{e}_2 + \widetilde{s}_0^T \underline{\kappa}_2) \\
2\varepsilon_{23} &= \underline{g}_3^{*T} (\underline{e}_2 + \widetilde{s}_0^T \underline{\kappa}_2) \\
\varepsilon_{33} &= 0
\end{aligned} \tag{3.167}$$

### 3.2.3 *Constitutive Laws for Shell*

For the easy description of shell constitutive relations, the Green-Lagrange strain tensor in its vector-dyadic form needs to be decomposed into

$$\widehat{\underline{\varepsilon}} = \left| \begin{matrix} \varepsilon_{11} \\ 2\varepsilon_{12} \\ 2\varepsilon_{13} \\ \varepsilon_{22} \\ 2\varepsilon_{23} \\ \varepsilon_{33} \end{matrix} \right| = \left| \begin{matrix} \varepsilon_{11} \\ \varepsilon_{21} \\ \varepsilon_{31} \\ 0 \\ 0 \\ 0 \end{matrix} \right| + \left| \begin{matrix} 0 \\ \varepsilon_{12} \\ 0 \\ \varepsilon_{22} \\ \varepsilon_{32} \\ 0 \end{matrix} \right| + \left| \begin{matrix} 0 \\ 0 \\ \varepsilon_{13} \\ 0 \\ \varepsilon_{23} \\ \varepsilon_{33} \end{matrix} \right| = \widehat{\underline{\varepsilon}}_1 + \widehat{\underline{\varepsilon}}_2 + \widehat{\underline{\varepsilon}}_3 \tag{3.168}$$

It is noted that the symbols $\varepsilon_{12}$, $\varepsilon_{21}$, $\varepsilon_{23}$, $\varepsilon_{32}$, $\varepsilon_{31}$ and $\varepsilon_{13}$ do not represent the components of Green-Lagrange strain tensor, but are defined in the following relations

$$2\varepsilon_{12} = \varepsilon_{21} + \varepsilon_{12}, \quad 2\varepsilon_{13} = \varepsilon_{31} + \varepsilon_{13}, \quad 2\varepsilon_{23} = \varepsilon_{32} + \varepsilon_{23} \tag{3.169}$$

After filtered out the zero items, the specific vectorial expressions are reduced to

$$\underline{\varepsilon}_1 = \begin{vmatrix} \varepsilon_{11} \\ \varepsilon_{21} \\ \varepsilon_{31} \end{vmatrix}, \ \underline{\varepsilon}_2 = \begin{vmatrix} \varepsilon_{12} \\ \varepsilon_{22} \\ \varepsilon_{32} \end{vmatrix}, \ \underline{\varepsilon}_3 = \begin{vmatrix} \varepsilon_{13} \\ \varepsilon_{23} \\ \varepsilon_{33} \end{vmatrix} = \underline{0} \tag{3.170}$$

and in detail

$$\begin{aligned} \underline{\varepsilon}_1 &= \begin{vmatrix} \varepsilon_{11} \\ \varepsilon_{21} \\ \varepsilon_{31} \end{vmatrix} = \begin{vmatrix} (\underline{x}_{0,1}^T R_0 + \underline{\kappa}_1^{0T} \widetilde{s}_0)(\underline{e}_1 + \widetilde{s}_0^T \underline{\kappa}_1) \\ (\underline{x}_{0,2}^T R_0 + \underline{\kappa}_2^{0T} \widetilde{s}_0)(\underline{e}_1 + \widetilde{s}_0^T \underline{\kappa}_1) \\ \bar{\imath}_3^T (\underline{e}_1 + \widetilde{s}_0^T \underline{\kappa}_1) \end{vmatrix} \\ \underline{\varepsilon}_2 &= \begin{vmatrix} \varepsilon_{12} \\ \varepsilon_{22} \\ \varepsilon_{32} \end{vmatrix} = \begin{vmatrix} (\underline{x}_{0,1}^T R_0 + \underline{\kappa}_1^{0T} \widetilde{s}_0)(\underline{e}_2 + \widetilde{s}_0^T \underline{\kappa}_2) \\ (\underline{x}_{0,2}^T R_0 + \underline{\kappa}_2^{0T} \widetilde{s}_0)(\underline{e}_2 + \widetilde{s}_0^T \underline{\kappa}_2) \\ \bar{\imath}_3^T (\underline{e}_2 + \widetilde{s}_0^T \underline{\kappa}_2) \end{vmatrix} \end{aligned} \tag{3.171}$$

The above formulas could be assembled into a two-dimensional array

$$\begin{bmatrix} \varepsilon_{11}, \ \varepsilon_{12} \\ \varepsilon_{21}, \ \varepsilon_{22} \\ \varepsilon_{31}, \ \varepsilon_{32} \end{bmatrix} = \begin{bmatrix} \underline{x}_{0,1}^T R_0 + \underline{\kappa}_1^{0T} \widetilde{s}_0 \\ \underline{x}_{0,2}^T R_0 + \underline{\kappa}_2^{0T} \widetilde{s}_0 \\ \bar{\imath}_3 \end{bmatrix} \begin{bmatrix} I, \ \widetilde{s}_0^T \end{bmatrix} \begin{bmatrix} \underline{e}_1, \ \underline{e}_2 \\ \underline{\kappa}_1, \ \underline{\kappa}_2 \end{bmatrix} \tag{3.172}$$

Since the normal vector of shell mid-surface is selected as the director, $\underline{s}_0 = \alpha_3 \bar{\imath}_3$, it must be careful when evaluating the components of Green-Lagrange strain tensor by Eq. (3.172), where the transpose of skew-symmetric matrix $\widetilde{s}_0^T$ needs to be constructed. Observations show that

$$\widetilde{s}_0^T = \alpha_3 \begin{bmatrix} 0 \ 1 \ 0 \\ -1 \ 0 \ 0 \\ 0 \ 0 \ 0 \end{bmatrix} \tag{3.173}$$

is a singular matrix, such that the product $\widetilde{s}_0^T \underline{\kappa}_i$ for $i = 1, 2$ could be truncated

$$\widetilde{s}_0^T \underline{\kappa}_i = \widehat{s}_0^T \widehat{\underline{\kappa}}_i \tag{3.174}$$

where the truncated matrix and curvature vectors are defined as

$$\widehat{s}_0^T = \alpha_3 \begin{bmatrix} 0 \ 1 \\ -1 \ 0 \\ 0 \ 0 \end{bmatrix}, \ \ \widehat{\underline{\kappa}}_1 = \begin{vmatrix} \kappa_{11} \\ \kappa_{21} \end{vmatrix}, \ \ \widehat{\underline{\kappa}}_2 = \begin{vmatrix} \kappa_{12} \\ \kappa_{22} \end{vmatrix} \tag{3.175}$$

As a result, the Green-Lagrange strain measures is recasted to a truncated form like

$$\lfloor \hat{\underline{\varepsilon}}_1, \ \hat{\underline{\varepsilon}}_2 \rfloor = \widehat{S}_0^T \lfloor \widehat{\underline{\epsilon}}_1, \ \widehat{\underline{\varepsilon}}_2 \rfloor \tag{3.176}$$

with the definitions of truncated symbols

$$\widehat{S}_0^T = \begin{bmatrix} \underline{x}_{0,1}^T R_0 + \underline{\kappa}_1^{0T} \widetilde{s}_0 \\ \underline{x}_{0,2}^T R_0 + \underline{\kappa}_2^{0T} \widetilde{s}_0 \\ \bar{\imath}_3 \end{bmatrix} [\, I,\ \widetilde{s}_0^T \,],\quad \widehat{\underline{\epsilon}}_1 = \begin{vmatrix} \underline{\epsilon}_1 \\ \widehat{\underline{\kappa}}_1 \end{vmatrix},\quad \widehat{\underline{\epsilon}}_2 = \begin{vmatrix} \underline{\epsilon}_2 \\ \widehat{\underline{\kappa}}_2 \end{vmatrix} \tag{3.177}$$

The strain energy $\mathcal{U}_e$ can be represented by using the non-zero components of Green-Lagrange strain tensor like

$$2\mathcal{U}_e = \int_{V_0} (\hat{\underline{\varepsilon}}_1^T \mathcal{D}_{11} \hat{\underline{\varepsilon}}_1 + \hat{\underline{\varepsilon}}_2^T \mathcal{D}_{22} \hat{\underline{\varepsilon}}_2 + 2\hat{\underline{\varepsilon}}_2^T \mathcal{D}_{21} \hat{\underline{\varepsilon}}_1)\, dV_0 \tag{3.178}$$

where the following constitutive matrix blocks are introduced

$$\mathcal{D}_{11} = \begin{bmatrix} d_{11},\ d_{12},\ d_{13} \\ d_{21},\ d_{22},\ d_{23} \\ d_{31},\ d_{32},\ d_{33} \end{bmatrix},\ \mathcal{D}_{22} = \begin{bmatrix} d_{22},\ d_{24},\ d_{25} \\ d_{42},\ d_{44},\ d_{45} \\ d_{52},\ d_{54},\ d_{55} \end{bmatrix},\ \mathcal{D}_{12} = \begin{bmatrix} d_{12},\ d_{14},\ d_{15} \\ d_{22},\ d_{24},\ d_{25} \\ d_{32},\ d_{34},\ d_{35} \end{bmatrix} \tag{3.179}$$

Since the constitutive matrix is a symmetric matrix, it is readily to verify that $\mathcal{D}_{12} = \mathcal{D}_{21}^T$. Referring the relations of Green-Lagrange strain and Lagrangian strain measures, Eq. (3.176), the above strain energy is recasted to

$$2\mathcal{U}_e = \int_{S_0} (\widehat{\underline{\epsilon}}_1^T \mathcal{C}_{11} \widehat{\underline{\epsilon}}_1 + \widehat{\underline{\epsilon}}_2^T \mathcal{C}_{22} \widehat{\underline{\epsilon}}_2 + 2\widehat{\underline{\epsilon}}_2^T \mathcal{C}_{21} \widehat{\underline{\epsilon}}_1)\, dS_0 \tag{3.180}$$

where $S_0$ is the area of the shell mid-surface, and $\mathcal{C}_{11}, \mathcal{C}_{22}, \mathcal{C}_{12}$ are the constitutive matrices of shell defined as

$$\mathcal{C}_{11} = \int_h \widehat{S}_0 \mathcal{D}_{11} \widehat{S}_0^T\, d\alpha_3,\ \mathcal{C}_{22} = \int_h \widehat{S}_0 \mathcal{D}_{22} \widehat{S}_0^T\, d\alpha_3,\ \mathcal{C}_{21} = \int_h \widehat{S}_0 \mathcal{D}_{21} \widehat{S}_0^T\, d\alpha_3 \tag{3.181}$$

Note that the symmetric properties of constitutive matrices are still hold, such that $\mathcal{C}_{12} = \mathcal{C}_{21}^T$. Apparently, the shell constitutive matrices are determined by the characteristic length $h$ in addition to the material parameters.

### *3.2.4 Variation of Strain Energy of Shell*

The second Piola-Kirchhoff stress and Green-Lagrange strain measures are selected so that their product gives an accurate energy density that is related to the undeformed volume of reference configuration. Referring the expressions of strain energy, Eq. (3.180), the strain energy density for a shell is given as

$$\mathcal{U}_\rho = \frac{1}{2}(\widehat{\underline{\epsilon}}_1^T C_{11}\widehat{\underline{\epsilon}}_1 + \widehat{\underline{\epsilon}}_2^T C_{22}\widehat{\underline{\epsilon}}_2) + \widehat{\underline{\epsilon}}_2^T C_{21}\widehat{\underline{\epsilon}}_1 \tag{3.182}$$

The variation of $\mathcal{U}_\rho$ is readily verified to be

$$\delta\mathcal{U}_\rho = \delta\widehat{\underline{\epsilon}}_1^T C_{11}\widehat{\underline{\epsilon}}_1 + \delta\widehat{\underline{\epsilon}}_2^T C_{22}\widehat{\underline{\epsilon}}_2 + \delta\widehat{\underline{\epsilon}}_2^T C_{21}\widehat{\underline{\epsilon}}_1 + \delta\widehat{\underline{\epsilon}}_1^T C_{12}\widehat{\underline{\epsilon}}_2 \tag{3.183}$$

After merging the similar terms at the right-hand-side of above equations, the variation of strain energy density can be rearranged into the compact form

$$\delta\mathcal{U}_\rho = \delta\widehat{\underline{\epsilon}}_1^T \underline{F}_{\epsilon\epsilon}^{11} + \delta\widehat{\underline{\epsilon}}_2^T \underline{F}_{\epsilon\epsilon}^{22} \tag{3.184}$$

where the stress measures of a Cosserat continuum are defined in the vectorial formulas like

$$\underline{F}_{\epsilon\epsilon}^{11} = C_{11}\widehat{\underline{\epsilon}}_1 + C_{12}\widehat{\underline{\epsilon}}_2, \quad \underline{F}_{\epsilon\epsilon}^{22} = C_{22}\widehat{\underline{\epsilon}}_2 + C_{21}\widehat{\underline{\epsilon}}_1 \tag{3.185}$$

Finally, the variation of strain energy can be represented by

$$\delta\mathcal{U}_e = \int_{S_0} (\delta\widehat{\underline{\epsilon}}_1^T \underline{F}_{\epsilon\epsilon}^{11} + \delta\widehat{\underline{\epsilon}}_2^T \underline{F}_{\epsilon\epsilon}^{22})dS_0 \tag{3.186}$$

### *3.2.5 Virtual Work of External Forces for a Shell*

The virtual work done by a corresponding virtual displacement operating on the external forces is evaluated in this section. According to Eq. (3.140), the virtual displacement of an arbitrary particle in deformed configuration is obtained by

$$\delta\underline{P} = \delta(\underline{X}_0 + R_1\underline{s}_0) = \delta(\underline{x}_0 + \underline{u}_s + R_1\underline{s}_0) \tag{3.187}$$

where $\underline{u}_s$ is the displacement of reference particle **P** on the mid-surface of the shell and $R_1$ the total rotation tensor that describes the orientation of local vector $\underline{s}_0$ in current deformed configuration. It will be readily verified the virtual displacement $\delta\underline{P}$, also the position variation, is determined to be

$$\delta\underline{P} = \delta\underline{u}_s + \delta R_1\underline{s}_0 \tag{3.188}$$

After resolved the above equations into the body attached frame, the variation of position vector will relate to the virtual motion of reference particle **P** like

$$R_1^T\delta\underline{P} = R_1^T\delta\underline{u}_s + \widetilde{s}_0^T\delta\underline{\psi} \tag{3.189}$$

where the virtual rotations are defined, $\widetilde{\delta\psi} = R_1^T \delta R_1$. Once again, due to the singularity of $\widetilde{s}_0^T$, the truncation must be applied to obtain

$$R_1^T \delta \underline{P} = R_1^T \delta \underline{u}_s + \widehat{s}_0^T \underline{\widehat{\delta\psi}} \tag{3.190}$$

and only two components of the virtual rotations are retained, $\underline{\widehat{\delta\psi}}^T = \lfloor \delta\psi_1,\ \delta\psi_2 \rfloor$. In current implementation, two types of external forces are considered, one is body force $\underline{g}$, the other surface pressure $\underline{p}_s$, both measured in body attached frame. Then, the virtual work of external forces can be computed from

$$\delta \mathcal{W}_e = \int_{V_0} \delta \underline{P}^T R_1 \underline{g}\, dV_0 + \int_{S_0} \delta \underline{P}^T R_1 \underline{p}_s\, dS_0 \tag{3.191}$$

Referring the formula of position variation, Eq. (3.190), the virtual work can be expressed in the following detailed formula as

$$\begin{aligned} \delta \mathcal{W}_e &= \int_{V_0} (R_1^T \delta \underline{u}_s + \widehat{s}_0^T \underline{\widehat{\delta\psi}})^T \underline{g}\, dV_0 + \int_{S_0} (R_1^T \delta \underline{u}_s + \widehat{s}_0^T \underline{\widehat{\delta\psi}})^T \underline{p}_s\, dS_0 \\ &= \int_{S_0} \left[ \delta \underline{u}_s^T R_1 (\int_h \underline{g}\, d\alpha_3 + \underline{p}_s) + \underline{\widehat{\delta\psi}}^T \left( \int_h \widehat{s}_0 \underline{g}\, d\alpha_3 + \widehat{s}_0 \underline{p}_s \right) \right] dS_0 \end{aligned} \tag{3.192}$$

For the easy description of virtual work, the external force and torque vectors are defined at first by

$$\underline{F} = \int_h \underline{g}\, d\alpha_3 + \underline{p}_s, \quad \underline{\widehat{T}} = \int_h \widehat{s}_0 \underline{g}\, d\alpha_3 + \widehat{s}_0 \underline{p}_s \tag{3.193}$$

then the virtual work can be rewritten as

$$\delta \mathcal{W}_e = \int_{S_0} (\delta \underline{u}_s^T R_1 \underline{F} + \underline{\widehat{\delta\psi}}^T \underline{\widehat{T}})\, dS_0 \tag{3.194}$$

or in the motion formalism like

$$\delta \mathcal{W}_e = \int_{S_0} \underline{\widehat{\delta\mathcal{U}}}^T \underline{\widehat{F}}_{\epsilon\alpha} dS_0 \tag{3.195}$$

where the reduced dual brackets are introduced

$$\underline{\widehat{F}}_{\epsilon\alpha} = \begin{vmatrix} \underline{F} \\ \underline{\widehat{T}} \end{vmatrix}, \quad \underline{\widehat{\delta\mathcal{U}}} = \begin{vmatrix} R_1^T \delta \underline{u}_s \\ \underline{\widehat{\delta\psi}} \end{vmatrix} \tag{3.196}$$

### 3.2.6 Governing Equations of Shell in Motion Formalism

Based on the principle of virtual work, which states

$$\delta \mathcal{U}_e - \delta \mathcal{W}_e = 0 \tag{3.197}$$

the governing equations for a shell problem could be obtained. Introducing Eqs. (3.186) and (3.194) leads to

$$\int_{S_0} (\delta \underline{\widehat{\epsilon}}_1^T \underline{F}_{\epsilon\epsilon}^{11} + \delta \underline{\widehat{\epsilon}}_2^T \underline{F}_{\epsilon\epsilon}^{22} - \underline{\widehat{\delta \mathcal{U}}}^T \underline{\widehat{F}}_{\epsilon\alpha}) dS_0 = 0 \tag{3.198}$$

In view of the compatibility attribute, Eq. (2.161), the variations of Lagrangian strains are truncated into

$$\delta \underline{\widehat{\epsilon}}_1 = \underline{\widehat{\delta \mathcal{U}}}_{,1} + \widetilde{\widetilde{\mathcal{K}}}_1^1 \underline{\widehat{\delta \mathcal{U}}}, \quad \delta \underline{\widehat{\epsilon}}_2 = \underline{\widehat{\delta \mathcal{U}}}_{,2} + \widetilde{\widetilde{\mathcal{K}}}_2^1 \underline{\widehat{\delta \mathcal{U}}} \tag{3.199}$$

where $\widetilde{\widetilde{\mathcal{K}}}_i^1$ are the dual skew-symmetric tensor of the curvatures with the last row and column eliminated, and the associated dual parts $\widetilde{\kappa}_i^1$ will be simplified to $\widetilde{\kappa}_i^1 = \kappa_{1i}\widetilde{\imath}_1 + \kappa_{2i}\widetilde{\imath}_2$ for $i = 1, 2$. As a result, the principle of virtual work is represented by

$$\int_{S_0} [(\underline{\widehat{\delta \mathcal{U}}}_{,1} + \widetilde{\widetilde{\mathcal{K}}}_1^1 \underline{\widehat{\delta \mathcal{U}}})^T \underline{F}_{\epsilon\epsilon}^{11} + (\underline{\widehat{\delta \mathcal{U}}}_{,2} + \widetilde{\widetilde{\mathcal{K}}}_2^1 \underline{\widehat{\delta \mathcal{U}}})^T \underline{F}_{\epsilon\epsilon}^{22} - \underline{\widehat{\delta \mathcal{U}}}^T \underline{\widehat{F}}_{\epsilon\alpha}] dS_0 = 0 \tag{3.200}$$

With the application of finite element discretization, the variations of motion vectors and its derivatives can be interpolated directly by using the shape functions $\mathcal{H}_\epsilon$ to obtain

$$\underline{\widehat{\delta \mathcal{U}}} = \mathcal{H}_\epsilon \underline{\widehat{\delta \mathcal{U}}}_n, \ \underline{\widehat{\delta \mathcal{U}}}_{,i} = \mathcal{H}_{\epsilon,i} \underline{\widehat{\delta \mathcal{U}}}_n \tag{3.201}$$

With the aid of these discrete interpolations, the principle of virtual work becomes

$$\underline{\widehat{\delta \mathcal{U}}}_n^T \int_{S_0} [(\mathcal{H}_{\epsilon,1}^T + \mathcal{H}_\epsilon^T \widetilde{\widetilde{\mathcal{K}}}_1^{1T}) \underline{F}_{\epsilon\epsilon}^{11} + (\mathcal{H}_{\epsilon,2}^T + \mathcal{H}_\epsilon^T \widetilde{\widetilde{\mathcal{K}}}_2^{1T}) \underline{F}_{\epsilon\epsilon}^{22} - \mathcal{H}_\epsilon^T \underline{\widehat{F}}_{\epsilon\alpha}] dS_0 = 0 \tag{3.202}$$

From which, the governing equations are obtained to be

$$\int_{S_0} [(\mathcal{H}_{\epsilon,1}^T + \mathcal{H}_\epsilon^T \widetilde{\widetilde{\mathcal{K}}}_1^{1T}) \underline{F}_{\epsilon\epsilon}^{11} + (\mathcal{H}_{\epsilon,2}^T + \mathcal{H}_\epsilon^T \widetilde{\widetilde{\mathcal{K}}}_2^{1T}) \underline{F}_{\epsilon\epsilon}^{22} - \mathcal{H}_\epsilon^T \underline{\widehat{F}}_{\epsilon\alpha}] dS_0 = 0 \tag{3.203}$$

After introduced the elastic force vector

$$\underline{\mathcal{F}_e} = \int_{S_0} [(\mathcal{H}_{\epsilon,1}^T + \mathcal{H}_\epsilon^T \tilde{\tilde{\mathcal{K}}}_1^{1T})\underline{F_{\epsilon\epsilon}^{11}} + (\mathcal{H}_{\epsilon,2}^T + \mathcal{H}_\epsilon^T \tilde{\tilde{\mathcal{K}}}_2^{1T})\underline{F_{\epsilon\epsilon}^{22}}] dS_0 \tag{3.204}$$

and the external force vector

$$\underline{\mathcal{F}_a} = \int_{S_0} \mathcal{H}_\epsilon^T \underline{\widehat{F}_{\epsilon\alpha}} dS_0 \tag{3.205}$$

the governing equations can be described by the following compact formula

$$\underline{\mathcal{F}_e} - \underline{\mathcal{F}_a} = 0 \tag{3.206}$$

### 3.2.7 Kinematic Energy of Shell

When governing equation of the shell is extended to the dynamic problem, the kinematic energy of shell needs to be evaluated. At first, the velocity of an arbitrary material particle in a shell can be evaluated by taking time derivative of position vector $\underline{P}$, defined by Eq. (3.140), to get

$$\underline{\dot{P}} = (\underline{x_0} + \underline{u_s} + R_1 \underline{s_0})^{\cdot} = \underline{\dot{u}_s} + \dot{R}_1 \underline{s_0} \tag{3.207}$$

After resolved the time derivatives into the body attached frame $\mathcal{F}_1$, the velocity of material particle will relate to the linear and angular velocities of reference particle **P** like

$$R_1^T \underline{\dot{P}} = R_1^T \underline{\dot{u}_s} + \widetilde{s}_0^T \underline{\omega} \tag{3.208}$$

where the angular velocity is defined as $\widetilde{\omega} = R_1^T \dot{R}_1$. By truncated the last column of $\widetilde{s}_0^T$, the velocity vector becomes

$$R_1^T \underline{\dot{P}} = R_1^T \underline{\dot{u}_s} - \widehat{s}_0^T \underline{\widehat{\omega}}. \tag{3.209}$$

and the last item of angular velocity is eliminated with only two components are retained, $\underline{\widehat{\omega}}^T = \lfloor \omega_1, \ \omega_2 \rfloor$. If denoting the material density as $\rho$, the kinematic energy of a shell can be computed from the integral related to the initial volume $V_0$ like

$$\begin{aligned} \mathcal{K}_e &= \int_{V_0} \frac{1}{2} \rho (R_1^T \underline{\dot{P}})^T R_1^T \underline{\dot{P}} dV_0 \\ &= \int_{V_0} \frac{1}{2} \lfloor \underline{\dot{u}_s}^T R_1, \ \underline{\omega}^T \rfloor \begin{bmatrix} \rho I, & \rho \widetilde{s}_0^T \\ \rho \widetilde{s}_0, & \rho \widetilde{s}_0 \widetilde{s}_0^T \end{bmatrix} \begin{vmatrix} R_1^T \underline{\dot{u}_s} \\ \underline{\omega} \end{vmatrix} dV_0 \end{aligned} \tag{3.210}$$

After truncated the last zero column of $\widetilde{s}_0^T$, the motion formalism of kinematic energy is readily obtained to be

$$\mathcal{K}_e = \int_{S_0} \frac{1}{2}\underline{\widehat{\mathcal{V}}}^T \mathcal{M}_{\epsilon\epsilon}\underline{\widehat{\mathcal{V}}} dS_0 \tag{3.211}$$

with the aid of following definitions

$$\mathcal{M}_{\epsilon\epsilon} = \int_h \begin{bmatrix} \rho I, & \rho\widehat{s}_0^T \\ \rho\widehat{s}_0, & \rho\widehat{s}_0\widehat{s}_0^T \end{bmatrix} d\alpha_3, \quad \underline{\widehat{\mathcal{V}}} = \begin{vmatrix} R_1^T \underline{\dot{u}}_s \\ \underline{\widehat{\omega}} \end{vmatrix} \tag{3.212}$$

Referring Eq. (3.211), the variation of kinematic energy is readily obtained to be

$$\delta\mathcal{K}_e = \int_{V_0} \delta\underline{\widehat{\mathcal{V}}}^T \mathcal{M}_{\epsilon\epsilon}\underline{\widehat{\mathcal{V}}} dV_0 \tag{3.213}$$

In view of compatibility properties, Eq. (2.160), the virtual dual velocity is related to the virtual motion and its time derivative in the truncated formula like

$$\delta\underline{\widehat{\mathcal{V}}} = \underline{\dot{\widehat{\delta\mathcal{U}}}} + \widetilde{\widetilde{\mathcal{V}}}\underline{\widehat{\delta\mathcal{U}}} \tag{3.214}$$

where $\widetilde{\widetilde{\mathcal{V}}}$ is the dual skew-symmetric tensor of velocities with the last row and column eliminated, and the associated dual part is approximated to $\widetilde{\omega} = \omega_1\widetilde{\iota}_1 + \omega_2\widetilde{\iota}_2$. The variation of kinematic energy will be represented by the weak form like

$$\delta\mathcal{K}_e = \int_{S_0} (\underline{\dot{\widehat{\delta\mathcal{U}}}}^T \mathcal{M}_{\epsilon\epsilon}\underline{\widehat{\mathcal{V}}} + \underline{\widehat{\delta\mathcal{U}}}^T \widetilde{\widetilde{\mathcal{V}}}^T \mathcal{M}_{\epsilon\epsilon}\underline{\widehat{\mathcal{V}}})\, dS_0 \tag{3.215}$$

Integrating the variation of kinematic energy by part in time domain will produce

$$\int_t \delta\mathcal{K}_e dt = \int_t \int_{S_0} (-\underline{\widehat{\delta\mathcal{U}}}^T \mathcal{M}_{\epsilon\epsilon}\underline{\dot{\widehat{\mathcal{V}}}} + \underline{\widehat{\delta\mathcal{U}}}^T \widetilde{\widetilde{\mathcal{V}}}^T \mathcal{M}_{\epsilon\epsilon}\underline{\widehat{\mathcal{V}}})\, dS_0 dt \tag{3.216}$$

From which the variation of kinematic energy is reformulated to

$$\delta\mathcal{K}_e = \int_{S_0} (-\underline{\widehat{\delta\mathcal{U}}}^T \mathcal{M}_{\epsilon\epsilon}\underline{\dot{\widehat{\mathcal{V}}}} + \underline{\widehat{\delta\mathcal{U}}}^T \widetilde{\widetilde{\mathcal{V}}}^T \mathcal{M}_{\epsilon\epsilon}\underline{\widehat{\mathcal{V}}})\, dS_0 \tag{3.217}$$

With the application of finite element method, the velocity and acceleration of reference particle are approximated by the following polynomials of nodal velocities $\underline{\widehat{\mathcal{V}}}_n$ and accelerations $\underline{\dot{\widehat{\mathcal{V}}}}_n$, respectively

$$\hat{\underline{\mathcal{V}}} = \mathcal{H}_\epsilon \hat{\underline{\mathcal{V}}}_n, \quad \dot{\hat{\underline{\mathcal{V}}}} = \mathcal{H}_\epsilon \dot{\hat{\underline{\mathcal{V}}}}_n \tag{3.218}$$

and the weak form of kinematic energy variation is found to be

$$\delta \mathcal{K}_e = \widehat{\underline{\delta \mathcal{U}}}_n^T \left( - \int_{S_0} \mathcal{H}_\epsilon^T \mathcal{M}_{\epsilon\epsilon} \mathcal{H}_\epsilon d S_0 \dot{\hat{\underline{\mathcal{V}}}}_n + \int_{S_0} \mathcal{H}_\epsilon^T \tilde{\hat{\mathcal{V}}}^T \mathcal{M}_{\epsilon\epsilon} \hat{\underline{\mathcal{V}}} d S_0 \right) \tag{3.219}$$

After defined the mass matrix and the Coriolis force vector

$$\mathcal{M}_n = \int_{S_0} \mathcal{H}_\epsilon^T \mathcal{M}_{\epsilon\epsilon} \mathcal{H}_\epsilon d S_0, \quad \underline{\mathcal{F}}_g = \int_{S_0} \mathcal{H}_\epsilon^T \tilde{\hat{\mathcal{V}}}^T \mathcal{M}_{\epsilon\epsilon} \hat{\underline{\mathcal{V}}} d S_0 \tag{3.220}$$

the variation of original kinematic energy can be rewritten into the compact formula

$$\delta \mathcal{K}_e = \underline{\delta \mathcal{U}}_n^T (-\mathcal{M}_n \dot{\hat{\underline{\mathcal{V}}}}_n + \underline{\mathcal{F}}_g) \tag{3.221}$$

### *3.2.8 Extended to Shell Dynamic Problem*

When considering the inertial force and the Coriolis force, the governing equations of motion for a shell can be obtained from the Hamilton's principle. Equivalently, the inertial force and the associated Coriolis force are required to be added into the Eq. (3.206) to get

$$\mathcal{M}_n \dot{\hat{\underline{\mathcal{V}}}}_n - \underline{\mathcal{F}}_g + \underline{\mathcal{F}}_e = \underline{\mathcal{F}}_a \tag{3.222}$$

## 3.3 General Shell-Like Theory of 6 DOFS

As discussed in Sect. 3.2, the director of a shell is selected as the vector along the unit vector $\bar{b}_3$. As a result, the sixth degree of freedom for a shell, called "drilling rotation," is immaterial and always be abandoned. Modifications is performed in this section with the purpose of affording material properties for the "drilling rotation", such that the shell theory of 6 DOFS could be implemented.

### *3.3.1 Modification of Material Properties for a Shell*

Referring Eq. (3.173), the elements of last column of $\tilde{s}_0^T$ are all zero. It will then lead to the last zero raw and column of material matrices

$$\mathcal{C}_{11} = \int_h \widetilde{S}_0 \mathcal{D}_{11} \widetilde{S}_0^T d\alpha_3, \ \mathcal{C}_{22} = \int_h \widetilde{S}_0 \mathcal{D}_{22} \widetilde{S}_0^T d\alpha_3, \ \mathcal{C}_{21} = \int_h \widetilde{S}_0 \mathcal{D}_{21} \widetilde{S}_0^T d\alpha_3 \tag{3.223}$$

and mass matrix $\mathcal{M}_{\epsilon\epsilon}$ with the definition of

$$\mathcal{M}_{\epsilon\epsilon} = \int_h \begin{bmatrix} \rho I, & \rho \widetilde{s}_0^T \\ \rho \widetilde{s}_0, & \rho \widetilde{s}_0 \widetilde{s}_0^T \end{bmatrix} d\alpha_3 \tag{3.224}$$

Here, the characteristic volume of three-dimensional Cosserat continuum with the magnitude of $V_c$ is introduced to evaluate approximately the above matrices, thereby replacing the use of the thickness $h$ of a shell as the characteristic length. As described in Fig. 3.3, the director $\underline{s}_{\ell_0}$ of the local three-dimensional volume is introduced. Hence, the original one-dimensional integrals along the thickness direction, Eqs. (3.223) and (3.224), can be approximated by the integrals on the characteristic volume averaged by the characteristic area $S_c$ like

$$\mathcal{C}_{11} = \frac{\int_{V_c} \widetilde{S}_0 \mathcal{D}_{11} \widetilde{S}_0^T dV_c}{S_c}, \ \mathcal{C}_{22} = \frac{\int_{V_c} \widetilde{S}_0 \mathcal{D}_{22} \widetilde{S}_0^T dV_c}{S_c}, \ \mathcal{C}_{21} = \frac{\int_{V_c} \widetilde{S}_0 \mathcal{D}_{21} \widetilde{S}_0^T dV_c}{S_c} \tag{3.225}$$

and

$$\mathcal{M}_{\epsilon\epsilon} = \frac{\int_{V_c} \begin{bmatrix} \rho I, & \rho \widetilde{s}_0^T \\ \rho \widetilde{s}_0, & \rho \widetilde{s}_0 \widetilde{s}_0^T \end{bmatrix} dV_c}{S_c} \tag{3.226}$$

to eliminate the singularity of material matrices and mass matrix, where the magnitude of characteristic area $S_c$ is determined from $S_c = V_c / h$, and the director $\underline{s}_0 = \alpha_3 \bar{\imath}_3$ will be replaced by using the director $\underline{s}_{\ell_0}$ of the local three-dimensional volume.

**Fig. 3.3** The characteristic volume $V_c$ replacing the use of the thickness $h$ as the characteristic length

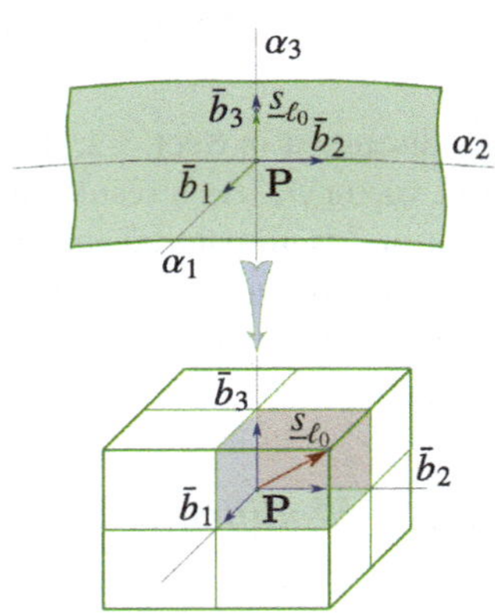

### 3.3.2 Variation of Strain Energy and Virtual Work of External Forces

The second Piola-Kirchhoff stress and Green-Lagrange strain measures are selected so that their product gives an accurate energy density that is related to the undeformed volume of reference configuration. Then, the strain energy of a shell can be written as

$$2\mathcal{U}_e = \int_{S_0} (\underline{\epsilon}_1^T \mathcal{C}_{11}\underline{\epsilon}_1 + \underline{\epsilon}_2^T \mathcal{C}_{22}\underline{\epsilon}_2 + 2\underline{\epsilon}_2^T \mathcal{C}_{21}\underline{\epsilon}_1)\, dS_0 \tag{3.227}$$

where $S_0$ is the area of the shell mid-surface. The variation of $\mathcal{U}_e$ is readily verified to be

$$\delta\mathcal{U}_e = \int_{S_0} \delta\underline{\epsilon}_1^T \mathcal{C}_{11}\underline{\epsilon}_1 + \delta\underline{\epsilon}_2^T \mathcal{C}_{22}\underline{\epsilon}_2 + \delta\underline{\epsilon}_2^T \mathcal{C}_{21}\underline{\epsilon}_1 + \delta\underline{\epsilon}_1^T \mathcal{C}_{12}\underline{\epsilon}_2\, dS_0 \tag{3.228}$$

After merging the similar terms at the right-hand-side of above equations, the variations can be rearranged into the compact form

$$\delta\mathcal{U} = \int_{S_0} \delta\underline{\epsilon}_1^T \underline{F}_{\epsilon\epsilon}^{11} + \delta\underline{\epsilon}_2^T \underline{F}_{\epsilon\epsilon}^{22}\, dS_0 \tag{3.229}$$

where the stress measures of a shell are defined in the vectorial formulas like

$$\underline{F}_{\epsilon\epsilon}^{11} = \mathcal{C}_{11}\underline{\epsilon}_1 + \mathcal{C}_{12}\underline{\epsilon}_2, \quad \underline{F}_{\epsilon\epsilon}^{22} = \mathcal{C}_{22}\underline{\epsilon}_2 + \mathcal{C}_{21}\underline{\epsilon}_1 \tag{3.230}$$

Then, the variation of strain energy can be represented by

$$\delta\mathcal{U}_e = \int_{S_0} (\delta\underline{\epsilon}_1^T \underline{F}_{\epsilon\epsilon}^{11} + \delta\underline{\epsilon}_2^T \underline{F}_{\epsilon\epsilon}^{22})dS_0 \tag{3.231}$$

Next, the virtual work of external forces can be computed from

$$\delta\mathcal{W}_e = \int_{V_0} \delta\underline{P}^T R_1 \underline{g}\, dV_0 + \int_{S_0} \delta\underline{P}^T R_1 \underline{p}_s\, dS_0 \tag{3.232}$$

Referring the formula of position variation, Eq. (3.189), the virtual work can be expressed in the following detailed formula as

$$\delta\mathcal{W}_e = \int_{V_0} (R_1^T \delta\underline{u}_s + \widetilde{s}_0^T \delta\underline{\psi})^T \underline{g}\, dV_0 + \int_{S_0} (R_1^T \delta\underline{u}_s + \widetilde{s}_0^T \delta\underline{\psi})^T \underline{p}_s\, dS_0$$
$$= \int_{S_0} \left[ \delta\underline{u}_s^T R_1 (\int_h \underline{g}\, d\alpha_3 + \underline{p}_s) + \delta\underline{\psi}^T (\int_h \widetilde{s}_0 \underline{g}\, d\alpha_3 + \widetilde{s}_0 \underline{p}_s) \right] dS_0 \tag{3.233}$$

For the easy description of virtual work, the external force and torque vectors are defined at first by

$$\underline{F} = \int_h \underline{g}\, d\alpha_3 + \underline{p}_s, \quad \underline{T} = \int_h \widetilde{s}_0 \underline{g}\, d\alpha_3 + \widetilde{s}_0 \underline{p}_s \tag{3.234}$$

and then the virtual work can be rewritten as

$$\delta\mathcal{W}_e = \int_{S_0} (\delta\underline{u}_s^T R_1 \underline{F} + \delta\underline{\psi}^T \underline{T})\, dS_0 \tag{3.235}$$

or in the motion formalism like

$$\delta\mathcal{W}_e = \int_{S_0} \underline{\delta\mathcal{U}}^T \underline{F}_{\epsilon\alpha} dS_0 \tag{3.236}$$

where the following dual brackets are introduced

$$\underline{F}_{\epsilon\alpha} = \begin{vmatrix} \underline{F} \\ \underline{T} \end{vmatrix}, \quad \underline{\delta\mathcal{U}} = \begin{vmatrix} R_1^T \delta\underline{u}_s \\ \delta\underline{\psi} \end{vmatrix} \tag{3.237}$$

### 3.3.3 Governing Equations of Shell with 6 DOFs in Motion Formalism

The governing equations for a shell problem could be obtained from the principle of virtual work, which states

$$\int_{S_0} (\delta\underline{\epsilon}_1^T \underline{F}_{\epsilon\epsilon}^{11} + \delta\underline{\epsilon}_2^T \underline{F}_{\epsilon\epsilon}^{22} - \underline{\delta\mathcal{U}}^T \underline{F}_{\epsilon\alpha}) dS_0 = 0 \tag{3.238}$$

In view of compatibility attributes, Eq. (2.161), the variations of Lagrangian strains are found to be

$$\delta\underline{\epsilon}_1 = \underline{\delta\mathcal{U}}_{,1} + \widetilde{\mathcal{K}}_1^1 \underline{\delta\mathcal{U}}, \quad \delta\underline{\epsilon}_2 = \underline{\delta\mathcal{U}}_{,2} + \widetilde{\mathcal{K}}_2^1 \underline{\delta\mathcal{U}} \tag{3.239}$$

then the principle of virtual work is recasted to

$$\int_{S_0} [(\underline{\delta \mathcal{U}}_{,1} + \widetilde{\mathcal{K}}_1^1 \underline{\delta \mathcal{U}})^T \underline{F}_{\epsilon\epsilon}^{11} + (\underline{\delta \mathcal{U}}_{,2} + \widetilde{\mathcal{K}}_2^1 \underline{\delta \mathcal{U}})^T \underline{F}_{\epsilon\epsilon}^{22} - \underline{\delta \mathcal{U}}^T \underline{F}_{\epsilon\alpha}] d S_0 = 0 \tag{3.240}$$

With the application of finite element discretization, the virtual motions and its spatial derivatives are interpolated by using the shape functions $\mathcal{H}_\epsilon$ to get

$$\underline{\delta \mathcal{U}} = \mathcal{H}_\epsilon \underline{\delta \mathcal{U}}_n, \ \underline{\delta \mathcal{U}}_{,i} = \mathcal{H}_{\epsilon,i} \underline{\delta \mathcal{U}}_n \tag{3.241}$$

With the aid of the above discretizations, the weak formula of the principle of virtual work becomes

$$\underline{\delta \mathcal{U}}_n^T \int_{S_0} [(\mathcal{H}_{\epsilon,1}^T + \mathcal{H}_\epsilon^T \widetilde{\mathcal{K}}_1^{1T}) \underline{F}_{\epsilon\epsilon}^{11} + (\mathcal{H}_{\epsilon,2}^T + \mathcal{H}_\epsilon^T \widetilde{\mathcal{K}}_2^{1T}) \underline{F}_{\epsilon\epsilon}^{22} - \mathcal{H}_\epsilon^T \underline{F}_{\epsilon\alpha}] d S_0 = 0 \tag{3.242}$$

From which, the governing equations are obtained to be

$$\int_{S_0} [(\mathcal{H}_{\epsilon,1}^T + \mathcal{H}_\epsilon^T \widetilde{\mathcal{K}}_1^{1T}) \underline{F}_{\epsilon\epsilon}^{11} + (\mathcal{H}_{\epsilon,2}^T + \mathcal{H}_\epsilon^T \widetilde{\mathcal{K}}_2^{1T}) \underline{F}_{\epsilon\epsilon}^{22} - \mathcal{H}_\epsilon^T \underline{F}_{\epsilon\alpha}] d S_0 = 0 \tag{3.243}$$

After introduced the elastic force vector

$$\underline{\mathcal{F}}_e = \int_{S_0} [(\mathcal{H}_{\epsilon,1}^T + \mathcal{H}_\epsilon^T \widetilde{\mathcal{K}}_1^{1T}) \underline{F}_{\epsilon\epsilon}^{11} + (\mathcal{H}_{\epsilon,2}^T + \mathcal{H}_\epsilon^T \widetilde{\mathcal{K}}_2^{1T}) \underline{F}_{\epsilon\epsilon}^{22}] d S_0 \tag{3.244}$$

and the external force vector

$$\underline{\mathcal{F}}_a = \int_{S_0} \mathcal{H}_\epsilon^T \underline{F}_{\epsilon\alpha} d S_0 \tag{3.245}$$

the governing equations can be described by the following compact formula

$$\underline{\mathcal{F}}_e - \underline{\mathcal{F}}_a = 0 \tag{3.246}$$

### 3.3.4 Kinematic Energy of Shell with 6 DOFs

When governing equation of the shell is extended to the dynamic problem, the kinematic energy of shell needs to be evaluated, which can be computed from the integral related to the initial volume $V_0$ like

$$\mathcal{K}_e = \int_{V_0} \frac{1}{2}\rho(R_1^T\underline{\dot{P}})^T R_1^T\underline{\dot{P}}dV_0$$
$$= \int_{V_0} \frac{1}{2}\lfloor \underline{\dot{u}}_s^T R_1,\ \underline{\omega}^T \rfloor \begin{bmatrix} \rho I, & \rho\tilde{s}_0^T \\ \rho\tilde{s}_0, & \rho\tilde{s}_0\tilde{s}_0^T \end{bmatrix} \begin{vmatrix} R_1^T\underline{\dot{u}}_s \\ \underline{\omega} \end{vmatrix} dV_0 \tag{3.247}$$

In view of Eq. (3.226), the motion formalism of kinematic energy is readily obtained to be

$$\mathcal{K}_e = \frac{1}{2}\int_{S_0} \underline{\mathcal{V}}^T\mathcal{M}_{\epsilon\epsilon}\underline{\mathcal{V}}dS_0 \tag{3.248}$$

the variation of kinematic energy is readily found from the above equations as

$$\delta\mathcal{K}_e = \int_{S_0} \delta\underline{\mathcal{V}}^T\mathcal{M}_{\epsilon\epsilon}\underline{\mathcal{V}}dS_0 \tag{3.249}$$

According to the associated compatibility attributes, Eq. (2.160), the variation of kinematic energy can be represented by the weak form like

$$\delta\mathcal{K}_e = \int_{S_0} (\underline{\delta\dot{\mathcal{U}}}^T\mathcal{M}_{\epsilon\epsilon}\underline{\mathcal{V}} + \underline{\delta\mathcal{U}}^T\tilde{\mathcal{V}}^T\mathcal{M}_{\epsilon\epsilon}\underline{\mathcal{V}})\, dS_0 \tag{3.250}$$

Integrating the variation of kinematic energy by part in time domain produces

$$\int_t \delta\mathcal{K}_e dt = \int_t\int_{S_0} (-\underline{\delta\mathcal{U}}^T\mathcal{M}_{\epsilon\epsilon}\underline{\dot{\mathcal{V}}} + \underline{\delta\mathcal{U}}^T\tilde{\mathcal{V}}^T\mathcal{M}_{\epsilon\epsilon}\underline{\mathcal{V}})\, dS_0 dt \tag{3.251}$$

From which the variation of kinematic energy is reformulated to

$$\delta\mathcal{K}_e = \int_{S_0} (-\underline{\delta\mathcal{U}}^T\mathcal{M}_{\epsilon\epsilon}\underline{\dot{\mathcal{V}}} + \underline{\delta\mathcal{U}}^T\tilde{\mathcal{V}}^T\mathcal{M}_{\epsilon\epsilon}\underline{\mathcal{V}})\, dS_0 \tag{3.252}$$

With the application of finite element discretization, the velocity and acceleration of reference particle is approximated by the following polynomials of nodal velocities $\underline{\mathcal{V}}_n$ and nodal accelerations $\underline{\dot{\mathcal{V}}}_n$, respectively

$$\underline{\mathcal{V}} = \mathcal{H}_\epsilon\underline{\mathcal{V}}_n, \quad \underline{\dot{\mathcal{V}}} = \mathcal{H}_\epsilon\underline{\dot{\mathcal{V}}}_n \tag{3.253}$$

and the weak form of kinematic energy variation is found to be

$$\delta\mathcal{K}_e = \underline{\delta\mathcal{U}}_n^T\left(-\int_{S_0} \mathcal{H}_\epsilon^T\mathcal{M}_{\epsilon\epsilon}\mathcal{H}_\epsilon dS_0\underline{\dot{\mathcal{V}}}_n + \int_{S_0} \mathcal{H}_\epsilon^T\tilde{\mathcal{V}}^T\mathcal{M}_{\epsilon\epsilon}\underline{\mathcal{V}}dS_0\right) \tag{3.254}$$

After defined the mass matrix and the Coriolis force vector

$$\mathcal{M}_n = \int_{S_0} \mathcal{H}_\epsilon^T \mathcal{M}_{\epsilon\epsilon} \mathcal{H}_\epsilon dS_0, \quad \underline{\mathcal{F}}_g = \int_{S_0} \mathcal{H}_\epsilon^T \widetilde{\mathcal{V}}^T \mathcal{M}_{\epsilon\epsilon} \underline{\mathcal{V}} dS_0 \tag{3.255}$$

the variation of original kinematic energy can be rewritten into the compact formula

$$\delta \mathcal{K}_e = \delta \underline{\mathcal{U}}_n^T (-\mathcal{M}_n \dot{\underline{\mathcal{V}}}_n + \underline{\mathcal{F}}_g) \tag{3.256}$$

For dynamic problem of a shell, the governing equations of shell motion can be obtained from the Hamilton's principle. Equivalently, the inertial force and the associated Coriolis force are required to be added into the Eq. (3.246) to obtain

$$\mathcal{M}_n \dot{\underline{\mathcal{V}}}_n - \underline{\mathcal{F}}_g + \underline{\mathcal{F}}_e = \underline{\mathcal{F}}_a \tag{3.257}$$

## 3.4 General Beam-Like Theory

Geometrically exact beam that has been well developed over the decades is the most appropriate model that can accurately describe arbitrary motions and deformations of slender beam-like structures. As mentioned in Sect. 3.2, the concept of *geometrically exact* [56, 57], stems from the identification of the analytical models, in which states that once the kinetic assumptions, such as small strain, have been made, the ensuing derivations of governing equations are strictly kept within the framework of these assumptions and no more kinematic assumptions are introduced. Note that the strain-displacement relations defined in this book are all exactly described under the small strain assumption.

Beginning with the work of Reissner [58], the exact intrinsic equations for beam static equilibrium are derived under the limitation of unrestrained warping. Following with the work of Simo and Vu-Quoc [59, 60], the research as to geometrically exact beam theories and numerical implementations [61] has yielded significant advances. Typically, Bauchau and his co-workers [62, 63] successively developed several beam models, which undergoing arbitrarily large deflections and rotations meanwhile the strain level remains low. In their latest works, the motion was manipulated via dual entities [64], and the global interpolation of motion [65] was proposed also. Hodges [66] developed a geometrically exact intrinsic dynamics model of beams that can be initially curved and twisted. The beam constitutive law is based on a separate finite element analysis [67, 68] and valid for anisotropic beams with nonhomogeneous cross-sections.

Since the geometrically exact beam theory has been well developed, few concentrations are paid to the theoretical fundamentals of the beam. Hence, it often makes the beginners fell confused as to what the relations of beam theory to three-dimensional solid mechanics. This confusion can be readily clarified by deriving the beam model from the theory of three-dimensional Cosserat continuum. In view of the geometric dimensions of beam-like structures, it is apparently observed that two dimensions associated with the cross-section of a beam are usually much smaller than the dimension of beam axis. Due to this observation, beam-like structure can be exactly treated as one-dimensional Cosserat continuum [2]. Apparently, the characteristic area of a beam will be the area of the cross-section. In current implementation, the curvilinear coordinate $\alpha_2$ and $\alpha_3$ are selected as the cross-section dimensions, such that the derivatives with respected to $\alpha_2$ and $\alpha_3$ are all neglected. Consequently, under certain conditions, it is possible to model a beam-like structure with governing equations that depend on only one spatial coordinate $\alpha_1$, and the associated theory is called general beam-like theory in this book.

### 3.4.1 Kinematics of Beam

As shown in Fig. 3.4, an arbitrary material particle **B** with infinitesimal volume in a beam is depicted by its position and director vectors in its reference configuration and current deformed configuration, respectively. In the original reference configuration, the position vector $\underline{p}(\alpha_i)$ of the material particle relative to a point **O**, the original of reference frame, $\mathcal{F}_{\mathcal{I}} = [\mathbf{O}, \mathcal{I}]$, can be written as

$$\underline{p}(\alpha_1, \alpha_2, \alpha_3) = \underline{x}_0(\alpha_1) + R_0(\alpha_1)\underline{s}_0(\alpha_2, \alpha_3) \tag{3.258}$$

where the general curvilinear coordinates, $(\alpha_1, \alpha_2, \alpha_3)$, are introduced to label the material particles in the original continuum, and $R_0$ the initial rotation tensor measures the orientation of the reference basis $\mathcal{B}_0$. It has been verified in Sect. 2.2.3 the rotation tensor $R_0$ is identical to the body attached basis $\mathcal{B}_0$, and in detail

$$R_0 = \lfloor \bar{b}_1, \bar{b}_2, \bar{b}_3 \rfloor = \mathcal{B}_0 \tag{3.259}$$

Note that the initial basis $\mathcal{B}_0$ is bounded by the referenced material particle **P** to produce the reference frame $\mathcal{F}_0 = [\mathbf{P}, \mathcal{B}_0]$. The local vector $\underline{s}_0$, defined in this reference frame $\mathcal{F}_0$, describes the director of material particle **B** and

$$\underline{s}_0 = \alpha_2 \bar{\imath}_2 + \alpha_3 \bar{\imath}_3 \tag{3.260}$$

Apparently, the director $\underline{s}_0$ locates in the plane of beam's cross-section. According to above definitions, it is readily verified

$$\underline{s}_{0,1} = \underline{0}, \quad \underline{s}_{0,2} = \bar{\imath}_2, \quad \underline{s}_{0,3} = \bar{\imath}_3 \tag{3.261}$$

It is very important to emphasize the area magnitude of cross-section is exactly the characteristic area in a one-dimensional Cosserat continuum associated with torsional and bending deformations. Obviously, Eq. (3.258) decomposes the position vector

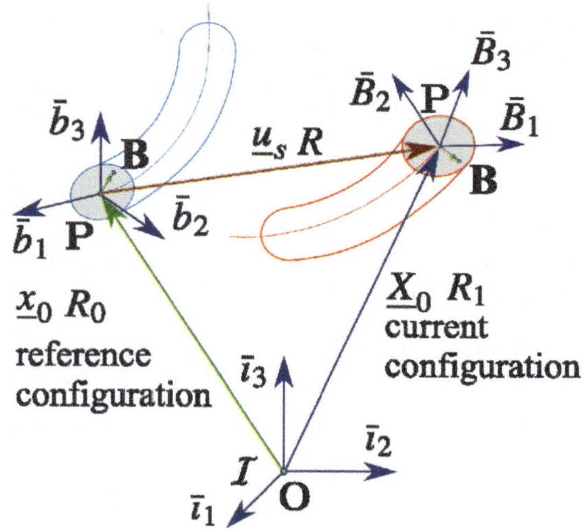


**Fig. 3.4** Kinematics of a material particle with infinitesimal volume in a beam

$\underline{p}$ of a material particle **B** into a reference position $\underline{x}_0$ and a local vector $\underline{s}_0$. The reference particle **P** with position $\underline{x}_0$ can translate freely, while the rigid director $\underline{s}_0$ attached to the reference particle **P** can only rotate without deformations.

For the easy description of beam deformations, the covariant base vectors,

$$\underline{g}_i = \frac{\partial \underline{p}}{\partial \alpha_i} \tag{3.262}$$

in the reference configuration are required, and the specific expressions can be obtained by taking derivatives of Eq. (3.258) with respected to the curvilinear coordinates $\alpha_i$ like

$$\begin{aligned} \underline{g}_1 &= \underline{x}_{0,1} + R_{0,1}\underline{s}_0 \\ \underline{g}_2 &= R_0\bar{\imath}_2 \\ \underline{g}_3 &= R_0\bar{\imath}_3 \end{aligned} \tag{3.263}$$

where the symbol $(\cdot)_{,i}$ for $i = 1, 2, 3$ denotes the spatial derivatives with respected to $\alpha_i$ and

$$(\cdot)_{,i} = \frac{\partial(\cdot)}{\partial \alpha_i} \tag{3.264}$$

Further operations are taken to resolve the base vectors into the local reference frame $\mathcal{F}_0$ to get

$$\begin{aligned} \underline{g}_1^* &= R_0^T\underline{x}_{0,1} + R_0^T R_{0,1}\underline{s}_0 \\ \underline{g}_2^* &= \bar{\imath}_2 \\ \underline{g}_3^* &= \bar{\imath}_3 \end{aligned} \tag{3.265}$$

According to the definitions of initial curvature tensors, Eq. (1.314), the base vectors can be represented by the following formulas like

$$\begin{aligned} \underline{g}_1^* &= R_0^T \underline{x}_{0,1} + \widetilde{\kappa}_1^0 \underline{s}_0 \\ \underline{g}_2^* &= \bar{\imath}_2 \\ \underline{g}_3^* &= \bar{\imath}_3 \end{aligned} \tag{3.266}$$

where the symbol $\widetilde{\kappa}_1^0 = R_0^T R_{0,1}$ denotes the initial curvature tensor. All these base vectors form the covariant bases $\mathcal{B}_g$ like

$$\mathcal{B}_g = \lfloor \underline{g}_1^*,\ \underline{g}_2^*,\ \underline{g}_3^* \rfloor \tag{3.267}$$

and then the metric tensor in the reference configuration is computed from

$$g = \mathcal{B}_g^T \mathcal{B}_g \tag{3.268}$$

For the easy description of global and local motions of the material particle, the covariant bases $\mathcal{B}_g$ are decomposed into two parts

$$\mathcal{B}_g = \mathcal{B}_g^s + \mathcal{B}_g^\alpha \tag{3.269}$$

with the aid of introduced symbols as

$$\mathcal{B}_g^s = \lfloor \underline{g}_1^s,\ \underline{g}_2^s,\ \underline{g}_3^s \rfloor,\ \mathcal{B}_g^\alpha = \lfloor \underline{g}_1^\alpha,\ \underline{g}_2^\alpha,\ \underline{g}_3^\alpha \rfloor \tag{3.270}$$

where the base vectors are given in detail for global part like

$$\begin{aligned} \underline{g}_1^s &= R_0^T \underline{x}_{0,1} + \widetilde{\kappa}_1^0 \underline{s}_0 \\ \underline{g}_2^s &= \underline{0} \\ \underline{g}_3^s &= \underline{0} \end{aligned} \tag{3.271}$$

and for local part

$$\underline{g}_1^\alpha = \underline{0},\ \underline{g}_2^\alpha = \bar{\imath}_2,\ \underline{g}_3^\alpha = \bar{\imath}_3 \tag{3.272}$$

In view of this, the metric tensor $g$ will be consisted of

$$g = g^{ss} + g^{s\alpha} + g^{\alpha s} + g^{\alpha\alpha} \tag{3.273}$$

and in detail

$$g^{ss} = \mathcal{B}_g^{sT} \mathcal{B}_g^s,\ g^{s\alpha} = \mathcal{B}_g^{sT} \mathcal{B}_g^\alpha,\ g^{\alpha s} = \mathcal{B}_g^{\alpha T} \mathcal{B}_g^s,\ g^{\alpha\alpha} = \mathcal{B}_g^{\alpha T} \mathcal{B}_g^\alpha \tag{3.274}$$

When the beam-like structure deforms and moves in space, the material particle **B** that had position vector $\underline{p}$ in the undeformed reference configuration now has position vector $\underline{P}$ in the deformed configuration, such that

$$\underline{P}(\alpha_1, \alpha_2, \alpha_3) = \underline{X}_0(\alpha_1) + R_1(\alpha_1)\underline{s}_0(\alpha_2, \alpha_3) \tag{3.275}$$

where $R_1$ is the total rotation tensor that describes the orientation of basis $\mathcal{B}_1$ attached to the deformed configuration and

$$R_1 = \lfloor \bar{B}_1, \bar{B}_2, \bar{B}_3 \rfloor = \mathcal{B}_1 \tag{3.276}$$

The associated frame $\mathcal{F}_1 = [\mathbf{P}, \mathcal{B}_1]$ will be constructed when bounding the current deformed basis $\mathcal{B}_1$ by the referenced material particle **P**. The position vector $\underline{X}_0$ of reference point **P** in current deformed configuration can be recasted to

$$\underline{X}_0(\alpha_1, \alpha_2, \alpha_3) = \underline{x}_0(\alpha_1) + \underline{u}_s(\alpha_1) \tag{3.277}$$

where the displacement vector $\underline{u}_s$ describes arbitrary translation of reference point. Alternatively, the position vector $\underline{P}$ of the material particle in its deformed configuration can be represented by

$$\underline{P}(\alpha_1, \alpha_2, \alpha_3) = \underline{p}(\alpha_1, \alpha_2, \alpha_3) + \underline{u}(\alpha_1, \alpha_2, \alpha_3) \tag{3.278}$$

where $\underline{u}$ is the displacement of material particle **B** in space. It is readily to find that the displacement $\underline{u}$ are related to

$$\underline{u} = \underline{u}_s + (R_1 - R_0)\underline{s}_0 \tag{3.279}$$

with a very clear physical meaning: the displacement $\underline{u}$ is composed of the translation $\underline{u}_s$ of reference particle and the pure rotation $(R_1 - R_0)\underline{s}_0$ of local director.

According to the definitions of covariant base vectors in deformed configuration

$$\underline{G}_i = \frac{\partial \underline{P}}{\partial \alpha_i} \tag{3.280}$$

the specific formulas can be obtained by taking spatial derivatives of Eq. (3.275) with respected to the curvilinear coordinates $\alpha_i$ for $i = 1, 2, 3$ like

$$\begin{aligned} \underline{G}_1 &= \underline{X}_{0,1} + R_{1,1}\underline{s}_0 \\ \underline{G}_2 &= R_1\bar{\imath}_2 \\ \underline{G}_3 &= R_1\bar{\imath}_3 \end{aligned} \tag{3.281}$$

Once again, the base vectors are resolved into the current local frame $\mathcal{F}_1$ to obtain

$$\begin{aligned} \underline{G}_1^* &= R_1^T \underline{X}_{0,1} + R_1^T R_{1,1} \underline{s}_0 \\ \underline{G}_2^* &= \bar{\imath}_2 \\ \underline{G}_3^* &= \bar{\imath}_3 \end{aligned} \tag{3.282}$$

In view of the definition of current curvature tensor $\widetilde{\kappa}_1^1 = R_1^T R_{1,1}$, the covariant base vectors are simplified to

$$\begin{aligned} \underline{G}_1^* &= R_1^T \underline{X}_{0,1} + \widetilde{\kappa}_1^1 \underline{s}_0 \\ \underline{G}_2^* &= \bar{\imath}_2 \\ \underline{G}_3^* &= \bar{\imath}_3 \end{aligned} \tag{3.283}$$

All these base vectors could be assembled into the covariant bases $\mathcal{B}_G$ like

$$\mathcal{B}_G = \lfloor \underline{G}_1^*,\ \underline{G}_2^*,\ \underline{G}_3^* \rfloor \tag{3.284}$$

which is decomposed into two parts

$$\mathcal{B}_G = \mathcal{B}_G^s + \mathcal{B}_G^\alpha \tag{3.285}$$

with the help of new symbols defined in detail as

$$\mathcal{B}_G^s = \lfloor \underline{G}_1^s,\ \underline{G}_2^s,\ \underline{G}_3^s \rfloor,\ \mathcal{B}_G^\alpha = \lfloor \underline{G}_1^\alpha,\ \underline{G}_2^\alpha,\ \underline{G}_3^\alpha \rfloor \tag{3.286}$$

where the global and local base vectors are introduced to be

$$\begin{aligned} \underline{G}_1^s &= R_1^T \underline{X}_{0,1} + \widetilde{\kappa}_1^1 \underline{s}_0 \\ \underline{G}_2^s &= \underline{0} \\ \underline{G}_3^s &= \underline{0} \end{aligned} \tag{3.287}$$

and

$$\underline{G}_1^\alpha = \underline{0},\ \underline{G}_2^\alpha = \bar{\imath}_2,\ \underline{G}_3^\alpha = \bar{\imath}_3 \tag{3.288}$$

In view of above expressions of covariant basis vectors in the deformed configuration, the associated metric tensor

$$G = \mathcal{B}_G^T \mathcal{B}_G \tag{3.289}$$

can be found to be composed of

$$G = G^{ss} + G^{s\alpha} + G^{\alpha s} + G^{\alpha\alpha} \tag{3.290}$$

with the following definitions

$$G^{ss} = \mathcal{B}_G^{sT} \mathcal{B}_G^s,\ G^{s\alpha} = \mathcal{B}_G^{sT} \mathcal{B}_G^\alpha,\ G^{\alpha s} = \mathcal{B}_G^{\alpha T} \mathcal{B}_G^s,\ G^{\alpha\alpha} = \mathcal{B}_G^{\alpha T} \mathcal{B}_G^\alpha \tag{3.291}$$

### 3.4.2 Strain Tensor of Beam

The rotation-free Green-Lagrange strain tensor [22] will be applied to measure the deformation of Cosserat continuum. In doing so, the Lagrangian stretch tensor and the Lagrangian curvature strain tensor [23, 24]

$$\underline{e}_1 = R_1^T \underline{X}_{0,1} - R_0^T \underline{x}_{0,1}, \; \widetilde{\kappa}_1 = \widetilde{\kappa}_1^1 - \widetilde{\kappa}_1^0 = R_1^T R_{1,1} - R_0^T R_{0,1} \tag{3.292}$$

are required to reformulate the deformed covariant base vector $\underline{G}_1^s$ in terms of

$$\underline{G}_1^s = \underline{e}_1 + R_0^T \underline{x}_{0,1} + \widetilde{\kappa}_1^1 \underline{s}_0 \tag{3.293}$$

Note that the Lagrangian stretch $\underline{e}_1$ and curvatures $\underline{\kappa}_1$ are nothing but the components of Lagrangian strain measures, Eq. (2.152), defined as the difference between the curvatures in the dual vector form, such that

$$\underline{\epsilon}_1 = \underline{\mathcal{K}}_1^1 - \underline{\mathcal{K}}_1^0 = \left| \begin{matrix} R_1^T \underline{X}_{0,1} \\ \underline{\kappa}_1^1 \end{matrix} \right| - \left| \begin{matrix} R_0^T \underline{x}_{0,1} \\ \underline{\kappa}_1^0 \end{matrix} \right| = \left| \begin{matrix} \underline{e}_1 \\ \underline{\kappa}_1 \end{matrix} \right| \tag{3.294}$$

Referring the decompositions of metric tensors in reference configuration and deformed configuration, Eqs. (3.273) and (3.290), respectively, the rotation-free Green-Lagrangian strain tensor measured in the curvilinear basis

$$\widehat{\varepsilon} = \frac{1}{2}(G - g) \tag{3.295}$$

can be regarded as a combination of two contributions, $\widehat{\varepsilon} = \varepsilon^s + \varepsilon^c$, one from the global deformation

$$\varepsilon^s = \frac{1}{2}(G^{ss} - g^{ss}) \tag{3.296}$$

and the other from the global-local coupling deformation

$$\varepsilon^c = \frac{1}{2}(G^{\alpha s} + G^{s\alpha} - g^{\alpha s} - g^{s\alpha}) \tag{3.297}$$

The contribution of local deformation is zero in classical Cosserat theory

$$\varepsilon^\alpha = \frac{1}{2}(G^{\alpha\alpha} - g^{\alpha\alpha}) = 0 \tag{3.298}$$

Substituting the specific formulations of metric tensors in terms of covariant basis vectors, Eq. (3.293), into the definitions of strain tensor, Eqs. (3.296) and (3.297), the components of rotation-free Green-Lagrange strain tensor are found to be related to the components of Lagrangian stretch strains and curvature strains. In the vector-dyadic form, the relations are written in detail like

$$\begin{aligned}
\varepsilon_{11} &= (\underline{x}_{0,1}^{T} + \underline{\kappa}_1^{0T}\widetilde{s}_0) R_0 (\underline{e}_1 + \widetilde{s}_0^T \underline{\kappa}_1) \\
2\varepsilon_{12} &= \bar{\imath}_2^T (\underline{e}_1 + \widetilde{s}_0^T \underline{\kappa}_1) \\
2\varepsilon_{13} &= \bar{\imath}_3^T (\underline{e}_1 + \widetilde{s}_0^T \underline{\kappa}_1) \\
\varepsilon_{22} &= 0 \\
2\varepsilon_{23} &= 0 \\
\varepsilon_{33} &= 0
\end{aligned} \tag{3.299}$$

Note that the assumptions of small stretch strain $\underline{e}_1 \ll 0$ and small curvature strain $\underline{\kappa}_1 \ll 0$ are applied such that the higher order terms as to $\underline{e}_1$ and $\underline{\kappa}_1$ are all neglected.

Furthermore, another straight forward approach is presented also to obtain the same expressions of Green-Lagrange strain components, Eq. (3.299), with the aid of base vector differences

$$\begin{aligned}
\Delta\underline{G}_1^* &= \underline{G}_1^* - \underline{g}_1^* = R_1^T \underline{X}_{0,1} - R_0^T \underline{x}_{0,1} + (\widetilde{\kappa}_1^1 - \widetilde{\kappa}_1^0)\underline{s}_0 \\
\Delta\underline{G}_2^* &= \underline{G}_2^* - \underline{g}_2^* = \underline{0} \\
\Delta\underline{G}_3^* &= \underline{G}_3^* - \underline{g}_3^* = \underline{0}
\end{aligned} \tag{3.300}$$

Referring the definitions of Lagrangian stretch and Lagrange curvature measures, Eq. (3.292), the above definitions can be rewritten in the compact form as

$$\Delta\underline{G}_1^* = \underline{e}_1 + \widetilde{s}_0^T \underline{\kappa}_1, \quad \Delta\underline{G}_2^* = \underline{0}, \quad \Delta\underline{G}_3^* = \underline{0} \tag{3.301}$$

The components of Green-Lagrange strain tensor are then approximated to be

$$\begin{aligned}
\varepsilon_{11} &= \frac{1}{2}[(\underline{g}_1^* + \Delta\underline{G}_1^*)^T(\underline{g}_1^* + \Delta\underline{G}_1^*) - \underline{g}_1^{*T}\underline{g}_1^*] \approx \underline{g}_1^{*T}\Delta\underline{G}_1^* \\
2\varepsilon_{12} &= \quad (\underline{g}_1^* + \Delta\underline{G}_1^*)^T(\underline{g}_2^* + \Delta\underline{G}_2^*) - \underline{g}_1^{*T}\underline{g}_2^* \quad = \underline{g}_2^{*T}\Delta\underline{G}_1^* \\
2\varepsilon_{13} &= \quad (\underline{g}_1^* + \Delta\underline{G}_1^*)^T(\underline{g}_3^* + \Delta\underline{G}_3^*) - \underline{g}_1^{*T}\underline{g}_3^* \quad = \underline{g}_3^{*T}\Delta\underline{G}_1^* \\
\varepsilon_{22} &= \frac{1}{2}[(\underline{g}_2^* + \Delta\underline{G}_2^*)^T(\underline{g}_2^* + \Delta\underline{G}_2^*) - \underline{g}_2^{*T}\underline{g}_2^*] = 0 \\
2\varepsilon_{23} &= \quad (\underline{g}_2^* + \Delta\underline{G}_2^*)^T(\underline{g}_3^* + \Delta\underline{G}_3^*) - \underline{g}_2^{*T}\underline{g}_3^* \quad = 0 \\
\varepsilon_{33} &= \frac{1}{2}[(\underline{g}_3^* + \Delta\underline{G}_3^*)^T(\underline{g}_3^* + \Delta\underline{G}_3^*) - \underline{g}_3^{*T}\underline{g}_3^*] = 0
\end{aligned} \tag{3.302}$$

under the same small strain assumption as mentioned before, such that the higher order terms $\Delta\underline{G}_1^{*T}\Delta\underline{G}_1^*$ are all truncated. Once again, the linearized strain measures are obtained to be

$$\begin{aligned}
\varepsilon_{11} &= \underline{g}_1^{*T}(\underline{e}_1 + \widetilde{s}_0^T \underline{\kappa}_1) \\
2\varepsilon_{12} &= \underline{g}_2^{*T}(\underline{e}_1 + \widetilde{s}_0^T \underline{\kappa}_1) \\
2\varepsilon_{13} &= \underline{g}_3^{*T}(\underline{e}_1 + \widetilde{s}_0^T \underline{\kappa}_1) \\
\varepsilon_{22} &= 0 \\
2\varepsilon_{23} &= 0 \\
\varepsilon_{33} &= 0
\end{aligned} \tag{3.303}$$

### 3.4.3 Constitutive Laws for Beam

For the easy description of one-dimensional Cosserat constitutive relations, the Green-Lagrange strain tensor in its vector-dyadic form needs to be decomposed into

$$\underline{\widehat{\varepsilon}} = \begin{vmatrix} \varepsilon_{11} \\ 2\varepsilon_{12} \\ 2\varepsilon_{13} \\ \varepsilon_{22} \\ 2\varepsilon_{23} \\ \varepsilon_{33} \end{vmatrix} = \begin{vmatrix} \varepsilon_{11} \\ \varepsilon_{21} \\ \varepsilon_{31} \\ 0 \\ 0 \\ 0 \end{vmatrix} + \begin{vmatrix} 0 \\ \varepsilon_{12} \\ 0 \\ \varepsilon_{22} \\ \varepsilon_{32} \\ 0 \end{vmatrix} + \begin{vmatrix} 0 \\ 0 \\ \varepsilon_{13} \\ 0 \\ \varepsilon_{23} \\ \varepsilon_{33} \end{vmatrix} = \underline{\varepsilon}_1 + \underline{\varepsilon}_2 + \underline{\varepsilon}_3 \tag{3.304}$$

It is noted that the symbols $\varepsilon_{12}$, $\varepsilon_{21}$, $\varepsilon_{23}$, $\varepsilon_{32}$, and $\varepsilon_{31}$, $\varepsilon_{13}$ do not represent the components of Green-Lagrange strain tensor, but are defined in the following relations

$$2\varepsilon_{12} = \varepsilon_{21} + \varepsilon_{12}, \quad 2\varepsilon_{13} = \varepsilon_{31} + \varepsilon_{13}, \quad 2\varepsilon_{23} = \varepsilon_{32} + \varepsilon_{23} \tag{3.305}$$

After filtered out the zero items, the specific vectorial expressions are reduced to

$$\underline{\hat{\varepsilon}}_1 = \begin{vmatrix} \varepsilon_{11} \\ \varepsilon_{21} \\ \varepsilon_{31} \end{vmatrix}, \ \underline{\hat{\varepsilon}}_2 = \begin{vmatrix} \varepsilon_{12} \\ \varepsilon_{22} \\ \varepsilon_{32} \end{vmatrix} = \underline{\hat{\varepsilon}}_3 = \begin{vmatrix} \varepsilon_{13} \\ \varepsilon_{23} \\ \varepsilon_{33} \end{vmatrix} = \underline{0} \tag{3.306}$$

and in detail

$$\begin{vmatrix} \varepsilon_{11} \\ \varepsilon_{21} \\ \varepsilon_{31} \end{vmatrix} = \begin{vmatrix} (\underline{x}_{0,1}^T R_0 + \underline{\kappa}_1^{0T} \widetilde{s}_0)(\underline{e}_1 + \widetilde{s}_0^T \underline{\kappa}_1) \\ \bar{\imath}_2^T (\underline{e}_1 + \widetilde{s}_0^T \underline{\kappa}_1) \\ \bar{\imath}_3^T (\underline{e}_1 + \widetilde{s}_0^T \underline{\kappa}_1) \end{vmatrix} \tag{3.307}$$

The above formulas could be resolved into

$$\begin{vmatrix} \varepsilon_{11} \\ \varepsilon_{21} \\ \varepsilon_{31} \end{vmatrix} = \begin{bmatrix} \underline{x}_{0,1}^T R_0 + \underline{\kappa}_1^{0T} \widetilde{s}_0 \\ \bar{\imath}_2 \\ \bar{\imath}_3 \end{bmatrix} \left[\, I, \ \widetilde{s}_0^T \,\right] \begin{vmatrix} \underline{e}_1 \\ \underline{\kappa}_1 \end{vmatrix} \tag{3.308}$$

and the associated compact form is found to be

$$\underline{\hat{\varepsilon}}_1 = \mathcal{S}_0^T \underline{\epsilon}_1 \tag{3.309}$$

with the introduced symbol

$$\mathcal{S}_0^T = \mathcal{B}_g \left[\, I, \ \widetilde{s}_0^T \,\right] \tag{3.310}$$

The strain energy $\mathcal{U}_e$ can be represented by using the Green-Lagrange strain vectors $\underline{\varepsilon}_1$ like

$$\mathcal{U}_e = \frac{1}{2}\int\limits_{V_0} \underline{\varepsilon}_1^T \mathcal{D} \underline{\varepsilon}_1 d V_0 \tag{3.311}$$

When the zero components of Green-Lagrange strain vector $\underline{\varepsilon}_1$ are filtered out from above formula, the strain energy will reduce to

$$\mathcal{U}_e = \frac{1}{2}\int\limits_{V_0} \hat{\underline{\varepsilon}}_1^T \mathcal{D}_{11} \hat{\underline{\varepsilon}}_1 \, d V_0 \tag{3.312}$$

where the following constitutive matrix block is defined as

$$\mathcal{D}_{11} = \begin{bmatrix} d_{11}, & d_{12}, & d_{13} \\ d_{21}, & d_{22}, & d_{23} \\ d_{31}, & d_{32}, & d_{33} \end{bmatrix} \tag{3.313}$$

Referring the relations of Green-Lagrange strain and Lagrangian strain measures, Eq. (3.309), the above strain energy is recasted to

$$\mathcal{U}_e = \frac{1}{2}\int\limits_{\ell_0} \underline{\epsilon}_1^T \mathcal{C}_{11} \underline{\epsilon}_1 \, d\alpha_1 \tag{3.314}$$

where $\ell_0$ is the length of the beam, and $\mathcal{C}_{11}$ the constitutive matrix of beam defined to be

$$\mathcal{C}_{11} = \int\limits_{s_0} \mathcal{S}_0 \mathcal{D}_{11} \mathcal{S}_0^T d S_0 \tag{3.315}$$

Apparently, the constitutive matrices of beam are determined by the characteristic area with the magnitude of $S_0$, which is exactly the area of beam cross-section. In a special case of straight beam, the initial curvature $\underline{\kappa}_1^0$ becomes zero, and the covariant bases $\mathcal{B}_g$ recover an identity matrix $I$ of $3 \times 3$. The constitutive matrix of beam can be given in the explicit form of

$$\mathcal{C}_{11} = \int\limits_{s_0} \begin{bmatrix} \mathcal{D}_{11}, & \mathcal{D}_{11}\widetilde{s}_0^T \\ \widetilde{s}_0\mathcal{D}_{11}, & \widetilde{s}_0\mathcal{D}_{11}\widetilde{s}_0^T \end{bmatrix} d S_0 \tag{3.316}$$

This formula is exactly the classical definition of beam cross-sectional stiffness matrix [69], containing six diagonal terms, axial stiffness, shear stiffness, torsional stiffness and bending stiffness, together with the off-diagonal coupling terms.

### 3.4.4 Variation of Strain Energy of Beam

The second Piola-Kirchhoff stress and Green-Lagrange strain measures are selected so that their product gives an accurate energy density that is related to the undeformed volume of reference configuration. Referring the expressions of strain energy, Eq. (3.314), the variation of strain energy for a beam is given as

$$\delta \mathcal{U}_e = \int_{\ell_0} \delta \underline{\epsilon}_1^T \underline{F}_{\epsilon\epsilon}^{11} d\alpha_1 \tag{3.317}$$

where the stress resultants of a beam are defined in the vectorial formula like

$$\underline{F}_{\epsilon\epsilon}^{11} = \mathcal{C}_{11} \underline{\epsilon}_1 \tag{3.318}$$

### 3.4.5 Virtual Work of External Forces for Beam

The virtual work done by a corresponding virtual displacement operating on the external forces is evaluated in this section. According to Eq. (3.275), the virtual displacement of an arbitrary particle **B** in deformed configuration can be expressed in terms of

$$\delta \underline{P} = \delta(\underline{X}_0 + R_1 \underline{s}_0) = \delta(\underline{x}_0 + \underline{u}_s + R_1 \underline{s}_0) \tag{3.319}$$

where $\underline{u}_s$ is the displacement of reference particle **P** and $R_1$ the total rotation tensor that describes the orientation of local vector $\underline{s}_0$ in current deformed configuration. It is readily verified the virtual displacement $\delta \underline{P}$, also the position variation, is determined to be

$$\delta \underline{P} = \delta \underline{u}_s + \delta R_1 \underline{s}_0 \tag{3.320}$$

After resolved the above equations into the body attached frame, the variation of position vector relates the virtual displacements of the reference particle in the form of

$$R_1^T \delta \underline{P} = R_1^T \delta \underline{u}_s + \tilde{s}_0^T \underline{\delta \psi} \tag{3.321}$$

where the virtual rotations are defined, $\widetilde{\delta \psi} = R_1^T \delta R_1$. In current implementation, two types of external forces are considered, one is body force $\underline{g}$, the other surface pressure $\underline{p}_s$, both measured in body attached frame. Then, the virtual work of the external forces can be computed from

$$\delta \mathcal{W}_e = \int_{V_0} \delta \underline{P}^T R_1 \underline{g} \, dV_0 + \int_{S_2} \delta \underline{P}^T R_1 \underline{p}_s \, dS_2 \tag{3.322}$$

where $S_2$ represents the area magnitude of a beam surface where the surface pressure is applied to, and $S_2 = \ell_0\ell_2$ and where $\ell_2$ the width magnitude along $\alpha_2$ direction. Referring the formula of position variation, Eq. (3.321), the virtual work can be expressed in the following detailed formula as

$$\begin{aligned}\delta\mathcal{W}_e &= \int_{V_0} (R_1^T\delta\underline{u}_s + \widetilde{s}_0^T\underline{\delta\psi})^T\underline{g}\,dV_0 + \int_{S_2} (R_1^T\delta\underline{u}_s + \widetilde{s}_0^T\underline{\delta\psi})^T\underline{p}_s\,dS_2 \\ &= \int_{\ell_0}\left[\delta\underline{u}_s^T R_1(\int_{S_0}\underline{g}\,dS_0 + \int_{\ell_2}\underline{p}_s\,d\alpha_2) + \underline{\delta\psi}^T(\int_{S_0}\widetilde{s}_0\underline{g}\,dS_0 + \int_{\ell_2}\widetilde{s}_0\underline{p}_s\,d\alpha_2)\right]d\alpha_1\end{aligned} \tag{3.323}$$

For the easy description of virtual work, the external force and torque vectors are defined at first by

$$\underline{F} = \int_{S_0}\underline{g}\,dS_0 + \int_{\ell_2}\underline{p}_s\,d\alpha_2,\quad \underline{T} = \int_{S_0}\widetilde{s}_0\underline{g}\,dS_0 + \int_{\ell_2}\widetilde{s}_0\underline{p}_s\,d\alpha_2 \tag{3.324}$$

then the virtual work can be approximated to be

$$\delta\mathcal{W}_e = \int_{\ell_0}(\delta\underline{u}_s^T R_1\underline{F} + \underline{\delta\psi}^T\underline{T})\,d\alpha_1 \tag{3.325}$$

or in the motion formalism like

$$\delta\mathcal{W}_e = \int_{\ell_0}\underline{\delta\mathcal{U}}^T\underline{F}_{\epsilon\alpha}d\alpha_1 \tag{3.326}$$

where the following dual brackets are introduced

$$\underline{F}_{\epsilon\alpha} = \begin{vmatrix}\underline{F}\\ \underline{T}\end{vmatrix},\quad \underline{\delta\mathcal{U}} = \begin{vmatrix}R_1^T\delta\underline{u}_s\\ \underline{\delta\psi}\end{vmatrix} \tag{3.327}$$

### 3.4.6 Governing Equations of Beam in Motion Formalism

Based on the principle of virtual work, which states

$$\delta\mathcal{U}_e - \delta\mathcal{W}_e = 0 \tag{3.328}$$

the governing equations for a beam problem could be obtained. Introducing Eqs. (3.317) and (3.325) leads to

$$\int_{\ell_0} (\delta \underline{\epsilon}_1^T \underline{F}_{\epsilon\epsilon}^{11} - \delta \underline{\mathcal{U}}^T \underline{F}_{\epsilon\alpha}) d\alpha_1 = 0 \tag{3.329}$$

In view of compatibility properties, Eq. (2.161), the variations of Lagrangian strains can be expressed in terms of

$$\delta \underline{\epsilon}_1 = \delta \underline{\mathcal{U}}_{,1} + \widetilde{\mathcal{K}}_1^1 \delta \underline{\mathcal{U}} \tag{3.330}$$

and then the principle of virtual work is recasted to

$$\int_{\ell_0} \left[ (\delta \underline{\mathcal{U}}_{,1} + \widetilde{\mathcal{K}}_1^1 \delta \underline{\mathcal{U}})^T \underline{F}_{\epsilon\epsilon}^{11} - \delta \underline{\mathcal{U}}^T \underline{F}_{\epsilon\alpha} \right] d\alpha_1 = 0 \tag{3.331}$$

Since the variations of motion vectors and its spatial derivatives form a linear six-dimensional space, they can be interpolated directly by using the shape functions $\mathcal{H}_\epsilon$ under the framework of finite element method. The following interpolations

$$\delta \underline{\mathcal{U}} = \mathcal{H}_\epsilon \delta \underline{\mathcal{U}}_n, \ \delta \underline{\mathcal{U}}_{,1} = \mathcal{H}_{\epsilon,1} \delta \underline{\mathcal{U}}_n \tag{3.332}$$

reformulate the principle of virtual work, Eq. (3.331), into the weak Galerkin form

$$\delta \underline{\mathcal{U}}_n^T \int_{\ell_0} \left[ (\mathcal{H}_{\epsilon,1}^T + \mathcal{H}_\epsilon^T \widetilde{\mathcal{K}}_1^{1T}) \underline{F}_{\epsilon\epsilon}^{11} - \mathcal{H}_\epsilon^T \underline{F}_{\epsilon\alpha} \right] d\alpha_1 = 0 \tag{3.333}$$

From which, the governing equations are obtained

$$\int_{\ell_0} \left[ (\mathcal{H}_{\epsilon,1}^T + \mathcal{H}_\epsilon^T \widetilde{\mathcal{K}}_1^{1T}) \underline{F}_{\epsilon\epsilon}^{11} - \mathcal{H}_\epsilon^T \underline{F}_{\epsilon\alpha} \right] d\alpha_1 = 0 \tag{3.334}$$

After introduced the elastic force vector

$$\underline{\mathcal{F}}_e = \int_{\ell_0} (\mathcal{H}_{\epsilon,1}^T + \mathcal{H}_\epsilon^T \widetilde{\mathcal{K}}_1^{1T}) \underline{F}_{\epsilon\epsilon}^{11} d\alpha_1 \tag{3.335}$$

and the discrete external force vector

$$\underline{\mathcal{F}}_a = \int_{\ell_0} \mathcal{H}_\epsilon^T \underline{F}_{\epsilon\alpha} d\alpha_1 \tag{3.336}$$

the governing equations can be described by the following compact formula

$$\underline{\mathcal{F}}_e - \underline{\mathcal{F}}_a = 0 \tag{3.337}$$

### 3.4.7 Kinematic Energy of Beam

When governing equation of the beam is extended to the dynamic problem, the kinematic energy of beam needs to be evaluated. At first, the velocity of an arbitrary material particle **B** in a beam can be evaluated by taking time derivatives of position vector $\underline{P}$, defined by Eq. (3.275), to get

$$\dot{\underline{P}} = (\underline{x}_0 + \underline{u}_s + R_1 \underline{s}_0)^{\cdot} = \dot{\underline{u}}_s + \dot{R}_1 \underline{s}_0 \tag{3.338}$$

After resolved the time derivatives into the body attached frame $\mathcal{F}_1$, the particle velocity will relate to the linear and angular velocities of reference particle like

$$R_1^T \dot{\underline{P}} = R_1^T \dot{\underline{u}}_s + \widetilde{s}_0^T \underline{\omega} \tag{3.339}$$

where the angular velocity is defined as $\widetilde{\omega} = R_1^T \dot{R}_1$. If denoting the material density as $\rho$, the kinematic energy of a beam can be computed from the integral related to the initial volume like

$$\begin{aligned} \mathcal{K}_e &= \int_{V_0} \frac{1}{2} \rho (R_1^T \dot{\underline{P}})^T R_1^T \dot{\underline{P}} dV_0 \\ &= \int_{V_0} \frac{1}{2} \lfloor \dot{\underline{u}}_s^T R_1, \ \underline{\omega}^T \rfloor \begin{bmatrix} \rho I, & \rho \widetilde{s}_0^T \\ \rho \widetilde{s}_0, & \rho \widetilde{s}_0 \widetilde{s}_0^T \end{bmatrix} \begin{vmatrix} R_1^T \dot{\underline{u}}_s \\ \underline{\omega} \end{vmatrix} dV_0 \end{aligned} \tag{3.340}$$

Its motion formalism

$$\mathcal{K}_e = \int_{\ell_0} \frac{1}{2} \underline{\mathcal{V}}^T \mathcal{M}_{\epsilon\epsilon} \underline{\mathcal{V}} d\alpha_1 \tag{3.341}$$

can be readily assembled with the aid of cross-sectional mass matrix $\mathcal{M}_{\epsilon\epsilon}$ and velocities $\underline{\mathcal{V}}$ in its dual bracket

$$\mathcal{M}_{\epsilon\epsilon} = \int_{S_0} \begin{bmatrix} \rho I, & \rho \widetilde{s}_0^T \\ \rho \widetilde{s}_0, & \rho \widetilde{s}_0 \widetilde{s}_0^T \end{bmatrix} dS_0, \quad \underline{\mathcal{V}} = \begin{vmatrix} R_1^T \dot{\underline{u}}_s \\ \underline{\omega} \end{vmatrix} \tag{3.342}$$

Referring Eq. (3.341), the variation of kinematic energy is found to be

$$\delta \mathcal{K}_e = \int_{\ell_0} \delta \underline{\mathcal{V}}^T \mathcal{M}_{\epsilon\epsilon} \underline{\mathcal{V}} d\alpha_1 \tag{3.343}$$

In view of the compatibility attributes, Eq. (2.160), the virtual velocity is related to virtual motion $\underline{\delta \mathcal{U}}$ and its time derivative like

$$\delta \underline{\mathcal{V}} = \underline{\delta \dot{\mathcal{U}}} + \widetilde{\mathcal{V}} \underline{\delta \mathcal{U}} \tag{3.344}$$

By introducing this relation, the variation of kinematic energy becomes

$$\delta\mathcal{K}_e = \int_{\ell_0} (\delta\underline{\dot{\mathcal{U}}}^T \mathcal{M}_{\epsilon\epsilon}\underline{\mathcal{V}} + \delta\underline{\mathcal{U}}^T \widetilde{\mathcal{V}}^T \mathcal{M}_{\epsilon\epsilon}\underline{\mathcal{V}})\, d\alpha_1 \tag{3.345}$$

Integrating the variation of kinematic energy by part in time domain will produce

$$\int_t \delta\mathcal{K}_e dt = \int_t \int_{\ell_0} (-\delta\underline{\mathcal{U}}^T \mathcal{M}_{\epsilon\epsilon}\underline{\dot{\mathcal{V}}} + \delta\underline{\mathcal{U}}^T \widetilde{\mathcal{V}}^T \mathcal{M}_{\epsilon\epsilon}\underline{\mathcal{V}})\, d\alpha_1 dt \tag{3.346}$$

From which the variation of kinematic energy is reformulated to

$$\delta\mathcal{K}_e = \int_{\ell_0} (-\delta\underline{\mathcal{U}}^T \mathcal{M}_{\epsilon\epsilon}\underline{\dot{\mathcal{V}}} + \delta\underline{\mathcal{U}}^T \widetilde{\mathcal{V}}^T \mathcal{M}_{\epsilon\epsilon}\underline{\mathcal{V}})\, d\alpha_1 \tag{3.347}$$

With the application of finite element discretization, the velocities and accelerations of reference particle **P** can be approximated by using the following polynomials of nodal velocities and its time derivatives like

$$\underline{\mathcal{V}} = \mathcal{H}_\epsilon \underline{\mathcal{V}}_n, \quad \underline{\dot{\mathcal{V}}} = \mathcal{H}_\epsilon \underline{\dot{\mathcal{V}}}_n \tag{3.348}$$

and the weak form of kinematic energy variation is found to be

$$\delta\mathcal{K}_e = \delta\underline{\mathcal{U}}_n^T \left( -\int_{\ell_0} \mathcal{H}_\epsilon^T \mathcal{M}_{\epsilon\epsilon}\mathcal{H}_\epsilon d\alpha_1 \underline{\dot{\mathcal{V}}}_n + \int_{\ell_0} \mathcal{H}_\epsilon^T \widetilde{\mathcal{V}}^T \mathcal{M}_{\epsilon\epsilon}\underline{\mathcal{V}} d\alpha_1 \right) \tag{3.349}$$

After introduced the discrete mass matrix and the Coriolis force vector

$$\mathcal{M}_n = \int_{\ell_0} \mathcal{H}_\epsilon^T \mathcal{M}_{\epsilon\epsilon}\mathcal{H}_\epsilon d\alpha_1, \quad \underline{\mathcal{F}}_g = \int_{\ell_0} \mathcal{H}_\epsilon^T \widetilde{\mathcal{V}}^T \mathcal{M}_{\epsilon\epsilon}\underline{\mathcal{V}}\, d\alpha_1 \tag{3.350}$$

the variation of original kinematic energy can be rewritten into the compact formula

$$\delta\mathcal{K}_e = \delta\underline{\mathcal{U}}_n^T (-\mathcal{M}_n \underline{\dot{\mathcal{V}}}_n + \underline{\mathcal{F}}_g) \tag{3.351}$$

### *3.4.8 Extended to Beam Dynamic Problem*

When considering the inertial and the Coriolis forces, the governing equations of motion for a beam is obtained from the Hamilton's principle. Equivalently, the inertial force and the associated Coriolis force are required to be added into the Eq. (3.337) to get

$$\mathcal{M}_n \dot{\underline{\mathcal{V}}}_n - \underline{\mathcal{F}}_g + \underline{\mathcal{F}}_e = \underline{\mathcal{F}}_a \tag{3.352}$$

## 3.5 General Rigid Body Theory

When the deformations of Cosserat continuum are all neglected in three dimensions $(\alpha_1, \alpha_2, \alpha_3)$, the Cosserat continuum will reduce to a rigid body suffering only the rigid motions. The governing equations of motion for a rigid body [21, 70] can be directly obtained by neglecting all the terms associated with the spatial derivatives in the governing equations for a three-dimensional Cosserat continuum [2].

### 3.5.1 *Kinematics of Rigid Body*

As shown in Fig. 3.5, an arbitrary material particle in a rigid body is depicted by its position and director vectors in its reference configuration and current deformed configuration, respectively. In the original reference configuration, the position vector $\underline{p}(\alpha_i)$ of the material particle relative to a point **O**, the original of reference frame, $\mathcal{F}_{\mathcal{I}} = [\mathbf{O}, \mathcal{I}]$, can be written as

$$\underline{p}(\alpha_1, \alpha_2, \alpha_3) = \underline{x}_0 + R_0 \underline{s}_0(\alpha_1, \alpha_2, \alpha_3) \tag{3.353}$$

where the general curvilinear coordinates, $(\alpha_1, \alpha_2, \alpha_3)$, are introduced to label the material particle in the original continuum, and $R_0$ the initial rotation tensor measures the orientation of reference basis $\mathcal{B}_0$, attached to the reference particle **P** with the position vector of $\underline{x}_0$. It means that the initial basis $\mathcal{B}_0$ is bounded by the particle **P** to produce the reference frame $\mathcal{F}_0 = [\mathbf{P}, \mathcal{B}_0]$. It has been verified in Sect. 2.2.3 the

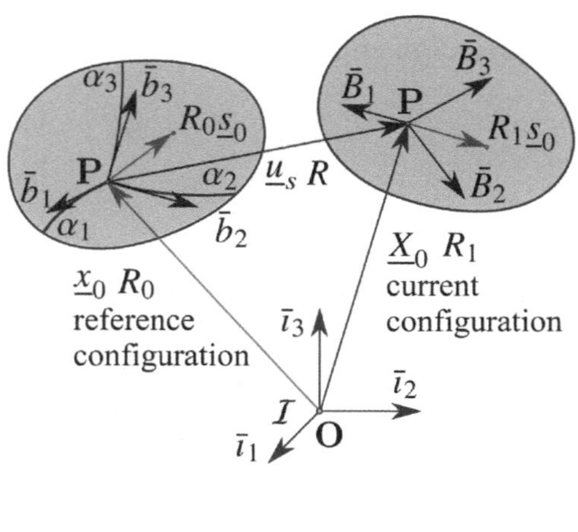


**Fig. 3.5** Kinematics of a three-dimensional rigid body

rotation tensor $R_0$ is identical to the body attached basis $\mathcal{B}_0$, and in detail

$$R_0 = \lfloor \bar{b}_1, \bar{b}_2, \bar{b}_3 \rfloor = \mathcal{B}_0 \tag{3.354}$$

The local vector $\underline{s}_0$, defined in this reference frame $\mathcal{F}_0$, describes the director of reference material particle **P**, and can be given directly as

$$\underline{s}_0 = \alpha_1 \bar{\imath}_1 + \alpha_2 \bar{\imath}_2 + \alpha_3 \bar{\imath}_3 \tag{3.355}$$

Obviously, Eq. (3.353) decomposes the position vector $\underline{p}$ of a material particle into a reference position $\underline{x}_0$ and a local vector $\underline{s}_0$. The reference particle **P** with position $\underline{x}_0$ can translate freely, while the rigid director $\underline{s}_0$ attached to the reference particle can only rotate. When the rigid body moves in space, the material particle that had position vector $\underline{p}$ in the undeformed reference configuration now has position vector $\underline{P}$ in the current configuration, such that

$$\underline{P}(\alpha_1, \alpha_2, \alpha_3) = \underline{X}_0 + R_1 \underline{s}_0(\alpha_1, \alpha_2, \alpha_3) \tag{3.356}$$

where $R_1$ is the total rotation tensor that describes the orientation of basis $\mathcal{B}_1$ attached to the deformed configuration and

$$R_1 = \lfloor \bar{B}_1, \bar{B}_2, \bar{B}_3 \rfloor = \mathcal{B}_1 \tag{3.357}$$

The associated frame $\mathcal{F}_1 = [\mathbf{P}, \mathcal{B}_1]$ can be constructed when bounding the current deformed basis $\mathcal{B}_1$ by the material particle **P**. The position vector $\underline{X}_0$ of reference point in current deformed configuration are defined as

$$\underline{X}_0 = \underline{x}_0 + \underline{u}_s \tag{3.358}$$

where the displacement vector $\underline{u}_s$ describes arbitrary translation of reference point.

### 3.5.2 *Work Done by External Forces for Rigid Body*

The governing equations of motion for a rigid body will be obtained from the principle of work and energy [21]. The differential work done by external forces are evaluated at first. By Eq. (3.356), the differential displacement of an arbitrary particle in current configuration is obtained by

$$d\underline{P} = d(\underline{X}_0 + R_1 \underline{s}_0) = d(\underline{x}_0 + \underline{u}_s + R_1 \underline{s}_0) \tag{3.359}$$

The explicit expressions of the differential displacement $d\underline{P}$ are readily verified to be

$$d\underline{P} = d\underline{u}_s + dR_1 \underline{s}_0 \tag{3.360}$$

After resolved the above equations into the body attached frame, the differential position vector $\underline{P}$ relates the differential displacements and rotations of reference particle like

$$R_1^T d\underline{P} = R_1^T d\underline{u}_s + \widetilde{s}_0^T \underline{d\psi} \tag{3.361}$$

where the differential rotations are defined, $\widetilde{d\psi} = R_1^T dR_1$. In current implementation, two types of external forces are considered, one is body force $\underline{g}$, and the other surface pressure $\underline{p}_s$, both measured in body attached frame. Then, the differential work done by the external forces can be computed from

$$d\mathcal{W}_e = \int_{V_0} d\underline{P}^T R_1 \underline{g}\, dV_0 + \int_{S_0} d\underline{P}^T R_1 \underline{p}_s\, dS_0 \tag{3.362}$$

where $S_0$ represents the area magnitude where the surface pressure is applied to. Referring the formula of differential position vector, Eq. (3.361), the differential work can be expressed in the following detailed formula as

$$\begin{aligned} d\mathcal{W}_e &= \int_{V_0} (R_1^T d\underline{u}_s + \widetilde{s}_0^T \underline{d\psi})^T \underline{g}\, dV_0 + \int_{S_0} (R_1^T d\underline{u}_s + \widetilde{s}_0^T \underline{d\psi})^T \underline{p}_s\, dS_0 \\ &= d\underline{u}_s^T R_1 (\int_{V_0} \underline{g}\, dV_0 + \int_{S_0} \underline{p}_s\, dS_0) + \underline{d\psi}^T (\int_{V_0} \widetilde{s}_0 \underline{g}\, dV_0 + \int_{S_0} \widetilde{s}_0 \underline{p}_s\, dS_0) \end{aligned} \tag{3.363}$$

For the easy description, the external force and torque are defined at first by

$$\underline{F} = \int_{V_0} \underline{g}\, dV_0 + \int_{S_0} \underline{p}_s\, dS_0, \quad \underline{T} = \int_{V_0} \widetilde{s}_0 \underline{g}\, dV_0 + \int_{S_0} \widetilde{s}_0 \underline{p}_s\, dS_0 \tag{3.364}$$

then the differential work done by the external force and torque can be recasted to

$$d\mathcal{W}_e = d\underline{u}_s^T R_1 \underline{F} + \underline{d\psi}^T \underline{T} \tag{3.365}$$

or in its motion formalism like

$$d\mathcal{W}_e = \underline{d\mathcal{U}}^T \underline{\mathcal{F}}_a \tag{3.366}$$

where the dual brackets of external forces and differential motion are given as

$$\underline{\mathcal{F}}_a = \begin{vmatrix} \underline{F} \\ \underline{T} \end{vmatrix}, \quad \underline{d\mathcal{U}} = \begin{vmatrix} R_1^T d\underline{u}_s \\ \underline{d\psi} \end{vmatrix} \tag{3.367}$$

### 3.5.3 Kinematic Energy of Rigid Body

Similar to the definition of differential position vector $d\underline{P}$, the velocity of an arbitrary material particle in a rigid body can be evaluated by taking time derivatives of position vector $\underline{P}$, defined by Eq. (3.356), to find

$$\dot{\underline{P}} = (\underline{x}_0 + \underline{u}_s + R_1 \underline{s}_0)^{\cdot} = \dot{\underline{u}}_s + \dot{R}_1 \underline{s}_0 \tag{3.368}$$

After resolved the time derivatives into the body attached frame $\mathcal{F}_1$, the particle velocity relates to the linear and angular velocities of reference particle **P** like

$$R_1^T \dot{\underline{P}} = R_1^T \dot{\underline{u}}_s + \widetilde{s}_0^T \underline{\omega} \tag{3.369}$$

where the angular velocity is defined as $\widetilde{\omega} = R_1^T \dot{R}_1$. If denoting the material density as $\rho$, the kinematic energy of a rigid body can be computed from the integral related to the initial volume like

$$\begin{aligned}\mathcal{K}_e &= \int_{V_0} \frac{1}{2} \rho (R_1^T \dot{\underline{P}})^T R_1^T \dot{\underline{P}} dV_0 \\ &= \int_{V_0} \frac{1}{2} \lfloor \dot{\underline{u}}_s^T R_1, \ \underline{\omega}^T \rfloor \begin{bmatrix} \rho I, & \rho \widetilde{s}_0^T \\ \rho \widetilde{s}_0, & \rho \widetilde{s}_0 \widetilde{s}_0^T \end{bmatrix} \begin{vmatrix} R_1^T \dot{\underline{u}}_s \\ \underline{\omega} \end{vmatrix} dV_0\end{aligned} \tag{3.370}$$

With the aid of following definitions

$$\mathcal{M} = \int_{V_0} \begin{bmatrix} \rho I, & \rho \widetilde{s}_0^T \\ \rho \widetilde{s}_0, & \rho \widetilde{s}_0 \widetilde{s}_0^T \end{bmatrix} dV_0, \quad \underline{\mathcal{V}} = \begin{vmatrix} R_1^T \dot{\underline{u}}_s \\ \underline{\omega} \end{vmatrix} \tag{3.371}$$

the motion formalism of kinematic energy is readily obtained

$$\mathcal{K}_e = \frac{1}{2} \underline{\mathcal{V}}^T \mathcal{M} \underline{\mathcal{V}} \tag{3.372}$$

and the differential kinematic energy becomes

$$d\mathcal{K}_e = d\underline{\mathcal{V}}^T \mathcal{M} \underline{\mathcal{V}} \tag{3.373}$$

Similar to the compatibility attributes, Eq. (2.160), the differential velocity is related to the differential motion vector $d\underline{\mathcal{U}}$ and its time derivative like

$$d\underline{\mathcal{V}} = d\dot{\underline{\mathcal{U}}} + \widetilde{\mathcal{V}} d\underline{\mathcal{U}} \tag{3.374}$$

and these relations are applied to specify the differential kinematic energy in the form of

$$d\mathcal{K}_e = \underline{d\dot{\mathcal{U}}}^T \mathcal{M}\underline{\mathcal{V}} + \underline{d\mathcal{U}}^T \widetilde{\mathcal{V}}^T \mathcal{M}\underline{\mathcal{V}} \tag{3.375}$$

Integrating the differential kinematic energy by part in time domain produces

$$\int_t d\mathcal{K}_e dt = \int_t (-\underline{d\mathcal{U}}^T \mathcal{M}\underline{\dot{\mathcal{V}}} + \underline{d\mathcal{U}}^T \widetilde{\mathcal{V}}^T \mathcal{M}\underline{\mathcal{V}})\, dt \tag{3.376}$$

From which the differential kinematic energy is reformulated to

$$d\mathcal{K}_e = -\underline{d\mathcal{U}}^T \mathcal{M}\underline{\dot{\mathcal{V}}} + \underline{d\mathcal{U}}^T \widetilde{\mathcal{V}}^T \mathcal{M}\underline{\mathcal{V}} \tag{3.377}$$

With the definition of the Coriolis force

$$\underline{\mathcal{F}}_g = \widetilde{\mathcal{V}}^T \mathcal{M}\underline{\mathcal{V}} \tag{3.378}$$

the differential kinematic energy can be rewritten into the compact form

$$d\mathcal{K}_e = \underline{d\mathcal{U}}^T (-\mathcal{M}\underline{\dot{\mathcal{V}}} + \underline{\mathcal{F}}_g) \tag{3.379}$$

### *3.5.4 Governing Equations of Motion in Motion Formalism*

Introducing the principle of work and energy for a rigid body

$$\int_t (d\mathcal{W}_e - d\mathcal{K}_e) dt = 0 \tag{3.380}$$

the governing equations of motion for dynamic problem are found to be

$$\mathcal{M}\underline{\dot{\mathcal{V}}} - \underline{\mathcal{F}}_g = \underline{\mathcal{F}}_a \tag{3.381}$$

## 3.6 General Cauchy Continuum Theory

As mentioned at the beginning of this chapter, the Cauchy continuum is a collection of material particles with only position vectors. Meanwhile, the Cosserat continuum includes a director vector at each material particle in addition to the position vector. Therefore, Cosserat continuum could be simplified to Cauchy continuum by ignoring the director vector at each material particle. In this section, the governing equations of motion for Cauchy continuum are presented by eliminating the terms associated with the directors in the governing equations of motion for a three-dimensional Cosserat continuum. All the theory details of Cauchy continuum are discussed below.

### *3.6.1 Kinematics of Cauchy Continuum*

As shown in Fig. 3.6, a material particle in a Cauchy continuum is depicted by its position vector in its reference configuration and current deformed configuration, respectively. In the original reference configuration, the position vector $\underline{p}(\alpha_i)$ of the material particle relative to a point **O**, the original of reference frame, $\mathcal{F}_{\mathcal{I}} = [\mathbf{O}, \mathcal{I}]$, can be written as

$$\underline{p}(\alpha_1, \alpha_2, \alpha_3) = \underline{x}_0(\alpha_1, \alpha_2, \alpha_3) \tag{3.382}$$

where the general curvilinear coordinates, $(\alpha_1, \alpha_2, \alpha_3)$, are introduced to label the material particle in the original continuum. The director vector of material particle for Cauchy continuum is ignored. Comparing Eqs. (3.1) and (3.382), it is apparent that an arbitrary material particle becomes coincide with the reference particle in a Cauchy continuum.

For the description of Cauchy continuum deformations, the covariant base vectors,

$$\underline{g}_i = \frac{\partial \underline{p}}{\partial \alpha_i} \tag{3.383}$$

in the reference configuration are still required, and the specific expressions can be obtained by taking derivatives of Eq. (3.382) with respected to the curvilinear coordinates $\alpha_i$ like

$$\underline{g}_1 = \underline{x}_{0,1}, \quad \underline{g}_2 = \underline{x}_{0,2}, \quad \underline{g}_3 = \underline{x}_{0,3} \tag{3.384}$$

where the symbol $(\cdot)_{,i}$ for $i = 1, 2, 3$ denotes the spatial derivatives with respected to $\alpha_i$ and

$$(\cdot)_{,i} = \frac{\partial (\cdot)}{\partial \alpha_i} \tag{3.385}$$

All these base vectors form the covariant bases $\mathcal{B}_g$ like

$$\mathcal{B}_g = \lfloor \underline{g}_1, \underline{g}_2, \underline{g}_3 \rfloor \tag{3.386}$$

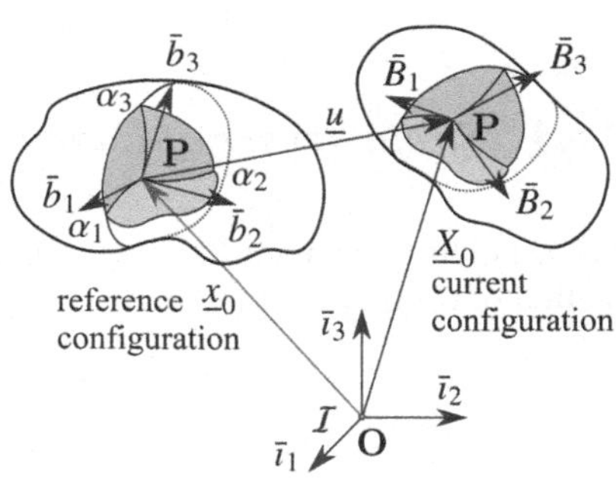


**Fig. 3.6** Kinematics of a material particle in a Cauchy continuum

and then the metric tensor in the reference configuration is computed from

$$g = \mathcal{B}_g^T \mathcal{B}_g \tag{3.387}$$

When the Cauchy continuum deforms with spatial motions, the material particle that had position vector $\underline{p}$ in the undeformed reference configuration now has position vector $\underline{P}$ in the deformed configuration, such that

$$\underline{P}(\alpha_1, \alpha_2, \alpha_3) = \underline{X}_0(\alpha_1, \alpha_2, \alpha_3) \tag{3.388}$$

The position vector $\underline{X}_0$ in current deformed configuration can be decomposed into

$$\underline{X}_0(\alpha_1, \alpha_2, \alpha_3) = \underline{x}_0(\alpha_1, \alpha_2, \alpha_3) + \underline{u}_s(\alpha_1, \alpha_2, \alpha_3) \tag{3.389}$$

where the displacement vector $\underline{u}_s$ describes arbitrary translation of material particle. According to the definitions of covariant base vectors in deformed configuration

$$\underline{G}_i = \frac{\partial \underline{P}}{\partial \alpha_i} \tag{3.390}$$

the specific formulas can be obtained by taking spatial derivatives of Eq. (3.17) with respected to the curvilinear coordinates $\alpha_i$ for $i = 1, 2, 3$ like

$$\underline{G}_1 = \underline{X}_{0,1}, \quad \underline{G}_2 = \underline{X}_{0,2}, \quad \underline{G}_3 = \underline{X}_{0,3} \tag{3.391}$$

All these base vectors could be assembled into the covariant bases $\mathcal{B}_G$ like

$$\mathcal{B}_G = \lfloor \underline{G}_1, \ \underline{G}_2, \ \underline{G}_3 \rfloor \tag{3.392}$$

In view of above expressions of covariant basis vectors in the deformed configuration, the associated metric tensor

$$G = \mathcal{B}_G^T \mathcal{B}_G \tag{3.393}$$

can be readily evaluated.

### 3.6.2 Strain Tensor of Cauchy Continuum

The rotation-free Green-Lagrange strain tensor [22] is applied to measure the deformation of Cauchy continuum. In doing so, the following relations

$$\underline{X}_{0,i} = \underline{x}_{0,i} + \underline{u}_{s,i} \tag{3.394}$$

for $i = 1, 2, 3$ are required to reformulate the deformed covariant base vectors in terms of

$$\underline{G}_1 = \underline{x}_{0,1} + \underline{u}_{s,1}, \quad \underline{G}_2 = \underline{x}_{0,2} + \underline{u}_{s,2}, \quad \underline{G}_3 = \underline{x}_{0,3} + \underline{u}_{s,3} \tag{3.395}$$

Referring the definitions of metric tensors in reference configuration and deformed configuration, Eqs. (3.387) and (3.393), respectively, the rotation-free Green-Lagrangian strain tensor can be measured in the curvilinear basis like

$$\widehat{\varepsilon} = \frac{1}{2}(G - g) \tag{3.396}$$

Substituting the specific formulations of metric tensors in terms of covariant basis vectors, Eq. (3.395), into the formulations of strain tensor, the components of rotation-free Green-Lagrange strain tensor in their vector-dyadic form are found to be

$$\begin{aligned} \varepsilon_{11} &= \underline{x}_{0,1}^T \underline{u}_{s,1} \\ 2\varepsilon_{12} &= \underline{x}_{0,1}^T \underline{u}_{s,2} + \underline{x}_{0,2}^T \underline{u}_{s,1} \\ 2\varepsilon_{13} &= \underline{x}_{0,1}^T \underline{u}_{s,3} + \underline{x}_{0,3}^T \underline{u}_{s,1} \\ \varepsilon_{22} &= \underline{x}_{0,2}^T \underline{u}_{s,2} \\ 2\varepsilon_{23} &= \underline{x}_{0,2}^T \underline{u}_{s,3} + \underline{x}_{0,3}^T \underline{u}_{s,2} \\ \varepsilon_{33} &= \underline{x}_{0,3}^T \underline{u}_{s,3} \end{aligned} \tag{3.397}$$

Note that the assumptions of small strains are applied such that the higher order terms as to $\underline{u}_{s,i}^T \underline{u}_{s,j}$ for $i, j = 1, 2, 3$ are all neglected.

### 3.6.3 Constitutive Laws for Cauchy Continuum

For general anisotropic nonlinear elastic materials, that are considered to be ideal materials, the components of second Piola-Kirchhoff stress tensor are correlated to the components of Green-Lagrange strain tensor by the constitutive equations. For the easy description of these constitutive relations, the Green-Lagrange strain tensor in its vector-dyadic form needs to be decomposed into

$$\underline{\widehat{\varepsilon}} = \begin{vmatrix} \varepsilon_{11} \\ 2\varepsilon_{12} \\ 2\varepsilon_{13} \\ \varepsilon_{22} \\ 2\varepsilon_{23} \\ \varepsilon_{33} \end{vmatrix} = \begin{vmatrix} \varepsilon_{11} \\ \varepsilon_{21} \\ \varepsilon_{31} \\ 0 \\ 0 \\ 0 \end{vmatrix} + \begin{vmatrix} 0 \\ \varepsilon_{12} \\ 0 \\ \varepsilon_{22} \\ \varepsilon_{32} \\ 0 \end{vmatrix} + \begin{vmatrix} 0 \\ 0 \\ \varepsilon_{13} \\ 0 \\ \varepsilon_{23} \\ \varepsilon_{33} \end{vmatrix} = \underline{\varepsilon}_1 + \underline{\varepsilon}_2 + \underline{\varepsilon}_3 \tag{3.398}$$

It is noted that the symbols $\varepsilon_{12}$, $\varepsilon_{21}$, $\varepsilon_{23}$, $\varepsilon_{32}$, and $\varepsilon_{31}$, $\varepsilon_{13}$ do not represent the components of Green-Lagrange strain tensor, but are defined in the following relations

$$2\varepsilon_{12} = \varepsilon_{21} + \varepsilon_{12}, \quad 2\varepsilon_{13} = \varepsilon_{31} + \varepsilon_{13}, \quad 2\varepsilon_{23} = \varepsilon_{32} + \varepsilon_{23} \tag{3.399}$$

After filtered out the zero items, the vectorial expressions are reduced to

$$\underline{\hat{\varepsilon}}_1 = \begin{vmatrix} \varepsilon_{11} \\ \varepsilon_{21} \\ \varepsilon_{31} \end{vmatrix}, \quad \underline{\hat{\varepsilon}}_2 = \begin{vmatrix} \varepsilon_{12} \\ \varepsilon_{22} \\ \varepsilon_{32} \end{vmatrix}, \quad \underline{\hat{\varepsilon}}_3 = \begin{vmatrix} \varepsilon_{13} \\ \varepsilon_{23} \\ \varepsilon_{33} \end{vmatrix} \tag{3.400}$$

and in detail

$$\begin{vmatrix} \varepsilon_{11} \\ \varepsilon_{21} \\ \varepsilon_{31} \end{vmatrix} = \begin{vmatrix} \underline{x}_{0,1}^T \underline{u}_{s,1} \\ \underline{x}_{0,2}^T \underline{u}_{s,1} \\ \underline{x}_{0,3}^T \underline{u}_{s,1} \end{vmatrix}, \quad \begin{vmatrix} \varepsilon_{12} \\ \varepsilon_{22} \\ \varepsilon_{32} \end{vmatrix} = \begin{vmatrix} \underline{x}_{0,1}^T \underline{u}_{s,2} \\ \underline{x}_{0,2}^T \underline{u}_{s,2} \\ \underline{x}_{0,3}^T \underline{u}_{s,2} \end{vmatrix}, \quad \begin{vmatrix} \varepsilon_{13} \\ \varepsilon_{23} \\ \varepsilon_{33} \end{vmatrix} = \begin{vmatrix} \underline{x}_{0,1}^T \underline{u}_{s,3} \\ \underline{x}_{0,2}^T \underline{u}_{s,3} \\ \underline{x}_{0,3}^T \underline{u}_{s,3} \end{vmatrix} \tag{3.401}$$

The above formulas could be assembled into

$$\begin{bmatrix} \varepsilon_{11}, \varepsilon_{12}, \varepsilon_{13} \\ \varepsilon_{21}, \varepsilon_{22}, \varepsilon_{23} \\ \varepsilon_{31}, \varepsilon_{32}, \varepsilon_{33} \end{bmatrix} = \begin{bmatrix} \underline{x}_{0,1}^T \\ \underline{x}_{0,2}^T \\ \underline{x}_{0,3}^T \end{bmatrix} \lfloor \underline{u}_{s,1}, \ \underline{u}_{s,2}, \ \underline{u}_{s,3} \rfloor \tag{3.402}$$

and the associated compact expressions are formulated

$$\lfloor \underline{\hat{\varepsilon}}_1, \ \underline{\hat{\varepsilon}}_2, \ \underline{\hat{\varepsilon}}_3 \rfloor = \mathcal{S}_0^T \lfloor \underline{\epsilon}_1, \ \underline{\epsilon}_2, \ \underline{\epsilon}_3 \rfloor \tag{3.403}$$

with the aid of following introduced transfer matrix

$$\mathcal{S}_0 = \lfloor \underline{x}_{0,1}, \ \underline{x}_{0,2}, \ \underline{x}_{0,3} \rfloor = \mathcal{B}_g \tag{3.404}$$

and reduced Lagrange stretch strains

$$\underline{\epsilon}_1 = \underline{u}_{s,1}, \quad \underline{\epsilon}_2 = \underline{u}_{s,2}, \quad \underline{\epsilon}_3 = \underline{u}_{s,3} \tag{3.405}$$

when the zero components of Green-Lagrange strain vectors $\underline{\varepsilon}_1$, $\underline{\varepsilon}_2$, and $\underline{\varepsilon}_3$ are filtered out, the strain energy can be evaluated in reference configuration from

$$\begin{aligned} 2\mathcal{U}_e = \int_{V_0} [&\underline{\hat{\varepsilon}}_1^T \mathcal{D}_{11} \underline{\hat{\varepsilon}}_1 + \underline{\hat{\varepsilon}}_2^T \mathcal{D}_{22} \underline{\hat{\varepsilon}}_2 + \underline{\hat{\varepsilon}}_3^T \mathcal{D}_{33} \underline{\hat{\varepsilon}}_3 \\ &+ 2(\underline{\hat{\varepsilon}}_2^T \mathcal{D}_{21} \underline{\hat{\varepsilon}}_1 + \underline{\hat{\varepsilon}}_3^T \mathcal{D}_{31} \underline{\hat{\varepsilon}}_1 + \underline{\hat{\varepsilon}}_3^T \mathcal{D}_{32} \underline{\hat{\varepsilon}}_2)] \, dV_0 \end{aligned} \tag{3.406}$$

for a given continuum volume with the magnitude of $V_0$. Referring the relations of Green-Lagrange strain and displacements derivatives, Eq. (3.403), the above strain energy is recasted to

$$2\mathcal{U}_e = \int_{V_0} [\underline{\epsilon}_1^T \mathcal{C}_{11} \underline{\epsilon}_1 + \underline{\epsilon}_2^T \mathcal{C}_{22} \underline{\epsilon}_2 + \underline{\epsilon}_3^T \mathcal{C}_{33} \underline{\epsilon}_3$$
$$+2(\underline{\epsilon}_2^T \mathcal{C}_{21} \underline{\epsilon}_1 + \underline{\epsilon}_3^T \mathcal{C}_{31} \underline{\epsilon}_1 + \underline{\epsilon}_3^T \mathcal{C}_{32} \underline{\epsilon}_2)]\, dV_0 \tag{3.407}$$

where the constitutive matrices of Cauchy continuum are defined

$$\begin{array}{lll} \mathcal{C}_{11} = \mathcal{S}_0 \mathcal{D}_{11} \mathcal{S}_0^T, & \mathcal{C}_{22} = \mathcal{S}_0 \mathcal{D}_{22} \mathcal{S}_0^T, & \mathcal{C}_{33} = \mathcal{S}_0 \mathcal{D}_{33} \mathcal{S}_0^T \\ \mathcal{C}_{21} = \mathcal{S}_0 \mathcal{D}_{21} \mathcal{S}_0^T, & \mathcal{C}_{31} = \mathcal{S}_0 \mathcal{D}_{31} \mathcal{S}_0^T, & \mathcal{C}_{32} = \mathcal{S}_0 \mathcal{D}_{32} \mathcal{S}_0^T \end{array} \tag{3.408}$$

and the symmetric properties of constitutive matrices are still hold, such that $\mathcal{C}_{12} = \mathcal{C}_{21}^T, \mathcal{C}_{13} = \mathcal{C}_{31}^T$ and $\mathcal{C}_{23} = \mathcal{C}_{32}^T$. Note for the Lagrangian representation, the transfer matrix $\mathcal{S}_0$ will reduce to identity matrix $I$ of $3 \times 3$, such that the constitutive matrices $\mathcal{C}_{ij}$ and $\mathcal{D}_{ij}$ for $i, j = 1, 2, 3$ become identical to each other.

### *3.6.4 Variation of Strain Energy of Cauchy Continuum*

Referring the expressions of strain energy, Eq. (3.407), the strain energy density for a Cauchy continuum is given as

$$\begin{aligned} \mathcal{U}_\rho = \frac{1}{2}(&\underline{\epsilon}_1^T \mathcal{C}_{11} \underline{\epsilon}_1 + \underline{\epsilon}_2^T \mathcal{C}_{22} \underline{\epsilon}_2 + \underline{\epsilon}_3^T \mathcal{C}_{33} \underline{\epsilon}_3) \\ &+\underline{\epsilon}_2^T \mathcal{C}_{21} \underline{\epsilon}_1 + \underline{\epsilon}_3^T \mathcal{C}_{31} \underline{\epsilon}_1 + \underline{\epsilon}_3^T \mathcal{C}_{32} \underline{\epsilon}_2 \end{aligned} \tag{3.409}$$

and the variation of $\mathcal{U}_\rho$ is readily verified to be

$$\begin{aligned} \delta\mathcal{U}_\rho = &\delta\underline{\epsilon}_1^T \mathcal{C}_{11} \underline{\epsilon}_1 + \delta\underline{\epsilon}_2^T \mathcal{C}_{22} \underline{\epsilon}_2 + \delta\underline{\epsilon}_3^T \mathcal{C}_{33} \underline{\epsilon}_3 \\ &+\delta\underline{\epsilon}_2^T \mathcal{C}_{21} \underline{\epsilon}_1 + \delta\underline{\epsilon}_3^T \mathcal{C}_{31} \underline{\epsilon}_1 + \delta\underline{\epsilon}_3^T \mathcal{C}_{32} \underline{\epsilon}_2 \\ &+\delta\underline{\epsilon}_1^T \mathcal{C}_{12} \underline{\epsilon}_2 + \delta\underline{\epsilon}_1^T \mathcal{C}_{13} \underline{\epsilon}_3 + \delta\underline{\epsilon}_2^T \mathcal{C}_{23} \underline{\epsilon}_3 \end{aligned} \tag{3.410}$$

After merging the similar terms at the right-hand-side of above equations, the variation of strain energy density can be rearranged into the compact form

$$\delta\mathcal{U}_\rho = \delta\underline{\epsilon}_1^T \underline{F}_{\epsilon\epsilon}^{11} + \delta\underline{\epsilon}_2^T \underline{F}_{\epsilon\epsilon}^{22} + \delta\underline{\epsilon}_3^T \underline{F}_{\epsilon\epsilon}^{33} \tag{3.411}$$

where the stress measures of a Cauchy continuum are defined in the vectorial formulas like

$$\begin{aligned} \underline{F}_{\epsilon\epsilon}^{11} &= \mathcal{C}_{11} \underline{\epsilon}_1 + \mathcal{C}_{12} \underline{\epsilon}_2 + \mathcal{C}_{13} \underline{\epsilon}_3 \\ \underline{F}_{\epsilon\epsilon}^{22} &= \mathcal{C}_{21} \underline{\epsilon}_1 + \mathcal{C}_{22} \underline{\epsilon}_2 + \mathcal{C}_{23} \underline{\epsilon}_3 \\ \underline{F}_{\epsilon\epsilon}^{33} &= \mathcal{C}_{31} \underline{\epsilon}_1 + \mathcal{C}_{32} \underline{\epsilon}_2 + \mathcal{C}_{33} \underline{\epsilon}_3 \end{aligned} \tag{3.412}$$

Finally, the variation of strain energy, the integral related to initial volume of magnitude $V_0$, can be represented by

$$\delta\mathcal{U}_e = \int\limits_{V_0} (\delta\underline{\epsilon}_1^T \underline{F}_{\epsilon\epsilon}^{11} + \delta\underline{\epsilon}_2^T \underline{F}_{\epsilon\epsilon}^{22} + \delta\underline{\epsilon}_3^T \underline{F}_{\epsilon\epsilon}^{33}) dV_0 \tag{3.413}$$

### 3.6.5 Nonlinear Strain Tensor of Cauchy Continuum

Referring the definitions of rotation-free Green-Lagrange strain tensor, Eq. (2.235), the associated components of this strain tensor can be written in terms of

$$\begin{aligned} \varepsilon_{11} &= \frac{1}{2}(\underline{G}_1^T \underline{G}_1 - \underline{g}_1^T \underline{g}_1) \\ 2\varepsilon_{12} &= \underline{G}_1^T \underline{G}_2 - \underline{g}_1^T \underline{g}_2 \\ 2\varepsilon_{13} &= \underline{G}_1^T \underline{G}_3 - \underline{g}_1^T \underline{g}_3 \\ \varepsilon_{22} &= \frac{1}{2}(\underline{G}_2^T \underline{G}_2 - \underline{g}_2^T \underline{g}_2) \\ 2\varepsilon_{23} &= \underline{G}_2^T \underline{G}_3 - \underline{g}_2^T \underline{g}_3 \\ \varepsilon_{33} &= \frac{1}{2}(\underline{G}_3^T \underline{G}_3 - \underline{g}_3^T \underline{g}_3) \end{aligned} \tag{3.414}$$

The above expressions are the complete definitions of strain components without neglecting any higher-order terms. By Eq. (3.405), the deformed covariant base vectors for Cauchy continuum are found to be

$$\underline{G}_1 = \underline{\epsilon}_1 + \underline{x}_{0,1}, \quad \underline{G}_2 = \underline{\epsilon}_2 + \underline{x}_{0,2}, \quad \underline{G}_3 = \underline{\epsilon}_3 + \underline{x}_{0,3} \tag{3.415}$$

and their variations become

$$\begin{aligned} \delta\underline{G}_1 &= \delta\underline{\epsilon}_1 = \delta\underline{u}_{s,1} \\ \delta\underline{G}_2 &= \delta\underline{\epsilon}_2 = \delta\underline{u}_{s,2} \\ \delta\underline{G}_3 &= \delta\underline{\epsilon}_3 = \delta\underline{u}_{s,3} \end{aligned} \tag{3.416}$$

### 3.6.6 Variation of Nonlinear Strain Energy of Cauchy Continuum

Given the initial volume of a Cauchy continuum with the magnitude of $V_0$, the strain energy $\mathcal{U}_e$ can be measured from the integral of stain and stress tensor product in a closed-form relation like

$$2\mathcal{U}_e = \int\limits_{V_0} \underline{\widehat{\varepsilon}}^T \mathcal{D} \underline{\widehat{\varepsilon}}\, dV_0 = \int\limits_{V_0} \underline{\widehat{\varepsilon}}^T \underline{\widehat{\sigma}}\, dV_0 \tag{3.417}$$

where the second Piola-Kirchhoff tress tensor are correlated to the Green-Lagrange strain tensor in the vector-dyadic form like $\underline{\widehat{\sigma}} = \mathcal{D}\underline{\widehat{\varepsilon}}$. The variation of strain energy is obtained by taking variation of Eq. (3.417) to get

$$\delta \mathcal{U}_e = \int_{V_0} \delta \underline{\widehat{\varepsilon}}^T \underline{\widehat{\sigma}} \, dV_0 \tag{3.418}$$

By Eq. (3.414), the variations of Green-Lagrange strain tensor components are found to be

$$\begin{aligned} \delta\varepsilon_{11} &= \underline{G}_1^T \delta \underline{G}_1 \\ \delta 2\varepsilon_{12} &= \underline{G}_1^T \delta \underline{G}_2 + \underline{G}_2^T \delta \underline{G}_1 \\ \delta 2\varepsilon_{13} &= \underline{G}_1^T \delta \underline{G}_3 + \underline{G}_3^T \delta \underline{G}_1 \\ \delta\varepsilon_{22} &= \underline{G}_2^T \delta \underline{G}_2 \\ \delta 2\varepsilon_{23} &= \underline{G}_2^T \delta \underline{G}_3 + \underline{G}_3^T \delta \underline{G}_2 \\ \delta\varepsilon_{33} &= \underline{G}_3^T \delta \underline{G}_3 \end{aligned} \tag{3.419}$$

For the easy description, the variation of Green-Lagrange strain tensor in its vector-dyadic form needs to be decomposed into

$$\delta \underline{\widehat{\varepsilon}} = \begin{vmatrix} \delta\varepsilon_{11} \\ \delta 2\varepsilon_{12} \\ \delta 2\varepsilon_{13} \\ \delta\varepsilon_{22} \\ \delta 2\varepsilon_{23} \\ \delta\varepsilon_{33} \end{vmatrix} = \begin{vmatrix} \delta\varepsilon_{11} \\ \delta\varepsilon_{21} \\ \delta\varepsilon_{31} \\ 0 \\ 0 \\ 0 \end{vmatrix} + \begin{vmatrix} 0 \\ \delta\varepsilon_{12} \\ 0 \\ \delta\varepsilon_{22} \\ \delta\varepsilon_{32} \\ 0 \end{vmatrix} + \begin{vmatrix} 0 \\ 0 \\ \delta\varepsilon_{13} \\ 0 \\ \delta\varepsilon_{23} \\ \delta\varepsilon_{33} \end{vmatrix} = \delta \underline{\widehat{\varepsilon}}_1 + \delta \underline{\widehat{\varepsilon}}_2 + \delta \underline{\widehat{\varepsilon}}_3 \tag{3.420}$$

After filtered out zero items, the specific vectorial expressions are found to be

$$\begin{aligned} \delta \underline{\varepsilon}_1 &= \begin{vmatrix} \delta\varepsilon_{11} \\ \delta\varepsilon_{21} \\ \delta\varepsilon_{31} \end{vmatrix} = \begin{vmatrix} \underline{G}_1^T \delta \underline{G}_1 \\ \underline{G}_2^T \delta \underline{G}_1 \\ \underline{G}_3^T \delta \underline{G}_1 \end{vmatrix} = \mathcal{B}_G^T \delta \underline{G}_1 \\ \delta \underline{\varepsilon}_2 &= \begin{vmatrix} \delta\varepsilon_{12} \\ \delta\varepsilon_{22} \\ \delta\varepsilon_{32} \end{vmatrix} = \begin{vmatrix} \underline{G}_1^T \delta \underline{G}_2 \\ \underline{G}_2^T \delta \underline{G}_2 \\ \underline{G}_3^T \delta \underline{G}_2 \end{vmatrix} = \mathcal{B}_G^T \delta \underline{G}_2 \\ \delta \underline{\varepsilon}_3 &= \begin{vmatrix} \delta\varepsilon_{13} \\ \delta\varepsilon_{23} \\ \delta\varepsilon_{33} \end{vmatrix} = \begin{vmatrix} \underline{G}_1^T \delta \underline{G}_3 \\ \underline{G}_2^T \delta \underline{G}_3 \\ \underline{G}_3^T \delta \underline{G}_3 \end{vmatrix} = \mathcal{B}_G^T \delta \underline{G}_3 \end{aligned} \tag{3.421}$$

With the aid of Eq. (3.416), the variation of strain vectors are further simplified to

$$\delta \underline{\varepsilon}_1 = \mathcal{B}_G^T \delta \underline{\epsilon}_1, \quad \delta \underline{\varepsilon}_2 = \mathcal{B}_G^T \delta \underline{\epsilon}_2, \quad \delta \underline{\varepsilon}_3 = \mathcal{B}_G^T \delta \underline{\epsilon}_3 \tag{3.422}$$

In view of the strain decomposition, Eq. (3.420), the variation of strain energy could be reformulated to

$$\delta \mathcal{U}_e = \int_{V_0} (\delta \underline{\widehat{\varepsilon}}_1 + \delta \underline{\widehat{\varepsilon}}_2 + \delta \underline{\widehat{\varepsilon}}_3)^T \underline{\widehat{\sigma}}\, d V_0$$
$$= \int_{V_0} (\delta \underline{\varepsilon}_1^T \underline{\sigma}_1 + \delta \underline{\varepsilon}_2^T \underline{\sigma}_2 + \delta \underline{\varepsilon}_3^T \underline{\sigma}_3)\, d V_0 \tag{3.423}$$

where $\underline{\sigma}_i$ for $i = 1, 2, 3$ are the reduced vectorial stresses by filtering out the components related to the zero terms of $\delta \underline{\widehat{\varepsilon}}_i$ defined by Eq. (3.420). Referring Eq. (3.422), the variation of strain energy can be recasted to

$$\delta \mathcal{U}_e = \int_{V_0} (\delta \underline{\epsilon}_1^T \mathcal{B}_G \underline{\sigma}_1 + \delta \underline{\epsilon}_2^T \mathcal{B}_G \underline{\sigma}_2 + \delta \underline{\epsilon}_3^T \mathcal{B}_G \underline{\sigma}_3)\, d V_0 \tag{3.424}$$

and then is rearranged into the compact form

$$\delta \mathcal{U} = \int_{V_0} (\delta \underline{\epsilon}_1^T \underline{F}_{\epsilon\epsilon}^{11} + \delta \underline{\epsilon}_2^T \underline{F}_{\epsilon\epsilon}^{22} + \delta \underline{\epsilon}_3^T \underline{F}_{\epsilon\epsilon}^{33})\, d V_0 \tag{3.425}$$

where the stress measures of a Cauchy continuum are defined in the vectorial formulas like

$$\underline{F}_{\epsilon\epsilon}^{11} = \mathcal{B}_G \underline{\sigma}_1, \quad \underline{F}_{\epsilon\epsilon}^{22} = \mathcal{B}_G \underline{\sigma}_2, \quad \underline{F}_{\epsilon\epsilon}^{33} = \mathcal{B}_G \underline{\sigma}_3 \tag{3.426}$$

### 3.6.7 Virtual Work of External Forces for Cauchy Continuum

The principle of virtual work is applied to express the Cauchy continuum mechanics problem in current implementation. The virtual work done by a corresponding virtual displacement operating on the external forces is then required. By Eq. (3.388), the virtual displacement of material particle **P** in deformed configuration is verified to be

$$\delta \underline{P} = \delta \underline{X}_0 = \delta \underline{u}_s \tag{3.427}$$

where $\underline{u}_s$ is the associated displacement. In current implementation, two types of external forces are considered, one is body force $\underline{g}$, the other surface pressure $\underline{p}_s$, both measured in inertial frame. By expression of above virtual displacement, the virtual work of external forces can be computed from

$$\delta \mathcal{W}_e = \int_{V_0} \delta \underline{u}_s^T \underline{g}\, d V_0 + \int_{S_0} \delta \underline{u}_s^T \underline{p}_s\, d S_0 \tag{3.428}$$

where $S_0$ represents the area magnitude of a Cauchy continuum where the surface pressure is applied to. For the easy description of virtual work, the external force vectors are averaged at first by the characteristic volume $V_c$ like

$$\underline{F}_{\epsilon\alpha} = \frac{\int_{V_c} \underline{g}\, dV_c + \int_{S_c} \underline{p}_s\, dS_c}{V_c} \tag{3.429}$$

then the virtual work can be approximated to be

$$\delta \mathcal{W}_e = \int_{V_0} \delta \underline{u}_s^T \underline{F}_{\epsilon\alpha}\, dV_0 \tag{3.430}$$

### 3.6.8 Governing Equations of Cauchy Continuum

The principle of virtual work states

$$\delta \mathcal{U}_e - \delta \mathcal{W}_e = 0 \tag{3.431}$$

which answers the virtual work done by the infinitesimal strain operating on the stress measures equals to the virtual work done by a corresponding virtual displacement operating on the external forces. Introducing Eqs. (3.413) and (3.430) leads to

$$\int_{V_0} (\delta \underline{\epsilon}_1^T \underline{F}_{\epsilon\epsilon}^{11} + \delta \underline{\epsilon}_2^T \underline{F}_{\epsilon\epsilon}^{22} + \delta \underline{\epsilon}_3^T \underline{F}_{\epsilon\epsilon}^{33} - \delta \underline{u}_s^T \underline{F}_{\epsilon\alpha}) dV_0 = 0 \tag{3.432}$$

In view of the relations, $\delta \underline{\epsilon}_1 = \delta \underline{u}_{s,1}$, $\delta \underline{\epsilon}_2 = \delta \underline{u}_{s,2}$ and $\delta \underline{\epsilon}_3 = \delta \underline{u}_{s,3}$, the principle of virtual work can be recasted to

$$\int_{V_0} (\delta \underline{u}_{s,1}^T \underline{F}_{\epsilon\epsilon}^{11} + \delta \underline{u}_{s,2}^T \underline{F}_{\epsilon\epsilon}^{22} + \delta \underline{u}_{s,3}^T \underline{F}_{\epsilon\epsilon}^{33} - \delta \underline{u}_s^T \underline{F}_{\epsilon\alpha}) dV_0 = 0 \tag{3.433}$$

Apparently, the virtual displacement and its spatial derivatives form a linear three-dimensional space, they can be interpolated directly by using the shape functions $\mathcal{H}_\epsilon$ of finite element method. The following interpolations

$$\delta \underline{u}_s = \mathcal{H}_\epsilon \delta \underline{u}_n,\ \delta \underline{u}_{s,i} = \mathcal{H}_{\epsilon,i} \delta \underline{u}_n \tag{3.434}$$

for $i = 1, 2, 3$ reformulate the principle of virtual work into the weak Galerkin form

$$\delta \underline{u}_n^T \int_{V_0} (\mathcal{H}_{\epsilon,1}^T \underline{F}_{\epsilon\epsilon}^{11} + \mathcal{H}_{\epsilon,2}^T \underline{F}_{\epsilon\epsilon}^{22} + \mathcal{H}_{\epsilon,3}^T \underline{F}_{\epsilon\epsilon}^{33} - \mathcal{H}_\epsilon^T \underline{F}_{\epsilon\alpha}) dV_0 = 0 \tag{3.435}$$

From which, the governing equations are found to be

$$\int_{V_0} (\mathcal{H}^T_{\epsilon,1}\underline{F}^{11}_{\epsilon\epsilon} + \mathcal{H}^T_{\epsilon,2}\underline{F}^{22}_{\epsilon\epsilon} + \mathcal{H}^T_{\epsilon,3}\underline{F}^{33}_{\epsilon\epsilon} - \mathcal{H}^T_{\epsilon}\underline{F}_{\epsilon\alpha})dV_0 = 0 \tag{3.436}$$

After introduced the elastic force vector

$$\underline{\mathcal{F}}_e = \int_{V_0} (\mathcal{H}^T_{\epsilon,1}\underline{F}^{11}_{\epsilon\epsilon} + \mathcal{H}^T_{\epsilon,2}\underline{F}^{22}_{\epsilon\epsilon} + \mathcal{H}^T_{\epsilon,3}\underline{F}^{33}_{\epsilon\epsilon})dV_0 \tag{3.437}$$

and the external force vector

$$\underline{\mathcal{F}}_a = \int_{V_0} \mathcal{H}^T_{\epsilon}\underline{F}_{\epsilon\alpha} dV_0 \tag{3.438}$$

the governing equations can be described by the following compact expression

$$\underline{\mathcal{F}}_e - \underline{\mathcal{F}}_a = 0 \tag{3.439}$$

### 3.6.9 Kinematic Energy of Cauchy Continuum

Similar to the definitions of position variation $\delta\underline{P}$, the velocity of an arbitrary material particle in a Cauchy continuum can be evaluated by taking time derivatives of position vector $\underline{P}$, defined by Eq. (3.388), to get

$$\dot{\underline{P}} = (\underline{x}_0 \dot{+} \underline{u}_s) = \dot{\underline{u}}_s \tag{3.440}$$

If denoting the material density as $\rho$, the kinematic energy of a Cauchy continuum can be computed from the integral related to the initial volume like

$$\mathcal{K}_e = \int_{V_0} \frac{1}{2}\rho\dot{\underline{u}}_s^T\dot{\underline{u}}_s dV_0 \tag{3.441}$$

or in compact formula alternatively

$$\mathcal{K} = \int_{V_0} \frac{1}{2}\dot{\underline{u}}_s^T\mathcal{M}_{\epsilon\epsilon}\dot{\underline{u}}_s dV_0 \tag{3.442}$$

where the mass matrix is defined as

$$\mathcal{M}_{\epsilon\epsilon} = \rho I \tag{3.443}$$

By Eq. (3.442), the variation of kinematic energy is found to be

$$\delta\mathcal{K}_e = \int_{V_0} \delta\dot{\underline{u}}_s^T \mathcal{M}_{\epsilon\epsilon}\dot{\underline{u}}_s \, dV_0 \tag{3.444}$$

Integrating the variation of kinematic energy by part in time domain produces

$$\int_t \delta\mathcal{K}_e dt = \int_t \int_{V_0} -\delta\underline{u}_s^T \mathcal{M}_{\epsilon\epsilon}\ddot{\underline{u}}_s \, dV_0 dt \tag{3.445}$$

From which the variation of kinematic energy is reformulated to

$$\delta\mathcal{K}_e = \int_{V_0} -\delta\underline{u}_s^T \mathcal{M}_{\epsilon\epsilon}\ddot{\underline{u}}_s \, dV_0 \tag{3.446}$$

With the application of finite element discretization, the velocity and accelerations of material particle can be approximated by using the polynomials

$$\dot{\underline{u}}_s = \mathcal{H}_\epsilon \dot{\underline{u}}_n, \quad \ddot{\underline{u}}_s = \mathcal{H}_\epsilon \ddot{\underline{u}}_n \tag{3.447}$$

and the weak form of kinematic energy variation is found to be

$$\delta\mathcal{K}_e = -\delta\underline{u}_n^T \int_{V_0} \mathcal{H}_\epsilon^T \mathcal{M}_{\epsilon\epsilon}\mathcal{H}_\epsilon dV_0 \, \ddot{\underline{u}}_n \tag{3.448}$$

After defined the mass matrix

$$\mathcal{M}_n = \int_{V_0} \mathcal{H}_\epsilon^T \mathcal{M}_{\epsilon\epsilon}\mathcal{H}_\epsilon dV_0 \tag{3.449}$$

the variation of original kinematic energy can be rewritten into the compact formula

$$\delta\mathcal{K}_e = -\delta\underline{u}_n^T \mathcal{M}_n \ddot{\underline{u}}_n \tag{3.450}$$

### *3.6.10 Extended to Cauchy Continuum Dynamic Problem*

When considering the inertial forces, the governing equations of motion for a Cauchy continuum can be obtained from the Hamilton's principle. Equivalently, the inertial force are required to be added into Eq. (3.439) to get

$$\mathcal{M}_n \ddot{\underline{u}}_n + \underline{\mathcal{F}}_e = \underline{\mathcal{F}}_a \tag{3.451}$$

## 3.7 General Membrane Theory

When the thickness of shell becomes infinitesimal, such that the bending stiffness and torsional stiffness can be ignored. In this special situation, the shell could be approximated by the membrane that only takes into account the in-plane strain measures.

### 3.7.1 *Kinematics of Membrane*

As depicted in Fig. 3.7, a material particle in a membrane is depicted by its position vector in the reference configuration and current deformed configuration, respectively. In the original reference configuration, the position vector $\underline{p}(\alpha_1, \alpha_2)$ of the material particle relative to a point $\mathbf{O}$, the original of reference frame, $\mathcal{F}_\mathcal{I} = [\mathbf{O}, \mathcal{I}]$, can be written as

$$\underline{p}(\alpha_1, \alpha_2) = \underline{x}_0(\alpha_1, \alpha_2) \tag{3.452}$$

where only two curvilinear coordinates, $(\alpha_1, \alpha_2)$, are enough to label the material particle in the original membrane. For the description of membrane deformations, the covariant base vectors

$$\underline{g}_i = \frac{\partial \underline{p}}{\partial \alpha_i} \tag{3.453}$$

for $i = 1, 2, 3$ in the reference configuration are required, and the specific expressions can be obtained by taking derivatives of Eq. (3.452) with respected to the curvilinear coordinates $\alpha_i$ as

$$\underline{g}_1 = \underline{x}_{0,1}, \quad \underline{g}_2 = \underline{x}_{0,2}, \quad \underline{g}_3 = \underline{0} \tag{3.454}$$

All these base vectors form the covariant bases $\mathcal{B}_g$ like

$$\mathcal{B}_g = \lfloor \underline{g}_1, \ \underline{g}_2, \ \underline{g}_3 \rfloor \tag{3.455}$$

and then computing the metric tensor in the reference configuration from

$$g = \mathcal{B}_g^T \mathcal{B}_g \tag{3.456}$$

when the membrane deforms with spatial motions, the material particle that had position vector $\underline{P}$ in the undeformed reference configuration now has position vector $\underline{P}$ in the deformed configuration, such that

$$\underline{P}(\alpha_1, \alpha_2) = \underline{x}_0(\alpha_1, \alpha_2) + \underline{u}_s(\alpha_1, \alpha_2) \tag{3.457}$$

where the displacement vector $\underline{u}_s$ describes arbitrary translation of material particle. According to the definitions of covariant base vectors in deformed configuration

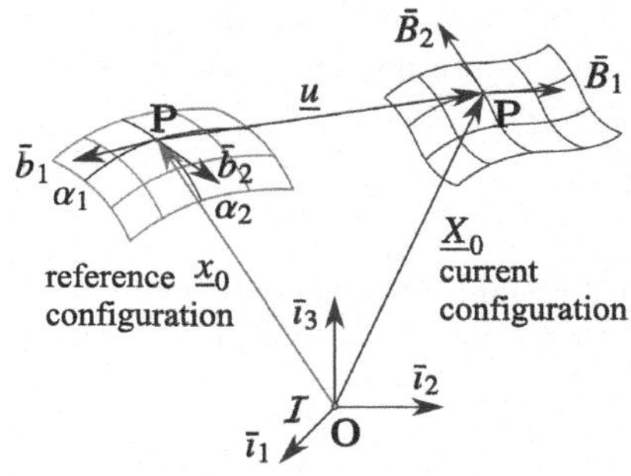


**Fig. 3.7** Kinematics of a material particle in a membrane

$$\underline{G}_i = \frac{\partial \underline{P}}{\partial \alpha_i} \tag{3.458}$$

the specific formulas can be obtained by taking spatial derivatives of Eq. (3.457) with respected to the curvilinear coordinates $\alpha_i$ for $i = 1, 2, 3$ like

$$\underline{G}_1 = \underline{P}_{0,1}, \quad \underline{G}_2 = \underline{P}_{0,2}, \quad \underline{G}_3 = \underline{0} \tag{3.459}$$

All these base vectors could be assembled into the covariant bases $\mathcal{B}_G$ like

$$\mathcal{B}_G = \lfloor \underline{G}_1, \ \underline{G}_2, \ \underline{G}_3 \rfloor \tag{3.460}$$

In view of above expressions of covariant basis vectors in the deformed configuration, the associated metric tensor can be computed from its definitions

$$G = \mathcal{B}_G^T \mathcal{B}_G \tag{3.461}$$

### *3.7.2 Strain Tensor of Membrane*

The rotation-free Green-Lagrange strain tensor [22] will be applied to measure the membrane deformation. In doing so, the Lagrangian stretch tensor, only containing two nonzero spatial derivatives

$$\underline{\epsilon}_1 = \underline{u}_{s,1}, \quad \underline{\epsilon}_2 = \underline{u}_{s,2} \tag{3.462}$$

is required to write the deformed covariant base vectors in terms of

$$\underline{G}_1 = \underline{\epsilon}_1 + \underline{x}_{0,1}, \quad \underline{G}_2 = \underline{\epsilon}_2 + \underline{x}_{0,2} \tag{3.463}$$

According to above definitions of covariant basis vectors, the components of rotation-free Green-Lagrange strain tensor

$$\hat{\varepsilon} = \frac{1}{2}(G - g) \tag{3.464}$$

are found to be closely related to the components of Lagrangian stretch strains. In specific, these relations in their vector-dyadic form are verified

$$\begin{aligned} \varepsilon_{11} &= \underline{x}_{0,1}^T \underline{\epsilon}_1 \\ 2\varepsilon_{12} &= \underline{x}_{0,1}^T \underline{\epsilon}_2 + \underline{x}_{0,2}^T \underline{\epsilon}_1 \\ 2\varepsilon_{13} &= 0 \\ \varepsilon_{22} &= \underline{x}_{0,2}^T \underline{\epsilon}_2 \\ 2\varepsilon_{23} &= 0 \\ \varepsilon_{33} &= 0 \end{aligned} \tag{3.465}$$

Note that the assumptions of small stretch strain $\underline{\epsilon}_i \ll 0$ for $i = 1, 2$ are applied such that the higher order terms $\underline{\epsilon}_i^T \underline{\epsilon}_j$ for $i, j = 1, 2$ are all neglected.

### 3.7.3 *Constitutive Laws for Membrane*

For the easy description of membrane constitutive relations, the Green-Lagrange strain tensor in its vector-dyadic form is decomposed into

$$\underline{\widehat{\varepsilon}} = \begin{vmatrix} \varepsilon_{11} \\ 2\varepsilon_{12} \\ 2\varepsilon_{13} \\ \varepsilon_{22} \\ 2\varepsilon_{23} \\ \varepsilon_{33} \end{vmatrix} = \begin{vmatrix} \varepsilon_{11} \\ \varepsilon_{21} \\ 0 \\ 0 \\ 0 \\ 0 \end{vmatrix} + \begin{vmatrix} 0 \\ \varepsilon_{12} \\ 0 \\ \varepsilon_{22} \\ 0 \\ 0 \end{vmatrix} = \underline{\varepsilon}_1 + \underline{\varepsilon}_2 \tag{3.466}$$

It is noted that the symbols $\varepsilon_{12}$ and $\varepsilon_{21}$ do not represent the components of Green-Lagrange strain tensor, but are defined in the following relations

$$2\varepsilon_{12} = \varepsilon_{21} + \varepsilon_{12} \tag{3.467}$$

After filtered out the zero terms, the specific vectorial expressions are reduced to

$$\hat{\underline{\varepsilon}}_1 = \begin{vmatrix} \varepsilon_{11} \\ \varepsilon_{21} \end{vmatrix} = \begin{vmatrix} \underline{x}_{0,1}^T \underline{\epsilon}_1 \\ \underline{x}_{0,2}^T \underline{\epsilon}_1 \end{vmatrix}, \quad \hat{\underline{\varepsilon}}_2 = \begin{vmatrix} \varepsilon_{12} \\ \varepsilon_{22} \end{vmatrix} = \begin{vmatrix} \underline{x}_{0,1}^T \underline{\epsilon}_2 \\ \underline{x}_{0,2}^T \underline{\epsilon}_2 \end{vmatrix} \tag{3.468}$$

and then assembled into

$$\begin{bmatrix} \varepsilon_{11}, & \varepsilon_{12} \\ \varepsilon_{21}, & \varepsilon_{22} \end{bmatrix} = \begin{bmatrix} \underline{x}_{0,1}^T \\ \underline{x}_{0,2}^T \end{bmatrix} \lfloor \underline{\epsilon}_1, \ \underline{\epsilon}_2 \rfloor \tag{3.469}$$

As a result, the Green-Lagrange strain measures is recasted to a truncated form like

$$\lfloor \hat{\underline{\varepsilon}}_1, \ \hat{\underline{\varepsilon}}_2 \rfloor = \mathcal{S}_0^T \lfloor \underline{\epsilon}_1, \ \underline{\epsilon}_2 \rfloor \tag{3.470}$$

with the definitions of transformation matrix $\mathcal{S}_0$ like

$$\mathcal{S}_0 = \lfloor \underline{x}_{0,1}, \ \underline{x}_{0,2} \rfloor \tag{3.471}$$

The strain energy $\mathcal{U}_e$ can be represented by using the non-zero components of Green-Lagrange strain tensor like

$$2\mathcal{U}_e = \int_{V_0} (\hat{\underline{\varepsilon}}_1^T \mathcal{D}_{11} \hat{\underline{\varepsilon}}_1 + \hat{\underline{\varepsilon}}_2^T \mathcal{D}_{22} \hat{\underline{\varepsilon}}_2 + 2\hat{\underline{\varepsilon}}_2^T \mathcal{D}_{21} \hat{\underline{\varepsilon}}_1) \, dV_0 \tag{3.472}$$

where the following constitutive matrix blocks are introduced

$$\mathcal{D}_{11} = \begin{bmatrix} d_{11}, & d_{12} \\ d_{21}, & d_{22} \end{bmatrix}, \ \mathcal{D}_{22} = \begin{bmatrix} d_{22}, & d_{24} \\ d_{42}, & d_{44} \end{bmatrix}, \ \mathcal{D}_{12} = \begin{bmatrix} d_{12}, & d_{14} \\ d_{22}, & d_{24} \end{bmatrix} \tag{3.473}$$

Since the constitutive matrix is a symmetric matrix, it is readily to verify that $\mathcal{D}_{12} = \mathcal{D}_{21}^T$. Referring the relations of Green-Lagrange strain and Lagrangian strain measures, Eq. (3.470), the above expression of strain energy is recasted to

$$2\mathcal{U}_e = \int_{S_0} (\underline{\epsilon}_1^T \mathcal{C}_{11} \underline{\epsilon}_1 + \underline{\epsilon}_2^T \mathcal{C}_{22} \underline{\epsilon}_2 + 2\underline{\epsilon}_2^T \mathcal{C}_{21} \underline{\epsilon}_1) \, dS_0 \tag{3.474}$$

where $S_0$ is the membrane area, and $\mathcal{C}_{ij}$ for $i, j = 1, 2$ are the constitutive matrices of membrane defined as

$$\mathcal{C}_{11} = \int_h \mathcal{S}_0 \mathcal{D}_{11} \mathcal{S}_0^T \, d\alpha_3, \ \mathcal{C}_{22} = \int_h \mathcal{S}_0 \mathcal{D}_{22} \mathcal{S}_0^T \, d\alpha_3, \ \mathcal{C}_{21} = \int_h \mathcal{S}_0 \mathcal{D}_{21} \mathcal{S}_0^T \, d\alpha_3 \tag{3.475}$$

It is necessary to note that the symmetric properties of constitutive matrices are still hold and $\mathcal{C}_{21} = \mathcal{C}_{12}^T$.

### 3.7.4 Variation of Strain Energy of Membrane

The second Piola-Kirchhoff stress and Green-Lagrange strain measures are selected so that their product gives an accurate energy density that is related to the undeformed volume of reference configuration. Referring the expressions of strain energy, Eq. (3.474), the strain energy density for a membrane is given as

$$\mathcal{U}_\rho = \frac{1}{2}(\underline{\epsilon}_1^T \mathcal{C}_{11}\underline{\epsilon}_1 + \underline{\epsilon}_2^T \mathcal{C}_{22}\underline{\epsilon}_2) + \underline{\epsilon}_2^T \mathcal{C}_{21}\underline{\epsilon}_1 \tag{3.476}$$

The variation of $\mathcal{U}_\rho$ is readily verified to be

$$\delta\mathcal{U}_\rho = \delta\underline{\epsilon}_1^T \mathcal{C}_{11}\underline{\epsilon}_1 + \delta\underline{\epsilon}_2^T \mathcal{C}_{22}\underline{\epsilon}_2 + \delta\underline{\epsilon}_2^T \mathcal{C}_{21}\underline{\epsilon}_1 + \delta\underline{\epsilon}_1^T \mathcal{C}_{12}\underline{\epsilon}_2 \tag{3.477}$$

After merging the similar terms at the right-hand-side of above equations, the variation of strain energy density can be rearranged into the compact form

$$\delta\mathcal{U}_\rho = \delta\underline{\epsilon}_1^T \underline{F}_{\epsilon\epsilon}^{11} + \delta\underline{\epsilon}_2^T \underline{F}_{\epsilon\epsilon}^{22} \tag{3.478}$$

where the stress measures of a membrane are defined in the vectorial formulas like

$$\underline{F}_{\epsilon\epsilon}^{11} = \mathcal{C}_{11}\underline{\epsilon}_1 + \mathcal{C}_{12}\underline{\epsilon}_2, \quad \underline{F}_{\epsilon\epsilon}^{22} = \mathcal{C}_{21}\underline{\epsilon}_1 + \mathcal{C}_{22}\underline{\epsilon}_2 \tag{3.479}$$

Finally, the variation of strain energy can be represented by

$$\delta\mathcal{U}_e = \int_{S_0} (\delta\underline{\epsilon}_1^T \underline{F}_{\epsilon\epsilon}^{11} + \delta\underline{\epsilon}_2^T \underline{F}_{\epsilon\epsilon}^{22}) dS_0 \tag{3.480}$$

### 3.7.5 Virtual Work of External Forces for a Membrane

The virtual work done by a corresponding virtual displacement operating on the external forces is evaluated in this section. According to Eq. (3.457), the virtual displacement of an arbitrary particle **P** of the membrane in deformed configuration is obtained by

$$\delta\underline{P} = \delta(\underline{x}_0 + \underline{u}_s) = \delta\underline{u}_s \tag{3.481}$$

where $\underline{u}_s$ is the associated displacement. In current implementation, only the surface pressure $\underline{p}_s$, measured in inertial frame, is considered. The virtual work of external forces can be computed from

$$\delta\mathcal{W}_e = \int_{S_0} \delta\underline{P}^T \underline{p}_s \, dS_0 = \int_{S_0} \delta\underline{u}_s^T \underline{p}_s \, dS_0 \tag{3.482}$$

### 3.7.6 Governing Equations of Membrane

Based on the principle of virtual work, which states, $\delta\mathcal{U}_e - \delta\mathcal{W}_e = 0$, the governing equations for a membrane problem could be obtained. Introducing Eqs. (3.480) and (3.482) leads to

$$\int_{S_0} (\delta\underline{\epsilon}_1^T \underline{F}_{\epsilon\epsilon}^{11} + \delta\underline{\epsilon}_2^T \underline{F}_{\epsilon\epsilon}^{22} - \delta\underline{u}_s^T \underline{p}_s) dS_0 = 0 \tag{3.483}$$

With the aid of relations, $\delta\underline{\epsilon}_1 = \delta\underline{u}_{s,1}$ and $\delta\underline{\epsilon}_2 = \delta\underline{u}_{s,2}$, the principle of virtual work can be represented by

$$\int_{S_0} (\delta\underline{u}_{s,1}^T \underline{F}_{\epsilon\epsilon}^{11} + \delta\underline{u}_{s,2} \underline{F}_{\epsilon\epsilon}^{22} - \delta\underline{u}_s^T \underline{p}_s) dS_0 = 0 \tag{3.484}$$

With the application of finite element discretization, the virtual displacement and its variations are interpolated by using the shape functions $H_\epsilon$. The following polynomial interpolations

$$\delta\underline{u}_s = H_\epsilon \delta\underline{u}_n, \ \delta\underline{u}_{s,i} = H_{\epsilon,i} \delta\underline{u}_n \tag{3.485}$$

for $i = 1, 2$, reformulate the principle of virtual work for a membrane into the weak Galerkin form

$$\delta\underline{u}_n^T \int_{S_0} (H_{\epsilon,1}^T \underline{F}_{\epsilon\epsilon}^{11} + H_{\epsilon,2}^T \underline{F}_{\epsilon\epsilon}^{22} - H_\epsilon^T \underline{p}_s) dS_0 = 0 \tag{3.486}$$

From which, the governing equations are obtained

$$\int_{S_0} (H_{\epsilon,1}^T \underline{F}_{\epsilon\epsilon}^{11} + H_{\epsilon,2}^T \underline{F}_{\epsilon\epsilon}^{22} - H_\epsilon^T \underline{p}_s) dS_0 = 0 \tag{3.487}$$

After introduced the elastic force vector

$$\underline{\mathcal{F}}_e = \int_{S_0} (H_{\epsilon,1}^T \underline{F}_{\epsilon\epsilon}^{11} + H_{\epsilon,2}^T \underline{F}_{\epsilon\epsilon}^{22}) dS_0 \tag{3.488}$$

and the external force vector

$$\underline{\mathcal{F}}_a = \int_{S_0} H_\epsilon^T \underline{p}_s \, dS_0 \tag{3.489}$$

the governing equations can be described by the following compact formula

$$\underline{\mathcal{F}}_e - \underline{\mathcal{F}}_a = 0 \tag{3.490}$$

### 3.7.7 Kinematic Energy of Membrane

When the governing equation of a membrane is extended to the dynamic problem, the kinematic energy of membrane needs to be evaluated. At first, the velocity of an arbitrary material particle in a membrane can be evaluated by taking time derivatives of position vector $\underline{P}$, defined by Eq. (3.457), to get

$$\underline{\dot{P}} = (\underline{x}_0 \dot{+} \underline{u}_s) = \underline{\dot{u}}_s \tag{3.491}$$

If denoting the material density as $\rho$, the kinematic energy of a membrane can be computed from the integral related to the initial volume $V_0$ like

$$\mathcal{K}_e = \int_{V_0} \frac{1}{2} \rho \underline{\dot{u}}_s^T \underline{\dot{u}}_s dV_0 \tag{3.492}$$

After introduced the constant mass matrix

$$\mathcal{M}_{\epsilon\epsilon} = \int_h \rho I \, d\alpha_3 = \rho h I, \tag{3.493}$$

where $h$ is the thickness of membrane, the kinematic energy can be readily recasted to

$$\mathcal{K}_e = \int_{S_0} \frac{1}{2} \underline{\dot{u}}_s^T \mathcal{M}_{\epsilon\epsilon} \underline{\dot{u}}_s \, dS_0 \tag{3.494}$$

By this expression, the variation of kinematic energy is found to be

$$\delta\mathcal{K}_e = \int_{S_0} \delta\underline{\dot{u}}_s^T \mathcal{M}_{\epsilon\epsilon} \underline{\dot{u}}_s \, dS_0 \tag{3.495}$$

Integrating the variation of kinematic energy by part in time domain produces

$$\int_t \delta\mathcal{K}_e dt = -\int_t \int_{S_0} \delta\underline{u}_s^T \mathcal{M}_{\epsilon\epsilon} \underline{\ddot{u}}_s \, dS_0 \tag{3.496}$$

From which the variation of kinematic energy is reformulated to

$$\delta\mathcal{K}_e = -\int_{S_0} \delta\underline{u}_s^T \mathcal{M}_{\epsilon\epsilon} \underline{\ddot{u}}_s \, dS_0 \tag{3.497}$$

With the application of finite element method, the velocity and acceleration of material particle is approximated by using the following polynomials of nodal velocities $\dot{\underline{u}}_n$ and accelerations $\ddot{\underline{u}}_n$ like

$$\dot{\underline{u}}_s = H_\epsilon \dot{\underline{u}}_n, \quad \ddot{\underline{u}}_s = H_\epsilon \ddot{\underline{u}}_n \tag{3.498}$$

and the weak form of kinematic energy variation is found to be

$$\delta \mathcal{K}_e = -\delta \underline{u}_n^T \int_{S_0} H_\epsilon^T \mathcal{M}_{\epsilon\epsilon} H_\epsilon \, dS_0 \, \ddot{\underline{u}}_n \tag{3.499}$$

After defined the mass matrix

$$\mathcal{M}_n = \int_{S_0} \mathcal{H}_\epsilon^T \mathcal{M}_{\epsilon\epsilon} \mathcal{H}_\epsilon dS_0 \tag{3.500}$$

the variation of original kinematic energy can be rewritten into the compact formula

$$\delta \mathcal{K}_e = -\delta \underline{u}_n^T \mathcal{M}_n \ddot{\underline{u}}_n \tag{3.501}$$

### 3.7.8 Extended to Membrane Dynamic Problem

When considering the inertial force, the governing equations of motion for a membrane can be obtained from the Hamilton's principle. Equivalently, the inertial force are required to be added into Eq. (3.490) to find

$$\mathcal{M}_n \ddot{\underline{u}}_n + \underline{\mathcal{F}}_e = \underline{\mathcal{F}}_a \tag{3.502}$$

## 3.8 General Cable Theory

If the stiffness properties of shearing, bending and torsion of a beam are all ignored, the beam could be treated as a cable suffering only extension. In other words, the governing equations of motion for a cable can be obtained directly from the reduction of beam equations by retaining the terms associated with the extension only.

### 3.8.1 Kinematics of Cable

As shown in Fig. 3.8, a material particle in a cable is depicted by its position vector in the reference configuration and current configuration, respectively. In the original reference configuration, the position vector $\underline{p}(\alpha_i)$ of the material particle relative to a point $\mathbf{O}$, the original of reference frame, $\bar{\mathcal{F}}_{\mathcal{I}} = [\mathbf{O}, \mathcal{I}]$, can be written as

$$\underline{p}(\alpha_1) = \underline{x}_0(\alpha_1) \tag{3.503}$$

where the curvilinear coordinate $\alpha_1$ is introduced to label the material particle in the cable. For the easy description of cable deformations, the covariant base vector,

$$\underline{g}_1 = \frac{\partial \underline{p}}{\partial \alpha_1} \tag{3.504}$$

in the reference configuration are required, and the specific expression can be obtained by taking derivative of Eq. (3.503) with respected to the curvilinear coordinate $\alpha_1$ like

$$\underline{g}_1 = \underline{x}_{0,1} \tag{3.505}$$

and other two covariant base vectors, $\underline{g}_2$ and $\underline{g}_3$, apparently become zero, $\underline{g}_2 = \underline{0}$ and $\underline{g}_3 = \underline{0}$. All these base vectors form the covariant bases $\mathcal{B}_g$ like

$$\mathcal{B}_g = \lfloor \underline{g}_1,\ \underline{g}_2,\ \underline{g}_3 \rfloor \tag{3.506}$$

and then the metric tensor in the reference configuration is computed from

$$g = \mathcal{B}_g^T \mathcal{B}_g \tag{3.507}$$

When the cable deforms with spatial motions, the material particle that had position vector $\underline{p}$ in the undeformed reference configuration now has position vector $\underline{P}$ in the deformed configuration, such that

**Fig. 3.8** Kinematics of a material particle in a cable

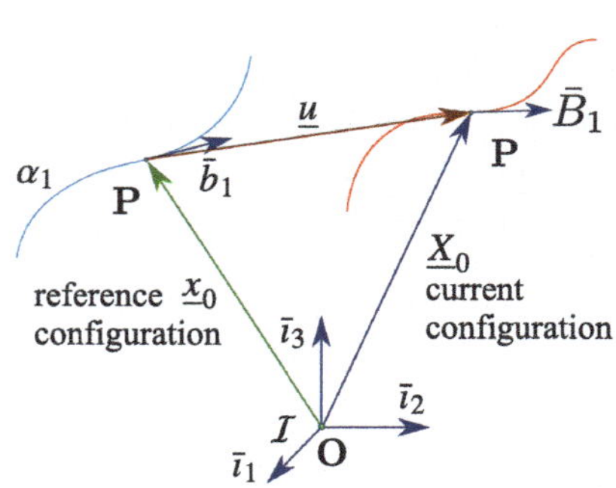

$$\underline{P}(\alpha_1) = \underline{x}_0(\alpha_1) + \underline{u}_s(\alpha_1) \tag{3.508}$$

where the displacement vector $\underline{u}_s$ describes arbitrary translation of material particle. According to the definition of covariant base vector in deformed configuration

$$\underline{G}_1 = \frac{\partial \underline{P}}{\partial \alpha_1} \tag{3.509}$$

its specific expression can be obtained by taking spatial derivative of Eq. (3.508) with respected to the curvilinear coordinate $\alpha_1$ like

$$\underline{G}_1 = \underline{P}_{,1} = \underline{x}_{0,1} + \underline{u}_{s,1} \tag{3.510}$$

This base vector with other two zero covariant base vectors, $\underline{G}_2$ and $\underline{G}_3$, could be assembled into the covariant bases

$$\mathcal{B}_G = \lfloor \underline{G}_1,\ \underline{G}_2,\ \underline{G}_3 \rfloor \tag{3.511}$$

In view of above expressions, the associated metric tensor $G = \mathcal{B}_G^T \mathcal{B}_G$ can be computed from its definition.

### 3.8.2 Strain Tensor of Cable

The rotation-free Green-Lagrange strain tensor [22] will be applied to measure the deformation of Cosserat continuum. In doing so, the Lagrangian stretch tensor, only containing one nonzero spatial derivatives

$$\underline{\epsilon}_1 = \underline{u}_{s,1} \tag{3.512}$$

is required to rewrite the deformed covariant base vector $\underline{G}_1$ in terms of

$$\underline{G}_1 = \underline{\epsilon}_1 + \underline{x}_{0,1} \tag{3.513}$$

Substituting the specific expressions of metric tensors in terms of covariant basis vector, Eq. (3.513), into the definitions of strain tensor, $\widehat{\varepsilon} = \frac{1}{2}(G - g)$, the component $\varepsilon_{11}$ of rotation-free Green-Lagrange strain tensor is related to

$$\varepsilon_{11} = \underline{x}_{0,1}^T \underline{\epsilon}_1 + \frac{1}{2} \underline{\epsilon}_1^T \underline{\epsilon}_1 \tag{3.514}$$

which is a complete definition of elongation without introducing any kind of assumptions.

### 3.8.3 Constitutive Laws for Cable

The strain energy $\mathcal{U}_e$ can be represented by using the component $\varepsilon_{11}$ of Green-Lagrange strain tensor like

$$\mathcal{U}_e = \frac{1}{2} \int_{V_0} \varepsilon_{11} d_{11} \varepsilon_1 d V_0 \tag{3.515}$$

Since the strain component $\varepsilon_{11}$ is not a function of $\alpha_2$ and $\alpha_3$, the above integral reduces to

$$\mathcal{U}_e = \frac{1}{2} \int_{\ell_1} \varepsilon_{11} E A \, \varepsilon_1 \, d\alpha_1 \tag{3.516}$$

where $\ell_1$ is the length of the cable, and $EA$ the axial stiffness of cable defined as

$$EA = \int_{S_0} d_{11} d S_0 \tag{3.517}$$

with the area magnitude of cable cross-section given as $S_0$.

### 3.8.4 Variation of Strain Energy of Cable

The second Piola-Kirchhoff stress and Green-Lagrange strain measures are selected so that their product gives an accurate energy density that is related to the undeformed length of reference configuration. Referring the expressions of strain energy, Eq. (3.516), the variation of strain energy for a cable is defined as

$$\delta \mathcal{U}_e = \int_{\ell_1} \delta \varepsilon_{11} F_{\epsilon\epsilon}^{11} d\alpha_1 \tag{3.518}$$

where the stress resultant of a cable are introduced with the definition of

$$F_{\epsilon\epsilon}^{11} = E A \, \varepsilon_{11} \tag{3.519}$$

### 3.8.5 Virtual Work of External Forces for Cable

The virtual work done by a corresponding virtual displacement operating on the external forces is evaluated in this section. According to Eq. (3.508), the virtual displacement of an arbitrary particle in deformed configuration is obtained by

$$\delta \underline{P} = \delta(\underline{x}_0 + \underline{u}_s) = \delta \underline{u}_s \tag{3.520}$$

and $\underline{u}_s$ is the displacement of material particle **P**. In current implementation, only the traction $\underline{t}_\sigma$ per infinitesimal area of cable cross-section is considered. Then, the virtual work of external forces can be computed from

$$\delta \mathcal{W}_e = \int_{V_0} \delta \underline{u}_s^T \underline{t}_\sigma \, dV_0 = \int_{\ell_0} \delta \underline{u}_s^T \underline{F} \, d\alpha_1 \tag{3.521}$$

where $\underline{F}$ represents the traction force acted on the cross-section of cable

$$\underline{F} = \int_{S_0} \underline{t}_\sigma dS_0 \tag{3.522}$$

## 3.8.6 Governing Equations of Cable

Based on the principle of virtual work, which states, $\delta \mathcal{U}_e - \delta \mathcal{W}_e = 0$, the governing equations for a cable problem could be obtained. Introducing Eqs. (3.518) and (3.521) leads to

$$\int_{\ell_1} (\delta \varepsilon_{11} F_{\epsilon\epsilon}^{11} - \delta \underline{u}_s^T \underline{F}) d\alpha_1 = 0 \tag{3.523}$$

In view of Eq. (3.514), the variation of $\varepsilon_{11}$ can be expressed in terms of

$$\delta \varepsilon_{11} = (\underline{x}_{0,1} + \underline{\epsilon}_1)^T \delta \underline{\epsilon}_1 = \underline{P}_{,1}^T \delta \underline{u}_{s,1} \tag{3.524}$$

and then the principle of virtual work is recasted to

$$\int_{\ell_1} (\delta \underline{u}_{s,1}^T \underline{P}_{,1} F_{\epsilon\epsilon}^{11} - \delta \underline{u}_s^T \underline{F}) d\alpha_1 = 0 \tag{3.525}$$

With the application of finite element method, the virtual displacement and its spatial derivative are interpolated by using the shape functions $H_\epsilon$. The following interpolations

$$\delta \underline{u}_s = H_\epsilon \delta \underline{u}_n, \ \delta \underline{u}_{s,1} = H_{\epsilon,1} \delta \underline{u}_n \tag{3.526}$$

reformulate the principle of virtual work, Eq. (3.525), into the weak Galerkin form

$$\delta \underline{u}_n^T \int_{\ell_1} (H_{\epsilon,1}^T \underline{P}_{,1} F_{\epsilon\epsilon}^{11} - H_\epsilon^T \underline{F}) d\alpha_1 = 0 \tag{3.527}$$

From which, the governing equations are obtained to be

$$\int_{\ell_1} (H^T_{\epsilon,1}\underline{P}_{,1}F^{11}_{\epsilon\epsilon} - H^T_\epsilon \underline{F})d\alpha_1 = 0 \tag{3.528}$$

After introduced the elastic force vector

$$\underline{\mathcal{F}_e} = \int_{\ell_1} H^T_{\epsilon,1}\underline{P}_{,1}F^{11}_{\epsilon\epsilon}d\alpha_1 \tag{3.529}$$

and the external force vector

$$\underline{\mathcal{F}_a} = \int_{\ell_1} H^T_\epsilon \underline{F}d\alpha_1 \tag{3.530}$$

the governing equations can be described by the following compact formula

$$\underline{\mathcal{F}_e} - \underline{\mathcal{F}_a} = 0 \tag{3.531}$$

## 3.8.7 Kinematic Energy of Cable

When governing equation of the cable is extended to the dynamic problem, the kinematic energy of cable needs to be evaluated. At first, the velocity of an arbitrary material particle in a cable can be evaluated by taking time derivatives of position vector $\underline{P}$, defined by Eq. (3.508), to get

$$\dot{\underline{P}} = (\underline{x}_0 \dot{+} \underline{u}_s) = \dot{\underline{u}}_s \tag{3.532}$$

If denoting the material density as $\rho$, the kinematic energy of a cable can be computed from the integral related to the initial volume like

$$\mathcal{K}_e = \int_{V_0} \frac{1}{2}\rho\dot{\underline{u}}_s^T\dot{\underline{u}}_s dV_0 \tag{3.533}$$

Here again, the observation that $\underline{u}_s$ is only the function of $\alpha_1$ leads to a one-dimensional integral of the kinematic energy like

$$\mathcal{K}_e = \int_{\ell_1} \frac{1}{2}\dot{\underline{u}}_s^T\mathcal{M}_{\epsilon\epsilon}\dot{\underline{u}}_s d\ell_1 \tag{3.534}$$

with the aid of following definition of mass matrix

$$\mathcal{M}_{\epsilon\epsilon} = \int\limits_{S_0} \rho I \, dS_0 \tag{3.535}$$

Referring Eq. (3.534), the variation of kinematic energy is readily obtained to be

$$\delta\mathcal{K}_e = \int\limits_{\ell_1} \delta\underline{\dot{u}}_s^T \mathcal{M}_{\epsilon\epsilon}\underline{\dot{u}}_s \, d\alpha_1 \tag{3.536}$$

Integrating the variation of kinematic energy by part in time domain produces

$$\int\limits_t \delta\mathcal{K}_e dt = -\int\limits_t \int\limits_{\ell_1} \delta\underline{u}_s^T \mathcal{M}_{\epsilon\epsilon}\underline{\ddot{u}}_s \, d\alpha_1 dt \tag{3.537}$$

From which the variation of kinematic energy can be approximated by

$$\delta\mathcal{K}_e = -\int\limits_{\ell_1} \delta\underline{u}_s^T \mathcal{M}_{\epsilon\epsilon}\underline{\ddot{u}}_s \, d\alpha_1 \tag{3.538}$$

With the application of finite element discretization, the acceleration of material particle is interpolated by using the follow polynomials of nodal accelerations $\underline{\ddot{u}}_n$ like

$$\underline{\ddot{u}}_s = H_\epsilon \underline{\ddot{u}}_n \tag{3.539}$$

and the weak form of kinematic energy variation is recasted to

$$\delta\mathcal{K}_e = -\delta\underline{u}_n^T \int\limits_{\ell_1} H_\epsilon^T \mathcal{M}_{\epsilon\epsilon} H_\epsilon d\alpha_1 \, \underline{\ddot{u}}_n \tag{3.540}$$

After defined the mass matrix

$$\mathcal{M}_n = \int\limits_{\ell_1} H_\epsilon^T \mathcal{M}_{\epsilon\epsilon} H_\epsilon d\alpha_1 \tag{3.541}$$

the variation of original kinematic energy can be rewritten into the compact formula

$$\delta\mathcal{K}_e = -\delta\underline{u}_n^T \mathcal{M}_n \underline{\ddot{u}}_n \tag{3.542}$$

### 3.8.8 Extended to Cable Dynamic Problem

When considering the inertial forces, the governing equations of motion for a cable can be obtained from the Hamilton's principle. Equivalently, the inertial force is required to be added into the Eq. (3.531) to get

$$\mathcal{M}_n \underline{\ddot{u}}_n + \underline{\mathcal{F}}_e = \underline{\mathcal{F}}_a \tag{3.543}$$

## 3.9 Summary of General Cosserat Continuum Theory

In this chapter, the general Cosserat continuum theory is discussed in detail, and the motion formalism of the governing equations is represented with the combination of finite element method. Since the geometric dimensions of most Cosserat continua feature different scales, for example, the shell has a "thin" dimension in the direction of thickness, these Cosserat continua can be reduced from the original three-dimensional ones to two-dimensional or one-dimensional continua by neglecting the deformations in the dimensions with small scales. In view of this, the general shell-like theory and beam-like theory, which are the typical two-dimensional and one-dimensional Cosserat continua, can be readily obtained from the general three-dimensional Cosserat continuum theory. When all the deformations of Cosserat continuum are ignored, the general Cosserat continuum theory will be simplified to the theory of rigid body. Furthermore, the general Cauchy continuum theory can be obtained also from Cosserat theory by neglecting the directors of material particles or neglecting the curvature strains equivalently of the Cosserat continuum. If

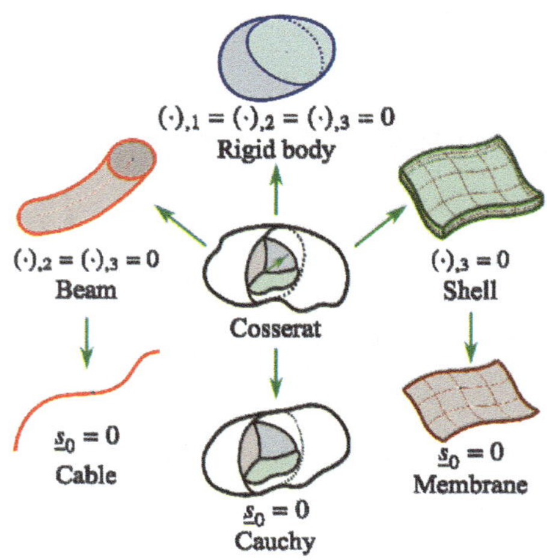


**Fig. 3.9** The unified descriptions for Cosserat continua and others

ignored the curvature strains or ignored the fiber directors in the theories of shell and beam, respectively, the membrane theory and cable theory are achieved. In this sense, as depicted in the Fig. 3.9, the general Cosserat continuum theory affords the unified descriptions for three-dimensional Cosserat continuum, shell, beam, rigid body, three-dimensional Cauchy continuum, membrane and cable, all of them.

## References


1. Cosserat, E., Cosserat, F.: Théorie des Corps Déformables. Hermann, Paris (1909)
2. Rubin, M.B.: Cosserat Theories: Shells. Rods and Points, Springer, Dordrecht (2000)
3. Toupin, R.A.: Theories of elasticity with couple-stress. Arch. Ration. Mech. Anal. **17**, 85–112 (1964)
4. Truesdell, C., Noll, W.: The nonlinear field theories of mechanics. In: Flügge, S. (ed.) Handbuch der Physik, III(3), pp. 1–602. Springer-Verlag, Berlin (1965)
5. Kafadar, C.B., Eringen, A.C.: Micropolar media-I. The classical theory. Int. J. Eng. Sci. **9**, 271-305 (1971)
6. Pabst, W.: Micropolar materials. Ceram. Silik. **49**, 170–180 (2005)
7. Nicholas, A.C., Ioannis, S., Jean, S., Itai, E.: A Cosserat breakage mechanics model for brittle granular media. J. Mech. Phys. Sol. (2020). https://doi.org/10.1016/j.jmps.2020.103975
8. Tornabene, F., Fantuzzi, N., Bacciocchi, M.: Mechanical behaviour of composite Cosserat solids in elastic problems with holes and discontinuities. Compos. Struct. (2017). https://doi.org/10.1016/j.compstruct.2017.07.087
9. Xi, Y., Yoshihiro, T.: Effective properties of Cosserat composites with periodic microstructure. Mech. Res. Commun. (2001). https://doi.org/10.1016/S0093-6413(01)00172-0
10. Rueger, Z., Lakes, R.S.: Strong Cosserat elastic effects in a unidirectional composite. Zeitschrift für Angewandte mathematik und Physik. (2017). https://doi.org/10.1007/s00033-017-0796-6
11. Samuel, F., Fabrice, B., Georges, C.: Cosserat modelling of size effects in the mechanical behaviour of polycrystals and multi-phase materials. Int. J. Sol. Struct. **37**, 7105–7126 (2000)
12. Onck, P.R.: Cosserat modeling of cellular solids. C. R. Mecanique. **330**, 717722 (2002)
13. Eremeyev, V.A., Skrzat, A., Vinakurava, A.: Application of the micropolar theory to the strength analysis of bioceramic materials for bone reconstruction. Strength Mater. **48**, 573–582 (2016)
14. Ericksen, J.L.: Liquid crystals and cosserat surfaces. Q J. Mech. Appl. Math. (1974). https://doi.org/10.1093/qjmam/27.2.213
15. Roderic, S.L.: Softening of Cosserat sensitivity in a foam: warp effects. Int. J. Mech. Sci. (2021). https://doi.org/10.1016/j.ijmecsci.2020.106125
16. Kumar, A., Kumar, S., Gupta, P.: A helical Cauchy-born rule for special Cosserat rod modeling of nano and continuum rods. J. Elast. (2016). https://doi.org/10.1007/s10659-015-9562-1
17. Gomez, J., Basaran, C.: Computational implementation of Cosserat continuum. Int. J. Mater. Prod. Tech. (2009). https://doi.org/10.1504/IJMPT.2009.022401
18. Dyskin, A., Pasternak, E., Esin, M.: Cracks of classical modes in small-scale Cosserat continuum. In: Ye, L. (eds.) Recent advances in structural integrity analysis-proceedings of the international congress (APCF/SIF-2014), pp. 32–36. Woodhead Publishing, Oxford (2014)
19. Forest, S.: Cosserat Media. In: Jürgen Buschow, K.H., Cahn, R.W., Flemings, M.C., Ilschner, B., Kramer, E.J., Mahajan, S., Veyssiére, P. (eds.) Encyclopedia of Materials: Science and Technology, pp. 1715–1717. Elsevier, Oxford (2001)
20. Pietraszkiewicz, W., Eremeyev, V.A.: On natural strain measures of the non-linear micoropolor continuum. Int. J. Solids Struct. (2009). https://doi.org/10.1016/j.ijsolstr.2008.09.027
21. Bauchau, O.A.: Flexible Multibody Dynamics. Springer, Dordrecht (2011)
22. Bathe, K.J.: Finite Element Procedures. Prentice-Hall, New Jersey (1996)

23. Pietraszkiewicz, W., Eremeyev, V.A.: On vectorially parameterized natural strain measures of the non-linear Cosserat continuum. Int. J. Solids Struct. (2009). https://doi.org/10.1016/j.ijsolstr.2009.01.030
24. Yu, W.: A unified theory for constitutive modeling of composites. J. Mech. Mater. Struct. (2016). https://doi.org/10.2140/jomms.2016.11.379
25. Žalohar, J.: Chapter 2—cosserat continuum. In: Žalohar, J. (eds.) Developments in structural geology and tectonics. The Omega-Theory: A New Physics of Earthquakes, pp. 17–32. Elsevier, Oxford (2018)
26. Lakes, R.: Experimental methods for study of Cosserat elastic solids and other generalized elastic continua. In: Mühlhaus, H., Wiley, J. (eds.) Continuum Models for Materials with Micro-structure, pp. 1–22. Wiley, New York (1995)
27. Ahmad, S., Irons, B.M., Zienkiewicz, O.: Analysis of thick and thin shell structures by curved finite element. Int. J. Numer. Methods in Eng. **2**, 419–451 (1970)
28. Hughes, T.J.R.: The Finite Element Method: Linear Static and Dynamic Finite Element Analysis. Prentice Hall, New Jersey (1987)
29. Crisfield, M.: Finite Elements on Solution Procedures for Structural Analysis. I. Linear Analysis. Pineridge Press, Swansea (1986)
30. Neff, P.: A geometrically exact cosserat shell-model including size effects, avoiding degeneracy in the thin shell limit. Part I: formal dimensional reduction for elastic plates and existence of minimizers for positive Cosserat couple modulus. Continuum Mech. Thermodyn. (2004). https://doi.org/10.1007/s00161-004-0182-4
31. Bathe, K.J., Dvorkin, E.: A four-node plate bending element based on Mindlin/Reissner plate theory and a mixed interpolation. Int. J. Numer. Methods Eng. **21**, 367–383 (1985)
32. Lee, P.S., Bathe, K.J.: The quadratic MITC plate and MITC shell elements in plate bending. Adv Eng. Softw. (2010). https://doi.org/10.1016/j.advengsoft.2009.12.011
33. Hughes, T.J.R., Liu, W.K.: Nonlinear finite element analysis of shells: Part I. three dimensional shells. Comput. Methods Appl. Mech. Eng. **26**, 331–362 (1981)
34. Hughes, T.J.R., Liu,W.K.: Nonlinear finite element analysis of shells: Part II. three dimensional shells. Comput. Methods Appl. Mech. Eng. **27**, 167–181 (1981)
35. Hughes, T.J.R., Carnoy, E.: Nonlinear finite element analysis shell formulation accounting for large membrane strains. Comput. Methods Appl. Mech. Eng. **39**, 69–82 (1983)
36. Belytschko, T., Stolarski, H., Liu, W.K., Carpenter, N., Ong, J.S.J.: Stress projection for membrane and shear locking in shell finite elements. Comput. Methods Appl. Mech. Eng. (1985). https://doi.org/10.1016/0045-7825(85)90035-0
37. Liu, W.K., Law, E.S., Lam, D., Belytschko, T.: Resultant-stress degenerated-shell element. Comput. Methods Appl. Mech. Eng. **55**, 259–300 (1986)
38. Sussman, T., Bathe, K.J.R.: 3D-shell elements for structures in large strains. Comput. Struct. **122**, 2–12 (2013)
39. Hughes, T.J.R., Cottrell, J.A., Bazilevs, Y.: Isogeometric analysis: CAD, finite elements, NURBS, exact geometry and mesh refinement. Comput. Methods Appl. Mech. Eng. (2005). https://doi.org/10.1016/j.cma.2004.10.008
40. Benson, D.J., Bazilevs, Y., Hsu, M.C., Hughes, T.J.R.: A large deformation, rotation-free, isogeometric shell. Comput. Methods Appl. Mech. Eng. (2011). https://doi.org/10.1016/j.cma.2010.12.003
41. Nguyen-Thanh, N., Valizadeh, N., Nguyen, M.N., Nguyen-Xuan, H., Zhuang, X., Areias, P., Zi, G., Bazilevs, Y., De, Lorenzis, L., Rabczuk, T.: An extended isogeometric thin shell analysis based on Kirchhoff-Love theory. Comput. Methods Appl. Mech. Eng. **284**, 265–291 (2015)
42. Duong, T.X., Roohbakhshan, F., Sauer, Roger A., R.A.: A new rotation-free isogeometric thin shell formulation and a corresponding continuity constraint for patch boundaries. Comput. Methods Appl. Mech. Eng. **316**, 43–83 (2017)
43. Benson, D.J., Bazilevs, Y., Hsu, M.C., Hughes, T.J.R.: Isogeometric shell analysis: the Reissner-Mindlin shell. Comput. Methods Appl. Mech. Eng. (2010). https://doi.org/10.1016/j.cma.2009.05.011

44. Pimenta, P.M., Campello, E.M.B.: Finite rotation parameterizations for the nonlinear static and dynamic analysis of shells. In: Ramm, E., Wall, W.A., Bletzinger, K.-U., Bischoff, M. (eds.) Proceedings of the 5th International Conference on Computation of Shell and Spatial Structures, pp. 1–5. Salzburg, Austria (2005)
45. Argyris, J.H., Scharpf, D.W.: The sheba family of shell elements for the matrix displacement method. Part II. interpolation scheme and stiffness matrix. J. Royal Aeronaut. Soc. **72**, 878–883 (1968)
46. Ericksen, J.L., Truesdell, C.: Exact theory of stress and strain in rods and shells. Arch. Rational Mech. Anal. **1**, 295–323 (1958)
47. Martin, S., Kaufmann, P., Botsch, M., Grinspun, E., Gross, M.: Unified simulation of elastic rods, shells, and solids. ACM Transactions on Graphics. (2010). https://doi.org/10.1145/1833351.1778776
48. Simo, J.C., Fox, D.D., Rifai, M.S.: On a stress resultant geometrically exact shell model. part VI: conserving algorithms for nonlinear dynamics. Int. J. Numer. Methods Eng **34**, 117–164 (1992)
49. Simo, J.C., Tarnow, N.: A new energy and momentum conserving algorithm for the nonlinear dynamics of shells. Int. J. Numer. Methods Eng. **37**, 2527–2549 (1994)
50. Ibrahimbegovic, A., Brank, B., Courtois, P.: Stress resultant geometrically exact form of classical shell model and vector-like parameterization of constrained finite rotations. Int. J. Numer. Methods Eng. (2001). https://doi.org/10.1002/nme.247
51. Roller, M., Betsch, P., Gallrein, A., Linn J.: On the use of geometrically exact shells for dynamic tire simulation. In: Terze, Z. (eds.) Multibody Dynamics. Computational Methods in Applied Sciences, PP. 1135–1144. Springer, Cham, 205–236 (2014)
52. Roller, M., Betsch, P., Gallrein, A., Linn, J.: An enhanced tire model for dynamic simulation based on geometrically exact shells. Arch. Mech. Eng. **63**, 277–295 (2016)
53. Campello, E.M.B., Pimenta, P.M., Wriggers, P.: An exact conserving algorithm for nonlinear dynamics with rotational DOFs and general hyperelasticity. Part 2: shells. Comput. Mech. **48**, 195–211 (2011)
54. Pimenta, P.M., Campello, E.M.B., Wriggers, P.: A fully nonlinear multi-parameter shell model with thickness variation and a triangular shell finite element. Comput. Mech. **34**, 181–193 (2004)
55. Pimenta, P.M., Campello, E.M.B.: A unified approach for the nonlinear dynamics of rods and shells using an exact conserving integration algorithm. In: De Mattos Pimenta, P., Wriggers, P. (eds.) New Trends in Thin Structures: Formulation, Optimization and Coupled Problems, pp 99–132. Springer, Vienna (2010)
56. Mota, A.A.: A Class of Geometrically Exact Membrane and Cable Finite Elements Based on the Hu-Washizu Functional. Ph.D. Dissertation, School of Civil and Environmental Engineering, Cornell University, Ithaca, NY (2000)
57. Simo, J.C.: Mathematical Modeling and Numerical Simulation of the Dynamics of Flexible Structures Undergoing Large Overall Motions. Air Force Office of Scientific Research, Technical Report AFOSR-92-0287TR, Bolling AFB, DC (1992)
58. Reissner, E.: On one-dimensional large-displacement finite-strain beam theory. Stud. Appl. Math. **52**, 87–95 (1973)
59. Simo, J.C.: A finite strain beam formulation. The three-dimensional dynamic problem. Part I. Comput. Methods Appl. Mech. Eng. (1985). https://doi.org/10.1016/0045-7825(85)90050-7
60. Simo, J.C.: A three-dimensional finite-strain rod model. Part II: Comput. Aspects. Comput. Methods Appl. Mech. Eng. (1986). https://doi.org/10.1016/0045-7825(86)90079-4
61. Ibrahimbegović, A.: On finite element implementation of geometrically nonlinear Reissner's beam theory: three-dimensional curved beam elements. Comput. Methods Appl. Mech. Eng. (1995). https://doi.org/10.1016/0045-7825(95)00724-F
62. Bauchau, O.A., Hong, C.H.: Nonlinear composite beam theory. J. Appl. Mech. (1988). https://doi.org/10.1115/1.3173622
63. Han, S., Bauchau, O.A.: Spectral formulation for geometrically exact beams: a motion interpolation based approach. AIAA J. (2019). https://doi.org/10.2514/1.J057489

64. Han, S., Bauchau, O.A.: Manipulation of motion via dual entities. Nonlinear Dynam. (2016). https://doi.org/10.1007/s11071-016-2703-7
65. Han, S., Bauchau, O.A.: On the global interpolation of motion. Comput. Methods Appl. Mech. Engrg. (2018). https://doi.org/10.1016/j.cma.2018.04.002
66. Hodges, D.H.: A mixed variational formulation based on exact intrinsic equations for dynamics of moving beams. Int. J. Solids and Struct. (1990). https://doi.org/10.1016/0020-7683(90)90060-9
67. Yu, W., Hodges, D.H., Volovoi, V., Cesnik, C.E.S.: On timoshenko-like modeling of initially curved and twisted composite beams. Int. J. Solids Struct. (2002). https://doi.org/10.1016/S0020-7683(02)00399-2
68. Yu, W., Hodges, D.H.: Elasticity solutions versus asymptotic sectional analysis of homogeneous, isotropic, prismatic beams. J. Appl. Mech. (2004). https://doi.org/10.1115/1.1640367
69. Bauchau, O.A., Craig, J.I.: Structural Analysis With Applications to Aerospace Structures. Springer, Dordrecht (2009)
70. Bauchau, O.A., Xin, H., Dong, S.Y., Li, Z.H., Han, S.L.: Intrinsic time integration procedures for rigid body dynamics. J. Comput. Nonlinear Dynam. (2013). https://doi.org/10.1115/1.4006252

# Part III
# Multiscale Modeling Technology of Multibody System

# Chapter 4
# Multiscale Multibody Dynamics

The multibody system [1] can be regarded as a collection of rigid and flexible bodies connected by mechanical joints. According to the definition of Cosserat continuum [2], it can be concluded that the rigid body or flexible bodies, such as beams, plates and shells, are exactly the Cosserat continuum with dimensions varying from zero to three. Obviously, all these Cosserat continua with different geometric sizes could be arbitrarily combined each other into a complicated multibody system with multiscale dimensions. For example, the wings of the civil aircraft are typical slender box structures composed of different parts, the lower and upper covers reinforced by the stiffeners, the front and rear beams together with the ribs, also contains the control surfaces, such as flaps and ailerons. When only the global deformations of the wing under the aerodynamic pressure are concerned, the wing could be modeled as one-dimensional Cosserat continuum, or a beam model, equivalently. When the special attentions have been paid to the stress concentrations at the root of the wing, the modeling of local structures at the root of the wing becomes so complicated because the geometric details of all concerned parts must be taken into account. In this special case, the analytical model of the center wing used for stress analysis can be viewed as a set of Cosserat continua with different geometric sizes, called a model of multiscale Cosserat continua.

The concept of multiscale arises from geometric dimension of different scales. Given a three-dimensional body, geometric description includes the size of height, width and length in three dimensions. If the width and length are much larger than the height, the deformation of the body in the "height" dimension can be neglected as a small quantity versus the deformations in other two dimensions. This is a popular assumption that has been applied to derive the shell theory from the original three-dimensional Cosserat theory. But it should be kept in mind that the small scale is a relative quantity only meaningful when comparing the dimension of small scale with the other two dimensions. The small scale is not a concept of absolute small. For example, the wing span of a civil aircraft under designing is about 72 m, the root chord is about 6 m and the thickness of the wing root is larger than 2 m approximately.

 
J. Wang, *Multiscale Multibody Dynamics*,
https://doi.org/10.1007/978-981-19-8441-9_4

Since the chord and thickness of the wing root are much smaller than the wing span, the wing could be modeled by using the beam theory in the analysis of aeroelasticity. The deformations of cross-sections of the wing are all neglected. However, the aerodynamic pressure distributed over the surface of the wing will cause the large stress concentration at the root of the wing. Hence, a detailed stress analysis must be performed for the structural design of center part of the wing. The components of Cauchy stress in all the three directions must be estimated to satisfy the strength conditions. These engineering practices show that the assumption of small scale of geometric dimension depends on what kind of analysis should be performed to satisfy the actual requirements of structural design. It is necessary to emphasize again that the small scale does not mean the absolutely small quantity in the sense of measurement.

In recent decades, Multiscale analysis of heterogeneous materials [3–5] has always been a topic of main research for the computational mechanics. Based on the assumption of scale separation, many analysis approaches were proposed, such as the asymptotic homogenization approaches [6–9], and variational asymptotic methods [10–13] for unit cell homogenization [14] and representative volume element methods [15–19]. Instead of introducing the assumption mentioned above, the multiscale finite element methods [20–22], the widely used approaches raised from the works of Babuška etc. [23], has been extended now to failure analyses [24] and Crack Propagation[25] of heterogeneous materials. This method connected the macroscopic and microscopic material properties by the so-called numerical shape functions. However, the determination of numerical shape functions, a key step of the multiscale finite element methods, must be carefully constructed by using different boundary conditions suitable for different problems, varying from linear, and oversampling to periodic boundary conditions.

The assumption of scale separation is still valid for multiscale analysis of multibody systems [26]. For example, considering the relative smallness of the cross-section size with respect to the beam length, the original three dimensional nonlinear problems can be rigorously split into a one-dimensional global beam analysis and a local two-dimensional linear analysis over the cross-section [27–29]. The same operations applied to split a three dimensional problem with one "thin" dimension into a two-dimensional shell model and a local one-dimensional linear analysis along this "thin" direction [30]. More generally, this assumption was also later used to develop the approach of structure genome [32, 34] for multiscale constitutive modeling of structures including the linear elastic Euler-Bernoulli beam, the Kirchhoff plate, and three-dimensional Cauchy continuum.

In this chapter, the rigid body assumption of the local region occupied by the director in a Cosserat continuum, introduced in Chap. 3, is abandoned. The local region occupied by the director could be a micro-structure composed of micro-shell, micro-beam and micro Cauchy continuum etc. The deformable micro-structures can contribute to the strain energy of global macro-structures, a collection of one-dimensional, two-dimensional or three-dimensional Cosserat continua. The strict mapping of global macro-structural degrees of freedom to the local micro-structural degrees of freedom will be constructed such that the original structure can be decoupled into a global macro-structural model and a local micro-structural model. If the decoupling process is repeated recursively, three different continuum models in the level of scales from submicro-to-micro-to-macro could be mapped one by one with the significance of multiscale modeling. Multiscale modeling affords an efficient approach that makes it possible to perform the detailed analysis of micro-structures at the speed of quickly global analysis. The global analysis will predict the global deformation and stress distributions of macro-structures, which can be used to estimate the global mechanical behavior. Through micro-macro mapping, the detailed micro-analysis could be performed with the stress concentration determined in the global analysis as the external input, then a fine analysis in the level of micro-scale can be accomplished.

## 4.1 Multiscale Cosserat Continuum Theory

This section presents the theory of multiscale Cosserat continuum with the benefits mentioned as before. The microscopic to macroscopic mapping of three-dimensional Cosserat continuum is defined at beginning. Then, the decoupled governing equations for micro- and macro-continuum are derived according to the principle of virtual work and Hamilton's theory for static and dynamic problems, respectively. Even a micro continuum could be composed of arbitrary Cosserat continuum with dimensions varying from one to three. In view of the unified theory of Cosserat continuum introduced in Chap. 3, the microscopic model considered in the current implementation is still assumed to be a three-dimension Cosserat continuum to make the multiscale Cosserat continuum theory more general.

### *4.1.1 Kinematics of Multiscale Cosserat Continuum*

As shown in Fig. 4.1, a material particle with infinitesimal volume in a Cosserat continuum is depicted by its position and director vectors in its reference configuration and current deformed configuration, respectively. In the reference configuration, the position vector $\underline{p}$ of the material particle relative to a point **O**, the original of reference frame, $\mathcal{F}_\mathcal{I} = [\mathbf{O}, \mathcal{I}]$, is defined as a function of the general curvilinear coordinates $(\alpha_1, \alpha_2, \alpha_3)$ like

$$\underline{p} = \underline{p}(\alpha_1, \alpha_2, \alpha_3) \tag{4.1}$$

If the position vector of a reference material particle **P** relative to a point **O** is given as

$$\underline{x}_0 = \underline{x}_0(\alpha_1, \alpha_2, \alpha_3) \tag{4.2}$$

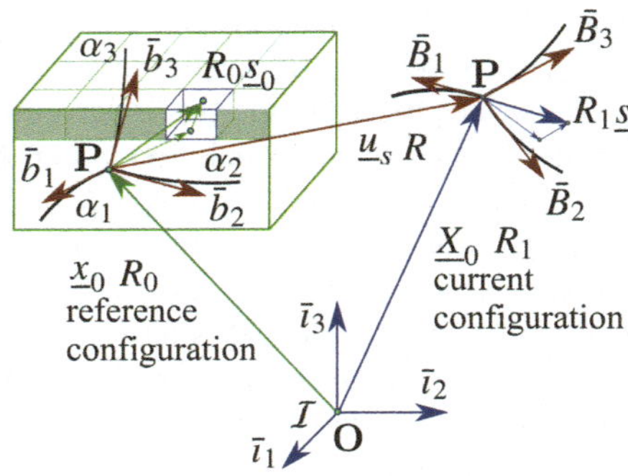


**Fig. 4.1** Kinematics of a material particle with infinitesimal volume in a multiscale Cosserat continuum

Taylor's expansion leads to

$$\underline{p}(\alpha_1, \alpha_2, \alpha_3) = \underline{x}_0(\alpha_1, \alpha_2, \alpha_3) + \frac{\partial \underline{x}_0}{\partial \alpha_1} d\alpha_1 + \frac{\partial \underline{x}_0}{\partial \alpha_2} d\alpha_2 + \frac{\partial \underline{x}_0}{\partial \alpha_3} d\alpha_3 \tag{4.3}$$

with the second and higher order terms truncated. According to the definitions of covariant base vectors, Eq. (1.274), it is readily verified that

$$\underline{p}(\alpha_1, \alpha_2, \alpha_3) = \underline{x}_0(\alpha_1, \alpha_2, \alpha_3) + \underline{g}_1 d\alpha_1 + \underline{g}_2 d\alpha_2 + \underline{g}_3 d\alpha_3 \tag{4.4}$$

If the material coordinated $(\alpha_1, \alpha_2, \alpha_3)$ are selected as the orthogonal curvilinear coordinates, the vectors $\underline{g}_1$, $\underline{g}_2$ and $\underline{g}_3$ will be mutually perpendicular. For this special case, the unit vectors $\bar{b}_1$, $\bar{b}_2$ and $\bar{b}_3$ of reference basis $\mathcal{B}_0$ can be determined from

$$\bar{b}_1 = \frac{\underline{g}_1}{h_1}, \; \bar{b}_2 = \frac{\underline{g}_2}{h_2}, \; \bar{b}_3 = \frac{\underline{g}_3}{h_3} \tag{4.5}$$

where $h_i$ for $i = 1, 2, 3$ are the Lame's constants, defined by Eq. (1.295). The position vector $\underline{p}$ is then represented by

$$\underline{p}(\alpha_1, \alpha_2, \alpha_3) = \underline{x}_0(\alpha_1, \alpha_2, \alpha_3) + R_0(\alpha_1, \alpha_2, \alpha_3)\underline{s}_0(\alpha_1, \alpha_2, \alpha_3) \tag{4.6}$$

where $R_0$ denotes the initial rotation tensor measuring the orientation of the reference basis $\mathcal{B}_0$. It has been verified in Sect. 2.2.3 the rotation tensor $R_0$ is identical to the body attached basis $\mathcal{B}_0$, and in detail

$$R_0 = \lfloor \bar{b}_1, \bar{b}_2, \bar{b}_3 \rfloor = \mathcal{B}_0 \tag{4.7}$$

Note that the initial basis $\mathcal{B}_0$ is bounded by the material particle **P** to produce the reference frame $\mathcal{F}_0 = [\mathbf{P}, \mathcal{B}_0]$. The local vector $\underline{s}_0$, defined in this reference frame $\mathcal{F}_0$, describes the director of a material particle like

$$\underline{s}_0^T = \lfloor h_1 d\alpha_1,\ h_2 d\alpha_2,\ h_3 d\alpha_3 \rfloor \tag{4.8}$$

or can be written in a more generic form like

$$\underline{s}_0^T(\alpha_1, \alpha_2, \alpha_3) = \lfloor s_{c1}\alpha_1,\ s_{c2}\alpha_2,\ s_{c3}\alpha_3 \rfloor \tag{4.9}$$

the scalar $s_c = \sqrt{s_{c1}^2 + s_{c2}^2 + s_{c3}^2}$, a small quantity, denotes the characteristic length of a microscale continuum. Obviously, Eq. (4.6) decomposes the position vector $\underline{p}$ of a material particle into a reference position $\underline{x}_0$ and a local vector $\underline{s}_0$. In classical Cosserat theory, director $\underline{s}_0$ is a rigid vector. Now it becomes deformable under the framework of multiscale Cosserat theory. If assuming the micro continuum still to be the Cosserat continuum, which is usually the case, the director $\underline{s}_0$ can be defined in a similar approach to the position vector, Eq. (4.6), like

$$\underline{s}_0(\alpha_1, \alpha_2, \alpha_3) = \underline{x}_0^\ell(\alpha_1, \alpha_2, \alpha_3) + R_0^\ell(\alpha_1, \alpha_2, \alpha_3)\underline{s}_0^\ell(\alpha_1, \alpha_2, \alpha_3) \tag{4.10}$$

where $\underline{x}_0^\ell$ is the position vector of reference particle $\mathbf{P}^\ell$ in local Cosserat continuum, and $R_0^\ell$ the local rotation tensor attached to the reference particle $\mathbf{P}^\ell$. Note the local director $\underline{s}_0^\ell$ is measured in the local basis spanned by $R_0^\ell$. In view of above formula, the position vector $\underline{p}$ can be determined from the following expression

$$\underline{p} = \underline{x}_0 + R_0(\underline{x}_0^\ell + R_0^\ell \underline{s}_0^\ell) \tag{4.11}$$

For the description of Cosserat continuum deformations, the covariant base vectors

$$\underline{g}_i = \frac{\partial \underline{p}}{\partial \alpha_i} \tag{4.12}$$

in the reference configuration are required, and the specific expressions can be obtained by taking derivatives of Eq. (4.6) with respected to the curvilinear coordinates $\alpha_i$ like

$$\begin{aligned}
\underline{g}_1 &= \underline{x}_{0,1} + R_{0,1}\underline{s}_0 + R_0\underline{s}_{0,1} \\
\underline{g}_2 &= \underline{x}_{0,2} + R_{0,2}\underline{s}_0 + R_0\underline{s}_{0,2} \\
\underline{g}_3 &= \underline{x}_{0,3} + R_{0,3}\underline{s}_0 + R_0\underline{s}_{0,3}
\end{aligned} \tag{4.13}$$

where the symbol $(\cdot)_{,i}$ for $i = 1, 2, 3$ denotes the spatial derivatives with respected to $\alpha_i$ and

$$(\cdot)_{,i} = \frac{\partial(\cdot)}{\partial \alpha_i} \tag{4.14}$$

Resolving the base vectors into the local reference frame $\mathcal{F}_0$ finds

$$\begin{aligned}\underline{g}_1^* &= R_0^T \underline{x}_{0,1} + R_0^T R_{0,1}\underline{s}_0 + \underline{s}_{0,1}\\ \underline{g}_2^* &= R_0^T \underline{x}_{0,2} + R_0^T R_{0,2}\underline{s}_0 + \underline{s}_{0,2}\\ \underline{g}_3^* &= R_0^T \underline{x}_{0,3} + R_0^T R_{0,3}\underline{s}_0 + \underline{s}_{0,3}\end{aligned} \tag{4.15}$$

In view of the definitions of initial curvature tensors, Eq. (1.314), the base vectors can be represented by the following formulas like

$$\begin{aligned}\underline{g}_1^* &= R_0^T \underline{x}_{0,1} + \widetilde{\kappa}_1^0\underline{s}_0 + \underline{s}_{0,1}\\ \underline{g}_2^* &= R_0^T \underline{x}_{0,2} + \widetilde{\kappa}_2^0\underline{s}_0 + \underline{s}_{0,2}\\ \underline{g}_3^* &= R_0^T \underline{x}_{0,3} + \widetilde{\kappa}_3^0\underline{s}_0 + \underline{s}_{0,3}\end{aligned} \tag{4.16}$$

where the symbols $\widetilde{\kappa}_i^0 = R_0^T R_{0,i}$ for $i = 1, 2, 3$ denote the skew-symmetric tensors of initial curvatures $\underline{\kappa}_i^0$. All these base vectors are assembled into the covariant bases $\mathcal{B}_g$ like

$$\mathcal{B}_g = \lfloor \underline{g}_1^*,\ \underline{g}_2^*,\ \underline{g}_3^* \rfloor \tag{4.17}$$

and then the metric tensor in the reference configuration can be computed from

$$g = \mathcal{B}_g^T \mathcal{B}_g \tag{4.18}$$

The similar manipulations can be performed repeatedly in the local Cosserat continuum to find the spatial derivatives of director

$$\begin{aligned}\underline{s}_{0,1} &= \underline{x}_{0,1}^\ell + R_{0,1}^\ell \underline{s}_0^\ell + R_0^\ell \underline{s}_{0,1}^\ell\\ \underline{s}_{0,2} &= \underline{x}_{0,2}^\ell + R_{0,2}^\ell \underline{s}_0^\ell + R_0^\ell \underline{s}_{0,2}^\ell\\ \underline{s}_{0,3} &= \underline{x}_{0,3}^\ell + R_{0,3}^\ell \underline{s}_0^\ell + R_0^\ell \underline{s}_{0,3}^\ell\end{aligned} \tag{4.19}$$

If defined the local curvature tensors as $\widetilde{\kappa}_i^{0\ell} = R_0^{\ell T} R_{0,i}^\ell$, the local base vectors can be resolved into the local basis $R_0^\ell$ like

$$\begin{aligned}R_0^{\ell T}\underline{s}_{0,1} &= R_0^{\ell T}\underline{x}_{0,1}^\ell + \widetilde{\kappa}_1^{0\ell}\underline{s}_0^\ell + \underline{s}_{0,1}^\ell\\ R_0^{\ell T}\underline{s}_{0,2} &= R_0^{\ell T}\underline{x}_{0,2}^\ell + \widetilde{\kappa}_2^{0\ell}\underline{s}_0^\ell + \underline{s}_{0,2}^\ell\\ R_0^{\ell T}\underline{s}_{0,3} &= R_0^{\ell T}\underline{x}_{0,3}^\ell + \widetilde{\kappa}_3^{0\ell}\underline{s}_0^\ell + \underline{s}_{0,3}^\ell\end{aligned} \tag{4.20}$$

When the Cosserat continuum changes its spatial configuration, the material particle that had position vector $\underline{p}$ in the undeformed reference configuration now has position vector $\underline{P}$ in the deformed configuration, such that

$$\underline{P}(\alpha_1, \alpha_2, \alpha_3) = \underline{X}_0(\alpha_1, \alpha_2, \alpha_3) + R_1(\alpha_1, \alpha_2, \alpha_3)\underline{S}_0(\alpha_1, \alpha_2, \alpha_3) \tag{4.21}$$

where $\underline{S}_0$ represents the deformed director in current configuration, and $R_1$ the total rotation tensor that describes the orientation of basis $\mathcal{B}_1$ attached to the deformed configuration and

$$R_1 = \lfloor \bar{B}_1, \bar{B}_2, \bar{B}_3 \rfloor = \mathcal{B}_1 \tag{4.22}$$

Note that bounding the current deformed basis $\mathcal{B}_1$ by the material particle **P**, the associated deformed frame $\mathcal{F}_1 = [\mathbf{P}, \mathcal{B}_1]$ can be constructed. Furthermore, if the displacement vector $\underline{u}_s$ is given to describe an arbitrary translation of reference particle, the reference position vector $\underline{X}_0$ in current deformed configuration is recasted to

$$\underline{X}_0(\alpha_1, \alpha_2, \alpha_3) = \underline{x}_0(\alpha_1, \alpha_2, \alpha_3) + \underline{u}_s(\alpha_1, \alpha_2, \alpha_3) \tag{4.23}$$

Similar to the expression, Eq. (4.21), the deformed director $\underline{S}_0$ can be decomposed into

$$\underline{S}_0(\alpha_1, \alpha_2, \alpha_3) = \underline{X}_0^\ell(\alpha_1, \alpha_2, \alpha_3) + R_1^\ell(\alpha_1, \alpha_2, \alpha_3)\underline{s}_0^\ell(\alpha_1, \alpha_2, \alpha_3) \tag{4.24}$$

where $\underline{X}_0^\ell$ is the deformed position vector of reference particle $\mathbf{P}^\ell$, and $R_1^\ell$ the deformed rotation tensor in local Cosserat continuum. Note that the local director $\underline{s}_0^\ell$ is assumed to be rigid. The position vector $\underline{X}_0^\ell$ can be related to its original position through

$$\underline{X}_0^\ell(\alpha_1, \alpha_2, \alpha_3) = \underline{x}_0^\ell(\alpha_1, \alpha_2, \alpha_3) + \underline{u}_s^\ell(\alpha_1, \alpha_2, \alpha_3) \tag{4.25}$$

where $\underline{u}_s^\ell$ is the displacement vector of local director $\underline{s}_0^\ell$ in micro continuum. In view of the definitions of covariant base vectors in deformed configuration

$$\underline{G}_i = \frac{\partial \underline{P}}{\partial \alpha_i} \tag{4.26}$$

the specific formulas can be obtained by taking spatial derivatives of Eq. (4.21) with respected to the curvilinear coordinates $\alpha_i$ for $i = 1, 2, 3$ like

$$\begin{aligned} \underline{G}_1 &= \underline{X}_{0,1} + R_{1,1}\underline{S}_0 + R_1\underline{S}_{0,1} \\ \underline{G}_2 &= \underline{X}_{0,2} + R_{1,2}\underline{S}_0 + R_1\underline{S}_{0,2} \\ \underline{G}_3 &= \underline{X}_{0,3} + R_{1,3}\underline{S}_0 + R_1\underline{S}_{0,3} \end{aligned} \tag{4.27}$$

Once again, the base vectors are resolved into the current local frame $\mathcal{F}_1$ to obtain

$$\begin{aligned} \underline{G}_1^* &= R_1^T\underline{X}_{0,1} + R_1^T R_{1,1}\underline{S}_0 + R_1^T R_1\underline{S}_{0,1} \\ \underline{G}_2^* &= R_1^T\underline{X}_{0,2} + R_1^T R_{1,2}\underline{S}_0 + R_1^T R_1\underline{S}_{0,2} \\ \underline{G}_3^* &= R_1^T\underline{X}_{0,3} + R_1^T R_{1,3}\underline{S}_0 + R_1^T R_1\underline{S}_{0,3} \end{aligned} \tag{4.28}$$

In view of the definitions of final curvature tensors $\widetilde{\kappa}_i^1 = R_1^T R_{1,i}$ for $i = 1, 2, 3$, the covariant base vectors are simplified to

$$\begin{aligned}
\underline{G}_1^* &= R_1^T \underline{X}_{0,1} + \widetilde{\kappa}_1^1 \underline{S}_0 + \underline{S}_{0,1} \\
\underline{G}_2^* &= R_1^T \underline{X}_{0,2} + \widetilde{\kappa}_2^1 \underline{S}_0 + \underline{S}_{0,2} \\
\underline{G}_3^* &= R_1^T \underline{X}_{0,3} + \widetilde{\kappa}_3^1 \underline{S}_0 + \underline{S}_{0,3}
\end{aligned} \tag{4.29}$$

All these base vectors could be assembled into the covariant bases $\mathcal{B}_G$ like

$$\mathcal{B}_G = \lfloor \underline{G}_1^*,\ \underline{G}_2^*,\ \underline{G}_3^* \rfloor \tag{4.30}$$

The similar expressions of deformed covariant base vectors $\underline{S}_{0,i}$ for $i = 1, 2, 3$ are found to be

$$\begin{aligned}
\underline{S}_{0,1} &= \underline{X}_{0,1}^\ell + R_{1,1}^\ell \underline{s}_0^\ell + R_1^\ell \underline{s}_{0,1}^\ell \\
\underline{S}_{0,2} &= \underline{X}_{0,2}^\ell + R_{1,2}^\ell \underline{s}_0^\ell + R_1^\ell \underline{s}_{0,2}^\ell \\
\underline{S}_{0,3} &= \underline{X}_{0,3}^\ell + R_{1,3}^\ell \underline{s}_0^\ell + R_1^\ell \underline{s}_{0,3}^\ell
\end{aligned} \tag{4.31}$$

and can be transferred to the local bases spanned by the rotation tensor $R_1^\ell$ like

$$\begin{aligned}
R_1^{\ell T} \underline{S}_{0,1} &= R_1^{\ell T} \underline{X}_{0,1}^\ell + \widetilde{\kappa}_1^{1\ell} \underline{s}_0^\ell + \underline{s}_{0,1}^\ell \\
R_1^{\ell T} \underline{S}_{0,2} &= R_1^{\ell T} \underline{X}_{0,2}^\ell + \widetilde{\kappa}_2^{1\ell} \underline{s}_0^\ell + \underline{s}_{0,2}^\ell \\
R_1^{\ell T} \underline{S}_{0,3} &= R_1^{\ell T} \underline{X}_{0,3}^\ell + \widetilde{\kappa}_3^{1\ell} \underline{s}_0^\ell + \underline{s}_{0,3}^\ell
\end{aligned} \tag{4.32}$$

where $\widetilde{\kappa}_i^{1\ell} = R_1^{\ell T} R_{1,i}^\ell$ for $i = 1, 2, 3$ give the definitions of current curvature tensors for local Cosserat continuum.

### 4.1.2 Strain Tensor of Multiscale Cosserat Continuum

As introduced in Sect. 3.1.2, the components of Green-Lagrange strain tensors [33], Eq. (3.44), can be approximated by using the difference of base vectors. The strain descriptions of a material particle for multiscale Cosserat continuum keep the same as before and are given as

$$\begin{aligned}
\varepsilon_{11} &= \underline{g}_1^{*T} \Delta \underline{G}_1^* \\
2\varepsilon_{12} &= \underline{g}_1^{*T} \Delta \underline{G}_2^* + \underline{g}_2^{*T} \Delta \underline{G}_1^* \\
2\varepsilon_{13} &= \underline{g}_1^{*T} \Delta \underline{G}_3^* + \underline{g}_3^{*T} \Delta \underline{G}_1^* \\
\varepsilon_{22} &= \underline{g}_2^{*T} \Delta \underline{G}_2^* \\
2\varepsilon_{23} &= \underline{g}_2^{*T} \Delta \underline{G}_3^* + \underline{g}_3^{*T} \Delta \underline{G}_2^* \\
\varepsilon_{33} &= \underline{g}_3^{*T} \Delta \underline{G}_3^*
\end{aligned} \tag{4.33}$$

Similarly, the vectorial expressions of Green-Lagrange strain tensor, Eq. (3.72), are readily verified to be

$$\underline{\varepsilon}_1 = \mathcal{B}_g^T \Delta \underline{G}_1^*, \ \underline{\varepsilon}_2 = \mathcal{B}_g^T \Delta \underline{G}_2^*, \ \underline{\varepsilon}_3 = \mathcal{B}_g^T \Delta \underline{G}_3^* \tag{4.34}$$

where the covariant bases $\mathcal{B}_g$ are assembled by the basis vectors for the reference configuration, Eq. (4.17). Before measuring the components of Green-Lagrange strain tensor, the differences between the base vectors $\underline{G}_i^*$ and $\underline{g}_i^*$ for $i = 1, 2, 3$ are evaluated according to the definitions, Eqs. (4.16) and (4.29), at first

$$\begin{aligned}
\Delta \underline{G}_1^* &= \underline{G}_1^* - \underline{g}_1^* = R_1^T \underline{X}_{0,1} - R_0^T \underline{x}_{0,1} + \widetilde{\kappa}_1^1 \underline{S}_0 - \widetilde{\kappa}_1^0 \underline{s}_0 + \underline{S}_{0,1} - \underline{s}_{0,1} \\
\Delta \underline{G}_2^* &= \underline{G}_2^* - \underline{g}_2^* = R_2^T \underline{X}_{0,2} - R_0^T \underline{x}_{0,2} + \widetilde{\kappa}_2^1 \underline{S}_0 - \widetilde{\kappa}_2^0 \underline{s}_0 + \underline{S}_{0,2} - \underline{s}_{0,2} \\
\Delta \underline{G}_3^* &= \underline{G}_3^* - \underline{g}_3^* = R_3^T \underline{X}_{0,3} - R_0^T \underline{x}_{0,3} + \widetilde{\kappa}_3^1 \underline{S}_0 - \widetilde{\kappa}_3^0 \underline{s}_0 + \underline{S}_{0,3} - \underline{s}_{0,3}
\end{aligned} \tag{4.35}$$

Rearranging the terms on the right hand side of above equations leads to

$$\begin{aligned}
\Delta \underline{G}_1^* &= R_1^T \underline{X}_{0,1} - R_0^T \underline{x}_{0,1} + (\widetilde{\kappa}_1^1 - \widetilde{\kappa}_1^0) \underline{s}_0 + \widetilde{\kappa}_1^1 (\underline{S}_0 - \underline{s}_0) + \underline{S}_{0,1} - \underline{s}_{0,1} \\
\Delta \underline{G}_2^* &= R_1^T \underline{X}_{0,2} - R_0^T \underline{x}_{0,2} + (\widetilde{\kappa}_2^1 - \widetilde{\kappa}_2^0) \underline{s}_0 + \widetilde{\kappa}_2^1 (\underline{S}_0 - \underline{s}_0) + \underline{S}_{0,2} - \underline{s}_{0,2} \\
\Delta \underline{G}_3^* &= R_1^T \underline{X}_{0,3} - R_0^T \underline{x}_{0,3} + (\widetilde{\kappa}_3^1 - \widetilde{\kappa}_3^0) \underline{s}_0 + \widetilde{\kappa}_3^1 (\underline{S}_0 - \underline{s}_0) + \underline{S}_{0,3} - \underline{s}_{0,3}
\end{aligned} \tag{4.36}$$

In view of the definitions of Lagrangian stretch and Lagrange curvature measures [31, 32], Eq. (2.152), the above equations can be further simplified to

$$\begin{aligned}
\Delta \underline{G}_1^* &= \lfloor I, \ \widetilde{s}_0^T \rfloor \underline{\epsilon}_1 + \widetilde{\kappa}_1^1 (\underline{S}_0 - \underline{s}_0) + \underline{S}_{0,1} - \underline{s}_{0,1} \\
\Delta \underline{G}_2^* &= \lfloor I, \ \widetilde{s}_0^T \rfloor \underline{\epsilon}_2 + \widetilde{\kappa}_2^1 (\underline{S}_0 - \underline{s}_0) + \underline{S}_{0,2} - \underline{s}_{0,2} \\
\Delta \underline{G}_3^* &= \lfloor I, \ \widetilde{s}_0^T \rfloor \underline{\epsilon}_3 + \widetilde{\kappa}_3^1 (\underline{S}_0 - \underline{s}_0) + \underline{S}_{0,3} - \underline{s}_{0,3}
\end{aligned} \tag{4.37}$$

The same relations hold for the local Cosserat continuum, such that

$$\begin{aligned}
R_1^{\ell T} \underline{S}_{0,1} - R_0^{\ell T} \underline{s}_{0,1} &= \lfloor I, \ \widetilde{s}_0^{\ell T} \rfloor \underline{\epsilon}_1^{\ell} \\
R_1^{\ell T} \underline{S}_{0,2} - R_0^{\ell T} \underline{s}_{0,2} &= \lfloor I, \ \widetilde{s}_0^{\ell T} \rfloor \underline{\epsilon}_2^{\ell} \\
R_1^{\ell T} \underline{S}_{0,3} - R_0^{\ell T} \underline{s}_{0,3} &= \lfloor I, \ \widetilde{s}_0^{\ell T} \rfloor \underline{\epsilon}_3^{\ell}
\end{aligned} \tag{4.38}$$

where $\underline{\epsilon}_i^{\ell}$ for $i = 1, 2, 3$ are the Lagrangian strain measures, Eq. (2.152), in the local Cosserat continuum

$$\underline{\epsilon}_i^{\ell} = \left| \begin{matrix} R_1^{\ell T} \underline{X}_{0,i}^{\ell} - R_0^{\ell T} \underline{x}_{0,i}^{\ell} \\ \underline{\kappa}_i^{1\ell} - \underline{\kappa}_i^{0\ell} \end{matrix} \right| \tag{4.39}$$

With the aid of expressions, Eq. (4.38), the difference of base vectors $\Delta \underline{G}_i^*$ for $i = 1, 2, 3$ are recasted to

$$
\begin{aligned}
\Delta\underline{G}_1^* &= \lfloor I,\ \widetilde{s}_0^T \rceil \underline{\epsilon}_1 + R_0^\ell \lfloor I,\ \widetilde{s}_0^{\ell T} \rceil \underline{\epsilon}_1^\ell + \widetilde{\kappa}_1^1(\underline{S}_0 - \underline{s}_0) + (I - R_0^\ell R_1^{\ell T})\underline{S}_{0,1} \\
\Delta\underline{G}_2^* &= \lfloor I,\ \widetilde{s}_0^T \rceil \underline{\epsilon}_2 + R_0^\ell \lfloor I,\ \widetilde{s}_0^{\ell T} \rceil \underline{\epsilon}_2^\ell + \widetilde{\kappa}_2^1(\underline{S}_0 - \underline{s}_0) + (I - R_0^\ell R_1^{\ell T})\underline{S}_{0,2} \\
\Delta\underline{G}_3^* &= \lfloor I,\ \widetilde{s}_0^T \rceil \underline{\epsilon}_3 + R_0^\ell \lfloor I,\ \widetilde{s}_0^{\ell T} \rceil \underline{\epsilon}_3^\ell + \widetilde{\kappa}_3^1(\underline{S}_0 - \underline{s}_0) + (I - R_0^\ell R_1^{\ell T})\underline{S}_{0,3}
\end{aligned}
\tag{4.40}
$$

Since the deformations and motions of micro Cosserat continuum is at least one order of magnitude smaller than those of macro Cosserat continuum, it can be concluded the local displacements are small $\underline{u}_s^\ell \ll 1$ and the local rotation $R^\ell = R_0^{\ell T} R_1^\ell$ is approximated by $R^\ell = I + \widetilde{\phi}^\ell$ corresponding to the small angle rotations for $\underline{\phi}^\ell \ll 1$. With the aid of these observations and the definitions of directors before and after deformations, $\underline{s}_0$ and $\underline{S}_0$, the coupling terms in above expressions, Eq. (4.40), can be further simplified to

$$
\begin{aligned}
\widetilde{\kappa}_1^1(\underline{S}_0 - \underline{s}_0) + (I - R_0^\ell R_1^{\ell T})\underline{S}_{0,1} &= \widetilde{\kappa}_1^1 \underline{u}^\ell + (R_1^\ell - R_0^\ell) R_1^{\ell T} \underline{S}_{0,1} \\
\widetilde{\kappa}_2^1(\underline{S}_0 - \underline{s}_0) + (I - R_0^\ell R_1^{\ell T})\underline{S}_{0,2} &= \widetilde{\kappa}_2^1 \underline{u}^\ell + (R_1^\ell - R_0^\ell) R_1^{\ell T} \underline{S}_{0,2} \\
\widetilde{\kappa}_3^1(\underline{S}_0 - \underline{s}_0) + (I - R_0^\ell R_1^{\ell T})\underline{S}_{0,3} &= \widetilde{\kappa}_3^1 \underline{u}^\ell + (R_1^\ell - R_0^\ell) R_1^{\ell T} \underline{S}_{0,3}
\end{aligned}
\tag{4.41}
$$

where $\underline{u}^\ell$ denotes the local displacement of material particle, $\underline{u}^\ell = \underline{S}_0 - \underline{s}_0$, and in detail, $\underline{u}^\ell = \underline{u}_s^\ell + R_0^\ell (R^\ell - I)\underline{s}_0^\ell = \underline{u}_s^\ell + R_0^\ell \widetilde{s}_0^{\ell T} \underline{\phi}^\ell$. Since $\underline{u}_s^\ell$ and $\underline{\phi}^\ell$ are all small, it answers that $\underline{u}^\ell$ is small also. Under the assumption of small stretch and shallow curvature strains both in the levels of micro and macro scales, further simplifications lead to

$$
\begin{aligned}
\widetilde{\kappa}_1^1(\underline{S}_0 - \underline{s}_0) + (I - R_0^\ell R_1^{\ell T})\underline{S}_{0,1} &\approx \widetilde{\kappa}_1^0(\underline{u}_s^\ell + R_0^\ell \widetilde{s}_0^{\ell T} \underline{\phi}^\ell) + \widetilde{s}_{0,1}^T R_0^\ell \underline{\phi}^\ell \\
\widetilde{\kappa}_2^1(\underline{S}_0 - \underline{s}_0) + (I - R_0^\ell R_1^{\ell T})\underline{S}_{0,2} &\approx \widetilde{\kappa}_2^0(\underline{u}_s^\ell + R_0^\ell \widetilde{s}_0^{\ell T} \underline{\phi}^\ell) + \widetilde{s}_{0,2}^T R_0^\ell \underline{\phi}^\ell \\
\widetilde{\kappa}_3^1(\underline{S}_0 - \underline{s}_0) + (I - R_0^\ell R_1^{\ell T})\underline{S}_{0,3} &\approx \widetilde{\kappa}_3^0(\underline{u}_s^\ell + R_0^\ell \widetilde{s}_0^{\ell T} \underline{\phi}^\ell) + \widetilde{s}_{0,3}^T R_0^\ell \underline{\phi}^\ell
\end{aligned}
\tag{4.42}
$$

where the higher order terms $\widetilde{\kappa}_i \underline{u}^\ell$ and $R_0^\ell \widetilde{\phi}^\ell \lfloor I,\ \widetilde{s}_0^{\ell T} \rceil \underline{\epsilon}_i^\ell$ for $i = 1, 2, 3$ have been truncated. By above approximations, the difference of base vectors for reference and current configurations are identified to be

$$
\begin{aligned}
\Delta\underline{G}_1^* &= \lfloor I,\ \widetilde{s}_0^T \rceil \underline{\epsilon}_1 + R_0^\ell \lfloor I,\ \widetilde{s}_0^{\ell T} \rceil \underline{\epsilon}_1^\ell + \widetilde{\kappa}_1^0(\underline{u}_s^\ell + R_0^\ell \widetilde{s}_0^{\ell T} \underline{\phi}^\ell) + \widetilde{s}_{0,1}^T R_0^\ell \underline{\phi}^\ell \\
\Delta\underline{G}_2^* &= \lfloor I,\ \widetilde{s}_0^T \rceil \underline{\epsilon}_2 + R_0^\ell \lfloor I,\ \widetilde{s}_0^{\ell T} \rceil \underline{\epsilon}_2^\ell + \widetilde{\kappa}_2^0(\underline{u}_s^\ell + R_0^\ell \widetilde{s}_0^{\ell T} \underline{\phi}^\ell) + \widetilde{s}_{0,2}^T R_0^\ell \underline{\phi}^\ell \\
\Delta\underline{G}_3^* &= \lfloor I,\ \widetilde{s}_0^T \rceil \underline{\epsilon}_3 + R_0^\ell \lfloor I,\ \widetilde{s}_0^{\ell T} \rceil \underline{\epsilon}_3^\ell + \widetilde{\kappa}_3^0(\underline{u}_s^\ell + R_0^\ell \widetilde{s}_0^{\ell T} \underline{\phi}^\ell) + \widetilde{s}_{0,3}^T R_0^\ell \underline{\phi}^\ell
\end{aligned}
\tag{4.43}
$$

If defining the micro displacements in its dual bracket

$$
\underline{\mathcal{U}}^\ell = \begin{vmatrix} R_0^{\ell T} \underline{u}_s^\ell \\ \underline{\phi}^\ell \end{vmatrix}
\tag{4.44}
$$

the differences of base vectors can be reformulated to

$$
\begin{aligned}
\Delta \underline{G}_1^* &= \lfloor I,\ \tilde{s}_0^T \rfloor \underline{\epsilon}_1 + R_0^\ell \lfloor I,\ \tilde{s}_0^{\ell T} \rfloor \underline{\epsilon}_1^\ell + \lfloor \tilde{\kappa}_1^0 R_0^\ell,\ \tilde{\kappa}_1^0 R_0^\ell \tilde{s}_0^{\ell T} + \tilde{s}_{0,1}^T R_0^\ell \rfloor \underline{\mathcal{U}}^\ell \\
\Delta \underline{G}_2^* &= \lfloor I,\ \tilde{s}_0^T \rfloor \underline{\epsilon}_2 + R_0^\ell \lfloor I,\ \tilde{s}_0^{\ell T} \rfloor \underline{\epsilon}_2^\ell + \lfloor \tilde{\kappa}_2^0 R_0^\ell,\ \tilde{\kappa}_2^0 R_0^\ell \tilde{s}_0^{\ell T} + \tilde{s}_{0,2}^T R_0^\ell \rfloor \underline{\mathcal{U}}^\ell \\
\Delta \underline{G}_3^* &= \lfloor I,\ \tilde{s}_0^T \rfloor \underline{\epsilon}_3 + R_0^\ell \lfloor I,\ \tilde{s}_0^{\ell T} \rfloor \underline{\epsilon}_3^\ell + \lfloor \tilde{\kappa}_3^0 R_0^\ell,\ \tilde{\kappa}_3^0 R_0^\ell \tilde{s}_0^{\ell T} + \tilde{s}_{0,3}^T R_0^\ell \rfloor \underline{\mathcal{U}}^\ell
\end{aligned}
\tag{4.45}
$$

and then are rewritten in the compact form

$$
\begin{aligned}
\Delta \underline{G}_1^* &= \lfloor I,\ \tilde{s}_0^T \rfloor \underline{\epsilon}_1 + R_0^\ell \lfloor I,\ \tilde{s}_0^{\ell T} \rfloor \underline{\epsilon}_1^\ell + C_1^\ell \underline{\mathcal{U}}^\ell \\
\Delta \underline{G}_2^* &= \lfloor I,\ \tilde{s}_0^T \rfloor \underline{\epsilon}_2 + R_0^\ell \lfloor I,\ \tilde{s}_0^{\ell T} \rfloor \underline{\epsilon}_2^\ell + C_2^\ell \underline{\mathcal{U}}^\ell \\
\Delta \underline{G}_3^* &= \lfloor I,\ \tilde{s}_0^T \rfloor \underline{\epsilon}_3 + R_0^\ell \lfloor I,\ \tilde{s}_0^{\ell T} \rfloor \underline{\epsilon}_3^\ell + C_3^\ell \underline{\mathcal{U}}^\ell
\end{aligned}
\tag{4.46}
$$

with the following definitions

$$
C_i^\ell = \lfloor \tilde{\kappa}_i^0 R_0^\ell,\ \tilde{\kappa}_i^0 R_0^\ell \tilde{s}_0^{\ell T} + \tilde{s}_{0,i}^T R_0^\ell \rfloor \tag{4.47}
$$

for $i = 1, 2, 3$.

In view of above expressions, the vectorial Green-Lagrange strain measures can be readily determined to be

$$
\begin{aligned}
\underline{\varepsilon}_1 &= \mathcal{B}_g^T \lfloor I,\ \tilde{s}_0^T \rfloor \underline{\epsilon}_1 + \mathcal{B}_g^T R_0^\ell \lfloor I,\ \tilde{s}_0^{\ell T} \rfloor \underline{\epsilon}_1^\ell + \mathcal{B}_g^T C_1^\ell \underline{\mathcal{U}}^\ell \\
\underline{\varepsilon}_2 &= \mathcal{B}_g^T \lfloor I,\ \tilde{s}_0^T \rfloor \underline{\epsilon}_2 + \mathcal{B}_g^T R_0^\ell \lfloor I,\ \tilde{s}_0^{\ell T} \rfloor \underline{\epsilon}_2^\ell + \mathcal{B}_g^T C_2^\ell \underline{\mathcal{U}}^\ell \\
\underline{\varepsilon}_3 &= \mathcal{B}_g^T \lfloor I,\ \tilde{s}_0^T \rfloor \underline{\epsilon}_3 + \mathcal{B}_g^T R_0^\ell \lfloor I,\ \tilde{s}_0^{\ell T} \rfloor \underline{\epsilon}_3^\ell + \mathcal{B}_g^T C_3^\ell \underline{\mathcal{U}}^\ell
\end{aligned}
\tag{4.48}
$$

With the definitions of following symbols

$$
\mathcal{S}_0^T = \mathcal{B}_g^T \lfloor I,\ \tilde{s}_0^T \rfloor,\ \mathcal{S}_0^{\ell T} = \mathcal{B}_g^T R_0^\ell \lfloor I,\ \tilde{s}_0^{\ell T} \rfloor \tag{4.49}
$$

and

$$
\mathcal{S}_i^{\ell T} = \mathcal{B}_g^T C_i^\ell = \mathcal{B}_g^T \lfloor \tilde{\kappa}_i^0 R_0^\ell,\ \tilde{\kappa}_i^0 R_0^\ell \tilde{s}_0^{\ell T} + \tilde{s}_{0,i}^T R_0^\ell \rfloor \tag{4.50}
$$

for $i = 1, 2, 3$, the Green-Lagrange strain tensor can be described as functions of Lagrangian stretch and curvature measures in the vectorial forms like

$$
\begin{aligned}
\underline{\varepsilon}_1 &= \mathcal{S}_0^T \underline{\epsilon}_1 + \mathcal{S}_0^{\ell T} \underline{\epsilon}_1^\ell + \mathcal{S}_1^{\ell T} \underline{\mathcal{U}}^\ell \\
\underline{\varepsilon}_2 &= \mathcal{S}_0^T \underline{\epsilon}_2 + \mathcal{S}_0^{\ell T} \underline{\epsilon}_2^\ell + \mathcal{S}_2^{\ell T} \underline{\mathcal{U}}^\ell \\
\underline{\varepsilon}_3 &= \mathcal{S}_0^T \underline{\epsilon}_3 + \mathcal{S}_0^{\ell T} \underline{\epsilon}_3^\ell + \mathcal{S}_3^{\ell T} \underline{\mathcal{U}}^\ell
\end{aligned}
\tag{4.51}
$$

In most of the cases, the initial curvatures are always zero, $\tilde{\kappa}_i^0 = 0$ for $i = 1, 2, 3$, the coefficient matrix $\mathcal{S}_i^\ell$ is then reduced to

$$
\mathcal{S}_i^{\ell T} = \mathcal{B}_g^T \lfloor 0,\ \tilde{s}_{0,i}^T R_0^\ell \rfloor \tag{4.52}
$$

and the last terms of vectorial strains, Eq. (4.51), are found to be

$$\mathcal{S}_i^{\ell T} \underline{\mathcal{U}}^{\ell} = \mathcal{B}_g^T \widetilde{s}_{0,i}^T R_0^{\ell} \underline{\phi}^{\ell} \tag{4.53}$$

By Eq. (4.9), the spatial derivatives $\underline{s}_{0,i}$ for $i = 1, 2, 3$ are verified to be the unit vectors scaled by a small quantity, $s_{ci}\bar{l}_i$, which makes the products $\widetilde{s}_{0,i}^T R_0^{\ell} \underline{\phi}^{\ell}$ to behavior like the small quantities of second order. This observation convinces the neglecting of the last terms in the expressions of vectorial strains

$$\begin{aligned} \underline{\varepsilon}_1 &= \mathcal{S}_0^T \underline{\epsilon}_1 + \mathcal{S}_0^{\ell T} \underline{\epsilon}_1^{\ell} \\ \underline{\varepsilon}_2 &= \mathcal{S}_0^T \underline{\epsilon}_2 + \mathcal{S}_0^{\ell T} \underline{\epsilon}_2^{\ell} \\ \underline{\varepsilon}_3 &= \mathcal{S}_0^T \underline{\epsilon}_3 + \mathcal{S}_0^{\ell T} \underline{\epsilon}_3^{\ell} \end{aligned} \tag{4.54}$$

It is interesting to note that the above expressions, Eq. (4.51), and its simplifications, Eq. (4.54), are presented in a very similar approach as described by the variational asymptotic methods [12]. However, the strain energy for multiscale Cosserat continuum, defined in the next section, will not be approximated in the asymptotic sense.

### *4.1.3 Constitutive Laws for Multiscale Cosserat Continua*

The strain energy $\mathcal{U}_e$ can be represented by using the vectorial Green-Lagrange strain measures like

$$\begin{aligned} 2\mathcal{U}_e = \int_{V_0} &[\underline{\varepsilon}_1^T \mathcal{D}_{11} \underline{\varepsilon}_1 + \underline{\varepsilon}_2^T \mathcal{D}_{22} \underline{\varepsilon}_2 + \underline{\varepsilon}_3^T \mathcal{D}_{33} \underline{\varepsilon}_3 \\ &+ 2(\underline{\varepsilon}_2^T \mathcal{D}_{21} \underline{\varepsilon}_1 + \underline{\varepsilon}_3^T \mathcal{D}_{31} \underline{\varepsilon}_1 + \underline{\varepsilon}_3^T \mathcal{D}_{32} \underline{\varepsilon}_2)] \, dV_0 \end{aligned} \tag{4.55}$$

where the material matrix blocks $\mathcal{D}_{ij}$ for $i, j = 1, 2, 3$ are introduced in Sect. 3.1.4. Referring Eq. (4.54), decomposed the vectorial Green-Lagrange strains $\underline{\varepsilon}_i$ into two parts, the Lagrangian stretch and curvature measures, $\underline{\epsilon}_i$ and $\underline{\epsilon}_i^{\ell}$, for the micro and macro models, respectively, each item of the above full descriptions of strain energy are found to be

$$\begin{aligned} \underline{\varepsilon}_i^T \mathcal{D}_{ij} \underline{\varepsilon}_j &= \lfloor \underline{\epsilon}_i^T, \ \underline{\epsilon}_i^{\ell T} \rfloor \begin{bmatrix} \mathcal{S}_0 \mathcal{D}_{ij} \mathcal{S}_0^T & \mathcal{S}_0 \mathcal{D}_{ij} \mathcal{S}_0^{\ell T} \\ \mathcal{S}_0^{\ell} \mathcal{D}_{ij} \mathcal{S}_0^T & \mathcal{S}_0^{\ell} \mathcal{D}_{ij} \mathcal{S}_0^{\ell T} \end{bmatrix} \begin{vmatrix} \underline{\epsilon}_j \\ \underline{\epsilon}_j^{\ell} \end{vmatrix} \\ &= \lfloor \underline{\epsilon}_i^T, \ \underline{\epsilon}_i^{\ell T} \rfloor \begin{bmatrix} \mathcal{D}_{\epsilon\epsilon}^{ij} & \mathcal{D}_{\epsilon\alpha}^{ij} \\ \mathcal{D}_{\alpha\epsilon}^{ij} & \mathcal{D}_{\alpha\alpha}^{ij} \end{bmatrix} \begin{vmatrix} \underline{\epsilon}_j \\ \underline{\epsilon}_j^{\ell} \end{vmatrix} \end{aligned} \tag{4.56}$$

where the constitutive matrices of Cosserat continuum are introduced

$$\begin{aligned} \mathcal{D}^{ij}_{\epsilon\epsilon} &= \mathcal{S}_0 \mathcal{D}_{ij} \mathcal{S}_0^T, \ \mathcal{D}^{ij}_{\epsilon\alpha} = \mathcal{S}_0 \mathcal{D}_{ij} \mathcal{S}_0^{\ell T} \\ \mathcal{D}^{ij}_{\alpha\epsilon} &= \mathcal{S}_0^{\ell} \mathcal{D}_{ij} \mathcal{S}_0^T, \ \mathcal{D}^{ij}_{\alpha\alpha} = \mathcal{S}_0^{\ell} \mathcal{D}_{ij} \mathcal{S}_0^{\ell T} \end{aligned} \tag{4.57}$$

for $i, j = 1, 2, 3$. Due to the symmetry of constitutive matrices, it can be verified that

$$\begin{aligned} \mathcal{D}^{ijT}_{\epsilon\epsilon} &= \mathcal{S}_0 \mathcal{D}^T_{ij} \mathcal{S}_0^T = \mathcal{S}_0 \mathcal{D}_{ji} \mathcal{S}_0^T = \mathcal{D}^{ji}_{\epsilon\epsilon} \\ \mathcal{D}^{ijT}_{\alpha\epsilon} &= \mathcal{S}_0 \mathcal{D}^T_{ij} \mathcal{S}_0^{\ell T} = \mathcal{S}_0 \mathcal{D}_{ji} \mathcal{S}_0^{\ell T} = \mathcal{D}^{ji}_{\epsilon\alpha} \end{aligned} \tag{4.58}$$

then all the terms of strain energy can be written explicitly

$$\begin{aligned} \underline{\varepsilon}_1^T \mathcal{D}_{11} \underline{\varepsilon}_1 &= \underline{\epsilon}_1^T \mathcal{D}^{11}_{\epsilon\epsilon} \underline{\epsilon}_1 + 2\underline{\epsilon}_1^T \mathcal{D}^{11}_{\epsilon\alpha} \underline{\epsilon}_1^{\ell} + \underline{\epsilon}_1^{\ell T} \mathcal{D}^{11}_{\alpha\alpha} \underline{\epsilon}_1^{\ell} \\ \underline{\varepsilon}_2^T \mathcal{D}_{22} \underline{\varepsilon}_2 &= \underline{\epsilon}_2^T \mathcal{D}^{22}_{\epsilon\epsilon} \underline{\epsilon}_2 + 2\underline{\epsilon}_2^T \mathcal{D}^{22}_{\epsilon\alpha} \underline{\epsilon}_2^{\ell} + \underline{\epsilon}_2^{\ell T} \mathcal{D}^{22}_{\alpha\alpha} \underline{\epsilon}_2^{\ell} \\ \underline{\varepsilon}_3^T \mathcal{D}_{33} \underline{\varepsilon}_3 &= \underline{\epsilon}_3^T \mathcal{D}^{33}_{\epsilon\epsilon} \underline{\epsilon}_3 + 2\underline{\epsilon}_3^T \mathcal{D}^{33}_{\epsilon\alpha} \underline{\epsilon}_3^{\ell} + \underline{\epsilon}_3^{\ell T} \mathcal{D}^{33}_{\alpha\alpha} \underline{\epsilon}_3^{\ell} \end{aligned} \tag{4.59}$$

and

$$\begin{aligned} \underline{\varepsilon}_2^T \mathcal{D}_{21} \underline{\varepsilon}_1 &= \underline{\epsilon}_2^T \mathcal{D}^{21}_{\epsilon\epsilon} \underline{\epsilon}_1 + \underline{\epsilon}_2^T \mathcal{D}^{21}_{\epsilon\alpha} \underline{\epsilon}_1^{\ell} + \underline{\epsilon}_1^T \mathcal{D}^{12}_{\epsilon\alpha} \underline{\epsilon}_2^{\ell} + \underline{\epsilon}_2^{\ell T} \mathcal{D}^{21}_{\alpha\alpha} \underline{\epsilon}_1^{\ell} \\ \underline{\varepsilon}_3^T \mathcal{D}_{31} \underline{\varepsilon}_1 &= \underline{\epsilon}_3^T \mathcal{D}^{31}_{\epsilon\epsilon} \underline{\epsilon}_1 + \underline{\epsilon}_3^T \mathcal{D}^{31}_{\epsilon\alpha} \underline{\epsilon}_1^{\ell} + \underline{\epsilon}_1^T \mathcal{D}^{13}_{\epsilon\alpha} \underline{\epsilon}_3^{\ell} + \underline{\epsilon}_3^{\ell T} \mathcal{D}^{31}_{\alpha\alpha} \underline{\epsilon}_1^{\ell} \\ \underline{\varepsilon}_3^T \mathcal{D}_{32} \underline{\varepsilon}_2 &= \underline{\epsilon}_3^T \mathcal{D}^{32}_{\epsilon\epsilon} \underline{\epsilon}_2 + \underline{\epsilon}_3^T \mathcal{D}^{32}_{\epsilon\alpha} \underline{\epsilon}_2^{\ell} + \underline{\epsilon}_2^T \mathcal{D}^{23}_{\epsilon\alpha} \underline{\epsilon}_3^{\ell} + \underline{\epsilon}_3^{\ell T} \mathcal{D}^{32}_{\alpha\alpha} \underline{\epsilon}_2^{\ell} \end{aligned} \tag{4.60}$$

If the characteristic volume with the magnitude of $V_c$ for a micro model is given, the strain energy can be written in detail like

$$\begin{aligned} \mathcal{U}_c = \int_{V_c} [\ & \frac{1}{2}(\underline{\epsilon}_1^T \mathcal{D}^{11}_{\epsilon\epsilon} \underline{\epsilon}_1 + \underline{\epsilon}_2^T \mathcal{D}^{22}_{\epsilon\epsilon} \underline{\epsilon}_2 + \underline{\epsilon}_3^T \mathcal{D}^{33}_{\epsilon\epsilon} \underline{\epsilon}_3) \\ & + \underline{\epsilon}_2^T \mathcal{D}^{21}_{\epsilon\epsilon} \underline{\epsilon}_1 + \underline{\epsilon}_3^T \mathcal{D}^{31}_{\epsilon\epsilon} \underline{\epsilon}_1 + \underline{\epsilon}_3^T \mathcal{D}^{32}_{\epsilon\epsilon} \underline{\epsilon}_2 \\ & + \frac{1}{2}(\underline{\epsilon}_1^{\ell T} \mathcal{D}^{11}_{\alpha\alpha} \underline{\epsilon}_1^{\ell} + \underline{\epsilon}_2^{\ell T} \mathcal{D}^{22}_{\alpha\alpha} \underline{\epsilon}_2^{\ell} + \underline{\epsilon}_3^{\ell T} \mathcal{D}^{33}_{\alpha\alpha} \underline{\epsilon}_3^{\ell}) \\ & + \underline{\epsilon}_2^{\ell T} \mathcal{D}^{21}_{\alpha\alpha} \underline{\epsilon}_1^{\ell} + \underline{\epsilon}_3^{\ell T} \mathcal{D}^{31}_{\alpha\alpha} \underline{\epsilon}_1^{\ell} + \underline{\epsilon}_3^{\ell T} \mathcal{D}^{32}_{\alpha\alpha} \underline{\epsilon}_2^{\ell} \\ & + \underline{\epsilon}_1^T \mathcal{D}^{11}_{\epsilon\alpha} \underline{\epsilon}_1^{\ell} + \underline{\epsilon}_1^T \mathcal{D}^{12}_{\epsilon\alpha} \underline{\epsilon}_2^{\ell} + \underline{\epsilon}_1^T \mathcal{D}^{13}_{\epsilon\alpha} \underline{\epsilon}_3^{\ell} \\ & + \underline{\epsilon}_2^T \mathcal{D}^{21}_{\epsilon\alpha} \underline{\epsilon}_1^{\ell} + \underline{\epsilon}_2^T \mathcal{D}^{22}_{\epsilon\alpha} \underline{\epsilon}_2^{\ell} + \underline{\epsilon}_2^T \mathcal{D}^{23}_{\epsilon\alpha} \underline{\epsilon}_3^{\ell} \\ & + \underline{\epsilon}_3^T \mathcal{D}^{31}_{\epsilon\alpha} \underline{\epsilon}_1^{\ell} + \underline{\epsilon}_3^T \mathcal{D}^{32}_{\epsilon\alpha} \underline{\epsilon}_2^{\ell} + \underline{\epsilon}_3^T \mathcal{D}^{33}_{\epsilon\alpha} \underline{\epsilon}_3^{\ell}\ ]\, dV_c \end{aligned} \tag{4.61}$$

According to the definitions of Lagrangian strain measures, Eq. (4.39), for the micro model, the Lagrangian stretch and curvature strains could be explicitly related to the displacement vector and its derivatives under the assumption of small micro translation and micro rotation like

$$\underline{\epsilon}_i^{\ell} = \underline{\mathcal{U}}^{\ell}_{,i} + \widetilde{\mathcal{K}}_i^{0\ell} \underline{\mathcal{U}}^{\ell} \tag{4.62}$$

for $i = 1, 2, 3$ with the following introduced symbols

$$\underline{\mathcal{U}}^{\ell} = \begin{vmatrix} R_0^{\ell T} \underline{u}_s^{\ell} \\ \underline{\phi}^{\ell} \end{vmatrix}, \ \underline{\mathcal{U}}_{,i}^{\ell} = \begin{vmatrix} (R_0^{\ell T} \underline{u}_s^{\ell})_{,i} \\ \underline{\phi}_{,i}^{\ell} \end{vmatrix}, \ \widetilde{\mathcal{K}}_i^{0\ell} = \begin{bmatrix} \widetilde{\kappa}_i^{0\ell} & \widetilde{R_0^{\ell T} \underline{x}_{0,i}} \\ 0 & \widetilde{\kappa}_i^{0\ell} \end{bmatrix} \tag{4.63}$$

Since it was assumed that the micro translation and rotation were all small, the micro motions form a linear space approximately. Hence, the classical finite element method can be directly applied to the micro model. Given the shape functions $\mathcal{H}_{\epsilon}^{\ell}$, the micro motion vectors and the spatial derivatives are interpolated by

$$\underline{\mathcal{U}}^{\ell} = \mathcal{H}_{\epsilon}^{\ell} \underline{\mathcal{U}}_n^{\ell}, \ \underline{\mathcal{U}}_{,i}^{\ell} = \mathcal{H}_{\epsilon,i}^{\ell} \underline{\mathcal{U}}_n^{\ell} \tag{4.64}$$

where $\underline{\mathcal{U}}_n^{\ell}$ denote the dual vectors of nodal motions for micro continuum. As a result, the Lagrangian stretch and curvature measures can be discretized into

$$\underline{\epsilon}_i^{\ell} = (\mathcal{H}_{\epsilon,i}^{\ell} + \widetilde{\mathcal{K}}_i^{0\ell} \mathcal{H}_{\epsilon}^{\ell}) \underline{\mathcal{U}}_n^{\ell} = \mathcal{Q}_i^{\ell} \underline{\mathcal{U}}_n^{\ell} \tag{4.65}$$

Under the assumption of scale separation, the global strains $\underline{\epsilon}_i$ for $i = 1, 2, 3$ are independent of micro variables. Hence, the strain energy for a micro characteristic volume with the amplitude of $V_c$ can be rewritten as

$$\begin{aligned}
\mathcal{U}_c = {} & \frac{1}{2}(\underline{\epsilon}_1^T \int_{V_c} \mathcal{D}_{\epsilon\epsilon}^{11}\, dV_c \underline{\epsilon}_1 + \underline{\epsilon}_2^T \int_{V_c} \mathcal{D}_{\epsilon\epsilon}^{22}\, dV_c \underline{\epsilon}_2 + \underline{\epsilon}_3^T \int_{V_c} \mathcal{D}_{\epsilon\epsilon}^{33}\, dV_c \underline{\epsilon}_3) \\
& + \underline{\epsilon}_2^T \int_{V_c} \mathcal{D}_{\epsilon\epsilon}^{21}\, dV_c \underline{\epsilon}_1 + \underline{\epsilon}_3^T \int_{V_c} \mathcal{D}_{\epsilon\epsilon}^{31}\, dV_c \underline{\epsilon}_1 + \underline{\epsilon}_3^T \int_{V_c} \mathcal{D}_{\epsilon\epsilon}^{32}\, dV_c \underline{\epsilon}_2 \\
& + \frac{1}{2}\underline{\mathcal{U}}_n^{\ell T} \int_{V_c} (\mathcal{Q}_1^{\ell T} \mathcal{D}_{\alpha\alpha}^{11} \mathcal{Q}_1^{\ell} + \mathcal{Q}_2^{\ell T} \mathcal{D}_{\alpha\alpha}^{22} \mathcal{Q}_2^{\ell} + \mathcal{Q}_3^{\ell T} \mathcal{D}_{\alpha\alpha}^{33} \mathcal{Q}_3^{\ell})\, dV_c \underline{\mathcal{U}}_n^{\ell} \\
& + \underline{\mathcal{U}}_n^{\ell T} \int_{V_c} (\mathcal{Q}_2^{\ell T} \mathcal{D}_{\alpha\alpha}^{21} \mathcal{Q}_1^{\ell} + \mathcal{Q}_3^{\ell T} \mathcal{D}_{\alpha\alpha}^{31} \mathcal{Q}_1^{\ell} + \mathcal{Q}_3^{\ell T} \mathcal{D}_{\alpha\alpha}^{32} \mathcal{Q}_2^{\ell})\, dV_c \underline{\mathcal{U}}_n^{\ell} \\
& + \underline{\epsilon}_1^T \int_{V_c} (\mathcal{D}_{\epsilon\alpha}^{11} \mathcal{Q}_1^{\ell} + \mathcal{D}_{\epsilon\alpha}^{12} \mathcal{Q}_2^{\ell} + \mathcal{D}_{\epsilon\alpha}^{13} \mathcal{Q}_3^{\ell})\, dV_c \underline{\mathcal{U}}_n^{\ell} \\
& + \underline{\epsilon}_2^T \int_{V_c} (\mathcal{D}_{\epsilon\alpha}^{21} \mathcal{Q}_1^{\ell} + \mathcal{D}_{\epsilon\alpha}^{22} \mathcal{Q}_2^{\ell} + \mathcal{D}_{\epsilon\alpha}^{23} \mathcal{Q}_3^{\ell})\, dV_c \underline{\mathcal{U}}_n^{\ell} \\
& + \underline{\epsilon}_3^T \int_{V_c} (\mathcal{D}_{\epsilon\alpha}^{31} \mathcal{Q}_1^{\ell} + \mathcal{D}_{\epsilon\alpha}^{32} \mathcal{Q}_2^{\ell} + \mathcal{D}_{\epsilon\alpha}^{33} \mathcal{Q}_3^{\ell})\, dV_c \underline{\mathcal{U}}_n^{\ell}
\end{aligned} \tag{4.66}$$

The strain energy per unit volume, called strain energy density, can be defined as the volume average of the above strain energy like

$$
\begin{aligned}
\mathcal{U}_\rho &= \frac{\mathcal{U}_C}{V_c} \\
&= \frac{1}{2}(\underline{\epsilon}_1^T \mathcal{C}_{\epsilon\epsilon}^{11} \underline{\epsilon}_1 + \underline{\epsilon}_2^T \mathcal{C}_{\epsilon\epsilon}^{22} \underline{\epsilon}_2 + \underline{\epsilon}_3^T \mathcal{C}_{\epsilon\epsilon}^{33} \underline{\epsilon}_3) + \underline{\epsilon}_2^T \mathcal{C}_{\epsilon\epsilon}^{21} \underline{\epsilon}_1 + \underline{\epsilon}_3^T \mathcal{C}_{\epsilon\epsilon}^{31} \underline{\epsilon}_1 + \underline{\epsilon}_3^T \mathcal{C}_{\epsilon\epsilon}^{32} \underline{\epsilon}_2 \\
&+ \frac{1}{2}\underline{\mathcal{U}}_n^{\ell T} \mathcal{C}_{\alpha\alpha} \underline{\mathcal{U}}_n^\ell + \underline{\epsilon}_1^T \mathcal{C}_{\epsilon\alpha}^{11} \underline{\mathcal{U}}_n^\ell + \underline{\epsilon}_2^T \mathcal{C}_{\epsilon\alpha}^{22} \underline{\mathcal{U}}_n^\ell + \underline{\epsilon}_3^T \mathcal{C}_{\epsilon\alpha}^{33} \underline{\mathcal{U}}_n^\ell
\end{aligned}
\tag{4.67}
$$

where the equivalent uniform stiffness matrices are computed from the weighted integration [34] over the characteristic volume with the magnitude of $V_c$ for a micro Cosserat continuum

$$
\begin{aligned}
\mathcal{C}_{\epsilon\epsilon}^{11} &= \frac{\int_{V_c} \mathcal{D}_{\epsilon\epsilon}^{11} dV_c}{V_c}, \quad \mathcal{C}_{\epsilon\epsilon}^{22} = \frac{\int_{V_c} \mathcal{D}_{\epsilon\epsilon}^{22} dV_c}{V_c}, \quad \mathcal{C}_{\epsilon\epsilon}^{33} = \frac{\int_{V_c} \mathcal{D}_{\epsilon\epsilon}^{33} dV_c}{V_c} \\
\mathcal{C}_{\epsilon\epsilon}^{21} &= \frac{\int_{V_c} \mathcal{D}_{\epsilon\epsilon}^{21} dV_c}{V_c}, \quad \mathcal{C}_{\epsilon\epsilon}^{31} = \frac{\int_{V_c} \mathcal{D}_{\epsilon\epsilon}^{31} dV_c}{V_c}, \quad \mathcal{C}_{\epsilon\epsilon}^{32} = \frac{\int_{V_c} \mathcal{D}_{\epsilon\epsilon}^{32} dV_c}{V_c}
\end{aligned}
\tag{4.68}
$$

$$
\begin{aligned}
\mathcal{C}_{\alpha\alpha} &= \frac{\int_{V_c} \mathcal{Q}_1^{\ell T} \mathcal{D}_{\alpha\alpha}^{11} \mathcal{Q}_1^\ell + \mathcal{Q}_2^{\ell T} \mathcal{D}_{\alpha\alpha}^{22} \mathcal{Q}_2^\ell + \mathcal{Q}_3^{\ell T} \mathcal{D}_{\alpha\alpha}^{33} \mathcal{Q}_3^\ell dV_c}{V_c} \\
&+ 2\frac{\int_{V_c} \mathcal{Q}_2^{\ell T} \mathcal{D}_{\alpha\alpha}^{21} \mathcal{Q}_1^\ell + \mathcal{Q}_3^{\ell T} \mathcal{D}_{\alpha\alpha}^{31} \mathcal{Q}_1^\ell + \mathcal{Q}_3^{\ell T} \mathcal{D}_{\alpha\alpha}^{32} \mathcal{Q}_2^\ell dV_c}{V_c}
\end{aligned}
\tag{4.69}
$$

$$
\begin{aligned}
\mathcal{C}_{\epsilon\alpha}^{11} &= \frac{\int_{V_c} \mathcal{D}_{\epsilon\alpha}^{11} \mathcal{Q}_1^\ell + \mathcal{D}_{\epsilon\alpha}^{12} \mathcal{Q}_2^\ell + \mathcal{D}_{\epsilon\alpha}^{13} \mathcal{Q}_3^\ell dV_c}{V_c} \\
\mathcal{C}_{\epsilon\alpha}^{22} &= \frac{\int_{V_c} \mathcal{D}_{\epsilon\alpha}^{21} \mathcal{Q}_1^\ell + \mathcal{D}_{\epsilon\alpha}^{22} \mathcal{Q}_2^\ell + \mathcal{D}_{\epsilon\alpha}^{23} \mathcal{Q}_3^\ell dV_c}{V_c} \\
\mathcal{C}_{\epsilon\alpha}^{33} &= \frac{\int_{V_c} \mathcal{D}_{\epsilon\alpha}^{31} \mathcal{Q}_1^\ell + \mathcal{D}_{\epsilon\alpha}^{32} \mathcal{Q}_2^\ell + \mathcal{D}_{\epsilon\alpha}^{33} \mathcal{Q}_3^\ell dV_c}{V_c}
\end{aligned}
\tag{4.70}
$$

The above volume averaging composes the general equivalence process of heterogeneous micro structures. Note that the symmetric properties of constitutive matrices are still hold, such that

$$
\begin{aligned}
\mathcal{C}_{\epsilon\epsilon}^{21T} &= \frac{\int_{V_c} \mathcal{D}_{\epsilon\epsilon}^{21T} dV_c}{V_c} = \frac{\int_{V_c} \mathcal{D}_{\epsilon\epsilon}^{12} dV_c}{V_c} = \mathcal{C}_{\epsilon\epsilon}^{12} \\
\mathcal{C}_{\epsilon\epsilon}^{31T} &= \frac{\int_{V_c} \mathcal{D}_{\epsilon\epsilon}^{31T} dV_c}{V_c} = \frac{\int_{V_c} \mathcal{D}_{\epsilon\epsilon}^{13} dV_c}{V_c} = \mathcal{C}_{\epsilon\epsilon}^{13} \\
\mathcal{C}_{\epsilon\epsilon}^{32T} &= \frac{\int_{V_c} \mathcal{D}_{\epsilon\epsilon}^{32T} dV_c}{V_c} = \frac{\int_{V_c} \mathcal{D}_{\epsilon\epsilon}^{23} dV_c}{V_c} = \mathcal{C}_{\epsilon\epsilon}^{23}
\end{aligned}
\tag{4.71}
$$

### 4.1.4 Variation of Strain Energy of Multiscale Cosserat Continuum

Since the constitutive matrix blocks are determined by the above weighted quantities on the micro Cosserat continuum, the full description of strain energy for a given volume with the magnitude of $V_0$ can be approximated by

$$\begin{aligned}\mathcal{U}_e &= \int_{V_0} \mathcal{U}_\rho dV_0 \\ &= \int_{V_0} [\frac{1}{2}(\underline{\epsilon}_1^T \mathcal{C}_{\epsilon\epsilon}^{11}\underline{\epsilon}_1 + \underline{\epsilon}_2^T \mathcal{C}_{\epsilon\epsilon}^{22}\underline{\epsilon}_2 + \underline{\epsilon}_3^T \mathcal{C}_{\epsilon\epsilon}^{33}\underline{\epsilon}_3) + \underline{\epsilon}_2^T \mathcal{C}_{\epsilon\epsilon}^{21}\underline{\epsilon}_1 + \underline{\epsilon}_3^T \mathcal{C}_{\epsilon\epsilon}^{31}\underline{\epsilon}_1 + \underline{\epsilon}_3^T \mathcal{C}_{\epsilon\epsilon}^{32}\underline{\epsilon}_2 \\ &+ \underline{\epsilon}_1^T \mathcal{C}_{\epsilon\alpha}^{11}\underline{\mathcal{U}}_n^\ell + \underline{\epsilon}_2^T \mathcal{C}_{\epsilon\alpha}^{22}\underline{\mathcal{U}}_n^\ell + \underline{\epsilon}_3^T \mathcal{C}_{\epsilon\alpha}^{33}\underline{\mathcal{U}}_n^\ell + \frac{1}{2}\underline{\mathcal{U}}_n^{\ell T} \mathcal{C}_{\alpha\alpha}\underline{\mathcal{U}}_n^\ell] dV_0\end{aligned} \tag{4.72}$$

Taking variation of averaged strain energy density leads to

$$\begin{aligned}\delta\mathcal{U}_\rho &= \delta\underline{\epsilon}_1^T (\mathcal{C}_{\epsilon\epsilon}^{11}\underline{\epsilon}_1 + \mathcal{C}_{\epsilon\epsilon}^{12}\underline{\epsilon}_2 + \mathcal{C}_{\epsilon\epsilon}^{13}\underline{\epsilon}_3) + \delta\underline{\epsilon}_1^T \mathcal{C}_{\epsilon\alpha}^{11}\underline{\mathcal{U}}_n^\ell \\ &+ \delta\underline{\epsilon}_2^T (\mathcal{C}_{\epsilon\epsilon}^{21}\underline{\epsilon}_1 + \mathcal{C}_{\epsilon\epsilon}^{22}\underline{\epsilon}_2 + \mathcal{C}_{\epsilon\epsilon}^{23}\underline{\epsilon}_3) + \delta\underline{\epsilon}_2^T \mathcal{C}_{\epsilon\alpha}^{22}\underline{\mathcal{U}}_n^\ell \\ &+ \delta\underline{\epsilon}_3^T (\mathcal{C}_{\epsilon\epsilon}^{31}\underline{\epsilon}_1 + \mathcal{C}_{\epsilon\epsilon}^{32}\underline{\epsilon}_2 + \mathcal{C}_{\epsilon\epsilon}^{33}\underline{\epsilon}_3) + \delta\underline{\epsilon}_3^T \mathcal{C}_{\epsilon\alpha}^{33}\underline{\mathcal{U}}_n^\ell \\ &+ \delta\underline{\mathcal{U}}_n^{\ell T} (\mathcal{C}_{\epsilon\alpha}^{11T}\underline{\epsilon}_1 + \mathcal{C}_{\epsilon\alpha}^{22T}\underline{\epsilon}_2 + \mathcal{C}_{\epsilon\alpha}^{33T}\underline{\epsilon}_3) + \delta\underline{\mathcal{U}}_n^{\ell T} \mathcal{C}_{\alpha\alpha}\underline{\mathcal{U}}_n^\ell\end{aligned} \tag{4.73}$$

After introducing the following stress measures

$$\begin{aligned}&\underline{F}_{\epsilon\epsilon}^{11} = \mathcal{C}_{\epsilon\epsilon}^{11}\underline{\epsilon}_1 + \mathcal{C}_{\epsilon\epsilon}^{12}\underline{\epsilon}_2 + \mathcal{C}_{\epsilon\epsilon}^{13}\underline{\epsilon}_3,\ \underline{F}_{\epsilon\epsilon}^{22} = \mathcal{C}_{\epsilon\epsilon}^{21}\underline{\epsilon}_1 + \mathcal{C}_{\epsilon\epsilon}^{22}\underline{\epsilon}_2 + \mathcal{C}_{\epsilon\epsilon}^{23}\underline{\epsilon}_3 \\ &\underline{F}_{\epsilon\epsilon}^{33} = \mathcal{C}_{\epsilon\epsilon}^{31}\underline{\epsilon}_1 + \mathcal{C}_{\epsilon\epsilon}^{32}\underline{\epsilon}_2 + \mathcal{C}_{\epsilon\epsilon}^{33}\underline{\epsilon}_3,\ \underline{F}_{\alpha\epsilon} = \mathcal{C}_{\epsilon\alpha}^{11T}\underline{\epsilon}_1 + \mathcal{C}_{\epsilon\alpha}^{22T}\underline{\epsilon}_2 + \mathcal{C}_{\epsilon\alpha}^{33T}\underline{\epsilon}_3\end{aligned} \tag{4.74}$$

The variation of strain energy density can be written in compact form like

$$\begin{aligned}\delta\mathcal{U}_\rho &= \delta\underline{\epsilon}_1^T (\underline{F}_{\epsilon\epsilon}^{11} + \mathcal{C}_{\epsilon\alpha}^{11}\underline{\mathcal{U}}_n^\ell) + \delta\underline{\epsilon}_2^T (\underline{F}_{\epsilon\epsilon}^{22} + \mathcal{C}_{\epsilon\alpha}^{22}\underline{\mathcal{U}}_n^\ell) + \delta\underline{\epsilon}_3^T (\underline{F}_{\epsilon\epsilon}^{33} + \mathcal{C}_{\epsilon\alpha}^{33}\underline{\mathcal{U}}_n^\ell) \\ &+ \delta\underline{\mathcal{U}}_n^{\ell T} (\underline{F}_{\alpha\epsilon} + \mathcal{C}_{\alpha\alpha}\underline{\mathcal{U}}_n^\ell)\end{aligned} \tag{4.75}$$

From which, the variation of strain energy $\mathcal{U}_e$ is readily evaluated from

$$\delta\mathcal{U}_e = \int_{V_0} \delta\mathcal{U}_\rho dV_0 \tag{4.76}$$

### 4.1.5 Virtual Work of External Forces for Multiscale Cosserat Continuum

The principle of virtual work is applied to express the problem of multiscale mechanics in current implementation. The virtual work done by a corresponding virtual displacement operating on the external forces is evaluated in this section. According to Eq. (4.21), the virtual displacement of an arbitrary particle in current configuration is obtained by taking variation of position vector $\underline{P}$ like

$$\delta \underline{P} = \delta(\underline{X}_0 + R_1 \underline{S}_0) = \delta(\underline{x}_0 + \underline{u}_s + R_1 \underline{S}_0) \tag{4.77}$$

where $\underline{u}_s$ is the displacement of reference particle **P** and $R_1$ the total rotation tensor that describes the orientation of local vector $\underline{S}_0$ in current deformed configuration. The virtual displacement $\delta \underline{P}$ can be readily verified to be

$$\delta \underline{P} = \delta \underline{u}_s + \delta R_1 \underline{S}_0 + R_1 \delta \underline{S}_0 \tag{4.78}$$

After resolved the above equations into the body attached frame, $\delta \underline{P}$ relates to the virtual displacements of reference particle like

$$R_1^T \delta \underline{P} = R_1^T \delta \underline{u}_s + \widetilde{S}_0^T \underline{\delta \psi} + \delta \underline{S}_0 \tag{4.79}$$

where the virtual rotations are defined $\widetilde{\delta \psi} = R_1^T \delta R_1$. The virtual displacement $R_1^T \delta \underline{P}$ needs to be simplified with the purposed of split the original multiscale continuum problem into a macro problem and a micro problem. Due to the fact that the scale of micro model is at least one order of magnitude smaller than the macro model, which also forms the fundamentals of scale separation assumption, it answers the displacement and deformation of micro continuum are small quantities, $\underline{u}^\ell = (\underline{S}_0 - \underline{s}_0) \ll 1$, and hence the higher order item $\widetilde{u}^{\ell T} \underline{\delta \psi}$ becomes much smaller and negligible. As a result, the cross-product item $\widetilde{S}_0^T \underline{\delta \psi}$ in Eq. (4.79) can be approximated by $\widetilde{s}_0^T \underline{\delta \psi}$, which leads to

$$R_1^T \delta \underline{P} = R_1^T \delta \underline{u}_s + \widetilde{s}_0^T \underline{\delta \psi} + \delta \underline{S}_0 \tag{4.80}$$

For the local position vector $\underline{S}_0$ of Cosserat continuum, its variation can be obtained in a similar manner under the assumptions of small displacements and rotations

$$R_1^{\ell T} \delta \underline{S}_0 = R_1^{\ell T} \delta \underline{u}_s^\ell + \widetilde{s}_0^{\ell T} \underline{\delta \psi}^\ell \tag{4.81}$$

with the application of virtual rotations $\widetilde{\delta \psi}^\ell = R_1^{\ell T} \delta R_1^\ell$.

In current implementation, two types of external forces are considered, one is body force $\underline{g}$ and the other surface pressure $\underline{p}_s$, both measured in body attached basis $R_1$. Then, the virtual work of the external forces applied to the characteristic volume of micro continuum can be computed from

$$\delta \mathcal{W}_c = \int_{V_c} \delta \underline{P}^T R_1 \underline{g}\, dV_c + \int_{S_c} \delta \underline{P}^T R_1 \underline{p}_s\, dS_c \tag{4.82}$$

where $S_c$ represents the area magnitude of a micro Cosserat continuum where the surface pressure is applied to. Referring the expressions of variations, Eqs. (4.80) and (4.81), the virtual work can be expressed in the following detailed formula as

$$\delta \mathcal{W}_c = \int_{V_c} \left( R_1^T \delta \underline{u}_s + \widetilde{s}_0^T \delta \underline{\psi} + \delta \underline{S}_0 \right)^T \underline{g}\, dV_c + \int_{S_c} \left( R_1^T \delta \underline{u}_s + \widetilde{s}_0^T \delta \underline{\psi} + \delta \underline{S}_0 \right)^T \underline{p}_s\, dS_c \tag{4.83}$$

Under the assumption of scale separation, the macro motion is independent of micro variables, the virtual work of external forces can be recasted to

$$\begin{aligned} \delta \mathcal{W}_c &= \delta \underline{\mathcal{U}}^T \left| \begin{matrix} \int_{V_c} \underline{g}\, dV_c + \int_{S_c} \underline{p}_s\, dS_c \\ \int_{V_c} \widetilde{s}_0 \underline{g}\, dV_c + \int_{S_c} \widetilde{s}_0 \underline{p}_s\, dS_c \end{matrix} \right| \\ &+ \int_{V_c} \delta \underline{\mathcal{U}}^{\ell T} \left| \begin{matrix} R_1^{\ell T} \underline{g} \\ \widetilde{s}_0^{\ell} R_1^{\ell T} \underline{g} \end{matrix} \right| dV_c + \int_{S_c} \delta \underline{\mathcal{U}}^{\ell T} \left| \begin{matrix} R_1^{\ell T} \underline{p}_s \\ \widetilde{s}_0^{\ell} R_1^{\ell T} \underline{p}_s \end{matrix} \right| dS_c \end{aligned} \tag{4.84}$$

where the following virtual motions in their Plucker coordinates are introduced

$$\delta \underline{\mathcal{U}} = \left| \begin{matrix} R_1^T \delta \underline{u}_s \\ \delta \underline{\psi} \end{matrix} \right|, \quad \delta \underline{\mathcal{U}}^{\ell} = \left| \begin{matrix} R_1^{\ell T} \delta \underline{u}_s^{\ell} \\ \delta \underline{\psi}^{\ell} \end{matrix} \right| \tag{4.85}$$

Since the virtual motion forms a six-dimensional linear space, the virtual motion $\delta \underline{\mathcal{U}}^{\ell}$ can be interpolated by using the finite element shape functions $\mathcal{H}_{\epsilon}^{\ell}$ directly

$$\delta \underline{\mathcal{U}}^{\ell} = \mathcal{H}_{\epsilon}^{\ell} \delta \underline{\mathcal{U}}_n^{\ell} \tag{4.86}$$

and then the virtual work decouples into

$$\begin{aligned} \delta \mathcal{W}_c &= \delta \underline{\mathcal{U}}^T \left| \begin{matrix} \int_{V_c} \underline{g}\, dV_c + \int_{S_c} \underline{p}_s\, dS_c \\ \int_{V_c} \widetilde{s}_0 \underline{g}\, dV_c + \int_{S_c} \widetilde{s}_0 \underline{p}_s\, dS_c \end{matrix} \right| \\ &+ \delta \underline{\mathcal{U}}_n^{\ell T} \left( \int_{V_c} \mathcal{H}_{\epsilon}^{\ell T} \left| \begin{matrix} R_1^{\ell T} \underline{g} \\ \widetilde{s}_0^{\ell} R_1^{\ell T} \underline{g} \end{matrix} \right| dV_c + \int_{S_c} \mathcal{H}_{\epsilon}^{\ell T} \left| \begin{matrix} R_1^{\ell T} \underline{p}_s \\ \widetilde{s}_0^{\ell} R_1^{\ell T} \underline{p}_s \end{matrix} \right| dS_c \right) \end{aligned} \tag{4.87}$$

For the easy description of virtual work, the external force and torque vectors are averaged [34] at first by the characteristic volume $V_c$ like

$$\underline{F}_{\epsilon\alpha} = \left| \begin{array}{c} \dfrac{\int_{V_c} \underline{g}\, dV_c + \int_{S_c} \underline{p}_s\, dS_c}{V_c} \\ \dfrac{\int_{V_c} \widetilde{s}_0 \underline{g}\, dV_c + \int_{S_c} \widetilde{s}_0 \underline{p}_s\, dS_c}{V_c} \end{array} \right|$$

$$\underline{F}_{\alpha\alpha} = \frac{1}{V_c} \left( \int_{V_c} \mathcal{H}_\epsilon^{\ell T} \left| \begin{array}{c} R_1^{\ell T} \underline{g} \\ \widetilde{s}_0^\ell R_1^{\ell T} \underline{g} \end{array} \right| dV_c + \int_{S_c} \mathcal{H}_\epsilon^{\ell T} \left| \begin{array}{c} R_1^{\ell T} \underline{p}_s \\ \widetilde{s}_0^\ell R_1^{\ell T} \underline{p}_s \end{array} \right| dS_c \right) \tag{4.88}$$

With the aid of above weighted integrations for external forces, $\underline{F}_{\epsilon\alpha}$ and $\underline{F}_{\alpha\alpha}$, the virtual work of external forces applied to the Cosserat continuum with the given volume $V_0$ can be written in the compact form as

$$\delta \mathcal{W}_e = \int_{V_0} (\delta \underline{\mathcal{U}}^T \underline{F}_{\epsilon\alpha} + \delta \underline{\mathcal{U}}_n^{\ell T} \underline{F}_{\alpha\alpha})\, dV_0 \tag{4.89}$$

### 4.1.6 Governing Equations of Multiscale Cosserat Continuum in Motion Formalism

By substituting Eqs. (4.75) and (4.89) into the principle of virtual work, which states

$$\delta \mathcal{U}_e - \delta \mathcal{W}_e = 0 \tag{4.90}$$

the governing equations of multiscale Cosserat continuum can be readily obtained like

$$\begin{aligned} \int_{V_0} [\, & \delta \underline{\epsilon}_1^T (\underline{F}_{\epsilon\epsilon}^{11} + C_{\epsilon\alpha}^{11} \underline{\mathcal{U}}_n^\ell) + \delta \underline{\epsilon}_2^T (\underline{F}_{\epsilon\epsilon}^{22} + C_{\epsilon\alpha}^{22} \underline{\mathcal{U}}_n^\ell) + \delta \underline{\epsilon}_3^T (\underline{F}_{\epsilon\epsilon}^{33} + C_{\epsilon\alpha}^{33} \underline{\mathcal{U}}_n^\ell) \\ & + \delta \underline{\mathcal{U}}_n^{\ell T} (\underline{F}_{\alpha\epsilon} + C_{\alpha\alpha} \underline{\mathcal{U}}_n^\ell) - \delta \underline{\mathcal{U}}^T \underline{F}_{\epsilon\alpha} - \delta \underline{\mathcal{U}}_n^{\ell T} \underline{F}_{\alpha\alpha}\, ]\, dV_0 = 0 \end{aligned} \tag{4.91}$$

Under the assumption of scale separation, the variations of macro strain measures are assumed to be independent of the variations of micro variables. Hence, the original multiscale problem can be decomposed into the macroscopic continuum problem

$$\begin{aligned} \int_{V_0} ( & \delta \underline{\epsilon}_1^T (\underline{F}_{\epsilon\epsilon}^{11} + C_{\epsilon\alpha}^{11} \underline{\mathcal{U}}_n^\ell) + \delta \underline{\epsilon}_2^T (\underline{F}_{\epsilon\epsilon}^{22} + C_{\epsilon\alpha}^{22} \underline{\mathcal{U}}_n^\ell) \\ & + \delta \underline{\epsilon}_3^T (\underline{F}_{\epsilon\epsilon}^{33} + C_{\epsilon\alpha}^{33} \underline{\mathcal{U}}_n^\ell) - \delta \underline{\mathcal{U}}^T \underline{F}_{\epsilon\alpha}) dV_0 = 0 \end{aligned} \tag{4.92}$$

and the microscopic continuum problem

$$C_{\alpha\alpha} \underline{\mathcal{U}}_n^\ell + \underline{F}_{\alpha\epsilon} - \underline{F}_{\alpha\alpha} = 0 \tag{4.93}$$

with appropriate boundary conditions. Once the elastic forces $\underline{F}_{\alpha\epsilon}$ and external forces $\underline{F}_{\alpha\alpha}$ are given, the micro motions could be solved from above equations

$$\underline{\mathcal{U}}_n^\ell = \mathcal{C}_{\alpha\alpha}^{-1}(\underline{F}_{\alpha\alpha} - \underline{F}_{\alpha\epsilon}) \tag{4.94}$$

The macro continuum problem becomes solvable after eliminating the micro unknowns $\underline{\mathcal{U}}_n^\ell$ by substituting Eq. (4.94) into Eq. (4.92). After merging similar terms, the macro continuum problem is recasted to

$$\begin{aligned}\int_{V_0} [\, & \delta\underline{\epsilon}_1^T(\widehat{\mathcal{C}}_{\epsilon\epsilon}^{11}\underline{\epsilon}_1 + \widehat{\mathcal{C}}_{\epsilon\epsilon}^{12}\underline{\epsilon}_2 + \widehat{\mathcal{C}}_{\epsilon\epsilon}^{13}\underline{\epsilon}_3 + \mathcal{C}_{\epsilon\alpha}^{11}\mathcal{C}_{\alpha\alpha}^{-1}\underline{F}_{\alpha\alpha}) \\ & + \delta\underline{\epsilon}_2^T(\widehat{\mathcal{C}}_{\epsilon\epsilon}^{21}\underline{\epsilon}_1 + \widehat{\mathcal{C}}_{\epsilon\epsilon}^{22}\underline{\epsilon}_2 + \widehat{\mathcal{C}}_{\epsilon\epsilon}^{23}\underline{\epsilon}_3 + \mathcal{C}_{\epsilon\alpha}^{22}\mathcal{C}_{\alpha\alpha}^{-1}\underline{F}_{\alpha\alpha}) \\ & + \delta\underline{\epsilon}_3^T(\widehat{\mathcal{C}}_{\epsilon\epsilon}^{31}\underline{\epsilon}_1 + \widehat{\mathcal{C}}_{\epsilon\epsilon}^{32}\underline{\epsilon}_2 + \widehat{\mathcal{C}}_{\epsilon\epsilon}^{33}\underline{\epsilon}_3 + \mathcal{C}_{\epsilon\alpha}^{33}\mathcal{C}_{\alpha\alpha}^{-1}\underline{F}_{\alpha\alpha}) - \underline{\delta\mathcal{U}}^T\underline{F}_{\epsilon\alpha}\,]\,dV_0 = 0\end{aligned} \tag{4.95}$$

where the constitutive matrix blocks are modified to

$$\begin{aligned}&\widehat{\mathcal{C}}_{\epsilon\epsilon}^{11} = \mathcal{C}_{\epsilon\epsilon}^{11} - \mathcal{C}_{\epsilon\alpha}^{11}\mathcal{C}_{\alpha\alpha}^{-1}\mathcal{C}_{\epsilon\alpha}^{11T},\ \widehat{\mathcal{C}}_{\epsilon\epsilon}^{12} = \mathcal{C}_{\epsilon\epsilon}^{12} - \mathcal{C}_{\epsilon\alpha}^{11}\mathcal{C}_{\alpha\alpha}^{-1}\mathcal{C}_{\epsilon\alpha}^{22T},\ \widehat{\mathcal{C}}_{\epsilon\epsilon}^{13} = \mathcal{C}_{\epsilon\epsilon}^{13} - \mathcal{C}_{\epsilon\alpha}^{11}\mathcal{C}_{\alpha\alpha}^{-1}\mathcal{C}_{\epsilon\alpha}^{33T} \\ &\widehat{\mathcal{C}}_{\epsilon\epsilon}^{21} = \mathcal{C}_{\epsilon\epsilon}^{21} - \mathcal{C}_{\epsilon\alpha}^{22}\mathcal{C}_{\alpha\alpha}^{-1}\mathcal{C}_{\epsilon\alpha}^{11T},\ \widehat{\mathcal{C}}_{\epsilon\epsilon}^{22} = \mathcal{C}_{\epsilon\epsilon}^{22} - \mathcal{C}_{\epsilon\alpha}^{22}\mathcal{C}_{\alpha\alpha}^{-1}\mathcal{C}_{\epsilon\alpha}^{22T},\ \widehat{\mathcal{C}}_{\epsilon\epsilon}^{23} = \mathcal{C}_{\epsilon\epsilon}^{23} - \mathcal{C}_{\epsilon\alpha}^{22}\mathcal{C}_{\alpha\alpha}^{-1}\mathcal{C}_{\epsilon\alpha}^{33T} \\ &\widehat{\mathcal{C}}_{\epsilon\epsilon}^{31} = \mathcal{C}_{\epsilon\epsilon}^{31} - \mathcal{C}_{\epsilon\alpha}^{33}\mathcal{C}_{\alpha\alpha}^{-1}\mathcal{C}_{\epsilon\alpha}^{11T},\ \widehat{\mathcal{C}}_{\epsilon\epsilon}^{32} = \mathcal{C}_{\epsilon\epsilon}^{32} - \mathcal{C}_{\epsilon\alpha}^{33}\mathcal{C}_{\alpha\alpha}^{-1}\mathcal{C}_{\epsilon\alpha}^{22T},\ \widehat{\mathcal{C}}_{\epsilon\epsilon}^{33} = \mathcal{C}_{\epsilon\epsilon}^{33} - \mathcal{C}_{\epsilon\alpha}^{33}\mathcal{C}_{\alpha\alpha}^{-1}\mathcal{C}_{\epsilon\alpha}^{33T}\end{aligned} \tag{4.96}$$

After introducing the stress measures as follow

$$\begin{aligned}\widehat{\underline{F}}_{\epsilon\epsilon}^{11} &= \widehat{\mathcal{C}}_{\epsilon\epsilon}^{11}\underline{\epsilon}_1 + \widehat{\mathcal{C}}_{\epsilon\epsilon}^{12}\underline{\epsilon}_2 + \widehat{\mathcal{C}}_{\epsilon\epsilon}^{13}\underline{\epsilon}_3 + \mathcal{C}_{\epsilon\alpha}^{11}\mathcal{C}_{\alpha\alpha}^{-1}\underline{F}_{\alpha\alpha} \\ \widehat{\underline{F}}_{\epsilon\epsilon}^{22} &= \widehat{\mathcal{C}}_{\epsilon\epsilon}^{22}\underline{\epsilon}_2 + \widehat{\mathcal{C}}_{\epsilon\epsilon}^{21}\underline{\epsilon}_1 + \widehat{\mathcal{C}}_{\epsilon\epsilon}^{23}\underline{\epsilon}_3 + \mathcal{C}_{\epsilon\alpha}^{22}\mathcal{C}_{\alpha\alpha}^{-1}\underline{F}_{\alpha\alpha} \\ \widehat{\underline{F}}_{\epsilon\epsilon}^{33} &= \widehat{\mathcal{C}}_{\epsilon\epsilon}^{33}\underline{\epsilon}_3 + \widehat{\mathcal{C}}_{\epsilon\epsilon}^{31}\underline{\epsilon}_1 + \widehat{\mathcal{C}}_{\epsilon\epsilon}^{32}\underline{\epsilon}_2 + \mathcal{C}_{\epsilon\alpha}^{33}\mathcal{C}_{\alpha\alpha}^{-1}\underline{F}_{\alpha\alpha}\end{aligned} \tag{4.97}$$

The macro continuum problem can be written in the compact formula like

$$\int_{V_0} (\delta\underline{\epsilon}_1^T\widehat{\underline{F}}_{\epsilon\epsilon}^{11} + \delta\underline{\epsilon}_2^T\widehat{\underline{F}}_{\epsilon\epsilon}^{22} + \delta\underline{\epsilon}_3^T\widehat{\underline{F}}_{\epsilon\epsilon}^{33} - \underline{\delta\mathcal{U}}^T\underline{F}_{\epsilon\alpha})dV_0 = 0 \tag{4.98}$$

Introducing Eq. (2.161) leads to

$$\begin{aligned}\int_{V_0} [&(\underline{\delta\mathcal{U}}_{,1} + \widetilde{\mathcal{K}}_1^1\underline{\delta\mathcal{U}})^T\widehat{\underline{F}}_{\epsilon\epsilon}^{11} + (\underline{\delta\mathcal{U}}_{,2} + \widetilde{\mathcal{K}}_2^1\underline{\delta\mathcal{U}})^T\widehat{\underline{F}}_{\epsilon\epsilon}^{22} \\ &+ (\underline{\delta\mathcal{U}}_{,3} + \widetilde{\mathcal{K}}_3^1\underline{\delta\mathcal{U}})^T\widehat{\underline{F}}_{\epsilon\epsilon}^{33} - \underline{\delta\mathcal{U}}^T\underline{F}_{\epsilon\alpha}]dV_0 = 0\end{aligned} \tag{4.99}$$

With the application of finite element discretization, the weak formula of the macro problem becomes

$$
\begin{aligned}
&\delta \underline{\mathcal{U}}_n^T \int_{V_0} [(\mathcal{H}_{\epsilon,1}^T + \mathcal{H}_\epsilon^T \widetilde{\mathcal{K}}_1^{1T})\underline{\widehat{F}}_{\epsilon\epsilon}^{11} + (\mathcal{H}_{\epsilon,2}^T + \mathcal{H}_\epsilon^T \widetilde{\mathcal{K}}_2^{1T})\underline{\widehat{F}}_{\epsilon\epsilon}^{22} \\
&+ (\mathcal{H}_{\epsilon,3}^T + \mathcal{H}_\epsilon^T \widetilde{\mathcal{K}}_3^{1T})\underline{\widehat{F}}_{\epsilon\epsilon}^{33} - \mathcal{H}_\epsilon^T \underline{F}_{\epsilon\alpha}] dV_0 = 0
\end{aligned}
\tag{4.100}
$$

From which, the governing equations are obtained to be

$$
\begin{aligned}
&\int_{V_0} [(\mathcal{H}_{\epsilon,1}^T + \mathcal{H}_\epsilon^T \widetilde{\mathcal{K}}_1^{1T})\underline{\widehat{F}}_{\epsilon\epsilon}^{11} + (\mathcal{H}_{\epsilon,2}^T + \mathcal{H}_\epsilon^T \widetilde{\mathcal{K}}_2^{1T})\underline{\widehat{F}}_{\epsilon\epsilon}^{22} \\
&+ (\mathcal{H}_{\epsilon,3}^T + \mathcal{H}_\epsilon^T \widetilde{\mathcal{K}}_3^{1T})\underline{\widehat{F}}_{\epsilon\epsilon}^{33} - \mathcal{H}_\epsilon^T \underline{F}_{\epsilon\alpha}] dV_0 = 0
\end{aligned}
\tag{4.101}
$$

After introduced the elastic force vector

$$
\begin{aligned}
\underline{\mathcal{F}}_e = &\int_{V_0} [(\mathcal{H}_{\epsilon,1}^T + \mathcal{H}_\epsilon^T \widetilde{\mathcal{K}}_1^{1T})\underline{\widehat{F}}_{\epsilon\epsilon}^{11} + (\mathcal{H}_{\epsilon,2}^T + \mathcal{H}_\epsilon^T \widetilde{\mathcal{K}}_2^{1T})\underline{\widehat{F}}_{\epsilon\epsilon}^{22} \\
&+ (\mathcal{H}_{\epsilon,3}^T + \mathcal{H}_\epsilon^T \widetilde{\mathcal{K}}_3^{1T})\underline{\widehat{F}}_{\epsilon\epsilon}^{33}] dV_0
\end{aligned}
\tag{4.102}
$$

and the external force vector

$$
\underline{\mathcal{F}}_a = \int_{V_0} \mathcal{H}_\epsilon^T \underline{F}_{\epsilon\alpha} dV_0 \tag{4.103}
$$

the governing equations can be described by the following compact formula

$$
\underline{\mathcal{F}}_e - \underline{\mathcal{F}}_a = 0 \tag{4.104}
$$

Even though the elastic forces, Eqs. (3.105) and (4.102), have the same expressions, the physical meaning is totally different. The first expressions, Eq. (3.105), are valid only for the Cosserat continuum, while the second expressions, Eq. (4.102), are the contributions of stain energy of multi-scale Cosserat continuum in the macro scale when the original model is decoupled into the macro and micro models in the sense of multiscale.

### 4.1.7 Extended to Multiscale Dynamic Problem

The velocity of an arbitrary material particle in a multiscale Cosserat continuum could be evaluated by taking time derivatives of position vector $\underline{P}$, defined by Eq. (4.21), to get

$$
\dot{\underline{P}} = (\underline{x}_0 + \underline{u}_s + R_1 \underline{S}_0)^{\cdot} = \dot{\underline{u}}_s + \dot{R}_1 \underline{S}_0 + R_1 \dot{\underline{S}}_0 \tag{4.105}
$$

Due to the fact that the scale of micro model are at least one order of magnitude smaller than that of the macro model, the velocity of micro continuum becomes negligible. Meanwhile, the centrifugal velocity is approximated to be

$$\dot{R}_1 \underline{S}_0 \approx \dot{R}_1 \underline{s}_0 \tag{4.106}$$

After resolved the time derivatives into the body attached frame $\mathcal{F}_1$, the particle velocity will relate to the linear and angular velocities of reference particle like

$$R_1^T \underline{\dot{P}} = R_1^T \underline{\dot{u}}_s + \tilde{s}_0^T \underline{\omega} \tag{4.107}$$

The kinematic energy of multiscale Cosserat continuum can be approximated by Eq. (3.111). The associated mass matrix is averaged by using the characteristic volume with the magnitude of $V_c$ in micro continuum like

$$\mathcal{M}_{\epsilon\epsilon} = \frac{1}{V_c} \int_{V_c} \begin{bmatrix} \rho I, & \rho \tilde{s}_0^T \\ \rho \tilde{s}_0, & \rho \tilde{s}_0 \tilde{s}_0^T \end{bmatrix} dV_c \tag{4.108}$$

The mass matrix and the Coriolis force vector are then defined by Eq. (3.121). When considering the inertial forces, the governing equations of motion for multiscale continuum will be decoupled into two sub-systems, one is dynamic macroscopic system

$$\mathcal{M}_n \underline{\dot{\mathcal{V}}}_n - \underline{\mathcal{F}}_g + \underline{\mathcal{F}}_e = \underline{\mathcal{F}}_a \tag{4.109}$$

and the other is the static microscopic system, Eq. (4.93).

## 4.2 Multiscale Shell-Like Theory

It has been clarified that the shell-like structure apparently has a "thin" dimension along the thickness significantly smaller than the other two geometric dimensions. Therefore, the shell-like structure is precisely defined as a two-dimensional Cosserat continuum with the deformation in the "thin" dimension ignored. The multiscale shell-like theory will take into account the deformation of the "thin" dimension at the micro scale, while keeping the deformation description of the original two-dimensional Cosserat continuum at the macro scale unchanged. The director of shell in the thickness direction will never be rigid under the framework of multiscale shell-like theory.

### *4.2.1 Kinematics of Multiscale Shell*

As shown in Fig. 4.2, a material particle with infinitesimal volume in a shell structure is depicted by its position and director vectors in its reference configuration and current deformed configuration, respectively. In the reference configuration, the

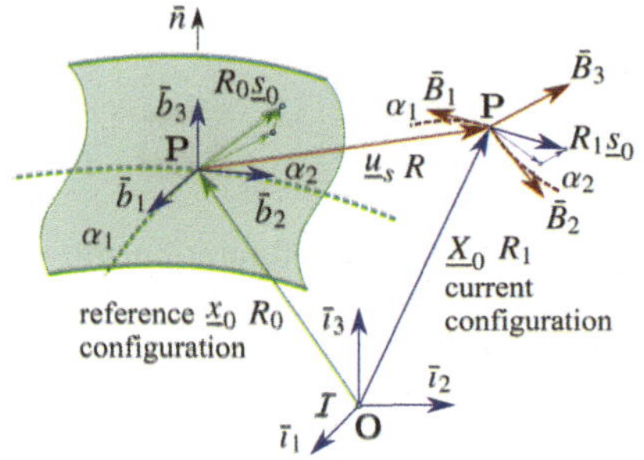


**Fig. 4.2** Kinematics of a material particle with infinitesimal volume in a multiscale shell

position vector $\underline{p}$ of the material particle relative to a point **O**, the original of reference frame, $\mathcal{F}_{\mathcal{I}} = [\bar{\mathbf{O}}, \mathcal{I}]$, is defined as a function of the general curvilinear coordinates, $(\alpha_1, \alpha_2, \alpha_3)$ like

$$\underline{p}(\alpha_1, \alpha_2, \alpha_3) = \underline{x}_0(\alpha_1, \alpha_2) + R_0(\alpha_1, \alpha_2)\underline{s}_0(\alpha_1, \alpha_2, \alpha_3) \tag{4.110}$$

where $\underline{x}_0$ is the position vector of a reference material particle on the middle surface of the shell relative to a point **O**, and $R_0$ denotes the initial rotation tensor measures the orientation of the reference basis $\mathcal{B}_0$. It has been verified in Sect. 2.2.3 the rotation tensor $R_0$ is identical to the body attached basis $\mathcal{B}_0$, and in detail

$$R_0 = \lfloor \bar{b}_1, \bar{b}_2, \bar{b}_3 \rfloor = \mathcal{B}_0 \tag{4.111}$$

Note that the initial basis $\mathcal{B}_0$ is bounded by the material particle **P** to produce the reference frame $\mathcal{F}_0 = [\mathbf{P}, \mathcal{B}_0]$. The local vector $\underline{s}_0$, defined in this reference frame $\mathcal{F}_0$, describes the director of a material particle. Obviously, Eq. (4.110) decomposes the position vector $\underline{p}$ of a material particle into a reference position $\underline{x}_0$ and a local vector $\underline{s}_0$. In classical shell theory, director $\underline{s}_0$ is a rigid vector. Now it becomes deformable under the framework of multiscale shell theory. If assuming the micro continuum to be the Cosserat continuum, which is usually the case, the director $\underline{s}_0$ can be defined in a similar approach to the definition of a position vector for a three-dimensional Cosserat continuum, Eq. (4.6), like

$$\underline{s}_0(\alpha_1, \alpha_2, \alpha_3) = \underline{x}_0^\ell(\alpha_1, \alpha_2, \alpha_3) + R_0^\ell(\alpha_1, \alpha_2, \alpha_3)\underline{s}_0^\ell(\alpha_1, \alpha_2, \alpha_3) \tag{4.112}$$

where $\underline{x}_0^\ell$ is the position vector of reference particle $\mathbf{P}^\ell$ in local Cosserat continuum, and $R_0^\ell$ the local rotation tensor attached to the reference particle $\mathbf{P}^\ell$ where the local director $\underline{s}_0^\ell$ is measured in the local basis spanned by $R_0^\ell$. In view of above formula, the position vector $\underline{p}$ can be determined from the following expression

$$\underline{p} = \underline{x}_0 + R_0(\underline{x}_0^\ell + R_0^\ell \underline{s}_0^\ell) \tag{4.113}$$

For the description of Cosserat continuum deformations, the covariant base vectors

$$\underline{g}_i = \frac{\partial \underline{p}}{\partial \alpha_i} \tag{4.114}$$

in the reference configuration are required, and the specific expressions can be obtained by taking derivatives of Eq. (4.110) with respected to the curvilinear coordinates $\alpha_i$ like

$$\begin{aligned}
\underline{g}_1 &= \underline{x}_{0,1} + R_{0,1}\underline{s}_0 + R_0\underline{s}_{0,1} \\
\underline{g}_2 &= \underline{x}_{0,2} + R_{0,2}\underline{s}_0 + R_0\underline{s}_{0,2} \\
\underline{g}_3 &= R_0\underline{s}_{0,3}
\end{aligned} \tag{4.115}$$

where the symbol $(\cdot)_{,i}$ for $i = 1, 2, 3$ denotes the spatial derivatives with respected to $\alpha_i$ and

$$(\cdot)_{,i} = \frac{\partial(\cdot)}{\partial \alpha_i} \tag{4.116}$$

Resolving the base vectors into the local reference frame $\mathcal{F}_0$ observes

$$\begin{aligned}
\underline{g}_1^* &= R_0^T \underline{x}_{0,1} + R_0^T R_{0,1}\underline{s}_0 + \underline{s}_{0,1} \\
\underline{g}_2^* &= R_0^T \underline{x}_{0,2} + R_0^T R_{0,2}\underline{s}_0 + \underline{s}_{0,2} \\
\underline{g}_3^* &= \underline{s}_{0,3}
\end{aligned} \tag{4.117}$$

According to the definitions of initial curvature tensors, Eq. (1.314), the base vectors can be represented by the following formulas like

$$\begin{aligned}
\underline{g}_1^* &= R_0^T \underline{x}_{0,1} + \widetilde{\kappa}_1^0\underline{s}_0 + \underline{s}_{0,1} \\
\underline{g}_2^* &= R_0^T \underline{x}_{0,2} + \widetilde{\kappa}_2^0\underline{s}_0 + \underline{s}_{0,2} \\
\underline{g}_3^* &= \underline{s}_{0,3}
\end{aligned} \tag{4.118}$$

where the symbols $\widetilde{\kappa}_i^0 = R_0^T R_{0,i}$ for $i = 1, 2$ denote the initial curvature tensors of a shell. All these base vectors form the covariant bases $\mathcal{B}_g$ like

$$\mathcal{B}_g = \lfloor \underline{g}_1^*,\ \underline{g}_2^*,\ \underline{g}_3^* \rfloor \tag{4.119}$$

and then the metric tensor in the reference configuration is determined from

$$g = \mathcal{B}_g^T \mathcal{B}_g \tag{4.120}$$

The similar manipulations can be performed repeatedly in the local Cosserat continuum to find the derivatives of director

$$\begin{aligned} \underline{s}_{0,1} &= \underline{x}^{\ell}_{0,1} + R^{\ell}_{0,1}\underline{s}^{\ell}_0 + R^{\ell}_0\underline{s}^{\ell}_{0,1} \\ \underline{s}_{0,2} &= \underline{x}^{\ell}_{0,2} + R^{\ell}_{0,2}\underline{s}^{\ell}_0 + R^{\ell}_0\underline{s}^{\ell}_{0,2} \\ \underline{s}_{0,3} &= \underline{x}^{\ell}_{0,3} + R^{\ell}_{0,3}\underline{s}^{\ell}_0 + R^{\ell}_0\underline{s}^{\ell}_{0,3} \end{aligned} \tag{4.121}$$

If defined the local curvature tensors as $\widetilde{\kappa}^{0\ell}_i = R^{\ell T}_0 R^{\ell}_{0,i}$ for $i = 1, 2, 3$, the local base vectors can be resolved into the local basis $R^{\ell}_0$ like

$$\begin{aligned} R^{\ell T}_0 \underline{s}_{0,1} &= R^{\ell T}_0 \underline{x}^{\ell}_{0,1} + \widetilde{\kappa}^{0\ell}_1 \underline{s}^{\ell}_0 + \underline{s}^{\ell}_{0,1} \\ R^{\ell T}_0 \underline{s}_{0,2} &= R^{\ell T}_0 \underline{x}^{\ell}_{0,2} + \widetilde{\kappa}^{0\ell}_2 \underline{s}^{\ell}_0 + \underline{s}^{\ell}_{0,2} \\ R^{\ell T}_0 \underline{s}_{0,3} &= R^{\ell T}_0 \underline{x}^{\ell}_{0,3} + \widetilde{\kappa}^{0\ell}_3 \underline{s}^{\ell}_0 + \underline{s}^{\ell}_{0,3} \end{aligned} \tag{4.122}$$

When the shell structure changes its spatial configuration, the material particle that had position vector $\underline{p}$ in the undeformed reference configuration now has position vector $\underline{P}$ in the deformed configuration, such that

$$\underline{P}(\alpha_1, \alpha_2, \alpha_3) = \underline{X}_0(\alpha_1, \alpha_2) + R_1(\alpha_1, \alpha_2)\underline{S}_0(\alpha_1, \alpha_2, \alpha_3) \tag{4.123}$$

where $\underline{S}_0$ represents the deformed director in current configuration, and $R_1$ the total rotation tensor that describes the orientation of basis $\mathcal{B}_1$ attached to the deformed configuration and

$$R_1 = \lfloor \bar{B}_1, \bar{B}_2, \bar{B}_3 \rfloor = \mathcal{B}_1 \tag{4.124}$$

If bounding the current deformed basis $\mathcal{B}_1$ by the material particle $\mathbf{P}$, the associated deformed frame $\mathcal{F}_1 = [\mathbf{P}, \mathcal{B}_1]$ can be constructed. Furthermore, if the displacement vector $\underline{u}_s$ is given to describe an arbitrary translation of reference particle, the reference position vector $\underline{X}_0$ in current deformed configuration can be recasted to

$$\underline{X}_0(\alpha_1, \alpha_2) = \underline{x}_0(\alpha_1, \alpha_2) + \underline{u}_s(\alpha_1, \alpha_2) \tag{4.125}$$

Similarly, the deformed director $\underline{S}_0$ is decomposed in the same approach like

$$\underline{S}_0(\alpha_1, \alpha_2, \alpha_3) = \underline{X}^{\ell}_0(\alpha_1, \alpha_2, \alpha_3) + R^{\ell}_1(\alpha_1, \alpha_2, \alpha_3)\underline{s}^{\ell}_0(\alpha_1, \alpha_2, \alpha_3) \tag{4.126}$$

where $\underline{X}^{\ell}_0$ is the deformed position vector of reference particle $\mathbf{P}^{\ell}$, and $R^{\ell}_1$ the deformed rotation tensor attached to the particle $\mathbf{P}$ in the local micro continuum. The position vector $\underline{X}^{\ell}_0$ can be related to its original position through

$$\underline{X}^{\ell}_0(\alpha_1, \alpha_2, \alpha_3) = \underline{x}^{\ell}_0(\alpha_1, \alpha_2, \alpha_3) + \underline{u}^{\ell}_s(\alpha_1, \alpha_2, \alpha_3) \tag{4.127}$$

where $\underline{u}^{\ell}_s$ is the displacement vector of local director in micro continuum. In view of the definitions of covariant base vectors in deformed configuration

$$\underline{G}_i = \frac{\partial \underline{P}}{\partial \alpha_i} \tag{4.128}$$

the specific formulas can be obtained by taking spatial derivatives of Eq. (4.123) with respected to the curvilinear coordinates $\alpha_i$ for $i = 1, 2, 3$ like

$$\begin{aligned} \underline{G}_1 &= \underline{X}_{0,1} + R_{1,1}\underline{S}_0 + R_1\underline{S}_{0,1} \\ \underline{G}_2 &= \underline{X}_{0,2} + R_{1,2}\underline{S}_0 + R_1\underline{S}_{0,2} \\ \underline{G}_3 &= R_1\underline{S}_{0,3} \end{aligned} \tag{4.129}$$

Once again, the base vectors are resolved into the current local frame $\mathcal{F}_1$ to obtain

$$\begin{aligned} \underline{G}_1^* &= R_1^T\underline{X}_{0,1} + R_1^T R_{1,1}\underline{S}_0 + R_1^T R_1\underline{S}_{0,1} \\ \underline{G}_2^* &= R_1^T\underline{X}_{0,2} + R_1^T R_{1,2}\underline{S}_0 + R_1^T R_1\underline{S}_{0,2} \\ \underline{G}_3^* &= R_1^T R_1\underline{S}_{0,3} \end{aligned} \tag{4.130}$$

In view of the definitions of curvature tensors $\widetilde{\kappa}_i^1 = R_1^T R_{1,i}$ for $i = 1, 2$ of a shell, the covariant base vectors are simplified to

$$\begin{aligned} \underline{G}_1^* &= R_1^T\underline{X}_{0,1} + \widetilde{\kappa}_1^1\underline{S}_0 + \underline{S}_{0,1} \\ \underline{G}_2^* &= R_1^T\underline{X}_{0,2} + \widetilde{\kappa}_2^1\underline{S}_0 + \underline{S}_{0,2} \\ \underline{G}_3^* &= \underline{S}_{0,3} \end{aligned} \tag{4.131}$$

All these base vectors could be assembled into the covariant bases $\mathcal{B}_G$ like

$$\mathcal{B}_G = \lfloor \underline{G}_1^*,\ \underline{G}_2^*,\ \underline{G}_3^* \rfloor \tag{4.132}$$

The expressions of the deformed covariant base vectors $\underline{S}_{0,i}$ for $i = 1, 2, 3$ can be obtained in a similar approach

$$\begin{aligned} \underline{S}_{0,1} &= \underline{X}_{0,1}^\ell + R_{1,1}^\ell\underline{s}_0^\ell + R_1^\ell\underline{s}_{0,1}^\ell \\ \underline{S}_{0,2} &= \underline{X}_{0,2}^\ell + R_{1,2}^\ell\underline{s}_0^\ell + R_1^\ell\underline{s}_{0,2}^\ell \\ \underline{S}_{0,3} &= \underline{X}_{0,3}^\ell + R_{1,3}^\ell\underline{s}_0^\ell + R_1^\ell\underline{s}_{0,3}^\ell \end{aligned} \tag{4.133}$$

Resolving these vectors of micro scale into the body attached frame produces

$$\begin{aligned} R_1^{\ell T}\underline{S}_{0,1} &= R_1^{\ell T}\underline{X}_{0,1}^\ell + \widetilde{\kappa}_1^{1\ell}\underline{s}_0^\ell + \underline{s}_{0,1}^\ell \\ R_1^{\ell T}\underline{S}_{0,2} &= R_1^{\ell T}\underline{X}_{0,2}^\ell + \widetilde{\kappa}_2^{1\ell}\underline{s}_0^\ell + \underline{s}_{0,2}^\ell \\ R_1^{\ell T}\underline{S}_{0,3} &= R_1^{\ell T}\underline{X}_{0,3}^\ell + \widetilde{\kappa}_3^{1\ell}\underline{s}_0^\ell + \underline{s}_{0,3}^\ell \end{aligned} \tag{4.134}$$

where $\widetilde{\kappa}_i^{1\ell} = R_1^{\ell T} R_{1,i}^{\ell}$ for $i = 1, 2, 3$ defines the deformed curvature tensors for local Cosserat continuum.

### 4.2.2 *Strain Tensor of Multiscale Shell*

Based on the assumption of small strains, the components of Green-Lagrange strain tensor can be approximated by using the difference of base vectors or so-called base vector increments. Referring the detailed expressions of Green-Lagrange strain tensor, Eq. (3.44), the same formulations are still valid for the shell structure

$$\begin{aligned} \varepsilon_{11} &= \underline{g}_1^{*T} \Delta \underline{G}_1^* \\ 2\varepsilon_{12} &= \underline{g}_1^{*T} \Delta \underline{G}_2^* + \underline{g}_2^{*T} \Delta \underline{G}_1^* \\ 2\varepsilon_{13} &= \underline{g}_1^{*T} \Delta \underline{G}_3^* + \underline{g}_3^{*T} \Delta \underline{G}_1^* \\ \varepsilon_{22} &= \underline{g}_2^{*T} \Delta \underline{G}_2^* \\ 2\varepsilon_{23} &= \underline{g}_2^{*T} \Delta \underline{G}_3^* + \underline{g}_3^{*T} \Delta \underline{G}_2^* \\ \varepsilon_{33} &= \underline{g}_3^{*T} \Delta \underline{G}_3^* \end{aligned} \tag{4.135}$$

where the assumption of small strains has been replaced by the assumption of negligible dot products, $\Delta \underline{G}_i^{*T} \Delta \underline{G}_i^* \approx 0$ for $i = 1, 2, 3$, equivalently. By merging the similar terms, above equations can be readily assembled into the vectorial expressions

$$\underline{\varepsilon}_1 = \mathcal{B}_g^T \Delta \underline{G}_1^*, \ \underline{\varepsilon}_2 = \mathcal{B}_g^T \Delta \underline{G}_2^*, \ \underline{\varepsilon}_3 = \mathcal{B}_g^T \Delta \underline{G}_3^* \tag{4.136}$$

where the vectorial strain measures $\underline{\varepsilon}_i$ for $i = 1, 2, 3$ are defined in Chap. 3 by Eq. (3.72). Before specified the explicit formulations of the Green-Lagrange strain tensor, it is required to evaluate the differences between the base vectors $\underline{G}_i^*$ and $\underline{g}_i^*$ for $i = 1, 2, 3$ at first. By Eqs. (4.118) and (4.131), the detailed formulas of $\Delta \underline{G}_i^*$ are found to be

$$\begin{aligned} \Delta \underline{G}_1^* &= \underline{G}_1^* - \underline{g}_1^* = R_1^T \underline{X}_{0,1} - R_0^T \underline{x}_{0,1} + \widetilde{\kappa}_1^1 \underline{S}_0 - \widetilde{\kappa}_1^0 \underline{s}_0 + \underline{S}_{0,1} - \underline{s}_{0,1} \\ \Delta \underline{G}_2^* &= \underline{G}_2^* - \underline{g}_2^* = R_2^T \underline{X}_{0,2} - R_0^T \underline{x}_{0,2} + \widetilde{\kappa}_2^1 \underline{S}_0 - \widetilde{\kappa}_2^0 \underline{s}_0 + \underline{S}_{0,2} - \underline{s}_{0,2} \\ \Delta \underline{G}_3^* &= \underline{G}_3^* - \underline{g}_3^* = \underline{S}_{0,3} - \underline{s}_{0,3} \end{aligned} \tag{4.137}$$

Minor modifications lead to

$$\begin{aligned} \Delta \underline{G}_1^* &= R_1^T \underline{X}_{0,1} - R_0^T \underline{x}_{0,1} + (\widetilde{\kappa}_1^1 - \widetilde{\kappa}_1^0) \underline{s}_0 + \widetilde{\kappa}_1^1 (\underline{S}_0 - \underline{s}_0) + \underline{S}_{0,1} - \underline{s}_{0,1} \\ \Delta \underline{G}_2^* &= R_1^T \underline{X}_{0,2} - R_0^T \underline{x}_{0,2} + (\widetilde{\kappa}_2^1 - \widetilde{\kappa}_2^0) \underline{s}_0 + \widetilde{\kappa}_2^1 (\underline{S}_0 - \underline{s}_0) + \underline{S}_{0,2} - \underline{s}_{0,2} \\ \Delta \underline{G}_2^* &= \underline{S}_{0,2} - \underline{s}_{0,2} \end{aligned} \tag{4.138}$$

By Eq. (2.152), the definitions of Lagrangian stretch and Lagrange curvature measures [31, 32], the base vector increments can be represented by

$$\begin{aligned}
\Delta \underline{G}_1^* &= \lfloor I,\ \widetilde{s}_0^T \rfloor \underline{\epsilon}_1 + \widetilde{\kappa}_1^1 (\underline{S}_0 - \underline{s}_0) + \underline{S}_{0,1} - \underline{s}_{0,1} \\
\Delta \underline{G}_2^* &= \lfloor I,\ \widetilde{s}_0^T \rfloor \underline{\epsilon}_2 + \widetilde{\kappa}_2^1 (\underline{S}_0 - \underline{s}_0) + \underline{S}_{0,2} - \underline{s}_{0,2} \\
\Delta \underline{G}_3^* &= \underline{S}_{0,3} - \underline{s}_{0,3}
\end{aligned} \tag{4.139}$$

The same relations hold in the case of micro Cosserat continuum, such that

$$\begin{aligned}
R_1^{\ell T} \underline{S}_{0,1} - R_0^{\ell T} \underline{s}_{0,1} &= \lfloor I,\ \widetilde{s}_0^{\ell T} \rfloor \underline{\epsilon}_1^\ell \\
R_1^{\ell T} \underline{S}_{0,2} - R_0^{\ell T} \underline{s}_{0,2} &= \lfloor I,\ \widetilde{s}_0^{\ell T} \rfloor \underline{\epsilon}_2^\ell \\
R_1^{\ell T} \underline{S}_{0,3} - R_0^{\ell T} \underline{s}_{0,3} &= \lfloor I,\ \widetilde{s}_0^{\ell T} \rfloor \underline{\epsilon}_3^\ell
\end{aligned} \tag{4.140}$$

where $\underline{\epsilon}_i^\ell$ for $i = 1, 2, 3$ are the Lagrangian stretch and curvature strain measures defined by Eq. (2.152) in the micro scale

$$\underline{\epsilon}_i^\ell = \left| \begin{matrix} R_1^{\ell T} \underline{X}_{0,i}^\ell - R_0^{\ell T} \underline{x}_{0,i}^\ell \\ \underline{\kappa}_i^{1\ell} - \underline{\kappa}_i^{0\ell} \end{matrix} \right| \tag{4.141}$$

With the aid of above relations, Eq. (4.140), the incremental base vectors $\Delta \underline{G}_i^*$ for $i = 1, 2, 3$ of a shell are recasted to

$$\begin{aligned}
\Delta \underline{G}_1^* &= \lfloor I,\ \widetilde{s}_0^T \rfloor \underline{\epsilon}_1 + R_0^\ell \lfloor I,\ \widetilde{s}_0^{\ell T} \rfloor \underline{\epsilon}_1^\ell + \widetilde{\kappa}_1^1 (\underline{S}_0 - \underline{s}_0) + (I - R_0^\ell R_1^{\ell T}) \underline{S}_{0,1} \\
\Delta \underline{G}_2^* &= \lfloor I,\ \widetilde{s}_0^T \rfloor \underline{\epsilon}_2 + R_0^\ell \lfloor I,\ \widetilde{s}_0^{\ell T} \rfloor \underline{\epsilon}_2^\ell + \widetilde{\kappa}_2^1 (\underline{S}_0 - \underline{s}_0) + (I - R_0^\ell R_1^{\ell T}) \underline{S}_{0,2} \\
\Delta \underline{G}_3^* &= R_0^\ell \lfloor I,\ \widetilde{s}_0^{\ell T} \rfloor \underline{\epsilon}_3^\ell + (I - R_0^\ell R_1^{\ell T}) \underline{S}_{0,3}
\end{aligned} \tag{4.142}$$

Due to the complexity of above expressions, the theory of multiscale shell is limited to linear strain description with the assumptions of small strain as mentioned before, such that $\underline{\epsilon}_i \ll 1$ for $i = 1, 2, 3$. This assumption is also suitable for micro Cosserat continuum. Observations also find that the microscopic motion, including displacement $\underline{u}_s^\ell$ and rotation $R^\ell = R_0^{\ell T} R_1^\ell$, are at least one order of magnitude smaller than the macroscopic motions, which also establishes the fundamental of scale separation. It shows that $\underline{u}_s^\ell \ll 1$ and $R^\ell = I + \widetilde{\phi}^\ell$, where $\underline{\phi}^\ell$ is the rotation angle of micro scale and $\underline{\phi}^\ell \ll 1$. According to these observations, the base vector increments finally can be approximated by

$$\begin{aligned}
\Delta \underline{G}_1^* &= \lfloor I,\ \widetilde{s}_0^T \rfloor \underline{\epsilon}_1 + R_0^\ell \lfloor I,\ \widetilde{s}_0^{\ell T} \rfloor \underline{\epsilon}_1^\ell + \widetilde{\kappa}_1^0 (\underline{u}_s^\ell + R_0^\ell \widetilde{s}_0^{\ell T} \underline{\phi}^\ell) + \widetilde{s}_{0,1}^T R_0^\ell \underline{\phi}^\ell \\
\Delta \underline{G}_2^* &= \lfloor I,\ \widetilde{s}_0^T \rfloor \underline{\epsilon}_2 + R_0^\ell \lfloor I,\ \widetilde{s}_0^{\ell T} \rfloor \underline{\epsilon}_2^\ell + \widetilde{\kappa}_2^0 (\underline{u}_s^\ell + R_0^\ell \widetilde{s}_0^{\ell T} \underline{\phi}^\ell) + \widetilde{s}_{0,2}^T R_0^\ell \underline{\phi}^\ell \\
\Delta \underline{G}_3^* &= R_0^\ell \lfloor I,\ \widetilde{s}_0^{\ell T} \rfloor \underline{\epsilon}_3^\ell + \widetilde{s}_{0,3}^T R_0^\ell \underline{\phi}^\ell
\end{aligned} \tag{4.143}$$

More operations as to these simplifications can be found in Sect. 4.1.2. Introduced the micro displacements in its Plucker coordinates

$$\underline{\mathcal{U}}^\ell = \begin{vmatrix} R_0^{\ell T} \underline{u}_s^\ell \\ \underline{\phi}^\ell \end{vmatrix} \tag{4.144}$$

the base vector increments can be rewritten in terms of

$$\begin{aligned}
\Delta \underline{G}_1^* &= \lfloor I,\ \widetilde{s}_0^T \rfloor \underline{\epsilon}_1 + R_0^\ell \lfloor I,\ \widetilde{s}_0^{\ell T} \rfloor \underline{\epsilon}_1^\ell + \lfloor \widetilde{\kappa}_1^0 R_0^\ell,\ \widetilde{\kappa}_1^0 R_0^\ell \widetilde{s}_0^{\ell T} + \widetilde{s}_{0,1}^T R_0^\ell \rfloor \underline{\mathcal{U}}^\ell \\
\Delta \underline{G}_2^* &= \lfloor I,\ \widetilde{s}_0^T \rfloor \underline{\epsilon}_2 + R_0^\ell \lfloor I,\ \widetilde{s}_0^{\ell T} \rfloor \underline{\epsilon}_2^\ell + \lfloor \widetilde{\kappa}_2^0 R_0^\ell,\ \widetilde{\kappa}_2^0 R_0^\ell \widetilde{s}_0^{\ell T} + \widetilde{s}_{0,2}^T R_0^\ell \rfloor \underline{\mathcal{U}}^\ell \\
\Delta \underline{G}_3^* &= R_0^\ell \lfloor I,\ \widetilde{s}_0^{\ell T} \rfloor \underline{\epsilon}_3^\ell + \lfloor 0,\ \widetilde{s}_{0,3}^T R_0^\ell \rfloor \underline{\mathcal{U}}^\ell
\end{aligned} \tag{4.145}$$

and then the compact expressions are obtained

$$\begin{aligned}
\Delta \underline{G}_1^* &= \lfloor I,\ \widetilde{s}_0^T \rfloor \underline{\epsilon}_1 + R_0^\ell \lfloor I,\ \widetilde{s}_0^{\ell T} \rfloor \underline{\epsilon}_1^\ell + C_1^\ell \underline{\mathcal{U}}^\ell \\
\Delta \underline{G}_2^* &= \lfloor I,\ \widetilde{s}_0^T \rfloor \underline{\epsilon}_2 + R_0^\ell \lfloor I,\ \widetilde{s}_0^{\ell T} \rfloor \underline{\epsilon}_2^\ell + C_2^\ell \underline{\mathcal{U}}^\ell \\
\Delta \underline{G}_3^* &= R_0^\ell \lfloor I,\ \widetilde{s}_0^{\ell T} \rfloor \underline{\epsilon}_3^\ell + C_3^\ell \underline{\mathcal{U}}^\ell
\end{aligned} \tag{4.146}$$

with the aid of the following definitions

$$C_i^\ell = \lfloor \widetilde{\kappa}_i^0 R_0^\ell,\ \widetilde{\kappa}_i^0 R_0^\ell \widetilde{s}_0^{\ell T} + \widetilde{s}_{0,i}^T R_0^\ell \rfloor,\ C_3^\ell = \lfloor 0,\ \widetilde{s}_{0,3}^T R_0^\ell \rfloor \tag{4.147}$$

for $i = 1, 2$. By above simplifications, the vectorial Green-Lagrange strain measures of a shell structure can be verified to be

$$\begin{aligned}
\underline{\varepsilon}_1 &= \mathcal{B}_g^T \lfloor I,\ \widetilde{s}_0^T \rfloor \underline{\epsilon}_1 + \mathcal{B}_g^T R_0^\ell \lfloor I,\ \widetilde{s}_0^{\ell T} \rfloor \underline{\epsilon}_1^\ell + \mathcal{B}_g^T C_1^\ell \underline{\mathcal{U}}^\ell \\
\underline{\varepsilon}_2 &= \mathcal{B}_g^T \lfloor I,\ \widetilde{s}_0^T \rfloor \underline{\epsilon}_2 + \mathcal{B}_g^T R_0^\ell \lfloor I,\ \widetilde{s}_0^{\ell T} \rfloor \underline{\epsilon}_2^\ell + \mathcal{B}_g^T C_2^\ell \underline{\mathcal{U}}^\ell \\
\underline{\varepsilon}_3 &= \mathcal{B}_g^T R_0^\ell \lfloor I,\ \widetilde{s}_0^{\ell T} \rfloor \underline{\epsilon}_3^\ell + \mathcal{B}_g^T C_3^\ell \underline{\mathcal{U}}^\ell
\end{aligned} \tag{4.148}$$

After introduced the following symbols

$$\mathcal{S}_0^T = \mathcal{B}_g^T \lfloor I,\ \widetilde{s}_0^T \rfloor,\ \mathcal{S}_0^{\ell T} = \mathcal{B}_g^T R_0^\ell \lfloor I,\ \widetilde{s}_0^{\ell T} \rfloor \tag{4.149}$$

and

$$\mathcal{S}_i^{\ell T} = \mathcal{B}_g^T C_i^\ell \tag{4.150}$$

for $i = 1, 2, 3$, the Green-Lagrange strain tensor can be rewritten as the functions of Lagrangian stretch and curvature strain measures together with the local motions in the vectorial forms like

$$\begin{aligned}\underline{\varepsilon}_1 &= \mathcal{S}_0^T \underline{\epsilon}_1 + \mathcal{S}_0^{\ell T} \underline{\epsilon}_1^\ell + \mathcal{S}_1^{\ell T} \underline{\mathcal{U}}^\ell \\ \underline{\varepsilon}_2 &= \mathcal{S}_0^T \underline{\epsilon}_2 + \mathcal{S}_0^{\ell T} \underline{\epsilon}_2^\ell + \mathcal{S}_2^{\ell T} \underline{\mathcal{U}}^\ell \\ \underline{\varepsilon}_3 &= \mathcal{S}_0^{\ell T} \underline{\epsilon}_3^\ell + \mathcal{S}_3^{\ell T} \underline{\mathcal{U}}^\ell\end{aligned} \tag{4.151}$$

In most of the case initial curvatures are zero, $\widetilde{\kappa}_i^0 = 0$ for $i = 1, 2, 3$, the above vectorial strain measures of a shell can be decoupled into two parts approximately

$$\begin{aligned}\underline{\varepsilon}_1 &= \mathcal{S}_0^T \underline{\epsilon}_1 + \mathcal{S}_0^{\ell T} \underline{\epsilon}_1^\ell \\ \underline{\varepsilon}_2 &= \mathcal{S}_0^T \underline{\epsilon}_2 + \mathcal{S}_0^{\ell T} \underline{\epsilon}_2^\ell \\ \underline{\varepsilon}_3 &= \mathcal{S}_0^{\ell T} \underline{\epsilon}_3^\ell\end{aligned} \tag{4.152}$$

after truncated the third terms containing the motions of micro scale. The detailed explanations as to the truncations can be found in Sect. 4.1.2.

### 4.2.3 Constitutive Laws for Multiscale Shell

Given the volume with the magnitude of $V_0$ for a shell structure, the strain energy $\mathcal{U}_e$ can be represented by using the vectorial Green-Lagrange strain measures like

$$\begin{aligned}2\mathcal{U}_e = \int_{V_0} &[\underline{\varepsilon}_1^T \mathcal{D}_{11} \underline{\varepsilon}_1 + \underline{\varepsilon}_2^T \mathcal{D}_{22} \underline{\varepsilon}_2 + \underline{\varepsilon}_3^T \mathcal{D}_{33} \underline{\varepsilon}_3 \\ &+ 2(\underline{\varepsilon}_2^T \mathcal{D}_{21} \underline{\varepsilon}_1 + \underline{\varepsilon}_3^T \mathcal{D}_{31} \underline{\varepsilon}_1 + \underline{\varepsilon}_3^T \mathcal{D}_{32} \underline{\varepsilon}_2)]\, dV_0\end{aligned} \tag{4.153}$$

where the constitutive matrix blocks $\mathcal{D}_{ij}$ for $i, j = 1, 2, 3$ are defined in Sect. 3.1.4. By Eq. (4.152), the vectorial Green-Lagrange strains $\underline{\varepsilon}_i$ is decomposed into two components, the Lagrangian stretch and curvature measures, $\underline{\epsilon}_i$ and $\underline{\epsilon}_i^\ell$ for the macroscopic and microscopic models respectively, each item of the above full descriptions of strain energy for a shell can be written in detail as

$$\begin{aligned}\underline{\varepsilon}_i^T \mathcal{D}_{ij} \underline{\varepsilon}_j &= \lfloor \underline{\epsilon}_i^T, \ \underline{\epsilon}_i^{\ell T} \rfloor \begin{bmatrix} \mathcal{S}_0 \mathcal{D}_{ij} \mathcal{S}_0^T & \mathcal{S}_0 \mathcal{D}_{ij} \mathcal{S}_0^{\ell T} \\ \mathcal{S}_0^\ell \mathcal{D}_{ij} \mathcal{S}_0^T & \mathcal{S}_0^\ell \mathcal{D}_{ij} \mathcal{S}_0^{\ell T} \end{bmatrix} \begin{vmatrix} \underline{\epsilon}_j \\ \underline{\epsilon}_j^\ell \end{vmatrix} \\ &= \lfloor \underline{\epsilon}_i^T, \ \underline{\epsilon}_i^{\ell T} \rfloor \begin{bmatrix} \mathcal{D}_{\epsilon\epsilon}^{ij} & \mathcal{D}_{\epsilon\alpha}^{ij} \\ \mathcal{D}_{\alpha\epsilon}^{ij} & \mathcal{D}_{\alpha\alpha}^{ij} \end{bmatrix} \begin{vmatrix} \underline{\epsilon}_j \\ \underline{\epsilon}_j^\ell \end{vmatrix}\end{aligned} \tag{4.154}$$

where the constitutive matrices for two-dimensional Cosserat continuum are defined

$$\begin{aligned}\mathcal{D}_{\epsilon\epsilon}^{ij} &= \mathcal{S}_0 \mathcal{D}_{ij} \mathcal{S}_0^T, \ \mathcal{D}_{\epsilon\alpha}^{ij} = \mathcal{S}_0 \mathcal{D}_{ij} \mathcal{S}_0^{\ell T} \\ \mathcal{D}_{\alpha\epsilon}^{ij} &= \mathcal{S}_0^\ell \mathcal{D}_{ij} \mathcal{S}_0^T, \ \mathcal{D}_{\alpha\alpha}^{ij} = \mathcal{S}_0^\ell \mathcal{D}_{ij} \mathcal{S}_0^{\ell T}\end{aligned} \tag{4.155}$$

for $i, j = 1, 2, 3$. When $i, j = 3$, a nonzero constitutive matrix block is clarified to be

$$\mathcal{D}_{\alpha\alpha}^{33} = \mathcal{S}_0^\ell \mathcal{D}_{33} \mathcal{S}_0^{\ell T} \tag{4.156}$$

Due to the symmetry of constitutive matrices, it is readily verified

$$\begin{aligned} \mathcal{D}_{\epsilon\epsilon}^{ijT} &= \mathcal{S}_0 \mathcal{D}_{ij}^T \mathcal{S}_0^T = \mathcal{S}_0 \mathcal{D}_{ji} \mathcal{S}_0^T = \mathcal{D}_{\epsilon\epsilon}^{ji} \\ \mathcal{D}_{\alpha\epsilon}^{ijT} &= \mathcal{S}_0 \mathcal{D}_{ij}^T \mathcal{S}_0^{\ell T} = \mathcal{S}_0 \mathcal{D}_{ji} \mathcal{S}_0^{\ell T} = \mathcal{D}_{\epsilon\alpha}^{ji} \end{aligned} \tag{4.157}$$

then all the terms of stain energy for a shell can be written explicitly

$$\begin{aligned} \underline{\varepsilon}_1^T \mathcal{D}_{11} \underline{\varepsilon}_1 &= \underline{\epsilon}_1^T \mathcal{D}_{\epsilon\epsilon}^{11} \underline{\epsilon}_1 + 2\underline{\epsilon}_1^T \mathcal{D}_{\epsilon\alpha}^{11} \underline{\epsilon}_1^\ell + \underline{\epsilon}_1^{\ell T} \mathcal{D}_{\alpha\alpha}^{11} \underline{\epsilon}_1^\ell \\ \underline{\varepsilon}_2^T \mathcal{D}_{22} \underline{\varepsilon}_2 &= \underline{\epsilon}_2^T \mathcal{D}_{\epsilon\epsilon}^{22} \underline{\epsilon}_2 + 2\underline{\epsilon}_2^T \mathcal{D}_{\epsilon\alpha}^{22} \underline{\epsilon}_2^\ell + \underline{\epsilon}_2^{\ell T} \mathcal{D}_{\alpha\alpha}^{22} \underline{\epsilon}_2^\ell \\ \underline{\varepsilon}_3^T \mathcal{D}_{33} \underline{\varepsilon}_3 &= \underline{\epsilon}_3^{\ell T} \mathcal{D}_{\alpha\alpha}^{33} \underline{\epsilon}_3^\ell \end{aligned} \tag{4.158}$$

and

$$\begin{aligned} \underline{\varepsilon}_2^T \mathcal{D}_{21} \underline{\varepsilon}_1 &= \underline{\epsilon}_2^T \mathcal{D}_{\epsilon\epsilon}^{21} \underline{\epsilon}_1 + \underline{\epsilon}_2^T \mathcal{D}_{\epsilon\alpha}^{21} \underline{\epsilon}_1^\ell + \underline{\epsilon}_1^T \mathcal{D}_{\epsilon\alpha}^{12} \underline{\epsilon}_2^\ell + \underline{\epsilon}_2^{\ell T} \mathcal{D}_{\alpha\alpha}^{21} \underline{\epsilon}_1^\ell \\ \underline{\varepsilon}_3^T \mathcal{D}_{31} \underline{\varepsilon}_1 &= \underline{\epsilon}_1^T \mathcal{D}_{\epsilon\alpha}^{13} \underline{\epsilon}_3^\ell + \underline{\epsilon}_3^{\ell T} \mathcal{D}_{\alpha\alpha}^{31} \underline{\epsilon}_1^\ell \\ \underline{\varepsilon}_3^T \mathcal{D}_{32} \underline{\varepsilon}_2 &= \underline{\epsilon}_2^T \mathcal{D}_{\epsilon\alpha}^{23} \underline{\epsilon}_3^\ell + \underline{\epsilon}_3^{\ell T} \mathcal{D}_{\alpha\alpha}^{32} \underline{\epsilon}_2^\ell \end{aligned} \tag{4.159}$$

If the characteristic volume with the magnitude of $V_c$ for a micro Cosserat continuum is given, the strain energy per characteristic volume can be written in detail like

$$\begin{aligned} \mathcal{U}_c = \int_{V_c} [\ & \frac{1}{2}(\underline{\epsilon}_1^T \mathcal{D}_{\epsilon\epsilon}^{11} \underline{\epsilon}_1 + \underline{\epsilon}_2^T \mathcal{D}_{\epsilon\epsilon}^{22} \underline{\epsilon}_2) + \underline{\epsilon}_2^T \mathcal{D}_{\epsilon\epsilon}^{21} \underline{\epsilon}_1 \\ & + \frac{1}{2}(\underline{\epsilon}_1^{\ell T} \mathcal{D}_{\alpha\alpha}^{11} \underline{\epsilon}_1^\ell + \underline{\epsilon}_2^{\ell T} \mathcal{D}_{\alpha\alpha}^{22} \underline{\epsilon}_2^\ell + \underline{\epsilon}_3^{\ell T} \mathcal{D}_{\alpha\alpha}^{33} \underline{\epsilon}_3^\ell) \\ & + \underline{\epsilon}_2^{\ell T} \mathcal{D}_{\alpha\alpha}^{21} \underline{\epsilon}_1^\ell + \underline{\epsilon}_3^{\ell T} \mathcal{D}_{\alpha\alpha}^{31} \underline{\epsilon}_1^\ell + \underline{\epsilon}_3^{\ell T} \mathcal{D}_{\alpha\alpha}^{32} \underline{\epsilon}_2^\ell \\ & + \underline{\epsilon}_1^T \mathcal{D}_{\epsilon\alpha}^{11} \underline{\epsilon}_1^\ell + \underline{\epsilon}_1^T \mathcal{D}_{\epsilon\alpha}^{12} \underline{\epsilon}_2^\ell + \underline{\epsilon}_1^T \mathcal{D}_{\epsilon\alpha}^{13} \underline{\epsilon}_3^\ell \\ & + \underline{\epsilon}_2^T \mathcal{D}_{\epsilon\alpha}^{21} \underline{\epsilon}_1^\ell + \underline{\epsilon}_2^T \mathcal{D}_{\epsilon\alpha}^{22} \underline{\epsilon}_2^\ell + \underline{\epsilon}_2^T \mathcal{D}_{\epsilon\alpha}^{23} \underline{\epsilon}_3^\ell\ ]\, dV_c \end{aligned} \tag{4.160}$$

The purpose of introducing the strain energy per characteristic volume is to produce the equivalent homogeneous stiffness matrices for a shell as discussed below. According to the definitions of Lagrangian stretch and curvature strain measures in the micro scale, Eq. (4.141), the local Lagrangian stretch and curvature strains could be explicitly related to the local motions and the derivatives under the assumption of small microscopic motion like

$$\underline{\epsilon}_i^\ell = \underline{\mathcal{U}}_{,i}^\ell + \widetilde{\mathcal{K}}_i^{0\ell} \underline{\mathcal{U}}^\ell \tag{4.161}$$

for $i = 1, 2, 3$ with the following introduced symbols

$$\underline{\mathcal{U}}^{\ell}_{,i} = \begin{vmatrix} (R_0^{\ell T}\underline{u}_s^{\ell})_{,i} \\ \underline{\phi}^{\ell}_{,i} \end{vmatrix}, \; \widetilde{\mathcal{K}}_i^{0\ell} = \begin{bmatrix} \widetilde{\kappa}_i^{0\ell} & R_0^{\ell T}\widetilde{\underline{x}_{0,i}} \\ 0 & \widetilde{\kappa}_i^{0\ell} \end{bmatrix} \tag{4.162}$$

Since the motions of micro scale are small, the micro motions can form a six-dimensional linear space approximately. Hence, the classical finite element method can be directly applied to the micro model. Given the shape functions $\mathcal{H}_{\epsilon}^{\ell}$, the micro motions and their derivatives are interpolated by

$$\underline{\mathcal{U}}^{\ell} = \mathcal{H}_{\epsilon}^{\ell}\underline{\mathcal{U}}_n^{\ell}, \; \underline{\mathcal{U}}^{\ell}_{,i} = \mathcal{H}_{\epsilon,i}^{\ell}\underline{\mathcal{U}}_n^{\ell} \tag{4.163}$$

where $\underline{\mathcal{U}}_n^{\ell}$ denotes nodal motions for micro Cosserat continuum. As a result, the Lagrangian strain measures can be discretized into

$$\underline{\epsilon}_i^{\ell} = (\mathcal{H}_{\epsilon,i}^{\ell} + \widetilde{\mathcal{K}}_i^{0\ell}\mathcal{H}_{\epsilon}^{\ell})\underline{\mathcal{U}}_n^{\ell} = \mathcal{Q}_i^{\ell}\underline{\mathcal{U}}_n^{\ell} \tag{4.164}$$

Under the assumption of scale separation, the global strains $\underline{\epsilon}_i$ for $i = 1, 2$ of a shell are independent of local variables. Hence, the strain energy per characteristic volume with the amplitude of $V_c$ for a micro Cosserat continuum can be rewritten as

$$\begin{aligned}
\mathcal{U}_c = {} & \frac{1}{2}(\underline{\epsilon}_1^T \int_{V_c} \mathcal{D}_{\epsilon\epsilon}^{11}\, dV_c \underline{\epsilon}_1 + \underline{\epsilon}_2^T \int_{V_c} \mathcal{D}_{\epsilon\epsilon}^{22}\, dV_c \underline{\epsilon}_2) + \underline{\epsilon}_2^T \int_{V_c} \mathcal{D}_{\epsilon\epsilon}^{21}\, dV_c \underline{\epsilon}_1 \\
& + \frac{1}{2}\underline{\mathcal{U}}_n^{\ell T} \int_{V_c} (\mathcal{Q}_1^{\ell T}\mathcal{D}_{\alpha\alpha}^{11}\mathcal{Q}_1^{\ell} + \mathcal{Q}_2^{\ell T}\mathcal{D}_{\alpha\alpha}^{22}\mathcal{Q}_2^{\ell} + \mathcal{Q}_3^{\ell T}\mathcal{D}_{\alpha\alpha}^{33}\mathcal{Q}_3^{\ell})\, dV_c \underline{\mathcal{U}}_n^{\ell} \\
& + \underline{\mathcal{U}}_n^{\ell T} \int_{V_c} (\mathcal{Q}_2^{\ell T}\mathcal{D}_{\alpha\alpha}^{21}\mathcal{Q}_1^{\ell} + \mathcal{Q}_3^{\ell T}\mathcal{D}_{\alpha\alpha}^{31}\mathcal{Q}_1^{\ell} + \mathcal{Q}_3^{\ell T}\mathcal{D}_{\alpha\alpha}^{32}\mathcal{Q}_2^{\ell})\, dV_c \underline{\mathcal{U}}_n^{\ell} \\
& + \underline{\epsilon}_1^T \int_{V_c} (\mathcal{D}_{\epsilon\alpha}^{11}\mathcal{Q}_1^{\ell} + \mathcal{D}_{\epsilon\alpha}^{12}\mathcal{Q}_2^{\ell} + \mathcal{D}_{\epsilon\alpha}^{13}\mathcal{Q}_3^{\ell})\, dV_c \underline{\mathcal{U}}_n^{\ell} \\
& + \underline{\epsilon}_2^T \int_{V_c} (\mathcal{D}_{\epsilon\alpha}^{21}\mathcal{Q}_1^{\ell} + \mathcal{D}_{\epsilon\alpha}^{22}\mathcal{Q}_2^{\ell} + \mathcal{D}_{\epsilon\alpha}^{23}\mathcal{Q}_3^{\ell})\, dV_c \underline{\mathcal{U}}_n^{\ell}
\end{aligned} \tag{4.165}$$

The strain energy per unit area, called strain energy density for a shell, can be defined as the area average of above strain energy

$$\begin{aligned}
\mathcal{U}_\rho &= \frac{\mathcal{U}_C}{S_c} \\
&= \frac{1}{2}(\underline{\epsilon}_1^T C_{\epsilon\epsilon}^{11}\underline{\epsilon}_1 + \underline{\epsilon}_2^T C_{\epsilon\epsilon}^{22}\underline{\epsilon}_2) + \underline{\epsilon}_2^T C_{\epsilon\epsilon}^{21}\underline{\epsilon}_1 \\
&\quad + \frac{1}{2}\underline{\mathcal{U}}_n^{\ell T} C_{\alpha\alpha}\underline{\mathcal{U}}_n^{\ell} + \underline{\epsilon}_1^T C_{\epsilon\alpha}^{11}\underline{\mathcal{U}}_n^{\ell} + \underline{\epsilon}_2^T C_{\epsilon\alpha}^{22}\underline{\mathcal{U}}_n^{\ell}
\end{aligned} \tag{4.166}$$

where the area is evaluated by $S_c = V_c/h$, and the equivalent homogeneous stiffness matrices of the shell structures are computed from the weighted integration [34] over the characteristic area $S_c$ of micro scale

$$\mathcal{C}^{11}_{\epsilon\epsilon} = \frac{\int_{V_c} \mathcal{D}^{11}_{\epsilon\epsilon} dV_c}{S_c}, \; \mathcal{C}^{22}_{\epsilon\epsilon} = \frac{\int_{V_c} \mathcal{D}^{22}_{\epsilon\epsilon} dV_c}{S_c}, \; \mathcal{C}^{21}_{\epsilon\epsilon} = \frac{\int_{V_c} \mathcal{D}^{21}_{\epsilon\epsilon} dV_c}{S_c} \tag{4.167}$$

$$\begin{aligned} \mathcal{C}_{\alpha\alpha} &= \frac{\int_{V_c} \mathcal{Q}^{\ell T}_1 \mathcal{D}^{11}_{\alpha\alpha} \mathcal{Q}^{\ell}_1 + \mathcal{Q}^{\ell T}_2 \mathcal{D}^{22}_{\alpha\alpha} \mathcal{Q}^{\ell}_2 + \mathcal{Q}^{\ell T}_3 \mathcal{D}^{33}_{\alpha\alpha} \mathcal{Q}^{\ell}_3 dV_c}{S_c} \\ &+ 2\frac{\int_{V_c} \mathcal{Q}^{\ell T}_2 \mathcal{D}^{21}_{\alpha\alpha} \mathcal{Q}^{\ell}_1 + \mathcal{Q}^{\ell T}_3 \mathcal{D}^{31}_{\alpha\alpha} \mathcal{Q}^{\ell}_1 + \mathcal{Q}^{\ell T}_3 \mathcal{D}^{32}_{\alpha\alpha} \mathcal{Q}^{\ell}_2 dV_c}{S_c} \end{aligned} \tag{4.168}$$

$$\begin{aligned} \mathcal{C}^{11}_{\epsilon\alpha} &= \frac{\int_{V_c} \mathcal{D}^{11}_{\epsilon\alpha} \mathcal{Q}^{\ell}_1 + \mathcal{D}^{12}_{\epsilon\alpha} \mathcal{Q}^{\ell}_2 + \mathcal{D}^{13}_{\epsilon\alpha} \mathcal{Q}^{\ell}_3 dV_c}{S_c} \\ \mathcal{C}^{22}_{\epsilon\alpha} &= \frac{\int_{V_c} \mathcal{D}^{21}_{\epsilon\alpha} \mathcal{Q}^{\ell}_1 + \mathcal{D}^{22}_{\epsilon\alpha} \mathcal{Q}^{\ell}_2 + \mathcal{D}^{23}_{\epsilon\alpha} \mathcal{Q}^{\ell}_3 dV_c}{S_c} \end{aligned} \tag{4.169}$$

Note that the symmetric properties of stiffness matrices are still hold, such that

$$\mathcal{C}^{21T}_{\epsilon\epsilon} = \frac{\int_{V_c} \mathcal{D}^{21T}_{\epsilon\epsilon} dV_c}{S_c} = \frac{\int_{V_c} \mathcal{D}^{12}_{\epsilon\epsilon} dV_c}{S_c} = \mathcal{C}^{12}_{\epsilon\epsilon} \tag{4.170}$$

### 4.2.4 Variation of Strain Energy of Multiscale Shell

Since the stiffness matrix blocks are determined by the above averaged quantities on the micro Cosserat continuum, the strain energy for a shell structure with the given area of $S_0$ can be written as

$$\begin{aligned} \mathcal{U}_e = \int_{S_0} \mathcal{U}_\rho dS_0 = \int_{S_0} [\frac{1}{2}(\underline{\epsilon}_1^T \mathcal{C}^{11}_{\epsilon\epsilon} \underline{\epsilon}_1 + \underline{\epsilon}_2^T \mathcal{C}^{22}_{\epsilon\epsilon} \underline{\epsilon}_2) + \underline{\epsilon}_2^T \mathcal{C}^{21}_{\epsilon\epsilon} \underline{\epsilon}_1 \\ + \underline{\epsilon}_1^T \mathcal{C}^{11}_{\epsilon\alpha} \underline{\mathcal{U}}^{\ell}_n + \underline{\epsilon}_2^T \mathcal{C}^{22}_{\epsilon\alpha} \underline{\mathcal{U}}^{\ell}_n + \frac{1}{2} \underline{\mathcal{U}}^{\ell T}_n \mathcal{C}_{\alpha\alpha} \underline{\mathcal{U}}^{\ell}_n ] dS_0 \end{aligned} \tag{4.171}$$

where the area $S_0$ relates to the volume $V_0$ by $S_0 = V_0/h$. Taking variation of averaged strain energy density leads to

$$\begin{aligned} \delta\mathcal{U}_\rho &= \delta\underline{\epsilon}_1^T (\mathcal{C}^{11}_{\epsilon\epsilon} \underline{\epsilon}_1 + \mathcal{C}^{12}_{\epsilon\epsilon} \underline{\epsilon}_2) + \delta\underline{\epsilon}_1^T \mathcal{C}^{11}_{\epsilon\alpha} \underline{\mathcal{U}}^{\ell}_n \\ &+ \delta\underline{\epsilon}_2^T (\mathcal{C}^{21}_{\epsilon\epsilon} \underline{\epsilon}_1 + \mathcal{C}^{22}_{\epsilon\epsilon} \underline{\epsilon}_2) + \delta\underline{\epsilon}_2^T \mathcal{C}^{22}_{\epsilon\alpha} \underline{\mathcal{U}}^{\ell}_n \\ &+ \delta\underline{\mathcal{U}}^{\ell T}_n (\mathcal{C}^{11T}_{\epsilon\alpha} \underline{\epsilon}_1 + \mathcal{C}^{22T}_{\epsilon\alpha} \underline{\epsilon}_2) + \delta\underline{\mathcal{U}}^{\ell T}_n \mathcal{C}_{\alpha\alpha} \underline{\mathcal{U}}^{\ell}_n \end{aligned} \tag{4.172}$$

After introducing the following stress measures

$$\begin{aligned} \underline{F}^{11}_{\epsilon\epsilon} &= \mathcal{C}^{11}_{\epsilon\epsilon}\underline{\epsilon}_1 + \mathcal{C}^{12}_{\epsilon\epsilon}\underline{\epsilon}_2, \\ \underline{F}^{22}_{\epsilon\epsilon} &= \mathcal{C}^{21}_{\epsilon\epsilon}\underline{\epsilon}_1 + \mathcal{C}^{22}_{\epsilon\epsilon}\underline{\epsilon}_2, \\ \underline{F}_{\alpha\epsilon} &= \mathcal{C}^{11T}_{\epsilon\alpha}\underline{\epsilon}_1 + \mathcal{C}^{22T}_{\epsilon\alpha}\underline{\epsilon}_2 \end{aligned} \tag{4.173}$$

The variations of strain energy density can be written in compact form like

$$\delta\mathcal{U}_\rho = \delta\underline{\epsilon}_1^T(\underline{F}^{11}_{\epsilon\epsilon} + \mathcal{C}^{11}_{\epsilon\alpha}\underline{\mathcal{U}}^\ell_n) + \delta\underline{\epsilon}_2^T(\underline{F}^{22}_{\epsilon\epsilon} + \mathcal{C}^{22}_{\epsilon\alpha}\underline{\mathcal{U}}^\ell_n) + \delta\underline{\mathcal{U}}^{\ell T}_n(\underline{F}_{\alpha\epsilon} + \mathcal{C}_{\alpha\alpha}\underline{\mathcal{U}}^\ell_n) \tag{4.174}$$

From which, the variation of strain energy $\mathcal{U}_e$ is readily evaluated

$$\delta\mathcal{U}_e = \int_{S_0} \delta\mathcal{U}_\rho dS_0 \tag{4.175}$$

## 4.2.5 Virtual Work of External Forces for Multiscale Shell

Different slightly form the definitions of external force, $\underline{F}_{\epsilon\alpha}$ and $\underline{F}_{\alpha\alpha}$, Eq. (4.88), the averaged quantities of external forces are modified to

$$\begin{aligned} \underline{F}_{\epsilon\alpha} &= \begin{vmatrix} \dfrac{\int_{V_c}\underline{g}\,dV_c + \int_{S_c}\underline{p}_s\,dS_c}{S_c} \\ \dfrac{\int_{V_c}\widetilde{s}_0\underline{g}\,dV_c + \int_{S_c}\widetilde{s}_0\underline{p}_s\,dS_c}{S_c} \end{vmatrix} \\ \underline{F}_{\alpha\alpha} &= \frac{1}{S_c}\left(\int_{V_c}\mathcal{H}^{\ell T}_\epsilon \begin{vmatrix} R_0^{\ell T}\underline{g} \\ \widetilde{s}_0^\ell R_0^{\ell T}\underline{g} \end{vmatrix} dV_c + \int_{S_c}\mathcal{H}^{\ell T}_\epsilon \begin{vmatrix} R_0^{\ell T}\underline{p}_s \\ \widetilde{s}_0^\ell R_0^{\ell T}\underline{p}_s \end{vmatrix} dS_c\right) \end{aligned} \tag{4.176}$$

With the application of averaged external force defined above, the work of external force applied to a shell can be written as

$$\delta\mathcal{W}_e = \int_{S_0} \underline{\delta\mathcal{U}}^T\underline{F}_{\epsilon\alpha} + \underline{\delta\mathcal{U}}^{\ell T}_n\underline{F}_{\alpha\alpha} dS_0 \tag{4.177}$$

### 4.2.6 Governing Equations of Multiscale Shell in Motion Formalism

By substituting Eqs. (4.174) and (4.177) into the principle of virtual work, which states

$$\delta\mathcal{U}_e - \delta\mathcal{W}_e = 0 \tag{4.178}$$

the governing equations of multiscale shell can be readily obtained like

$$\begin{aligned}\int_{S_0} [\, & \delta\underline{\epsilon}_1^T(\underline{F}_{\epsilon\epsilon}^{11} + \mathcal{C}_{\epsilon\alpha}^{11}\underline{\mathcal{U}}_n^{\ell}) + \delta\underline{\epsilon}_2^T(\underline{F}_{\epsilon\epsilon}^{22} + \mathcal{C}_{\epsilon\alpha}^{22}\underline{\mathcal{U}}_n^{\ell}) \\ & + \delta\underline{\mathcal{U}}_n^{\ell T}(\underline{F}_{\alpha\epsilon} + \mathcal{C}_{\alpha\alpha}\underline{\mathcal{U}}_n^{\ell}) - \delta\underline{\mathcal{U}}^T\underline{F}_{\epsilon\alpha} - \delta\underline{\mathcal{U}}_n^{\ell T}\underline{F}_{\alpha\alpha}\,]\,dS_0 = 0\end{aligned} \tag{4.179}$$

Since the variations of macro strain measures are independent of the variations of micro motions, the original problem can be decomposed into a macroscopic problem

$$\int_{S_0} (\delta\underline{\epsilon}_1^T(\underline{F}_{\epsilon\epsilon}^{11} + \mathcal{C}_{\epsilon\alpha}^{11}\underline{\mathcal{U}}_n^{\ell}) + \delta\underline{\epsilon}_2^T(\underline{F}_{\epsilon\epsilon}^{22} + \mathcal{C}_{\epsilon\alpha}^{22}\underline{\mathcal{U}}_n^{\ell}) - \delta\underline{\mathcal{U}}^T\underline{F}_{\epsilon\alpha})dS_0 = 0 \tag{4.180}$$

and a linearized microscopic problem

$$\mathcal{C}_{\alpha\alpha}\underline{\mathcal{U}}_n^{\ell} + \underline{F}_{\alpha\epsilon} - \underline{F}_{\alpha\alpha} = 0 \tag{4.181}$$

with appropriate boundary conditions. Once the elastic forces $\underline{F}_{\alpha\epsilon}$ and external forces $\underline{F}_{\alpha\alpha}$ are given, the micro motions could be solved from above equations as

$$\underline{\mathcal{U}}_n^{\ell} = \mathcal{C}_{\alpha\alpha}^{-1}(\underline{F}_{\alpha\alpha} - \underline{F}_{\alpha\epsilon}) \tag{4.182}$$

The macro continuum problem becomes solvable after eliminating the micro unknowns $\underline{\mathcal{U}}_n^{\ell}$ by substituting Eq. (4.182) into Eq. (4.180). After merging similar terms, the macro continuum problem is recasted to

$$\begin{aligned}\int_{S_0} [\, & \delta\underline{\epsilon}_1^T(\widehat{\mathcal{C}}_{\epsilon\epsilon}^{11}\underline{\epsilon}_1 + \widehat{\mathcal{C}}_{\epsilon\epsilon}^{12}\underline{\epsilon}_2 + \mathcal{C}_{\epsilon\alpha}^{11}\mathcal{C}_{\alpha\alpha}^{-1}\underline{F}_{\alpha\alpha}) \\ & + \delta\underline{\epsilon}_2^T(\widehat{\mathcal{C}}_{\epsilon\epsilon}^{21}\underline{\epsilon}_1 + \widehat{\mathcal{C}}_{\epsilon\epsilon}^{22}\underline{\epsilon}_2 + \mathcal{C}_{\epsilon\alpha}^{22}\mathcal{C}_{\alpha\alpha}^{-1}\underline{F}_{\alpha\alpha}) - \delta\underline{\mathcal{U}}^T\underline{F}_{\epsilon\alpha}\,]\,dS_0 = 0\end{aligned} \tag{4.183}$$

where the stiffness matrix blocks are modified to

$$\begin{aligned}\widehat{\mathcal{C}}_{\epsilon\epsilon}^{11} &= \mathcal{C}_{\epsilon\epsilon}^{11} - \mathcal{C}_{\epsilon\alpha}^{11}\mathcal{C}_{\alpha\alpha}^{-1}\mathcal{C}_{\epsilon\alpha}^{11T}, \quad \widehat{\mathcal{C}}_{\epsilon\epsilon}^{12} = \mathcal{C}_{\epsilon\epsilon}^{12} - \mathcal{C}_{\epsilon\alpha}^{11}\mathcal{C}_{\alpha\alpha}^{-1}\mathcal{C}_{\epsilon\alpha}^{22T} \\ \widehat{\mathcal{C}}_{\epsilon\epsilon}^{21} &= \mathcal{C}_{\epsilon\epsilon}^{21} - \mathcal{C}_{\epsilon\alpha}^{22}\mathcal{C}_{\alpha\alpha}^{-1}\mathcal{C}_{\epsilon\alpha}^{11T}, \quad \widehat{\mathcal{C}}_{\epsilon\epsilon}^{22} = \mathcal{C}_{\epsilon\epsilon}^{22} - \mathcal{C}_{\epsilon\alpha}^{22}\mathcal{C}_{\alpha\alpha}^{-1}\mathcal{C}_{\epsilon\alpha}^{22T}\end{aligned} \tag{4.184}$$

After introduced the following stress measures

$$\begin{aligned}\underline{\widehat{F}}^{11}_{\epsilon\epsilon} &= \widehat{\mathcal{C}}^{11}_{\epsilon\epsilon}\underline{\epsilon}_1 + \widehat{\mathcal{C}}^{12}_{\epsilon\epsilon}\underline{\epsilon}_2 + \mathcal{C}^{11}_{\epsilon\alpha}\mathcal{C}^{-1}_{\alpha\alpha}\underline{F}_{\alpha\alpha} \\ \underline{\widehat{F}}^{22}_{\epsilon\epsilon} &= \widehat{\mathcal{C}}^{22}_{\epsilon\epsilon}\underline{\epsilon}_2 + \widehat{\mathcal{C}}^{21}_{\epsilon\epsilon}\underline{\epsilon}_1 + \mathcal{C}^{22}_{\epsilon\alpha}\mathcal{C}^{-1}_{\alpha\alpha}\underline{F}_{\alpha\alpha}\end{aligned} \tag{4.185}$$

The macro continuum problem can be written in the compact formula like

$$\int_{S_0} (\delta\underline{\epsilon}_1^T \underline{\widehat{F}}^{11}_{\epsilon\epsilon} + \delta\underline{\epsilon}_2^T \underline{\widehat{F}}^{22}_{\epsilon\epsilon} - \underline{\delta\mathcal{U}}^T \underline{F}_{\epsilon\alpha}) dS_0 = 0 \tag{4.186}$$

Introducing compatibility conditions, Eq. (2.161), leads to

$$\int_{S_0} \left[ (\underline{\delta\mathcal{U}}_{,1} + \widetilde{\mathcal{K}}^1_1 \underline{\delta\mathcal{U}})^T \underline{\widehat{F}}^{11}_{\epsilon\epsilon} + (\underline{\delta\mathcal{U}}_{,2} + \widetilde{\mathcal{K}}^1_2 \underline{\delta\mathcal{U}})^T \underline{\widehat{F}}^{22}_{\epsilon\epsilon} - \underline{\delta\mathcal{U}}^T \underline{F}_{\epsilon\alpha} \right] dS_0 = 0 \tag{4.187}$$

With the application of finite element discretization, the weak formula of the macro problem becomes

$$\begin{aligned}\underline{\delta\mathcal{U}}_n^T \int_{S_0} [(\mathcal{H}^T_{\epsilon,1} + \mathcal{H}^T_\epsilon \widetilde{\mathcal{K}}^{1T}_1)\underline{\widehat{F}}^{11}_{\epsilon\epsilon} \\ + (\mathcal{H}^T_{\epsilon,2} + \mathcal{H}^T_\epsilon \widetilde{\mathcal{K}}^{1T}_2)\underline{\widehat{F}}^{22}_{\epsilon\epsilon} - \mathcal{H}^T_\epsilon \underline{F}_{\epsilon\alpha}] dS_0 = 0\end{aligned} \tag{4.188}$$

From which, the governing equations are obtained to be

$$\int_{S_0} [(\mathcal{H}^T_{\epsilon,1} + \mathcal{H}^T_\epsilon \widetilde{\mathcal{K}}^{1T}_1)\underline{\widehat{F}}^{11}_{\epsilon\epsilon} + (\mathcal{H}^T_{\epsilon,2} + \mathcal{H}^T_\epsilon \widetilde{\mathcal{K}}^{1T}_2)\underline{\widehat{F}}^{22}_{\epsilon\epsilon} - \mathcal{H}^T_\epsilon \underline{F}_{\epsilon\alpha}] dS_0 = 0 \tag{4.189}$$

After introduced the elastic force vector

$$\underline{\mathcal{F}}_e = \int_{S_0} \left[ (\mathcal{H}^T_{\epsilon,1} + \mathcal{H}^T_\epsilon \widetilde{\mathcal{K}}^{1T}_1)\underline{\widehat{F}}^{11}_{\epsilon\epsilon} + (\mathcal{H}^T_{\epsilon,2} + \mathcal{H}^T_\epsilon \widetilde{\mathcal{K}}^{1T}_2)\underline{\widehat{F}}^{22}_{\epsilon\epsilon} \right] dS_0 \tag{4.190}$$

and the external force vector

$$\underline{\mathcal{F}}_a = \int_{S_0} \mathcal{H}^T_\epsilon \underline{F}_{\epsilon\alpha} dS_0 \tag{4.191}$$

the governing equations can be described by the following compact formula

$$\underline{\mathcal{F}}_e - \underline{\mathcal{F}}_a = 0 \tag{4.192}$$

Even though the elastic forces, Eqs. (3.105) and (4.190), have the same expressions, the physical meaning is totally different. The first expression is valid only for the shell structure, the second expressions are the contributions of dummy macro model when the original shell problem is decoupled into the macroscopic problem and microscopic problem in the sense of multiscale.

### *4.2.7 Extended to Multiscale Shell Dynamic Problem*

The velocity of an arbitrary material particle in a multiscale shell structure could be evaluated by taking time derivatives of position vector $\underline{P}$, defined by Eq. (4.123), to get

$$\dot{\underline{P}} = (\underline{x}_0 + \underline{u}_s + R_1 \underline{S}_0)^{\cdot} = \dot{\underline{u}}_s + \dot{R}_1 \underline{S}_0 + R_1 \dot{\underline{S}}_0 \tag{4.193}$$

Under the assumption of scale separation, the scales of micro continuum are at least one order of magnitude smaller than those of macro continuum, the velocity of micro continuum becomes negligible. Meanwhile, the centrifugal velocity can be approximated by

$$\dot{R}_1 \underline{S}_0 \approx \dot{R}_1 \underline{s}_0 \tag{4.194}$$

After resolved the time derivatives into the body attached frame $\mathcal{F}_1$, the particle velocity will relate to the linear and angular velocities of reference particle like

$$R_1^T \dot{\underline{P}} = R_1^T \dot{\underline{u}}_s + \widetilde{s}_0^T \underline{\omega} \tag{4.195}$$

The kinematic energy of multiscale shell structure can be approximated by Eq. (3.248). The associated material matrix is averaged by the weighted integration [34] over the characteristic area with the magnitude of $S_c$ for a micro Cosserat continuum

$$\mathcal{M}_{\epsilon\epsilon} = \frac{1}{S_c} \int_{V_c} \begin{bmatrix} \rho I, & \rho \widetilde{s}_0^T \\ \rho \widetilde{s}_0, & \rho \widetilde{s}_0 \widetilde{s}_0^T \end{bmatrix} dV_c \tag{4.196}$$

The mass matrix and the Coriolis force vector are then defined by Eq. (3.255). When considering the inertial forces, the governing equations of motion for multiscale shell will be decoupled into two sub-systems, one is dynamic macroscopic system

$$\mathcal{M}_n \dot{\underline{\mathcal{V}}}_n - \underline{\mathcal{F}}_g + \underline{\mathcal{F}}_e = \underline{\mathcal{F}}_a \tag{4.197}$$

and the other is the static microscopic system, Eq. (4.181).

## 4.3 A Special Multiscale Shell

A special multiscale shell is presented in this section, which the micro Cosserat continuum is limited to one-dimensional Cosserat continuum. In the real situation, micro Cosserat continuum could be a combination of continua with dimensions varying from one to three. Typically, three-dimensional Cosserat will be the most representative continuum that could be reduced to six different types of bodies, as discussed in Chap. 3.

### *4.3.1 Kinematics of Multiscale Shell with 5 DOFS*

As shown in Fig. 4.3, a material particle with infinitesimal volume in a shell structure is depicted by its position and director vectors in its reference configuration and current deformed configuration, respectively. For the special multiscale shell, the normal vector of the shell mid-surface is selected as the director $\underline{s}_0$, such that

$$\underline{s}_0(\alpha_3) = \underline{x}_0^\ell(\alpha_3) + R_0^\ell(\alpha_3)\underline{s}_0^\ell(\alpha_3) \tag{4.198}$$

where $\underline{x}_0^\ell$ is the position vector of reference particle $\mathbf{P}^\ell$ in local Cosserat continuum, and $R_0^\ell$ the local rotation tensor attached to the reference particle $\mathbf{P}^\ell$ where the local director $\underline{s}_0^\ell$ is measured in the local basis spanned by $R_0^\ell$. In view of above formulation, the position vector $\underline{p}$ can be determined from

$$\underline{p}(\alpha_1, \alpha_2, \alpha_3) = \underline{x}_0(\alpha_1, \alpha_2) + R_0(\alpha_1, \alpha_2)\left[\underline{x}_0^\ell(\alpha_3) + R_0^\ell(\alpha_3)\underline{s}_0^\ell(\alpha_3)\right] \tag{4.199}$$

For the description of shell deformation, the covariant base vectors with the definition of

$$\underline{g}_i = \frac{\partial \underline{p}}{\partial \alpha_i} \tag{4.200}$$

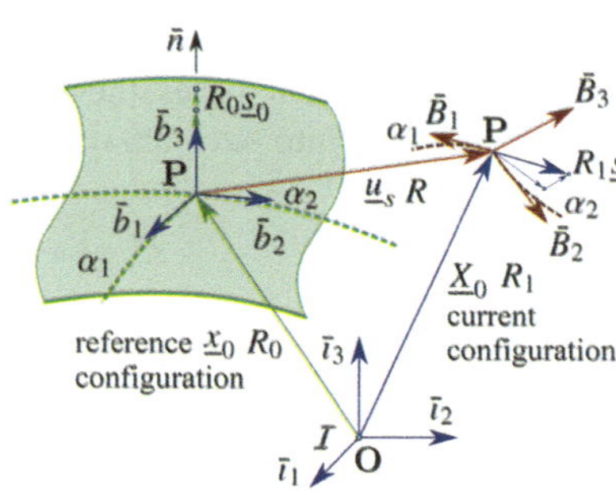


**Fig. 4.3** Kinematics of a material particle with infinitesimal volume in a special multiscale shell

in the reference configuration are required, and the specific expressions can be obtained by taking derivatives of position vector $\underline{p}$ with respected to the curvilinear coordinates $\alpha_i$ like

$$\begin{aligned} \underline{g}_1 &= \underline{x}_{0,1} + R_{0,1}\underline{s}_0 \\ \underline{g}_2 &= \underline{x}_{0,2} + R_{0,2}\underline{s}_0 \\ \underline{g}_3 &= R_0\underline{s}_{0,3} \end{aligned} \tag{4.201}$$

where the symbol $(\cdot)_{,i}$ for $i = 1, 2, 3$ denotes the spatial derivatives with respected to $\alpha_i$ and

$$(\cdot)_{,i} = \frac{\partial(\cdot)}{\partial\alpha_i} \tag{4.202}$$

Resolving the base vectors into the local reference frame $\mathcal{F}_0$ leads to

$$\begin{aligned} \underline{g}_1^* &= R_0^T\underline{x}_{0,1} + R_0^T R_{0,1}\underline{s}_0 \\ \underline{g}_2^* &= R_0^T\underline{x}_{0,2} + R_0^T R_{0,2}\underline{s}_0 \\ \underline{g}_3^* &= \underline{s}_{0,3} \end{aligned} \tag{4.203}$$

According to the definitions of initial curvature tensors, Eq. (1.314), the base vectors can be represented by the following formulas like

$$\begin{aligned} \underline{g}_1^* &= R_0^T\underline{x}_{0,1} + \widetilde{\kappa}_1^0\underline{s}_0 \\ \underline{g}_2^* &= R_0^T\underline{x}_{0,2} + \widetilde{\kappa}_2^0\underline{s}_0 \\ \underline{g}_3^* &= \underline{s}_{0,3} \end{aligned} \tag{4.204}$$

where the symbols $\widetilde{\kappa}_i^0 = R_0^T R_{0,i}$ for $i = 1, 2$ denote the initial curvature tensors of a shell. All these base vectors assemble into the covariant bases $\mathcal{B}_g$ like

$$\mathcal{B}_g = \lfloor \underline{g}_1^*,\ \underline{g}_2^*,\ \underline{g}_3^* \rfloor \tag{4.205}$$

and then the metric tensor in the reference configuration is computed from

$$g = \mathcal{B}_g^T\mathcal{B}_g \tag{4.206}$$

The similar manipulations can be performed repeatedly in the local Cosserat continuum to obtain the derivatives of director

$$\underline{s}_{0,3} = \underline{x}_{0,3}^\ell + R_{0,3}^\ell\underline{s}_0^\ell + R_0^\ell\underline{s}_{0,3}^\ell \tag{4.207}$$

If defined the local curvature tensors as $\widetilde{\kappa}_3^{0\ell} = R_0^{\ell T} R_{0,3}^\ell$, the local base vector $\underline{s}_{0,3}$ can be resolved into the local basis like

$$R_0^{\ell T}\underline{s}_{0,3} = R_0^{\ell T}\underline{x}_{0,3}^{\ell} + \widetilde{\kappa}_3^{0\ell}\underline{s}_0^{\ell} + \underline{s}_{0,3}^{\ell} \tag{4.208}$$

When the shell structure changes its spatial configuration, the material particle that had position vector $\underline{p}$ in the undeformed reference configuration now has position vector $\underline{P}$ in the deformed configuration, such that

$$\underline{P}(\alpha_1, \alpha_2, \alpha_3) = \underline{X}_0(\alpha_1, \alpha_2) + R_1(\alpha_1, \alpha_2)\underline{S}_0(\alpha_3) \tag{4.209}$$

where $\underline{S}_0$ represents the deformed director in current configuration, and $R_1$ the total rotation tensor that describes the orientation of basis $\mathcal{B}_1$ attached to the deformed configuration and

$$R_1 = \lfloor \bar{B}_1, \bar{B}_2, \bar{B}_3 \rfloor = \mathcal{B}_1 \tag{4.210}$$

After bounding the current deformed basis $\mathcal{B}_1$ by the material particle $\mathbf{P}$, the associated deformed frame $\mathcal{F}_1 = [\mathbf{P}, \mathcal{B}_1]$ is constructed. Furthermore, the deformed director $\underline{s}_0$ can be decomposed into

$$\underline{S}_0(\alpha_3) = \underline{X}_0^{\ell}(\alpha_3) + R_1^{\ell}(\alpha_3)\underline{S}_0^{\ell}(\alpha_3) \tag{4.211}$$

where $\underline{X}_0^{\ell}$ is the deformed position vector of reference particle $\mathbf{P}^{\ell}$, and $R_1^{\ell}$ the deformed rotation tensor in the micro continuum. The position vector $\underline{X}_0^{\ell}$ relates to its original position through

$$\underline{X}_0^{\ell}(\alpha_3) = \underline{x}_0^{\ell}(\alpha_3) + \underline{u}_s^{\ell}(\alpha_3) \tag{4.212}$$

where $\underline{u}_s^{\ell}$ is the displacement vector of local director in micro continuum. In view of the definitions of covariant base vectors in deformed configuration

$$\underline{G}_i = \frac{\partial \underline{P}}{\partial \alpha_i} \tag{4.213}$$

the specific formulas are obtained by taking spatial derivatives of Eq. (4.209) with respected to the curvilinear coordinates $\alpha_i$ for $i = 1, 2, 3$ like

$$\begin{aligned} \underline{G}_1 &= \underline{X}_{0,1} + R_{1,1}\underline{S}_0 \\ \underline{G}_2 &= \underline{X}_{0,2} + R_{1,2}\underline{S}_0 \\ \underline{G}_3 &= R_1\underline{S}_{0,3} \end{aligned} \tag{4.214}$$

Once again, the base vectors are resolved into the current local frame $\mathcal{F}_1$ to obtain

$$\begin{aligned} \underline{G}_1^* &= R_1^T\underline{X}_{0,1} + R_1^T R_{1,1}\underline{S}_0 \\ \underline{G}_2^* &= R_1^T\underline{X}_{0,2} + R_1^T R_{1,2}\underline{S}_0 \\ \underline{G}_3^* &= R_1^T R_1\underline{S}_{0,3} \end{aligned} \tag{4.215}$$

In view of the definitions of curvature tensors $\widetilde{\kappa}_i^1 = R_1^T R_{1,i}$ for $i = 1, 2$ in current configuration, the covariant base vectors are simplified to

$$\begin{aligned} \underline{G}_1^* &= R_1^T \underline{X}_{0,1} + \widetilde{\kappa}_1^1 \underline{S}_0 \\ \underline{G}_2^* &= R_1^T \underline{X}_{0,2} + \widetilde{\kappa}_2^1 \underline{S}_0 \\ \underline{G}_3^* &= \underline{S}_{0,3} \end{aligned} \tag{4.216}$$

All these base vectors could be assembled into the covariant bases $\mathcal{B}_G$ like

$$\mathcal{B}_G = \lfloor \underline{G}_1^*, \ \underline{G}_2^*, \ \underline{G}_3^* \rfloor \tag{4.217}$$

The similar expressions of the deformed covariant base vectors $\underline{S}_{0,3}$ can be obtained in a similar approach

$$R_1^{\ell T} \underline{S}_{0,3} = R_1^{\ell T} \underline{X}_{0,3}^{\ell} + \widetilde{\kappa}_3^{1\ell} \underline{s}_0^{\ell} + \underline{s}_{0,3}^{\ell} \tag{4.218}$$

## 4.3.2 Strain Tensor of Multiscale Shell with 5 DOFS

Before measuring the components of Green-Lagrange strain tensor, the differences between the base vectors $\underline{G}_i^*$ and $\underline{g}_i^*$ for $i = 1, 2, 3$ are evaluated at first. Referring Eqs. (4.204) and (4.216), the detailed expressions of $\Delta \underline{G}_i^*$ for $i = 1, 2, 3$ are found to be

$$\begin{aligned} \Delta \underline{G}_1^* &= \underline{G}_1^* - \underline{g}_1^* = R_1^T \underline{X}_{0,1} - R_0^T \underline{x}_{0,1} + \widetilde{\kappa}_1^1 \underline{S}_0 - \widetilde{\kappa}_1^0 \underline{s}_0 \\ \Delta \underline{G}_2^* &= \underline{G}_2^* - \underline{g}_2^* = R_2^T \underline{X}_{0,2} - R_0^T \underline{x}_{0,2} + \widetilde{\kappa}_2^1 \underline{S}_0 - \widetilde{\kappa}_2^0 \underline{s}_0 \\ \Delta \underline{G}_3^* &= \underline{G}_3^* - \underline{g}_3^* = \underline{S}_{0,3} - \underline{s}_{0,3} \end{aligned} \tag{4.219}$$

Minor modifications lead to

$$\begin{aligned} \Delta \underline{G}_1^* &= R_1^T \underline{X}_{0,1} - R_0^T \underline{x}_{0,1} + (\widetilde{\kappa}_1^1 - \widetilde{\kappa}_1^0) \underline{s}_0 + \widetilde{\kappa}_1^1 (\underline{S}_0 - \underline{s}_0) \\ \Delta \underline{G}_2^* &= R_1^T \underline{X}_{0,2} - R_0^T \underline{x}_{0,2} + (\widetilde{\kappa}_2^1 - \widetilde{\kappa}_2^0) \underline{s}_0 + \widetilde{\kappa}_2^1 (\underline{S}_0 - \underline{s}_0) \\ \Delta \underline{G}_3^* &= \underline{S}_{0,3} - \underline{s}_{0,3} \end{aligned} \tag{4.220}$$

In view of the definitions of Lagrangian stretch and Lagrange curvature measures, Eq. (3.177), the base vector increments can be further simplified to

$$\begin{aligned} \Delta \underline{G}_1^* &= \lfloor I, \ \widetilde{s}_0^T \rfloor \widehat{\underline{\epsilon}}_1 + \widetilde{\kappa}_1^1 (\underline{S}_0 - \underline{s}_0) \\ \Delta \underline{G}_2^* &= \lfloor I, \ \widetilde{s}_0^T \rfloor \widehat{\underline{\epsilon}}_2 + \widetilde{\kappa}_2^1 (\underline{S}_0 - \underline{s}_0) \\ \Delta \underline{G}_3^* &= \underline{S}_{0,3} - \underline{s}_{0,3} \end{aligned} \tag{4.221}$$

The same relations hold for the micro Cosserat continuum, such that

$$R_1^{\ell T}\underline{S}_{0,3} - R_0^{\ell T}\underline{s}_{0,3} = \lfloor I,\ \widetilde{s}_0^{\ell T}\rfloor \underline{\epsilon}_3^{\ell} = \lfloor I,\ \widetilde{s}_0^{\ell T}\rfloor \widehat{\underline{\epsilon}}_3^{\ell} \tag{4.222}$$

where $\underline{\epsilon}_3^{\ell}$ is the Lagrangian strain measures, Eq. (2.152), defined in the local micro Cosserat continuum like

$$\underline{\epsilon}_3^{\ell} = \left| \begin{array}{c} R_1^{\ell T}\underline{X}_{0,3}^{\ell} - R_0^{\ell T}\underline{x}_{0,3}^{\ell} \\ \underline{\kappa}_3^{1\ell} - \underline{\kappa}_3^{0\ell} \end{array} \right| \tag{4.223}$$

and $\widehat{\underline{\epsilon}}_3^{\ell}$ is the result of truncating the last component of the local curvature vector. With the aid of above definitions, the difference of base vectors $\Delta \underline{G}_i^*$ for $i = 1, 2, 3$ are recasted to

$$\begin{aligned}
\Delta \underline{G}_1^* &= \lfloor I,\ \widetilde{s}_0^T\rfloor \widehat{\underline{\epsilon}}_1 + \widetilde{\kappa}_1^1(\underline{S}_0 - \underline{s}_0) \\
\Delta \underline{G}_2^* &= \lfloor I,\ \widetilde{s}_0^T\rfloor \widehat{\underline{\epsilon}}_2 + \widetilde{\kappa}_2^1(\underline{S}_0 - \underline{s}_0) \\
\Delta \underline{G}_3^* &= R_0^{\ell}\lfloor I,\ \widetilde{s}_0^{\ell T}\rfloor \widehat{\underline{\epsilon}}_3^{\ell} + (I - R_0^{\ell}R_1^{\ell T})\underline{S}_{0,3}
\end{aligned} \tag{4.224}$$

Due to the complexity of the problem, the assumption of small strains has to be introduced to simplify the strain expressions in the theory of multiscale shell. The assumption is valid for both microscopic and macroscopic models. Under this assumption, the base vector increments are finally simplified to

$$\begin{aligned}
\Delta \underline{G}_1^* &= \lfloor I,\ \widetilde{s}_0^T\rfloor \widehat{\underline{\epsilon}}_1 + \widetilde{\kappa}_1^0(\underline{u}_s^{\ell} + R_0^{\ell}\widetilde{s}_0^{\ell T}\underline{\phi}^{\ell}) \\
\Delta \underline{G}_2^* &= \lfloor I,\ \widetilde{s}_0^T\rfloor \widehat{\underline{\epsilon}}_2 + \widetilde{\kappa}_2^0(\underline{u}_s^{\ell} + R_0^{\ell}\widetilde{s}_0^{\ell T}\underline{\phi}^{\ell}) \\
\Delta \underline{G}_3^* &= R_0^{\ell}\lfloor I,\ \widetilde{s}_0^{\ell T}\rfloor \widehat{\underline{\epsilon}}_3^{\ell} + \widetilde{s}_{0,3}^T R_0^{\ell}\underline{\phi}^{\ell}
\end{aligned} \tag{4.225}$$

More details of the simplifications can be found in the Sect. 4.1.2. When introducing the micro motion in its dual bracket, $\underline{\mathcal{U}}^{\ell}$, Eq. (4.144), the base vector increments can be reformulated to

$$\begin{aligned}
\Delta \underline{G}_1^* &= \lfloor I,\ \widetilde{s}_0^T\rfloor \widehat{\underline{\epsilon}}_1 + \widetilde{\kappa}_1^0 R_0^{\ell}\lfloor I,\ \widetilde{s}_0^{\ell T}\rfloor \underline{\mathcal{U}}^{\ell} \\
\Delta \underline{G}_2^* &= \lfloor I,\ \widetilde{s}_0^T\rfloor \widehat{\underline{\epsilon}}_2 + \widetilde{\kappa}_2^0 R_0^{\ell}\lfloor I,\ \widetilde{s}_0^{\ell T}\rfloor \underline{\mathcal{U}}^{\ell} \\
\Delta \underline{G}_3^* &= R_0^{\ell}\lfloor I,\ \widetilde{s}_0^{\ell T}\rfloor \widehat{\underline{\epsilon}}_3^{\ell} + \lfloor 0,\ \widetilde{s}_{0,3}^T R_0^{\ell}\rfloor \underline{\mathcal{U}}^{\ell}
\end{aligned} \tag{4.226}$$

and then recasted to the compact form

$$\begin{aligned}
\Delta \underline{G}_1^* &= \lfloor I,\ \widetilde{s}_0^T\rfloor \widehat{\underline{\epsilon}}_1 + C_1^{\ell}\underline{\mathcal{U}}^{\ell} \\
\Delta \underline{G}_2^* &= \lfloor I,\ \widetilde{s}_0^T\rfloor \widehat{\underline{\epsilon}}_2 + C_2^{\ell}\underline{\mathcal{U}}^{\ell} \\
\Delta \underline{G}_3^* &= R_0^{\ell}\lfloor I,\ \widetilde{s}_0^{\ell T}\rfloor \widehat{\underline{\epsilon}}_3^{\ell} + C_3^{\ell}\underline{\mathcal{U}}^{\ell}
\end{aligned} \tag{4.227}$$

with the definitions of

$$\mathcal{C}_i^\ell = \widetilde{\kappa}_i^0 R_0^\ell \lfloor I,\ \widetilde{s}_0^{\ell T} \rfloor,\ \mathcal{C}_3^\ell = \lfloor 0,\ \widetilde{s}_{0,3}^T R_0^\ell \rfloor \tag{4.228}$$

for $i = 1, 2$. In view of above expressions, the vectorial Green-Lagrange strain measures for a shell structure can be readily verified to be

$$\begin{aligned}
\underline{\varepsilon}_1 &= \mathcal{B}_g^T \lfloor I,\ \widetilde{s}_0^T \rfloor \widehat{\underline{\epsilon}}_1 + \mathcal{B}_g^T \mathcal{C}_1^\ell \underline{\mathcal{U}}^\ell \\
\underline{\varepsilon}_2 &= \mathcal{B}_g^T \lfloor I,\ \widetilde{s}_0^T \rfloor \widehat{\underline{\epsilon}}_2 + \mathcal{B}_g^T \mathcal{C}_2^\ell \underline{\mathcal{U}}^\ell \\
\underline{\varepsilon}_3 &= \mathcal{B}_g^T R_0^\ell \lfloor I,\ \widetilde{s}_0^{\ell T} \rfloor \widehat{\underline{\epsilon}}_3^\ell + \mathcal{B}_g^T \mathcal{C}_3^\ell \underline{\mathcal{U}}^\ell
\end{aligned} \tag{4.229}$$

Defining the following symbols

$$\widehat{\mathcal{S}}_0^T = \mathcal{B}_g^T \lfloor I,\ \widetilde{s}_0^T \rfloor,\ \widehat{\mathcal{S}}_0^{\ell T} = \mathcal{B}_g^T R_0^\ell \lfloor I,\ \widetilde{s}_0^{\ell T} \rfloor \tag{4.230}$$

and

$$\mathcal{S}_i^{\ell T} = \mathcal{B}_g^T \widetilde{\kappa}_i^0 R_0^\ell \lfloor I,\ \widetilde{s}_0^{\ell T} \rfloor,\ \mathcal{S}_3^{\ell T} = \mathcal{B}_g^T \lfloor 0,\ \widetilde{s}_{0,3}^T R_0^\ell \rfloor \tag{4.231}$$

for $i = 1, 2$, the components of Green-Lagrange strain tensor can be rewritten as the functions of Lagrangian stretch and curvature measures, $\widehat{\underline{\epsilon}}_i$ for $i = 1, 2$ and $\widehat{\underline{\epsilon}}_3^\ell$, together with the local motion like

$$\begin{aligned}
\underline{\varepsilon}_1 &= \widehat{\mathcal{S}}_0^T \widehat{\underline{\epsilon}}_1 + \mathcal{S}_1^{\ell T} \underline{\mathcal{U}}^\ell \\
\underline{\varepsilon}_2 &= \widehat{\mathcal{S}}_0^T \widehat{\underline{\epsilon}}_2 + \mathcal{S}_2^{\ell T} \underline{\mathcal{U}}^\ell \\
\underline{\varepsilon}_3 &= \widehat{\mathcal{S}}_0^{\ell T} \widehat{\underline{\epsilon}}_3^\ell + \mathcal{S}_3^{\ell T} \underline{\mathcal{U}}^\ell
\end{aligned} \tag{4.232}$$

In most of the case initial curvatures are zero, $\widetilde{\kappa}_i^0 = 0$ for $i = 1, 2$, the above equations can be simplified to

$$\underline{\varepsilon}_1 = \widehat{\mathcal{S}}_0^T \widehat{\underline{\epsilon}}_1,\ \underline{\varepsilon}_2 = \widehat{\mathcal{S}}_0^T \widehat{\underline{\epsilon}}_2,\ \underline{\varepsilon}_3 = \widehat{\mathcal{S}}_0^{\ell T} \widehat{\underline{\epsilon}}_3^\ell \tag{4.233}$$

after the truncation of third terms containing the motions of micro scale, as discussed in Sect. 4.1.2. It is necessary to note that the formulations, both Eqs. (4.232) and (4.233), are available for the full description of the strain energy for a multiscale shell. While, the later will make the full description clearer.

### *4.3.3 Constitutive Laws for Multiscale Shell with 5 DOFS*

Given the volume with the magnitude of $V_0$ for a shell, the strain energy $\mathcal{U}_e$ can be described by using the Green-Lagrange strains and its work conjugates, the second Piola-Kirchhoff stress measures. For general anisotropic nonlinear elastic materials, the strain energy can be written in the following vectorial form

$$2\mathcal{U}_e = \int\limits_{V_0} [\underline{\varepsilon}_1^T \mathcal{D}_{11} \underline{\varepsilon}_1 + \underline{\varepsilon}_2^T \mathcal{D}_{22} \underline{\varepsilon}_2 + \underline{\varepsilon}_3^T \mathcal{D}_{33} \underline{\varepsilon}_3 + 2(\underline{\varepsilon}_2^T \mathcal{D}_{21} \underline{\varepsilon}_1 + \underline{\varepsilon}_3^T \mathcal{D}_{31} \underline{\varepsilon}_1 + \underline{\varepsilon}_3^T \mathcal{D}_{32} \underline{\varepsilon}_2)]\, dV_0 \tag{4.234}$$

where the constitutive matrix blocks $\mathcal{D}_{ij}$ for $i, j = 1, 2, 3$ are defined in Sect. 3.1.4. After truncated the immaterial "drilling rotation," the strain energy reduces to

$$2\mathcal{U}_e = \int\limits_{S_0} (\widehat{\underline{\epsilon}}_1^T \mathcal{C}_{11} \widehat{\underline{\epsilon}}_1 + \widehat{\underline{\epsilon}}_2^T \mathcal{C}_{22} \widehat{\underline{\epsilon}}_2 + 2\widehat{\underline{\epsilon}}_2^T \mathcal{C}_{21} \widehat{\underline{\epsilon}}_1)\, dS_0 + \int\limits_{V_0} [\widehat{\underline{\epsilon}}_3^{\ell T} \mathcal{D}_{\alpha\alpha}^{33} \widehat{\underline{\epsilon}}_3^{\ell} + 2(\widehat{\underline{\epsilon}}_1^T \mathcal{D}_{\epsilon\alpha}^{13} \widehat{\underline{\epsilon}}_3^{\ell} + \widehat{\underline{\epsilon}}_2^T \mathcal{D}_{\epsilon\alpha}^{23} \widehat{\underline{\epsilon}}_3^{\ell})]\, dV_0 \tag{4.235}$$

where $S_0$ is the integral area computed from $S_0 = V_0/h$, and $h$ the thickness of shell. The stiffness matrices $\mathcal{C}_{ij}$ for $i, j = 1, 2$ are defined by Eq. (3.181), and the new involved material blocks are defined as follow

$$\mathcal{D}_{\alpha\alpha}^{33} = \widehat{S}_0^{\ell} \mathcal{D}_{33} \widehat{S}_0^{\ell T},\ \mathcal{D}_{\epsilon\alpha}^{13} = \widehat{S}_0 \mathcal{D}_{13} \widehat{S}_0^{\ell T},\ \mathcal{D}_{\epsilon\alpha}^{23} = \widehat{S}_0 \mathcal{D}_{23} \widehat{S}_0^{\ell T} \tag{4.236}$$

with the symmetry properties

$$\mathcal{D}_{\alpha\epsilon}^{31T} = \mathcal{D}_{\epsilon\alpha}^{13},\ \mathcal{D}_{\alpha\epsilon}^{32T} = \mathcal{D}_{\epsilon\alpha}^{23} \tag{4.237}$$

When the thickness $h$ is selected by default as the characteristic length of a shell, the integral of the micro scale part for the strain energy can be split into

$$2\mathcal{U}_e = \int\limits_{S_0} (\widehat{\underline{\epsilon}}_1^T \mathcal{C}_{11} \widehat{\underline{\epsilon}}_1 + \widehat{\underline{\epsilon}}_2^T \mathcal{C}_{22} \widehat{\underline{\epsilon}}_2 + 2\widehat{\underline{\epsilon}}_2^T \mathcal{C}_{21} \widehat{\underline{\epsilon}}_1)\, dS_0 + \int\limits_{S_0} [\int\limits_{h} \widehat{\underline{\epsilon}}_3^{\ell T} \mathcal{D}_{\alpha\alpha}^{33} \widehat{\underline{\epsilon}}_3^{\ell} + 2(\widehat{\underline{\epsilon}}_1^T \mathcal{D}_{\epsilon\alpha}^{13} \widehat{\underline{\epsilon}}_3^{\ell} + \widehat{\underline{\epsilon}}_2^T \mathcal{D}_{\epsilon\alpha}^{23} \widehat{\underline{\epsilon}}_3^{\ell})] d\alpha_3\, dS_0 \tag{4.238}$$

According to the definitions of Lagrangian stretch and curvature measures in the sense of micro scale, Eq. (4.223), these strain measures could be explicitly related to the motion of micro scale and its derivative

$$\underline{\epsilon}_3^{\ell} = \underline{\mathcal{U}}_{,3}^{\ell} + \widetilde{\mathcal{K}}_3^{0\ell} \underline{\mathcal{U}}^{\ell} \tag{4.239}$$

with the following introduced symbols

$$\underline{\mathcal{U}}_{,3}^{\ell} = \left| \begin{matrix} (R_0^{\ell T} \underline{u}_s^{\ell})_{,3} \\ \underline{\phi}_{,3}^{\ell} \end{matrix} \right|,\ \widetilde{\mathcal{K}}_3^{0\ell} = \begin{bmatrix} \widetilde{\kappa}_3^{0\ell} & R_0^{\ell T} \widetilde{\underline{x}_{0,3}} \\ 0 & \widetilde{\kappa}_3^{0\ell} \end{bmatrix} \tag{4.240}$$

under the assumption of small motion. As mentioned before, the micro motion with small magnitude can form a linear space. It means the classical finite element method can be applied directly to discretize the strain energy of micro model. Given the shape functions $\mathcal{H}_\epsilon^\ell$ of finite element method, the local motion and the derivative is interpolated by

$$\underline{\mathcal{U}}^\ell = \mathcal{H}_\epsilon^\ell \underline{\mathcal{U}}_n^\ell, \; \underline{\mathcal{U}}_{,3}^\ell = \mathcal{H}_{\epsilon,3}^\ell \underline{\mathcal{U}}_n^\ell \tag{4.241}$$

where $\underline{\mathcal{U}}^\ell$ denotes the nodal motions of discrete micro model. As a result, the Lagrangian stretch and curvature measures of micro scale can be approximated to

$$\underline{\epsilon}_3^\ell = (\mathcal{H}_{\epsilon,3}^\ell + \widetilde{\mathcal{K}}_3^{0\ell} \mathcal{H}_\epsilon^\ell) \underline{\mathcal{U}}_n^\ell = \mathcal{Q}_3^\ell \underline{\mathcal{U}}_n^\ell \tag{4.242}$$

After neglected the immaterial “drilling rotation” from each nodal velocities of micro scale, the truncated expression can be readily found

$$\widehat{\underline{\epsilon}}_3^\ell = \widehat{\mathcal{Q}}_3^\ell \widehat{\underline{\mathcal{U}}}_n^\ell \tag{4.243}$$

Under the assumption of scale separation, the macro strains $\widehat{\underline{\epsilon}}_i$ for $i = 1, 2$ are independent of local variables. Hence, the full description of strain energy on the area $S_0$ of the middle-surface for a shell can be determined to be

$$\begin{aligned} 2\mathcal{U}_e = \int_{S_0} [\, & \widehat{\underline{\epsilon}}_1^T C_{11} \widehat{\underline{\epsilon}}_1 + \widehat{\underline{\epsilon}}_2^T C_{22} \widehat{\underline{\epsilon}}_2 + 2\widehat{\underline{\epsilon}}_2^T C_{21} \widehat{\underline{\epsilon}}_1 \\ & + \widehat{\underline{\mathcal{U}}}_n^{\ell T} C_{\alpha\alpha} \widehat{\underline{\mathcal{U}}}_n^\ell + 2(\widehat{\underline{\epsilon}}_1^T C_{\epsilon\alpha}^{11} \widehat{\underline{\mathcal{U}}}_n^\ell + \widehat{\underline{\epsilon}}_2^T C_{\epsilon\alpha}^{22} \widehat{\underline{\mathcal{U}}}_n^\ell) \,] \, dS_0 \end{aligned} \tag{4.244}$$

where the new stiffness matrices are introduced

$$C_{\alpha\alpha} = \int_h \widehat{\mathcal{Q}}_3^{\ell T} \mathcal{D}_{\alpha\alpha}^{33} \widehat{\mathcal{Q}}_3 d\alpha_3, \; C_{\epsilon\alpha}^{11} = \int_h \mathcal{D}_{\epsilon\alpha}^{13} \widehat{\mathcal{Q}}_3 d\alpha_3, \; C_{\epsilon\alpha}^{22} = \int_h \mathcal{D}_{\epsilon\alpha}^{23} \widehat{\mathcal{Q}}_3 d\alpha_3 \tag{4.245}$$

### 4.3.4 *Variation of Strain Energy of Multiscale Shell with 5 DOFS*

After defined the density of strain energy for a shell as follow

$$\begin{aligned} \mathcal{U}_\rho = & \frac{1}{2} (\widehat{\underline{\epsilon}}_1^T C_{11} \widehat{\underline{\epsilon}}_1 + \widehat{\underline{\epsilon}}_2^T C_{22} \widehat{\underline{\epsilon}}_2) + \widehat{\underline{\epsilon}}_2^T C_{21} \widehat{\underline{\epsilon}}_1 \\ & + \widehat{\underline{\epsilon}}_1^T C_{\epsilon\alpha}^{11} \widehat{\underline{\mathcal{U}}}_n^\ell + \widehat{\underline{\epsilon}}_2^T C_{\epsilon\alpha}^{22} \widehat{\underline{\mathcal{U}}}_n^\ell + \frac{1}{2} \widehat{\underline{\mathcal{U}}}_n^{\ell T} C_{\alpha\alpha} \widehat{\underline{\mathcal{U}}}_n^\ell \end{aligned} \tag{4.246}$$

the strain energy $\mathcal{U}_e$ can be represented by the integration of the strain energy density in the form of $\mathcal{U}_e = \int_{S_0} \mathcal{U}_\rho dS_0$, which makes the variational operations of strain energy more convenient. Taking variation of averaged strain energy density leads to

$$\begin{aligned}\delta\mathcal{U}_\rho &= \delta\widehat{\underline{\epsilon}}_1^T(C_{11}\widehat{\underline{\epsilon}}_1 + C_{12}\widehat{\underline{\epsilon}}_2) + \delta\widehat{\underline{\epsilon}}_1^T C_{\epsilon\alpha}^{11}\widehat{\underline{\mathcal{U}}}_n^\ell \\ &+ \delta\widehat{\underline{\epsilon}}_2^T(C_{21}\widehat{\underline{\epsilon}}_1 + C_{22}\widehat{\underline{\epsilon}}_2) + \delta\widehat{\underline{\epsilon}}_2^T C_{\epsilon\alpha}^{22}\widehat{\underline{\mathcal{U}}}_n^\ell \\ &+ \delta\widehat{\underline{\mathcal{U}}}_n^{\ell T}(C_{\epsilon\alpha}^{11T}\widehat{\underline{\epsilon}}_1 + C_{\epsilon\alpha}^{22T}\widehat{\underline{\epsilon}}_2) + \delta\widehat{\underline{\mathcal{U}}}_n^{\ell T} C_{\alpha\alpha}\widehat{\underline{\mathcal{U}}}_n^\ell\end{aligned} \tag{4.247}$$

The compact expressions

$$\delta\mathcal{U}_\rho = \delta\widehat{\underline{\epsilon}}_1^T(\underline{F}_{\epsilon\epsilon}^{11} + C_{\epsilon\alpha}^{11}\widehat{\underline{\mathcal{U}}}_n^\ell) + \delta\widehat{\underline{\epsilon}}_2^T(\underline{F}_{\epsilon\epsilon}^{22} + C_{\epsilon\alpha}^{22}\widehat{\underline{\mathcal{U}}}_n^\ell) + \delta\widehat{\underline{\mathcal{U}}}_n^{\ell T}(\underline{F}_{\alpha\epsilon} + C_{\alpha\alpha}\widehat{\underline{\mathcal{U}}}_n^\ell) \tag{4.248}$$

are readily obtained after introducing the following stress measures

$$\underline{F}_{\epsilon\epsilon}^{11} = C_{11}\widehat{\underline{\epsilon}}_1 + C_{12}\widehat{\underline{\epsilon}}_2,\ \underline{F}_{\epsilon\epsilon}^{22} = C_{21}\widehat{\underline{\epsilon}}_1 + C_{22}\widehat{\underline{\epsilon}}_2,\ \widehat{\underline{F}}_{\alpha\epsilon} = C_{\epsilon\alpha}^{11T}\widehat{\underline{\epsilon}}_1 + C_{\epsilon\alpha}^{22T}\widehat{\underline{\epsilon}}_2 \tag{4.249}$$

Then, the variation of strain energy $\mathcal{U}_e$ is described by using the follow formula

$$\delta\mathcal{U}_e = \int_{S_0} \delta\mathcal{U}_\rho dS_0 \tag{4.250}$$

## 4.3.5 Virtual Work of External Forces for Multiscale Shell with 5 DOFS

According to the definitions of external force, $\underline{F}_{\epsilon\alpha}$ and $\underline{F}_{\alpha\alpha}$, Eq. (4.176), the last terms which corresponding to the immaterial "drilling rotation" must be truncated from the dual brackets of $\underline{F}_{\epsilon\alpha}$ and $\underline{F}_{\alpha\alpha}$, respectively, to obtain the reduced forces $\widehat{\underline{F}}_{\epsilon\alpha}$ and $\widehat{\underline{F}}_{\alpha\alpha}$. The virtual work of external force applied to the multiscale shell is then obtained readily

$$\delta\mathcal{W}_e = \int_{S_0} \delta\widehat{\underline{\mathcal{U}}}^T\widehat{\underline{F}}_{\epsilon\alpha} + \delta\widehat{\underline{\mathcal{U}}}_n^{\ell T}\widehat{\underline{F}}_{\alpha\alpha} dS_0 \tag{4.251}$$

## 4.3.6 Governing Equations of Multiscale Shell with 5 DOFS in Motion Formalism

By substituting Eqs. (4.248) and (4.251) into the principle of virtual work, which states

$$\delta\mathcal{U}_e - \delta\mathcal{W}_e = 0 \tag{4.252}$$

the governing equations of multiscale shell are readily obtained

$$\begin{aligned}\int_{S_0} [\, & \delta\underline{\epsilon}_1^T (\underline{F}_{\epsilon\epsilon}^{11} + C_{\epsilon\alpha}^{11}\underline{\widehat{\mathcal{U}}}_n^{\ell}) + \delta\underline{\epsilon}_2^T (\underline{F}_{\epsilon\epsilon}^{22} + C_{\epsilon\alpha}^{22}\underline{\widehat{\mathcal{U}}}_n^{\ell}) \\ & + \delta\underline{\widehat{\mathcal{U}}}_n^{\ell T} (\underline{\widehat{F}}_{\alpha\epsilon} + C_{\alpha\alpha}\underline{\widehat{\mathcal{U}}}_n^{\ell}) - \delta\underline{\widehat{\mathcal{U}}}^T \underline{\widehat{F}}_{\epsilon\alpha} - \delta\underline{\widehat{\mathcal{U}}}_n^{\ell T} \underline{\widehat{F}}_{\alpha\alpha} \,] d S_0 = 0\end{aligned} \tag{4.253}$$

Based on the assumption of scale separation, the variations of macro strain measures are independent of the variational variables for a micro model. As a result, the original problem can be decomposed into a macroscopic continuum problem

$$\int_{S_0} (\delta\underline{\epsilon}_1^T (\underline{F}_{\epsilon\epsilon}^{11} + C_{\epsilon\alpha}^{11}\underline{\widehat{\mathcal{U}}}_n^{\ell}) + \delta\underline{\epsilon}_2^T (\underline{F}_{\epsilon\epsilon}^{22} + C_{\epsilon\alpha}^{22}\underline{\widehat{\mathcal{U}}}_n^{\ell}) - \delta\underline{\widehat{\mathcal{U}}}^T \underline{\widehat{F}}_{\epsilon\alpha}) d S_0 = 0 \tag{4.254}$$

and a microscopic continuum problem

$$C_{\alpha\alpha}\underline{\widehat{\mathcal{U}}}_n^{\ell} + \underline{\widehat{F}}_{\alpha\epsilon} - \underline{\widehat{F}}_{\alpha\alpha} = 0 \tag{4.255}$$

with appropriate boundary conditions. Once the elastic forces $\underline{\widehat{F}}_{\alpha\epsilon}$ and external forces $\underline{\widehat{F}}_{\alpha\alpha}$ are given, the micro displacements could be solved from above equations

$$\underline{\mathcal{U}}_n^{\ell} = C_{\alpha\alpha}^{-1} (\underline{\widehat{F}}_{\alpha\alpha} - \underline{\widehat{F}}_{\alpha\epsilon}) \tag{4.256}$$

The macroscopic continuum problem becomes solvable after eliminating the micro unknowns $\underline{\mathcal{U}}_n^{\ell}$ by substituting Eq. (4.256) into Eq. (4.254). Merging the similar terms leads to

$$\begin{aligned}\int_{S_0} [\, & \delta\underline{\epsilon}_1^T (\widehat{C}_{\epsilon\epsilon}^{11}\underline{\epsilon}_1 + \widehat{C}_{\epsilon\epsilon}^{12}\underline{\epsilon}_2 + C_{\epsilon\alpha}^{11} C_{\alpha\alpha}^{-1} \underline{\widehat{F}}_{\alpha\alpha}) \\ & + \delta\underline{\epsilon}_2^T (\widehat{C}_{\epsilon\epsilon}^{21}\underline{\epsilon}_1 + \widehat{C}_{\epsilon\epsilon}^{22}\underline{\epsilon}_2 + C_{\epsilon\alpha}^{22} C_{\alpha\alpha}^{-1} \underline{\widehat{F}}_{\alpha\alpha}) - \delta\underline{\widehat{\mathcal{U}}}^T \underline{\widehat{F}}_{\epsilon\alpha} \,] d S_0 = 0\end{aligned} \tag{4.257}$$

where the stiffness matrix blocks are modified to

$$\begin{aligned}\widehat{C}_{\epsilon\epsilon}^{11} &= C_{11} - C_{\epsilon\alpha}^{11} C_{\alpha\alpha}^{-1} C_{\epsilon\alpha}^{11T}, \; \widehat{C}_{\epsilon\epsilon}^{12} = C_{12} - C_{\epsilon\alpha}^{11} C_{\alpha\alpha}^{-1} C_{\epsilon\alpha}^{22T} \\ \widehat{C}_{\epsilon\epsilon}^{21} &= C_{21} - C_{\epsilon\alpha}^{22} C_{\alpha\alpha}^{-1} C_{\epsilon\alpha}^{11T}, \; \widehat{C}_{\epsilon\epsilon}^{22} = C_{22} - C_{\epsilon\alpha}^{22} C_{\alpha\alpha}^{-1} C_{\epsilon\alpha}^{22T}\end{aligned} \tag{4.258}$$

After introduced new symbols of the following stress measures

$$\begin{aligned}\underline{\widehat{F}}_{\epsilon\epsilon}^{11} &= \widehat{C}_{\epsilon\epsilon}^{11}\underline{\epsilon}_1 + \widehat{C}_{\epsilon\epsilon}^{12}\underline{\epsilon}_2 + C_{\epsilon\alpha}^{11} C_{\alpha\alpha}^{-1} \underline{\widehat{F}}_{\alpha\alpha} \\ \underline{\widehat{F}}_{\epsilon\epsilon}^{22} &= \widehat{C}_{\epsilon\epsilon}^{22}\underline{\epsilon}_2 + \widehat{C}_{\epsilon\epsilon}^{21}\underline{\epsilon}_1 + C_{\epsilon\alpha}^{22} C_{\alpha\alpha}^{-1} \underline{\widehat{F}}_{\alpha\alpha}\end{aligned} \tag{4.259}$$

the compact formulation of macroscopic problem are further simplified to

$$\int_{S_0} (\delta \underline{\epsilon}_1^T \underline{\widehat{F}}_{\epsilon\epsilon}^{11} + \delta \underline{\epsilon}_2^T \underline{\widehat{F}}_{\epsilon\epsilon}^{22} - \delta \widehat{\mathcal{U}}^T \underline{\widehat{F}}_{\epsilon\alpha}) dS_0 = 0 \tag{4.260}$$

Referring the compatibility attributes, Eq. (2.161), it produces

$$\int_{S_0} \left[ (\delta \underline{\mathcal{U}}_{,1} + \widetilde{\widetilde{\mathcal{K}}}_1^1 \delta \underline{\mathcal{U}})^T \underline{\widehat{F}}_{\epsilon\epsilon}^{11} + (\delta \underline{\mathcal{U}}_{,2} + \widetilde{\widetilde{\mathcal{K}}}_2^1 \delta \underline{\mathcal{U}})^T \underline{\widehat{F}}_{\epsilon\epsilon}^{22} - \delta \widehat{\mathcal{U}}^T \underline{\widehat{F}}_{\epsilon\alpha} \right] dS_0 = 0 \tag{4.261}$$

With the application of finite element discretization, the weak formula of the macroscopic shell problem becomes

$$\begin{aligned} \delta \underline{\mathcal{U}}_n^T \int_{S_0} [(\mathcal{H}_{\epsilon,1}^T + \mathcal{H}_\epsilon^T \widetilde{\widetilde{\mathcal{K}}}_1^{1T}) \underline{\widehat{F}}_{\epsilon\epsilon}^{11} \\ + (\mathcal{H}_{\epsilon,2}^T + \mathcal{H}_\epsilon^T \widetilde{\widetilde{\mathcal{K}}}_2^{1T}) \underline{\widehat{F}}_{\epsilon\epsilon}^{22} - \mathcal{H}_\epsilon^T \underline{\widehat{F}}_{\epsilon\alpha}] dS_0 = 0 \end{aligned} \tag{4.262}$$

From which, the governing equations are readily obtained

$$\int_{S_0} [(\mathcal{H}_{\epsilon,1}^T + \mathcal{H}_\epsilon^T \widetilde{\widetilde{\mathcal{K}}}_1^{1T}) \underline{\widehat{F}}_{\epsilon\epsilon}^{11} + (\mathcal{H}_{\epsilon,2}^T + \mathcal{H}_\epsilon^T \widetilde{\widetilde{\mathcal{K}}}_2^{1T}) \underline{\widehat{F}}_{\epsilon\epsilon}^{22} - \mathcal{H}_\epsilon^T \underline{\widehat{F}}_{\epsilon\alpha}] dS_0 = 0 \tag{4.263}$$

Given the definitions of the elastic force vector

$$\underline{\mathcal{F}}_e = \int_{S_0} \left[ (\mathcal{H}_{\epsilon,1}^T + \mathcal{H}_\epsilon^T \widetilde{\widetilde{\mathcal{K}}}_1^{1T}) \underline{\widehat{F}}_{\epsilon\epsilon}^{11} + (\mathcal{H}_{\epsilon,2}^T + \mathcal{H}_\epsilon^T \widetilde{\widetilde{\mathcal{K}}}_2^{1T}) \underline{\widehat{F}}_{\epsilon\epsilon}^{22} \right] dS_0 \tag{4.264}$$

and the external force vector

$$\underline{\mathcal{F}}_a = \int_{S_0} \mathcal{H}_\epsilon^T \underline{\widehat{F}}_{\epsilon\alpha} dS_0 \tag{4.265}$$

the governing equations of macroscopic shell model can be described by the following compact formula

$$\underline{\mathcal{F}}_e - \underline{\mathcal{F}}_a = 0 \tag{4.266}$$

For the dynamic problem, the inertial forces must be taken into account. At first, the velocity of an arbitrary material particle in a multiscale shell is required, which could be evaluated by taking time derivatives of position vector $\underline{P}$, defined by Eq. (4.209), to find

$$\dot{\underline{P}} = (\underline{x}_0 + \underline{u}_s \dot{} + R_1 \underline{S}_0) = \dot{\underline{u}}_s + \dot{R}_1 \underline{S}_0 + R_1 \dot{\underline{S}}_0 \tag{4.267}$$

According to the assumption of scale separation, the dimensions of micro continuum are at least one order of magnitude smaller than that of the macro continuum. Hence, the velocity of micro continuum becomes negligible. Meanwhile, the centrifugal velocity is approximated by

$$\dot{R}_1 \underline{S}_0 \approx \dot{R}_1 \underline{s}_0 \tag{4.268}$$

After resolved the time derivatives into the body attached frame $\mathcal{F}_1$, the particle velocity will relate to the linear and angular velocities of reference particle like

$$R_1^T \underline{\dot{P}} = R_1^T \underline{\dot{u}}_s + \widehat{s}_0^T \underline{\widehat{\omega}} \tag{4.269}$$

with the truncation of "rolling" speed $\omega_3$. The kinematic energy for this special multiscale shell is defined by Eq. (3.211). The associated mass matrix is averaged by using the characteristic area with the magnitude of $S_c$ in micro model like

$$\mathcal{M}_{\epsilon\epsilon} = \frac{1}{S_c} \int_{V_c} \begin{bmatrix} \rho I, & \rho \widehat{s}_0^T \\ \rho \widehat{s}_0, & \rho \widehat{s}_0 \widehat{s}_0^T \end{bmatrix} dV_c \tag{4.270}$$

The definitions of mass matrix and the Coriolis force vector, Eq. (3.220), are still available. When considering the inertial forces, the governing equations of motion for multiscale shell are decoupled into two sub-systems, one is dynamic macroscopic system

$$\mathcal{M}_n \underline{\hat{\mathcal{V}}}_n - \underline{\mathcal{F}}_g + \underline{\mathcal{F}}_e = \underline{\mathcal{F}}_a \tag{4.271}$$

and the other is the static microscopic system, Eq. (4.255).

## 4.4 Multiscale Beam-Like Theory

The beam is regarded as a one-dimensional Cosserat continuum, in which the deformations of two dimensions defining the cross-section of the beam are ignored. In multiscale beam-like theory, the rigid assumption of the cross-section of the beam is not applicable. The characteristic scale of micro Cosserat continuum is determined according to the size of the cross-section and then the deformation of the cross-section could be described by using the theory of three-dimensional or two-dimensional Cosserat continuum in micro scale.

### *4.4.1 Kinematics of Multiscale Beam*

As shown in Fig. 4.4, a material particle with infinitesimal volume in a beam is depicted by its position and director vectors in its reference configuration and current

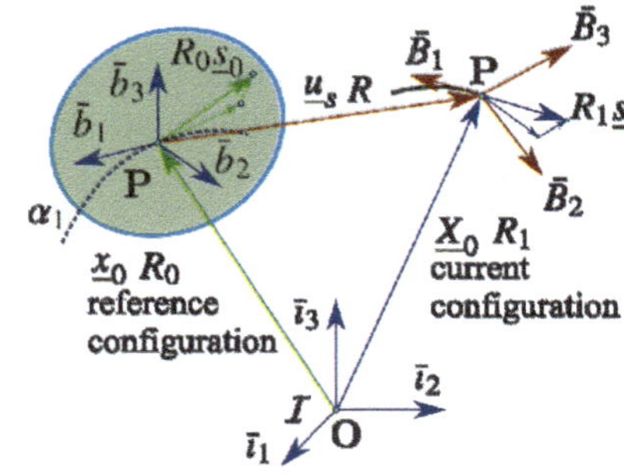


**Fig. 4.4** Kinematics of a material particle with infinitesimal volume in a multiscale beam

deformed configuration, respectively. In the reference configuration, the position vector $\underline{p}$ of the material particle relative to a point **O**, the original of reference frame, $\mathcal{F}_{\mathcal{I}} = [\mathbf{O}, \mathcal{I}]$, is defined as a function of the general curvilinear coordinates, $(\alpha_1, \alpha_2, \alpha_3)$ like

$$\underline{p}(\alpha_1, \alpha_2, \alpha_3) = \underline{x}_0(\alpha_1) + R_0(\alpha_1)\underline{s}_0(\alpha_1, \alpha_2, \alpha_3) \tag{4.272}$$

where $\underline{x}_0$ is the position vector of a reference material particle **P** relative to a point **O**, and $R_0$ denotes the initial rotation tensor measuring the orientation of the reference basis $\mathcal{B}_0$. It has been verified in Sect. 2.2.3 the rotation tensor $R_0$ is identical to the body attached basis $\mathcal{B}_0$, and in detail

$$R_0 = \lfloor \bar{b}_1, \bar{b}_2, \bar{b}_3 \rfloor = \mathcal{B}_0 \tag{4.273}$$

Note that the initial basis $\mathcal{B}_0$ is bounded by the reference particle **P** to produce the reference frame $\mathcal{F}_0 = [\mathbf{P}, \mathcal{B}_0]$. The local vector $\underline{s}_0$, defined in this reference frame $\mathcal{F}_0$, describes the director of a material particle. Obviously, Eq. (4.272) decomposes the position vector $\underline{p}$ of a material particle into a reference position $\underline{x}_0$ and a local vector $\underline{s}_0$. In classical beam theory, director $\underline{s}_0$ is a rigid vector. Now it becomes deformable under the framework of multiscale beam theory. If assuming the micro continuum to be the three-dimensional Cosserat continuum, which is usually the case, the director $\underline{s}_0$ can be defined in a similar approach to the position vector $\underline{p}$, Eq. (4.272), like

$$\underline{s}_0(\alpha_1, \alpha_2, \alpha_3) = \underline{x}_0^\ell(\alpha_1, \alpha_2, \alpha_3) + R_0^\ell(\alpha_1, \alpha_2, \alpha_3)\underline{s}_0^\ell(\alpha_1, \alpha_2, \alpha_3) \tag{4.274}$$

where $\underline{x}_0^\ell$ is the position vector of reference particle $\mathbf{P}^\ell$ in local Cosserat continuum, and $R_0^\ell$ the local rotation tensor attached to the reference particle $\mathbf{P}^\ell$ where the local director $\underline{s}_0^\ell$ is measured in the local basis spanned by $R_0^\ell$. In view of above formula, the position vector $\underline{p}$ can be determined from the following expression

$$\underline{p} = \underline{x}_0 + R_0(\underline{x}_0^\ell + R_0^\ell \underline{s}_0^\ell) \tag{4.275}$$

For the description of multiscale beam deformations, the covariant base vectors

$$\underline{g}_i = \frac{\partial \underline{p}}{\partial \alpha_i} \tag{4.276}$$

in the reference configuration are required, and the specific expressions can be obtained by taking derivatives of Eq. (4.272) with respected to the curvilinear coordinates $\alpha_i$ like

$$\begin{aligned} \underline{g}_1 &= \underline{x}_{0,1} + R_{0,1}\underline{s}_0 + R_0\underline{s}_{0,1} \\ \underline{g}_2 &= R_0\underline{s}_{0,2} \\ \underline{g}_3 &= R_0\underline{s}_{0,3} \end{aligned} \tag{4.277}$$

where the symbol $(\cdot)_{,i}$ for $i = 1, 2, 3$ denotes the spatial derivatives with respected to $\alpha_i$ and

$$(\cdot)_{,i} = \frac{\partial(\cdot)}{\partial \alpha_i} \tag{4.278}$$

After resolved the base vectors into the local reference frame $\mathcal{F}_0$, it can be obtained

$$\begin{aligned} \underline{g}_1^* &= R_0^T\underline{x}_{0,1} + R_0^T R_{0,1}\underline{s}_0 + \underline{s}_{0,1} \\ \underline{g}_2^* &= \underline{s}_{0,2} \\ \underline{g}_3^* &= \underline{s}_{0,3} \end{aligned} \tag{4.279}$$

According to the definitions of initial curvature tensors, Eq. (1.314), the base vectors are then represented by the following formulas like

$$\begin{aligned} \underline{g}_1^* &= R_0^T\underline{x}_{0,1} + \widetilde{\kappa}_1^0\underline{s}_0 + \underline{s}_{0,1} \\ \underline{g}_2^* &= \underline{s}_{0,2} \\ \underline{g}_3^* &= \underline{s}_{0,3} \end{aligned} \tag{4.280}$$

where the symbols $\widetilde{\kappa}_1^0 = R_0^T R_{0,1}$ denote the initial curvature tensor of a beam. All these base vectors assemble the covariant bases $\mathcal{B}_g$ like

$$\mathcal{B}_g = \lfloor \underline{g}_1^*,\ \underline{g}_2^*,\ \underline{g}_3^* \rfloor \tag{4.281}$$

and then the metric tensor in the reference configuration can be computed from

$$g = \mathcal{B}_g^T \mathcal{B}_g \tag{4.282}$$

The similar manipulations are performed repeatedly in the local Cosserat continuum to find that the derivatives of director $\underline{s}_0$ becomes

$$
\begin{aligned}
\underline{s}_{0,1} &= \underline{x}_{0,1}^{\ell} + R_{0,1}^{\ell}\underline{s}_{0}^{\ell} + R_{0}^{\ell}\underline{s}_{0,1}^{\ell} \\
\underline{s}_{0,2} &= \underline{x}_{0,2}^{\ell} + R_{0,2}^{\ell}\underline{s}_{0}^{\ell} + R_{0}^{\ell}\underline{s}_{0,2}^{\ell} \\
\underline{s}_{0,3} &= \underline{x}_{0,3}^{\ell} + R_{0,3}^{\ell}\underline{s}_{0}^{\ell} + R_{0}^{\ell}\underline{s}_{0,3}^{\ell}
\end{aligned}
\tag{4.283}
$$

If defined the local curvature tensors as $\widetilde{\kappa}_i^{0\ell} = R_0^{\ell T} R_{0,i}^{\ell}$ for $i = 1, 2, 3$, the local base vectors can be resolved into the local basis spanned by $R_0^{\ell}$ like

$$
\begin{aligned}
R_0^{\ell T}\underline{s}_{0,1} &= R_0^{\ell T}\underline{x}_{0,1}^{\ell} + \widetilde{\kappa}_1^{0\ell}\underline{s}_0^{\ell} + \underline{s}_{0,1}^{\ell} \\
R_0^{\ell T}\underline{s}_{0,2} &= R_0^{\ell T}\underline{x}_{0,2}^{\ell} + \widetilde{\kappa}_2^{0\ell}\underline{s}_0^{\ell} + \underline{s}_{0,2}^{\ell} \\
R_0^{\ell T}\underline{s}_{0,3} &= R_0^{\ell T}\underline{x}_{0,3}^{\ell} + \widetilde{\kappa}_3^{0\ell}\underline{s}_0^{\ell} + \underline{s}_{0,3}^{\ell}
\end{aligned}
\tag{4.284}
$$

When the beam changes its spatial configuration, the material particle that had position vector $\underline{p}$ in the undeformed reference configuration now has position vector $\underline{P}$ in the deformed configuration, such that

$$
\underline{P}(\alpha_1, \alpha_2, \alpha_3) = \underline{X}_0(\alpha_1, \alpha_2) + R_1(\alpha_1, \alpha_2)\underline{S}_0(\alpha_1, \alpha_2, \alpha_3) \tag{4.285}
$$

where $\underline{S}_0$ represents the deformed director in current configuration, and $R_1$ the total rotation tensor that describes the orientation of basis $\mathcal{B}_1$ attached to the deformed configuration and

$$
R_1 = \lfloor \bar{B}_1, \bar{B}_2, \bar{B}_3 \rfloor = \mathcal{B}_1 \tag{4.286}
$$

Note that bounding the current deformed basis $\mathcal{B}_1$ by the material particle $\mathbf{P}$, the associated deformed frame $\mathcal{F}_1 = [\mathbf{P}, \mathcal{B}_1]$ can be constructed. Furthermore, if the displacement vector $\underline{u}_s$ is given to describe an arbitrary translation of reference particle $\mathbf{P}$, the reference position vector $\underline{X}_0$ in current deformed configuration can be recasted to

$$
\underline{X}_0(\alpha_1) = \underline{x}_0(\alpha_1) + \underline{u}_s(\alpha_1) \tag{4.287}
$$

Similarly, the deformed director $\underline{S}_0$ is decomposed into

$$
\underline{S}_0(\alpha_1, \alpha_2, \alpha_3) = \underline{X}_0^{\ell}(\alpha_1, \alpha_2, \alpha_3) + R_1^{\ell}(\alpha_1, \alpha_2, \alpha_3)\underline{s}_0^{\ell}(\alpha_1, \alpha_2, \alpha_3) \tag{4.288}
$$

where $\underline{X}_0^{\ell}$ is the deformed position vector of reference particle $\mathbf{P}^{\ell}$, and $R_1^{\ell}$ the deformed rotation tensor in the micro three-dimensional Cosserat continuum. The position vector $\underline{X}_0^{\ell}$ can be related to its original position $\underline{x}_0^{\ell}$ through

$$
\underline{X}_0^{\ell}(\alpha_1, \alpha_2, \alpha_3) = \underline{x}_0^{\ell}(\alpha_1, \alpha_2, \alpha_3) + \underline{u}_s^{\ell}(\alpha_1, \alpha_2, \alpha_3) \tag{4.289}
$$

where $\underline{u}_s^{\ell}$ is the displacement vector of local director in micro continuum. In view of the definitions of covariant base vectors in deformed configuration

$$\underline{G}_i = \frac{\partial \underline{P}}{\partial \alpha_i} \tag{4.290}$$

the specific formulas can be obtained by taking spatial derivatives of Eq. (4.285) with respected to the curvilinear coordinates $\alpha_i$ for $i = 1, 2, 3$ like

$$\begin{aligned} \underline{G}_1 &= \underline{X}_{0,1} + R_{1,1}\underline{S}_0 + R_1\underline{S}_{0,1} \\ \underline{G}_2 &= R_1\underline{S}_{0,2} \\ \underline{G}_3 &= R_1\underline{S}_{0,3} \end{aligned} \tag{4.291}$$

Once again, the base vectors are resolved into the current frame $\mathcal{F}_1$ to obtain

$$\begin{aligned} \underline{G}_1^* &= R_1^T\underline{X}_{0,1} + R_1^T R_{1,1}\underline{S}_0 + R_1^T R_1\underline{S}_{0,1} \\ \underline{G}_2^* &= R_1^T R_1\underline{S}_{0,2} \\ \underline{G}_3^* &= R_1^T R_1\underline{S}_{0,3} \end{aligned} \tag{4.292}$$

In view of the definitions of curvature tensors $\widetilde{\kappa}_1^1 = R_1^T R_{1,1}$, the covariant base vectors are simplified to

$$\begin{aligned} \underline{G}_1^* &= R_1^T\underline{X}_{0,1} + \widetilde{\kappa}_1^1\underline{S}_0 + \underline{S}_{0,1} \\ \underline{G}_2^* &= \underline{S}_{0,2} \\ \underline{G}_3^* &= \underline{S}_{0,3} \end{aligned} \tag{4.293}$$

All these base vectors could be assembled into the covariant bases $\mathcal{B}_G$ like

$$\mathcal{B}_G = \lfloor \underline{G}_1^*,\ \underline{G}_2^*,\ \underline{G}_3^* \rfloor \tag{4.294}$$

The similar expressions of deformed covariant base vectors $\underline{S}_{0,i}$ for $i = 1, 2, 3$ can be obtained as

$$\begin{aligned} \underline{S}_{0,1} &= \underline{X}_{0,1}^\ell + R_{1,1}^\ell\underline{s}_0^\ell + R_1^\ell\underline{s}_{0,1}^\ell \\ \underline{S}_{0,2} &= \underline{X}_{0,2}^\ell + R_{1,2}^\ell\underline{s}_0^\ell + R_1^\ell\underline{s}_{0,2}^\ell \\ \underline{S}_{0,3} &= \underline{X}_{0,3}^\ell + R_{1,3}^\ell\underline{s}_0^\ell + R_1^\ell\underline{s}_{0,3}^\ell \end{aligned} \tag{4.295}$$

Resolving these vectors of micro three-dimensional continuum into the body attached frame spanned by $R_1^\ell$ produces

$$\begin{aligned} R_1^{\ell T}\underline{S}_{0,1} &= R_1^{\ell T}\underline{X}_{0,1}^\ell + \widetilde{\kappa}_1^{1\ell}\underline{s}_0^\ell + \underline{s}_{0,1}^\ell \\ R_1^{\ell T}\underline{S}_{0,2} &= R_1^{\ell T}\underline{X}_{0,2}^\ell + \widetilde{\kappa}_2^{1\ell}\underline{s}_0^\ell + \underline{s}_{0,2}^\ell \\ R_1^{\ell T}\underline{S}_{0,3} &= R_1^{\ell T}\underline{X}_{0,3}^\ell + \widetilde{\kappa}_3^{1\ell}\underline{s}_0^\ell + \underline{s}_{0,3}^\ell \end{aligned} \tag{4.296}$$

where $\widetilde{\kappa}_i^{1\ell} = R_1^{\ell T} R_{1,i}^\ell$ for $i = 1, 2, 3$ defines the deformed curvature tensors for micro Cosserat continuum.

### 4.4.2 Strain Tensor of Multiscale Beam

The Green-Lagrange strain tensor is still valid for the full description of deformations for a multiscale beam. According to the small strain assumptions or small base vector increments equivalently, the components of Green-Lagrange strain tensor can be approximated by using the base vector increments in an efficient manner

$$
\begin{aligned}
\varepsilon_{11} &= \underline{g}_1^{*T} \Delta \underline{G}_1^* \\
2\varepsilon_{12} &= \underline{g}_1^{*T} \Delta \underline{G}_2^* + \underline{g}_2^{*T} \Delta \underline{G}_1^* \\
2\varepsilon_{13} &= \underline{g}_1^{*T} \Delta \underline{G}_3^* + \underline{g}_3^{*T} \Delta \underline{G}_1^* \\
\varepsilon_{22} &= \underline{g}_2^{*T} \Delta \underline{G}_2^* \\
2\varepsilon_{23} &= \underline{g}_2^{*T} \Delta \underline{G}_3^* + \underline{g}_3^{*T} \Delta \underline{G}_2^* \\
\varepsilon_{33} &= \underline{g}_3^{*T} \Delta \underline{G}_3^*
\end{aligned}
\tag{4.297}
$$

where the higher order terms $\Delta \underline{G}_i^{*T} \Delta \underline{G}_j^*$ for $i, j = 1, 2, 3$ are all truncated. At first, the differences between the base vectors $\underline{G}_i^*$ and $\underline{g}_i^*$ for $i = 1, 2, 3$ are evaluated

$$
\begin{aligned}
\Delta \underline{G}_1^* &= \underline{G}_1^* - \underline{g}_1^* = R_1^T \underline{X}_{0,1} - R_0^T \underline{x}_{0,1} + \widetilde{\kappa}_1^1 \underline{S}_0 - \widetilde{\kappa}_1^0 \underline{s}_0 + \underline{S}_{0,1} - \underline{s}_{0,1} \\
\Delta \underline{G}_2^* &= \underline{G}_2^* - \underline{g}_2^* = \underline{S}_{0,2} - \underline{s}_{0,2} \\
\Delta \underline{G}_3^* &= \underline{G}_3^* - \underline{g}_3^* = \underline{S}_{0,3} - \underline{s}_{0,3}
\end{aligned}
\tag{4.298}
$$

and then minor modification leads to

$$
\begin{aligned}
\Delta \underline{G}_1^* &= R_1^T \underline{X}_{0,1} - R_0^T \underline{x}_{0,1} + (\widetilde{\kappa}_1^1 - \widetilde{\kappa}_1^0) \underline{s}_0 + \widetilde{\kappa}_1^1 (\underline{S}_0 - \underline{s}_0) + \underline{S}_{0,1} - \underline{s}_{0,1} \\
\Delta \underline{G}_2^* &= \underline{S}_{0,2} - \underline{s}_{0,2} \\
\Delta \underline{G}_3^* &= \underline{S}_{0,3} - \underline{s}_{0,3}
\end{aligned}
\tag{4.299}
$$

In view of the definitions of Lagrangian stretch and Lagrange curvature measures [31, 32], Eq. (2.152), the above equations can be further simplified to

$$
\begin{aligned}
\Delta \underline{G}_1^* &= \lfloor I,\ \widetilde{s}_0^T \rceil \underline{\epsilon}_1 + \widetilde{\kappa}_1^1 (\underline{S}_0 - \underline{s}_0) + \underline{S}_{0,1} - \underline{s}_{0,1} \\
\Delta \underline{G}_2^* &= \underline{S}_{0,2} - \underline{s}_{0,2} \\
\Delta \underline{G}_3^* &= \underline{S}_{0,3} - \underline{s}_{0,3}
\end{aligned}
\tag{4.300}
$$

The same relations hold in the case of micro Cosserat continuum, such that

$$
\begin{aligned}
R_1^{\ell T} \underline{S}_{0,1} - R_0^{\ell T} \underline{s}_{0,1} &= \lfloor I,\ \widetilde{s}_0^{\ell T} \rceil \underline{\epsilon}_1^{\ell} \\
R_1^{\ell T} \underline{S}_{0,2} - R_0^{\ell T} \underline{s}_{0,2} &= \lfloor I,\ \widetilde{s}_0^{\ell T} \rceil \underline{\epsilon}_2^{\ell} \\
R_1^{\ell T} \underline{S}_{0,3} - R_0^{\ell T} \underline{s}_{0,3} &= \lfloor I,\ \widetilde{s}_0^{\ell T} \rceil \underline{\epsilon}_3^{\ell}
\end{aligned}
\tag{4.301}
$$

where $\underline{\epsilon}_i^\ell$ for $i = 1, 2, 3$ are the Lagrangian strain measures, Eq. (2.152), defined in the local Cosserat continuum like

$$\underline{\epsilon}_i^\ell = \begin{vmatrix} R_1^{\ell T} \underline{X}_{0,i}^\ell - R_0^{\ell T} \underline{x}_{0,i}^\ell \\ \underline{\kappa}_i^{1\ell} - \underline{\kappa}_i^{0\ell} \end{vmatrix} \tag{4.302}$$

By above definitions, the difference of base vectors $\Delta \underline{G}_i^*$ for $i = 1, 2, 3$ are recasted to

$$\begin{aligned} \Delta \underline{G}_1^* &= R_0^\ell \lfloor I,\ \widetilde{s}_0^{\ell T} \rfloor \underline{\epsilon}_1^\ell + (I - R_0^\ell R_1^{\ell T}) \underline{S}_{0,1} + \lfloor I,\ \widetilde{s}_0^T \rfloor \underline{\epsilon}_1 + \widetilde{\kappa}_1^1 (\underline{S}_0 - \underline{s}_0) \\ \Delta \underline{G}_2^* &= R_0^\ell \lfloor I,\ \widetilde{s}_0^{\ell T} \rfloor \underline{\epsilon}_2^\ell + (I - R_0^\ell R_1^{\ell T}) \underline{S}_{0,2} \\ \Delta \underline{G}_3^* &= R_0^\ell \lfloor I,\ \widetilde{s}_0^{\ell T} \rfloor \underline{\epsilon}_3^\ell + (I - R_0^\ell R_1^{\ell T}) \underline{S}_{0,3} \end{aligned} \tag{4.303}$$

Due to the complexity of deformation measures, current implementation limits the theory of multiscale beam to linear strain description with the assumption of small strains, such that $\underline{\epsilon}_1 \ll 1$ and $\underline{\epsilon}_i^\ell \ll 1$ for $i = 1, 2, 3$. Note that the deformations and displacements of micro model is at least one order of magnitude smaller than those of macro model, which is constitute the fundamentals of scale separation assumption. In view of these two assumptions, the base vector increments finally can be approximated by

$$\begin{aligned} \Delta \underline{G}_1^* &= R_0^\ell \lfloor I,\ \widetilde{s}_0^{\ell T} \rfloor \underline{\epsilon}_1^\ell + \widetilde{s}_{0,1}^T R_0^\ell \underline{\phi}^\ell + \lfloor I,\ \widetilde{s}_0^T \rfloor \underline{\epsilon}_1 + \widetilde{\kappa}_1^0 (\underline{u}_s^\ell + R_0^\ell \widetilde{s}_0^{\ell T} \underline{\phi}^\ell) \\ \Delta \underline{G}_2^* &= R_0^\ell \lfloor I,\ \widetilde{s}_0^{\ell T} \rfloor \underline{\epsilon}_2^\ell + \widetilde{s}_{0,2}^T R_0^\ell \underline{\phi}^\ell \\ \Delta \underline{G}_3^* &= R_0^\ell \lfloor I,\ \widetilde{s}_0^{\ell T} \rfloor \underline{\epsilon}_3^\ell + \widetilde{s}_{0,3}^T R_0^\ell \underline{\phi}^\ell \end{aligned} \tag{4.304}$$

By introducing the micro motion $\underline{\mathcal{U}}^\ell$ in its dual bracket

$$\underline{\mathcal{U}}^\ell = \begin{vmatrix} R_0^{\ell T} \underline{u}_s^\ell \\ \underline{\phi}^\ell \end{vmatrix} \tag{4.305}$$

the differences of base vectors can be reformulated to

$$\begin{aligned} \Delta \underline{G}_1^* &= \lfloor I,\ \widetilde{s}_0^T \rfloor \underline{\epsilon}_1 + R_0^\ell \lfloor I,\ \widetilde{s}_0^{\ell T} \rfloor \underline{\epsilon}_1^\ell + \lfloor \widetilde{\kappa}_1^0 R_0^\ell,\ \widetilde{\kappa}_1^0 R_0^\ell \widetilde{s}_0^{\ell T} + \widetilde{s}_{0,1}^T R_0^\ell \rfloor \underline{\mathcal{U}}^\ell \\ \Delta \underline{G}_2^* &= R_0^\ell \lfloor I,\ \widetilde{s}_0^{\ell T} \rfloor \underline{\epsilon}_2^\ell + \lfloor 0,\ \widetilde{s}_{0,2}^T R_0^\ell \rfloor \underline{\mathcal{U}}^\ell \\ \Delta \underline{G}_3^* &= R_0^\ell \lfloor I,\ \widetilde{s}_0^{\ell T} \rfloor \underline{\epsilon}_3^\ell + \lfloor 0,\ \widetilde{s}_{0,3}^T R_0^\ell \rfloor \underline{\mathcal{U}}^\ell \end{aligned} \tag{4.306}$$

and then written in the compact form of

$$\begin{aligned} \Delta \underline{G}_1^* &= \lfloor I,\ \widetilde{s}_0^T \rfloor \underline{\epsilon}_1 + R_0^\ell \lfloor I,\ \widetilde{s}_0^{\ell T} \rfloor \underline{\epsilon}_1^\ell + C_1^\ell \underline{\mathcal{U}}^\ell \\ \Delta \underline{G}_2^* &= R_0^\ell \lfloor I,\ \widetilde{s}_0^{\ell T} \rfloor \underline{\epsilon}_2^\ell + C_2^\ell \underline{\mathcal{U}}^\ell \\ \Delta \underline{G}_3^* &= R_0^\ell \lfloor I,\ \widetilde{s}_0^{\ell T} \rfloor \underline{\epsilon}_3^\ell + C_3^\ell \underline{\mathcal{U}}^\ell \end{aligned} \tag{4.307}$$

with the help of symbols introduced as follow

$$\mathcal{C}_1^\ell = \lfloor \widetilde{\kappa}_1^0 R_0^\ell,\ \widetilde{\kappa}_1^0 R_0^\ell \widetilde{s}_0^{\ell T} + \widetilde{s}_{0,1}^T R_0^\ell \rfloor,\ \mathcal{C}_i^\ell = \lfloor 0,\ \widetilde{s}_{0,i}^T R_0^\ell \rfloor \tag{4.308}$$

for $i = 2, 3$. In view of above expressions, the vectorial Green-Lagrange strain measures can be readily found to be

$$\begin{aligned}
\underline{\varepsilon}_1 &= \mathcal{B}_g^T \lfloor I,\ \widetilde{s}_0^T \rfloor \underline{\epsilon}_1 + \mathcal{B}_g^T R_0^\ell \lfloor I,\ \widetilde{s}_0^{\ell T} \rfloor \underline{\epsilon}_1^\ell + \mathcal{B}_g^T \mathcal{C}_1^\ell \underline{\mathcal{U}}^\ell \\
\underline{\varepsilon}_2 &= \mathcal{B}_g^T R_0^\ell \lfloor I,\ \widetilde{s}_0^{\ell T} \rfloor \underline{\epsilon}_2^\ell + \mathcal{B}_g^T \mathcal{C}_2^\ell \underline{\mathcal{U}}^\ell \\
\underline{\varepsilon}_3 &= \mathcal{B}_g^T R_0^\ell \lfloor I,\ \widetilde{s}_0^{\ell T} \rfloor \underline{\epsilon}_3^\ell + \mathcal{B}_g^T \mathcal{C}_3^\ell \underline{\mathcal{U}}^\ell
\end{aligned} \tag{4.309}$$

where $\underline{\varepsilon}_i$ for $i = 1, 2, 3$ is defined by Eq. (3.72). With the definitions of following symbols

$$\mathcal{S}_0^T = \mathcal{B}_g^T \lfloor I,\ \widetilde{s}_0^T \rfloor,\ \mathcal{S}_0^{\ell T} = \mathcal{B}_g^T R_0^\ell \lfloor I,\ \widetilde{s}_0^{\ell T} \rfloor \tag{4.310}$$

and

$$\mathcal{S}_1^{\ell T} = \mathcal{B}_g^T \mathcal{C}_1^\ell = \mathcal{B}_g^T \lfloor \widetilde{\kappa}_1^0 R_0^\ell,\ \widetilde{\kappa}_1^0 R_0^\ell \widetilde{s}_0^{\ell T} + \widetilde{s}_{0,1}^T R_0^\ell \rfloor,\ \mathcal{S}_i^{\ell T} = \mathcal{B}_g^T \lfloor 0,\ \widetilde{s}_{0,i}^T R_0^\ell \rfloor \tag{4.311}$$

for $i = 2, 3$, the Green-Lagrange strain tensor in its vectorial form can be simplify described as a functions of Lagrangian stretch and curvature measures

$$\begin{aligned}
\underline{\varepsilon}_1 &= \mathcal{S}_0^T \underline{\epsilon}_1 + \mathcal{S}_0^{\ell T} \underline{\epsilon}_1^\ell + \mathcal{S}_1^{\ell T} \underline{\mathcal{U}}^\ell \\
\underline{\varepsilon}_2 &= \mathcal{S}_0^{\ell T} \underline{\epsilon}_2^\ell + \mathcal{S}_2^{\ell T} \underline{\mathcal{U}}^\ell \\
\underline{\varepsilon}_3 &= \mathcal{S}_0^{\ell T} \underline{\epsilon}_3^\ell + \mathcal{S}_3^{\ell T} \underline{\mathcal{U}}^\ell
\end{aligned} \tag{4.312}$$

In most of the analysis case, initial curvatures are zeros, $\widetilde{\kappa}_1^0 = 0$, the above equations are then simplified to

$$\begin{aligned}
\underline{\varepsilon}_1 &= \mathcal{S}_0^T \underline{\epsilon}_1 + \mathcal{S}_0^{\ell T} \underline{\epsilon}_1^\ell \\
\underline{\varepsilon}_2 &= \mathcal{S}_0^{\ell T} \underline{\epsilon}_2^\ell \\
\underline{\varepsilon}_3 &= \mathcal{S}_0^{\ell T} \underline{\epsilon}_3^\ell
\end{aligned} \tag{4.313}$$

Note that the third terms containing the motions of micro scale are truncated. The detailed explanations as to the truncations can be found in Sect. 4.1.2.

### *4.4.3 Constitutive Laws for Multiscale Beam*

Given the volume with the magnitude of $V_0$ for a beam, the strain energy $\mathcal{U}_e$ can be represented by using the vectorial Green-Lagrange strain measures

$$2\mathcal{U}_e = \int_{V_0} [\underline{\varepsilon}_1^T \mathcal{D}_{11} \underline{\varepsilon}_1 + \underline{\varepsilon}_2^T \mathcal{D}_{22} \underline{\varepsilon}_2 + \underline{\varepsilon}_3^T \mathcal{D}_{33} \underline{\varepsilon}_3 + 2(\underline{\varepsilon}_2^T \mathcal{D}_{21} \underline{\varepsilon}_1 + \underline{\varepsilon}_3^T \mathcal{D}_{31} \underline{\varepsilon}_1 + \underline{\varepsilon}_3^T \mathcal{D}_{32} \underline{\varepsilon}_2)]\, dV_0 \tag{4.314}$$

where the constitutive matrix blocks $\mathcal{D}_{ij}$ for $i, j = 1, 2, 3$ are introduced in Sect. 3.1.4. In view of the decomposition of vectorial Green-Lagrange strains into the Lagrange stretch and curvature measures for the macroscopic and microscopic models, Eq. (4.313), the first term of above full descriptions of strain energy for a beam are recasted to

$$\begin{aligned} \underline{\varepsilon}_1^T \mathcal{D}_{11} \underline{\varepsilon}_1 &= \lfloor \underline{\epsilon}_1^T, \ \underline{\epsilon}_1^{\ell T} \rfloor \begin{bmatrix} \mathcal{S}_0 \mathcal{D}_{11} \mathcal{S}_0^T & \mathcal{S}_0 \mathcal{D}_{11} \mathcal{S}_0^{\ell T} \\ \mathcal{S}_0^{\ell} \mathcal{D}_{11} \mathcal{S}_0^T & \mathcal{S}_0^{\ell} \mathcal{D}_{11} \mathcal{S}_0^{\ell T} \end{bmatrix} \begin{vmatrix} \underline{\epsilon}_1 \\ \underline{\epsilon}_1^{\ell} \end{vmatrix} \\ &= \lfloor \underline{\epsilon}_1^T, \ \underline{\epsilon}_1^{\ell T} \rfloor \begin{bmatrix} \mathcal{D}_{\epsilon\epsilon}^{11} & \mathcal{D}_{\epsilon\alpha}^{11} \\ \mathcal{D}_{\alpha\epsilon}^{11} & \mathcal{D}_{\alpha\alpha}^{11} \end{bmatrix} \begin{vmatrix} \underline{\epsilon}_1 \\ \underline{\epsilon}_1^{\ell} \end{vmatrix} \end{aligned} \tag{4.315}$$

where the following constitutive matrices are defined

$$\begin{aligned} \mathcal{D}_{\epsilon\epsilon}^{11} &= \mathcal{S}_0 \mathcal{D}_{11} \mathcal{S}_0^T, \ \mathcal{D}_{\epsilon\alpha}^{11} = \mathcal{S}_0 \mathcal{D}_{11} \mathcal{S}_0^{\ell T} \\ \mathcal{D}_{\alpha\epsilon}^{11} &= \mathcal{S}_0^{\ell} \mathcal{D}_{11} \mathcal{S}_0^T, \ \mathcal{D}_{\alpha\alpha}^{11} = \mathcal{S}_0^{\ell} \mathcal{D}_{11} \mathcal{S}_0^{\ell T} \end{aligned} \tag{4.316}$$

For $i, j = 1, 2, 3$, nonzero constitutive matrix block becomes

$$\mathcal{D}_{\epsilon\alpha}^{ij} = \mathcal{S}_0 \mathcal{D}_{ij} \mathcal{S}_0^{\ell T}, \ \mathcal{D}_{\alpha\epsilon}^{ij} = \mathcal{S}_0^{\ell} \mathcal{D}_{ij} \mathcal{S}_0^T, \ \mathcal{D}_{\alpha\alpha}^{ij} = \mathcal{S}_0^{\ell} \mathcal{D}_{ij} \mathcal{S}_0^{\ell T} \tag{4.317}$$

Due to the symmetry of constitutive matrices, it can be verified that

$$\mathcal{D}_{\alpha\epsilon}^{ijT} = \mathcal{S}_0 \mathcal{D}_{ij}^T \mathcal{S}_0^{\ell T} = \mathcal{S}_0 \mathcal{D}_{ji} \mathcal{S}_0^{\ell T} = \mathcal{D}_{\epsilon\alpha}^{ji} \tag{4.318}$$

then all the terms of strain energy for a multiscale beam can be written explicitly

$$\begin{aligned} \underline{\varepsilon}_1^T \mathcal{D}_{11} \underline{\varepsilon}_1 &= \underline{\epsilon}_1^T \mathcal{D}_{\epsilon\epsilon}^{11} \underline{\epsilon}_1 + 2 \underline{\epsilon}_1^T \mathcal{D}_{\epsilon\alpha}^{11} \underline{\epsilon}_1^{\ell} + \underline{\epsilon}_1^{\ell T} \mathcal{D}_{\alpha\alpha}^{11} \underline{\epsilon}_1^{\ell} \\ \underline{\varepsilon}_2^T \mathcal{D}_{22} \underline{\varepsilon}_2 &= \underline{\epsilon}_2^{\ell T} \mathcal{D}_{\alpha\alpha}^{22} \underline{\epsilon}_2^{\ell} \\ \underline{\varepsilon}_3^T \mathcal{D}_{33} \underline{\varepsilon}_3 &= \underline{\epsilon}_3^{\ell T} \mathcal{D}_{\alpha\alpha}^{33} \underline{\epsilon}_3^{\ell} \end{aligned} \tag{4.319}$$

and

$$\begin{aligned} \underline{\varepsilon}_2^T \mathcal{D}_{21} \underline{\varepsilon}_1 &= \underline{\epsilon}_1^T \mathcal{D}_{\epsilon\alpha}^{12} \underline{\epsilon}_2^{\ell} + \underline{\epsilon}_2^{\ell T} \mathcal{D}_{\alpha\alpha}^{21} \underline{\epsilon}_1^{\ell} \\ \underline{\varepsilon}_3^T \mathcal{D}_{31} \underline{\varepsilon}_1 &= \underline{\epsilon}_1^T \mathcal{D}_{\epsilon\alpha}^{13} \underline{\epsilon}_3^{\ell} + \underline{\epsilon}_3^{\ell T} \mathcal{D}_{\alpha\alpha}^{31} \underline{\epsilon}_1^{\ell} \\ \underline{\varepsilon}_3^T \mathcal{D}_{32} \underline{\varepsilon}_2 &= \underline{\epsilon}_3^{\ell T} \mathcal{D}_{\alpha\alpha}^{32} \underline{\epsilon}_2^{\ell} \end{aligned} \tag{4.320}$$

Given the characteristic volume with the magnitude of $V_c$ for a micro Cosserat continuum, the strain energy per characteristic volume can be written in detail like

$$\begin{aligned}\mathcal{U}_c = \int_{V_c} [\ & \frac{1}{2}\underline{\epsilon}_1^T \mathcal{D}_{\epsilon\epsilon}^{11}\underline{\epsilon}_1 + \underline{\epsilon}_1^T \mathcal{D}_{\epsilon\alpha}^{11}\underline{\epsilon}_1^\ell + \underline{\epsilon}_1^T \mathcal{D}_{\epsilon\alpha}^{12}\underline{\epsilon}_2^\ell + \underline{\epsilon}_1^T \mathcal{D}_{\epsilon\alpha}^{13}\underline{\epsilon}_3^\ell \\ & + \frac{1}{2}(\underline{\epsilon}_1^{\ell T} \mathcal{D}_{\alpha\alpha}^{11}\underline{\epsilon}_1^\ell + \underline{\epsilon}_2^{\ell T} \mathcal{D}_{\alpha\alpha}^{22}\underline{\epsilon}_2^\ell + \underline{\epsilon}_3^{\ell T} \mathcal{D}_{\alpha\alpha}^{33}\underline{\epsilon}_3^\ell) \\ & + \underline{\epsilon}_2^{\ell T} \mathcal{D}_{\alpha\alpha}^{21}\underline{\epsilon}_1^\ell + \underline{\epsilon}_3^{\ell T} \mathcal{D}_{\alpha\alpha}^{31}\underline{\epsilon}_1^\ell + \underline{\epsilon}_3^{\ell T} \mathcal{D}_{\alpha\alpha}^{32}\underline{\epsilon}_2^\ell\ ]\, dV_c\end{aligned} \tag{4.321}$$

The purpose of introducing the strain energy per characteristic volume is to produce the equivalent homogeneous stiffness matrices for a beam as discussed below. According to the definitions of Lagrangian stretch and curvature measures, Eq. (4.302), the vectorial Lagrangian measures could be explicitly related to the micro motion and its derivatives under the assumption of small motion like

$$\underline{\epsilon}_1^\ell = \underline{\mathcal{U}}_{,1}^\ell + \widetilde{\mathcal{K}}_1^{0\ell}\underline{\mathcal{U}}^\ell \tag{4.322}$$

with the following introduced symbols

$$\underline{\mathcal{U}}_{,1}^\ell = \begin{vmatrix} (R_0^{\ell T}\underline{u}_s^\ell)_{,1} \\ \underline{\phi}_{,1}^\ell \end{vmatrix},\ \widetilde{\mathcal{K}}_1^{0\ell} = \begin{bmatrix} \widetilde{\kappa}_1^{0\ell} & \widetilde{R_0^{\ell T}\underline{x}_{0,1}} \\ 0 & \widetilde{\kappa}_1^{0\ell} \end{bmatrix} \tag{4.323}$$

Due to the fact that the motions of micro Cosserat continuum are small, the micro motions can form a six-dimensional linear space approximately. Hence, the classical finite element method can be directly applied to the micro model. The local motions and the derivatives are interpolated by using the shape functions $\mathcal{H}_\epsilon^\ell$ to find

$$\underline{\mathcal{U}}^\ell = \mathcal{H}_\epsilon^\ell \underline{\mathcal{U}}_n^\ell,\ \underline{\mathcal{U}}_{,i}^\ell = \mathcal{H}_{\epsilon,i}^\ell \underline{\mathcal{U}}_n^\ell \tag{4.324}$$

where $\underline{\mathcal{U}}_n^\ell$ denotes nodal motions for micro Cosserat continuum. Finally, the Lagrangian stretch and curvature measures can be approximated by

$$\underline{\epsilon}_i^\ell = (\mathcal{H}_{\epsilon,i}^\ell + \widetilde{\mathcal{K}}_i^{0\ell}\mathcal{H}_\epsilon^\ell)\underline{\mathcal{U}}_n^\ell = \mathcal{Q}_i^\ell \underline{\mathcal{U}}_n^\ell \tag{4.325}$$

Under the assumption of scale separation, the global strain $\underline{\epsilon}_1$ of a beam is independent of local variables. Hence, the strain energy for a characteristic volume with the amplitude of $V_c$ can be rewritten as

$$\begin{aligned}\mathcal{U}_c = & \frac{1}{2}\underline{\epsilon}_1^T \int_{V_c} \mathcal{D}_{\epsilon\epsilon}^{11}\, dV_c \underline{\epsilon}_1 + \underline{\epsilon}_1^T \int_{V_c} (\mathcal{D}_{\epsilon\alpha}^{11}\mathcal{Q}_1^\ell + \mathcal{D}_{\epsilon\alpha}^{12}\mathcal{Q}_2^\ell + \mathcal{D}_{\epsilon\alpha}^{13}\mathcal{Q}_3^\ell)\, dV_c \underline{\mathcal{U}}_n^\ell \\ & + \frac{1}{2}\underline{\mathcal{U}}_n^{\ell T} \int_{V_c} (\mathcal{Q}_1^{\ell T}\mathcal{D}_{\alpha\alpha}^{11}\mathcal{Q}_1^\ell + \mathcal{Q}_2^{\ell T}\mathcal{D}_{\alpha\alpha}^{22}\mathcal{Q}_2^\ell + \mathcal{Q}_3^{\ell T}\mathcal{D}_{\alpha\alpha}^{33}\mathcal{Q}_3^\ell)\, dV_c \underline{\mathcal{U}}_n^\ell \\ & + \underline{\mathcal{U}}_n^{\ell T} \int_{V_c} (\mathcal{Q}_2^{\ell T}\mathcal{D}_{\alpha\alpha}^{21}\mathcal{Q}_1^\ell + \mathcal{Q}_3^{\ell T}\mathcal{D}_{\alpha\alpha}^{31}\mathcal{Q}_1^\ell + \mathcal{Q}_3^{\ell T}\mathcal{D}_{\alpha\alpha}^{32}\mathcal{Q}_2^\ell)\, dV_c \underline{\mathcal{U}}_n^\ell\end{aligned} \tag{4.326}$$

The strain energy per unit length called strain energy density can be obtained by using the above formulations of strain energy divided by $\ell_c$ like

$$\mathcal{U}_\rho = \frac{1}{2}\underline{\epsilon}_1^T \mathcal{C}_{\epsilon\epsilon}^{11}\underline{\epsilon}_1 + \underline{\epsilon}_1^T \mathcal{C}_{\epsilon\alpha}^{11}\underline{\mathcal{U}}_n^\ell + \frac{1}{2}\underline{\mathcal{U}}_n^{\ell T}\mathcal{C}_{\alpha\alpha}\underline{\mathcal{U}}_n^\ell \tag{4.327}$$

where the equivalent uniform stiffness matrices are computed from the averaged integral [34] as follow

$$\mathcal{C}_{\epsilon\epsilon}^{11} = \frac{\int_{V_c}\mathcal{D}_{\epsilon\epsilon}^{11}dV_c}{\ell_c},\ \mathcal{C}_{\epsilon\alpha}^{11} = \frac{\int_{V_c}(\mathcal{D}_{\epsilon\alpha}^{11}\mathcal{Q}_1^\ell + \mathcal{D}_{\epsilon\alpha}^{12}\mathcal{Q}_2^\ell + \mathcal{D}_{\epsilon\alpha}^{13}\mathcal{Q}_3^\ell)dV_c}{\ell_c} \tag{4.328}$$

$$\begin{aligned}\mathcal{C}_{\alpha\alpha} = &\frac{\int_{V_c}(\mathcal{Q}_1^{\ell T}\mathcal{D}_{\alpha\alpha}^{11}\mathcal{Q}_1^\ell + \mathcal{Q}_2^{\ell T}\mathcal{D}_{\alpha\alpha}^{22}\mathcal{Q}_2^\ell + \mathcal{Q}_3^{\ell T}\mathcal{D}_{\alpha\alpha}^{33}\mathcal{Q}_3^\ell)dV_c}{\ell_c}\\ &+2\frac{\int_{V_c}(\mathcal{Q}_2^{\ell T}\mathcal{D}_{\alpha\alpha}^{21}\mathcal{Q}_1^\ell + \mathcal{Q}_3^{\ell T}\mathcal{D}_{\alpha\alpha}^{31}\mathcal{Q}_1^\ell + \mathcal{Q}_3^{\ell T}\mathcal{D}_{\alpha\alpha}^{32}\mathcal{Q}_2^\ell)dV_c}{\ell_c}\end{aligned} \tag{4.329}$$

### 4.4.4 Variation of Strain Energy of Multiscale Beam

Since the constitutive matrix blocks are determined by the above averaged quantities on the micro Cosserat continuum, the strain energy for a beam with volume of $V_0$ can be written as

$$\mathcal{U}_e = \int_{\ell_0}\mathcal{U}_\rho d\alpha_1 = \int_{\ell_0}\left(\frac{1}{2}\underline{\epsilon}_1^T \mathcal{C}_{\epsilon\epsilon}^{11}\underline{\epsilon}_1 + \underline{\epsilon}_1^T \mathcal{C}_{\epsilon\alpha}^{11}\underline{\mathcal{U}}_n^\ell + \frac{1}{2}\underline{\mathcal{U}}_n^{\ell T}\mathcal{C}_{\alpha\alpha}\underline{\mathcal{U}}_n^\ell\right)d\alpha_1 \tag{4.330}$$

Taking variation of averaged strain energy density leads to

$$\delta\mathcal{U}_\rho = \delta\underline{\epsilon}_1^T \mathcal{C}_{\epsilon\epsilon}^{11}\underline{\epsilon}_1 + \delta\underline{\epsilon}_1^T \mathcal{C}_{\epsilon\alpha}^{11}\underline{\mathcal{U}}_n^\ell + \delta\underline{\mathcal{U}}_n^{\ell T}\mathcal{C}_{\epsilon\alpha}^{11T}\underline{\epsilon}_1 + \delta\underline{\mathcal{U}}_n^{\ell T}\mathcal{C}_{\alpha\alpha}\underline{\mathcal{U}}_n^\ell \tag{4.331}$$

After introducing the following stress measures

$$\underline{F}_{\epsilon\epsilon}^{11} = \mathcal{C}_{\epsilon\epsilon}^{11}\underline{\epsilon}_1,\ \underline{F}_{\alpha\epsilon} = \mathcal{C}_{\epsilon\alpha}^{11T}\underline{\epsilon}_1 \tag{4.332}$$

The variations of strain energy density can be written in compact form like

$$\delta\mathcal{U}_\rho = \delta\underline{\epsilon}_1^T(\underline{F}_{\epsilon\epsilon}^{11} + \mathcal{C}_{\epsilon\alpha}^{11}\underline{\mathcal{U}}_n^\ell) + \delta\underline{\mathcal{U}}_n^{\ell T}(\underline{F}_{\alpha\epsilon} + \mathcal{C}_{\alpha\alpha}\underline{\mathcal{U}}_n^\ell) \tag{4.333}$$

From which, the variation of strain energy $\mathcal{U}_e$ is readily evaluated

$$\delta\mathcal{U}_e = \int_{\ell_0}\delta\mathcal{U}_\rho d\alpha_1 \tag{4.334}$$

### 4.4.5 Virtual Work of External Forces for Multiscale Beam

Different form the definitions of external force applied to a beam, Eq. (3.324), the averaged quantities of external forces in the case of multiscale beam are introduced

$$\underline{F}_{\epsilon\alpha} = \begin{vmatrix} \dfrac{\int_{V_c} \underline{g}\, dV_c + \int_{S_c} \underline{p}_s\, dS_c}{\ell_c} \\ \dfrac{\int_{V_c} \widetilde{s}_0 \underline{g}\, dV_c + \int_{S_c} \widetilde{s}_0 \underline{p}_s\, dS_c}{\ell_c} \end{vmatrix}$$
$$\underline{F}_{\alpha\alpha} = \frac{1}{\ell_c}\left( \int\limits_{V_c} \mathcal{H}_\epsilon^{\ell T} \begin{vmatrix} R_0^{\ell T} \underline{g} \\ \widetilde{s}_0^{\ell} R_0^{\ell T} \underline{g} \end{vmatrix} dV_c + \int\limits_{S_c} \mathcal{H}_\epsilon^{\ell T} \begin{vmatrix} R_0^{\ell T} \underline{p}_s \\ \widetilde{s}_0^{\ell} R_0^{\ell T} \underline{p}_s \end{vmatrix} dS_c \right) \tag{4.335}$$

where $\ell_c$ is the characteristic length of a beam. Usually, $\ell_c$ is selected as the length of a beam element. With the application of averaged external force in above vectorial form, the work of external force applied to the macro model can be written as

$$\delta \mathcal{W}_e = \int\limits_{\ell_0} \underline{\delta\mathcal{U}}^T \underline{F}_{\epsilon\alpha} + \underline{\delta\mathcal{U}_n^{\ell T}} \underline{F}_{\alpha\alpha} d\alpha_1 \tag{4.336}$$

### 4.4.6 Governing Equations of Multiscale Beam in Motion Formalism

By substituting Eqs. (4.334) and (4.336) into the principle of virtual work, which states

$$\delta \mathcal{U}_e - \delta \mathcal{W}_e = 0 \tag{4.337}$$

the governing equations of multiscale beam can be readily obtained like

$$\begin{aligned} \int\limits_{\ell_0} [\, & \delta \underline{\epsilon}_1^T (\underline{F}_{\epsilon\epsilon}^{11} + C_{\epsilon\alpha}^{11} \underline{\mathcal{U}_n^\ell}) + \delta \underline{\mathcal{U}_n^{\ell T}} (\underline{F}_{\alpha\epsilon} + C_{\alpha\alpha} \underline{\mathcal{U}_n^\ell}) \\ & - \underline{\delta\mathcal{U}}^T \underline{F}_{\epsilon\alpha} - \underline{\delta\mathcal{U}_n^{\ell T}} \underline{F}_{\alpha\alpha} \,] d\alpha_1 = 0 \end{aligned} \tag{4.338}$$

Based on the assumption of scale separation, the variations of macro strain measures are independent of the variations of micro displacements. Hence, the original problem can be decomposed into the macroscopic beam problem

$$\int\limits_{\ell_0} \left[ \delta \underline{\epsilon}_1^T (\underline{F}_{\epsilon\epsilon}^{11} + C_{\epsilon\alpha}^{11} \underline{\mathcal{U}_n^\ell}) - \underline{\delta\mathcal{U}}^T \underline{F}_{\epsilon\alpha} \right] d\alpha_1 = 0 \tag{4.339}$$

and the microscopic Cosserat continuum problem

$$\mathcal{C}_{\alpha\alpha}\underline{\mathcal{U}}_n^\ell + \underline{F}_{\alpha\epsilon} - \underline{F}_{\alpha\alpha} = 0 \tag{4.340}$$

with appropriate boundary conditions. Once the elastic forces $\underline{F}_{\alpha\epsilon}$ and external forces $\underline{F}_{\alpha\alpha}$ are given, the micro displacements could be solved from above equations as

$$\underline{\mathcal{U}}_n^\ell = \mathcal{C}_{\alpha\alpha}^{-1}(\underline{F}_{\alpha\alpha} - \underline{F}_{\alpha\epsilon}) \tag{4.341}$$

The macro continuum problem becomes solvable after eliminating the micro unknowns $\underline{\mathcal{U}}_n^\ell$ by substituting Eq. (4.341) into Eq. (4.339). After merging similar terms, the macro continuum problem is recasted to

$$\int_{\ell_0} [\ \delta\underline{\epsilon}_1^T(\widehat{\mathcal{C}}_{\epsilon\epsilon}^{11}\underline{\epsilon}_1 + \mathcal{C}_{\epsilon\alpha}^{11}\mathcal{C}_{\alpha\alpha}^{-1}\underline{F}_{\alpha\alpha}) - \underline{\delta\mathcal{U}}^T\underline{F}_{\epsilon\alpha}\ ]d\alpha_1 = 0 \tag{4.342}$$

where the stiffness matrix of a beam is modified to

$$\widehat{\mathcal{C}}_{\epsilon\epsilon}^{11} = \mathcal{C}_{\epsilon\epsilon}^{11} - \mathcal{C}_{\epsilon\alpha}^{11}\mathcal{C}_{\alpha\alpha}^{-1}\mathcal{C}_{\epsilon\alpha}^{11T} \tag{4.343}$$

After introduced the stress measures as follow

$$\underline{\widehat{F}}_{\epsilon\epsilon}^{11} = \widehat{\mathcal{C}}_{\epsilon\epsilon}^{11}\underline{\epsilon}_1 + \mathcal{C}_{\epsilon\alpha}^{11}\mathcal{C}_{\alpha\alpha}^{-1}\underline{F}_{\alpha\alpha} \tag{4.344}$$

the macroscopic continuum problem can be written in the compact form

$$\int_{\ell_0} (\delta\underline{\epsilon}_1^T\underline{\widehat{F}}_{\epsilon\epsilon}^{11} - \underline{\delta\mathcal{U}}^T\underline{F}_{\epsilon\alpha})d\alpha_1 = 0 \tag{4.345}$$

In view of compatibility attribute, Eq. (2.161), it leads to

$$\int_{\ell_0} \left[(\underline{\delta\mathcal{U}}_{,1} + \widetilde{\mathcal{K}}_1^1\underline{\delta\mathcal{U}})^T\underline{\widehat{F}}_{\epsilon\epsilon}^{11} - \underline{\delta\mathcal{U}}^T\underline{F}_{\epsilon\alpha}\right]d\alpha_1 = 0 \tag{4.346}$$

With the application of finite element discretization, the weak formula of the macro problem becomes

$$\underline{\delta\mathcal{U}}_n^T \int_{\ell_0} [(\mathcal{H}_{\epsilon,1}^T + \mathcal{H}_\epsilon^T\widetilde{\mathcal{K}}_1^{1T})\underline{\widehat{F}}_{\epsilon\epsilon}^{11} - \mathcal{H}_\epsilon^T\underline{F}_{\epsilon\alpha}]d\alpha_1 = 0 \tag{4.347}$$

From which, the governing equations are obtained

$$\int_{\ell_0} [(\mathcal{H}_{\epsilon,1}^T + \mathcal{H}_\epsilon^T \widetilde{\mathcal{K}}_1^{1T})\underline{\widehat{F}}_{\epsilon\epsilon}^{11} - \mathcal{H}_\epsilon^T \underline{F}_{\epsilon\alpha}]d\alpha_1 = 0 \tag{4.348}$$

After introduced the elastic force vector

$$\underline{\mathcal{F}}_e = \int_{\ell_0} (\mathcal{H}_{\epsilon,1}^T + \mathcal{H}_\epsilon^T \widetilde{\mathcal{K}}_1^{1T})\underline{\widehat{F}}_{\epsilon\epsilon}^{11} d\alpha_1 \tag{4.349}$$

and the external force vector

$$\underline{\mathcal{F}}_a = \int_{S_0} \mathcal{H}_\epsilon^T \underline{F}_{\epsilon\alpha} dS_0 \tag{4.350}$$

the governing equations can be described by the following compact formula

$$\underline{\mathcal{F}}_e - \underline{\mathcal{F}}_a = 0 \tag{4.351}$$

### 4.4.7 Extended to Multiscale Beam Dynamic Problem

The velocity of an arbitrary material particle in a multiscale beam could be evaluated by taking time derivatives of position vector $\underline{P}$, defined by Eq. (4.285), to get

$$\dot{\underline{P}} = (\underline{x}_0 + \underline{u}_s \dot{} + R_1 \underline{S}_0) = \dot{\underline{u}}_s + \dot{R}_1 \underline{S}_0 + R_1 \dot{\underline{S}}_0 \tag{4.352}$$

Due to the fact that the motion of micro model is at least one order of magnitude smaller than that of the macro model, the velocity of micro model becomes negligible, $R_1 \dot{\underline{S}}_0 \approx 0$. Meanwhile, the centrifugal velocity is approximated by

$$\dot{R}_1 \underline{S}_0 \approx \dot{R}_1 \underline{s}_0 \tag{4.353}$$

After resolved the time derivatives into the body attached frame $\mathcal{F}_1$, the particle velocity relates to the linear and angular velocities of reference particle like

$$R_1^T \dot{\underline{P}} = R_1^T \dot{\underline{u}}_s + \widetilde{s}_0^T \underline{\omega} \tag{4.354}$$

The kinematic energy of multiscale beam can be approximated by Eq. (3.341). The associated mass matrix is evaluated by averaging the mass matrix of micro continuum with the characteristic volume of magnitude $V_c$ like

$$\mathcal{M}_{\epsilon\epsilon} = \frac{1}{\ell_c} \int_{V_c} \begin{bmatrix} \rho I, & \rho \widetilde{s}_0^T \\ \rho \widetilde{s}_0, & \rho \widetilde{s}_0 \widetilde{s}_0^T \end{bmatrix} dV_c \tag{4.355}$$

The mass matrix and the Coriolis force vector are then defined by Eq. (3.350). When considering the inertial forces, the governing equations of motion for multiscale beam will be decoupled into two sub-systems, one is dynamic macroscopic system

$$\mathcal{M}_n \dot{\underline{\mathcal{V}}}_n - \underline{\mathcal{F}}_g + \underline{\mathcal{F}}_e = \underline{\mathcal{F}}_a \tag{4.356}$$

and the other is the static microscopic system, Eq. (4.340).

## 4.5 A Special Formula of Multiscale Beam

This section will introduce a special formula of multiscale beam composed of a macro one-dimensional Cosserat continuum and a two-dimensional micro Cosserat continuum. According to the assumption of scale separation, the macro variables, such as positions, motions and strain measures of macro model are independent of the same variables of micro model. Therefore, the original multiscale problem can be decomposed into the macroscopic beam problem and the microscopic two-dimensional continuum problem. The macroscopic model is labeled only by a material coordinate $\alpha_1$, while the micro continuum will be labeled by using the rest material coordinates $\alpha_2$ and $\alpha_3$.

### 4.5.1 *Kinematics of a Special Multiscale Beam*

As shown in Fig. 4.5, a material particle with infinitesimal volume in a beam is depicted by its position and director vectors in its reference configuration and current deformed configuration, respectively. In the reference configuration, the position

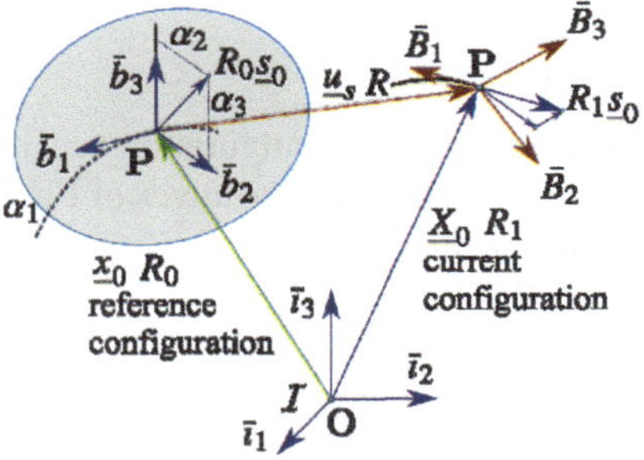


**Fig. 4.5** Kinematics of a material particle with infinitesimal volume in a multiscale beam

vector $\underline{p}$ of the material particle relative to a point **O**, the original of reference frame, $\mathcal{F}_{\mathcal{I}} = [\mathbf{O}, \mathcal{I}]$, is defined as a function of the general curvilinear coordinates $(\alpha_1, \alpha_2, \alpha_3)$ like

$$\underline{p}(\alpha_1, \alpha_2, \alpha_3) = \underline{x}_0(\alpha_1) + R_0(\alpha_1)\underline{s}_0(\alpha_2, \alpha_3) \tag{4.357}$$

where $\underline{x}_0$ is the position vector of a reference material particle **P** relative to a point **O**, and $R_0$ denotes the initial rotation tensor measures the orientation of the reference basis $\mathcal{B}_0$. It has been verified in Sect. 2.2.3 the rotation tensor $R_0$ is identical to the body attached basis $\mathcal{B}_0$, and in detail

$$R_0 = \lfloor \bar{b}_1, \bar{b}_2, \bar{b}_3 \rfloor = \mathcal{B}_0 \tag{4.358}$$

Note that the initial basis $\mathcal{B}_0$ is bounded by the material particle **P** to produce the reference frame $\mathcal{F}_0 = [\mathbf{P}, \mathcal{B}_0]$. The local vector $\underline{s}_0$, defined in this reference frame $\mathcal{F}_0$, describes the director of a material particle. Obviously, Eq. (4.357) decomposes the position vector $\underline{p}$ of a material particle into a reference position $\underline{x}_0$ and a local vector $\underline{s}_0$. In classical beam theory, director $\underline{s}_0$ is a rigid vector. Now it becomes deformable under the framework of multiscale beam theory. If assuming the micro continuum to be a two-dimensional Cosserat continuum, as depicted in Fig. 4.5, the director $\underline{s}_0$ can be described in a similar approach to the definition of position vector, Eq. (4.357), like

$$\underline{s}_0(\alpha_2, \alpha_3) = \underline{x}_0^\ell(\alpha_2, \alpha_3) + R_0^\ell(\alpha_2, \alpha_3)\underline{s}_0^\ell(\alpha_2, \alpha_3) \tag{4.359}$$

where $\underline{x}_0^\ell$ is the position vector of reference particle $\mathbf{P}^\ell$ in local Cosserat continuum, and $R_0^\ell$ the local rotation tensor attached to the reference particle $\mathbf{P}^\ell$ where the local director $\underline{s}_0^\ell$ is measured in the local basis spanned by $R_0^\ell$. In view of above formula, the position vector $\underline{p}$ can be determined from the following expression

$$\underline{p} = \underline{x}_0 + R_0(\underline{x}_0^\ell + R_0^\ell \underline{s}_0^\ell) \tag{4.360}$$

For the description of beam deformations, the covariant base vectors

$$\underline{g}_i = \frac{\partial \underline{p}}{\partial \alpha_i} \tag{4.361}$$

in the reference configuration are required, and the specific expressions can be obtained by taking derivatives of Eq. (4.357) with respected to the curvilinear coordinates $\alpha_i$ like

$$\begin{aligned} \underline{g}_1 &= \underline{x}_{0,1} + R_{0,1}\underline{s}_0 \\ \underline{g}_2 &= R_0\underline{s}_{0,2} \\ \underline{g}_3 &= R_0\underline{s}_{0,3} \end{aligned} \tag{4.362}$$

where the symbol $(\cdot)_{,i}$ for $i = 1, 2, 3$ denotes the spatial derivatives with respected to $\alpha_i$ and

$$(\cdot)_{,i} = \frac{\partial(\cdot)}{\partial\alpha_i} \tag{4.363}$$

Resolving the base vectors into the local reference frame $\mathcal{F}_0$ obtains

$$\begin{aligned}
\underline{g}_1^* &= R_0^T \underline{x}_{0,1} + R_0^T R_{0,1} \underline{s}_0 \\
\underline{g}_2^* &= \underline{s}_{0,2} \\
\underline{g}_3^* &= \underline{s}_{0,3}
\end{aligned} \tag{4.364}$$

According to the definitions of initial curvature tensors, Eq. (1.314), the base vectors can be represented by the following formulas

$$\begin{aligned}
\underline{g}_1^* &= R_0^T \underline{x}_{0,1} + \widetilde{\kappa}_1^0 \underline{s}_0 \\
\underline{g}_2^* &= \underline{s}_{0,2} \\
\underline{g}_3^* &= \underline{s}_{0,3}
\end{aligned} \tag{4.365}$$

where the symbol $\widetilde{\kappa}_1^0 = R_0^T R_{0,1}$ denotes the initial curvature tensor of a beam. All these base vectors form the covariant bases $\mathcal{B}_g$ like

$$\mathcal{B}_g = \lfloor \underline{g}_1^*,\ \underline{g}_2^*,\ \underline{g}_3^* \rfloor \tag{4.366}$$

and then the metric tensor in the reference configuration is computed from

$$g = \mathcal{B}_g^T \mathcal{B}_g \tag{4.367}$$

The similar manipulations can be performed repeatedly in the local Cosserat continuum to find that the derivatives of director becomes

$$\begin{aligned}
\underline{s}_{0,1} &= \underline{0} \\
\underline{s}_{0,2} &= \underline{x}_{0,2}^\ell + R_{0,2}^\ell \underline{s}_0^\ell + R_0^\ell \underline{s}_{0,2}^\ell \\
\underline{s}_{0,3} &= \underline{x}_{0,3}^\ell + R_{0,3}^\ell \underline{s}_0^\ell + R_0^\ell \underline{s}_{0,3}^\ell
\end{aligned} \tag{4.368}$$

If defined the local curvature tensors as $\widetilde{\kappa}_i^{0\ell} = R_0^{\ell T} R_{0,i}^\ell$ for $i = 2, 3$, the local base vectors can be resolved into the local basis spanned by $R_0^\ell$ like

$$\begin{aligned}
R_0^{\ell T} \underline{s}_{0,1} &= \underline{0} \\
R_0^{\ell T} \underline{s}_{0,2} &= R_0^{\ell T} \underline{x}_{0,2}^\ell + \widetilde{\kappa}_2^{0\ell} \underline{s}_0^\ell + \underline{s}_{0,2}^\ell \\
R_0^{\ell T} \underline{s}_{0,3} &= R_0^{\ell T} \underline{x}_{0,3}^\ell + \widetilde{\kappa}_3^{0\ell} \underline{s}_0^\ell + \underline{s}_{0,3}^\ell
\end{aligned} \tag{4.369}$$

When the beam changes its spatial configuration, the material particle that had position vector $\underline{p}$ in the undeformed reference configuration now has position vector $\underline{P}$ in the deformed configuration, such that

$$\underline{P}(\alpha_1, \alpha_2, \alpha_3) = \underline{X}_0(\alpha_1) + R_1(\alpha_1)\underline{S}_0(\alpha_2, \alpha_3) \tag{4.370}$$

where $\underline{S}_0$ represents the deformed director in current configuration, and $R_1$ the total rotation tensor that describes the orientation of basis $\mathcal{B}_1$ attached to the deformed configuration and

$$R_1 = \lfloor \bar{B}_1, \bar{B}_2, \bar{B}_3 \rfloor = \mathcal{B}_1 \tag{4.371}$$

Note that bounding the current deformed basis $\mathcal{B}_1$ by the material particle **P**, the associated deformed frame $\mathcal{F}_1 = [\mathbf{P}, \mathcal{B}_1]$ will be constructed. Furthermore, if the displacement vector $\underline{u}_s$ is given to describe an arbitrary translation of reference particle **P**, the reference position vector $\underline{X}_0$ in current deformed configuration can be recasted to

$$\underline{X}_0(\alpha_1) = \underline{x}_0(\alpha_1) + \underline{u}_s(\alpha_1) \tag{4.372}$$

In the similar manner, the deformed director $\underline{S}_0$ can be decomposed into

$$\underline{S}_0(\alpha_2, \alpha_3) = \underline{X}_0^\ell(\alpha_2, \alpha_3) + R_1^\ell(\alpha_2, \alpha_3)\underline{s}_0^\ell(\alpha_2, \alpha_3) \tag{4.373}$$

where $\underline{X}_0^\ell$ is the deformed position vector of reference particle $\mathbf{P}^\ell$, and $R_1^\ell$ the deformed rotation tensor in the local continuum. The position vector $\underline{X}_0^\ell$ can be related to its original position through

$$\underline{X}_0^\ell(\alpha_2, \alpha_3) = \underline{x}_0^\ell(\alpha_2, \alpha_3) + \underline{u}_s^\ell(\alpha_2, \alpha_3) \tag{4.374}$$

where $\underline{u}_s^\ell$ is the displacement vector of local director $\underline{s}_0^\ell$ in micro continuum. In view of the definitions of covariant base vectors in deformed configuration

$$\underline{G}_i = \frac{\partial \underline{P}}{\partial \alpha_i} \tag{4.375}$$

the specific formulas can be obtained by taking spatial derivatives of Eq. (4.370) with respected to the curvilinear coordinates $\alpha_i$ for $i = 1, 2, 3$ like

$$\begin{aligned} \underline{G}_1 &= \underline{X}_{0,1} + R_{1,1}\underline{S}_0 \\ \underline{G}_2 &= R_1\underline{S}_{0,2} \\ \underline{G}_3 &= R_1\underline{S}_{0,3} \end{aligned} \tag{4.376}$$

Once again, the base vectors are resolved into the current local frame $\mathcal{F}_1$ to obtain

$$\begin{aligned} \underline{G}_1^* &= R_1^T \underline{X}_{0,1} + R_1^T R_{1,1} \underline{S}_0 \\ \underline{G}_2^* &= R_1^T R_1 \underline{S}_{0,2} \\ \underline{G}_3^* &= R_1^T R_1 \underline{S}_{0,3} \end{aligned} \tag{4.377}$$

Introducing the definitions of current curvature tensors $\widetilde{\kappa}_1^1 = R_1^T R_{1,1}$, the covariant base vectors are simplified to

$$\begin{aligned} \underline{G}_1^* &= R_1^T \underline{X}_{0,1} + \widetilde{\kappa}_1^1 \underline{S}_0 \\ \underline{G}_2^* &= \underline{S}_{0,2} \\ \underline{G}_3^* &= \underline{S}_{0,3} \end{aligned} \tag{4.378}$$

All these base vectors could be assembled into the covariant bases $\mathcal{B}_G$ like

$$\mathcal{B}_G = \lfloor \underline{G}_1^*,\ \underline{G}_2^*,\ \underline{G}_3^* \rfloor \tag{4.379}$$

The similar expressions of the deformed covariant base vectors $\underline{S}_{0,i}$ for $i = 1, 2, 3$ are obtained

$$\begin{aligned} R_1^{\ell T} \underline{S}_{0,1} &= \underline{0} \\ R_1^{\ell T} \underline{S}_{0,2} &= R_1^{\ell T} \underline{X}_{0,2}^{\ell} + \widetilde{\kappa}_2^{1\ell} \underline{s}_0^{\ell} + \underline{s}_{0,2}^{\ell} \\ R_1^{\ell T} \underline{S}_{0,3} &= R_1^{\ell T} \underline{X}_{0,3}^{\ell} + \widetilde{\kappa}_3^{1\ell} \underline{s}_0^{\ell} + \underline{s}_{0,3}^{\ell} \end{aligned} \tag{4.380}$$

with the aid of the deformed curvature tensors $\widetilde{\kappa}_i^{1\ell} = R_1^{\ell T} R_{1,i}^{\ell}$ for $i = 2, 3$ defined in the local Cosserat continuum.

## 4.5.2 Strain Tensor of a Special Multiscale Beam

Before measuring the components of Green-Lagrange strain tensor, the differences between the base vectors $\underline{G}_i^*$ and $\underline{g}_i^*$ for $i = 1, 2, 3$ are evaluated at first

$$\Delta \underline{G}_i^* = \underline{G}_i^* - \underline{g}_i^* \tag{4.381}$$

and then the components of Green-Lagrange strain tensor are approximated to be

$$
\begin{aligned}
\varepsilon_{11} &= \frac{1}{2}[(\underline{g}_1^* + \Delta\underline{G}_1^*)^T(\underline{g}_1^* + \Delta\underline{G}_1^*) - \underline{g}_1^{*T}\underline{g}_1^*] \approx \underline{g}_1^{*T}\Delta\underline{G}_1^* \\
2\varepsilon_{12} &= (\underline{g}_1^* + \Delta\underline{G}_1^*)^T(\underline{g}_2^* + \Delta\underline{G}_2^*) - \underline{g}_1^{*T}\underline{g}_2^* \approx \underline{g}_1^{*T}\Delta\underline{G}_2^* + \underline{g}_2^{*T}\Delta\underline{G}_1^* \\
2\varepsilon_{13} &= (\underline{g}_1^* + \Delta\underline{G}_1^*)^T(\underline{g}_3^* + \Delta\underline{G}_3^*) - \underline{g}_1^{*T}\underline{g}_3^* \approx \underline{g}_1^{*T}\Delta\underline{G}_3^* + \underline{g}_3^{*T}\Delta\underline{G}_1^* \\
\varepsilon_{22} &= \frac{1}{2}[(\underline{g}_2^* + \Delta\underline{G}_2^*)^T(\underline{g}_2^* + \Delta\underline{G}_2^*) - \underline{g}_2^{*T}\underline{g}_2^*] \approx \underline{g}_2^{*T}\Delta\underline{G}_2^* \\
2\varepsilon_{23} &= (\underline{g}_2^* + \Delta\underline{G}_2^*)^T(\underline{g}_3^* + \Delta\underline{G}_3^*) - \underline{g}_2^{*T}\underline{g}_3^* \approx \underline{g}_2^{*T}\Delta\underline{G}_3^* + \underline{g}_3^{*T}\Delta\underline{G}_2^* \\
\varepsilon_{33} &= \frac{1}{2}[(\underline{g}_3^* + \Delta\underline{G}_3^*)^T(\underline{g}_3^* + \Delta\underline{G}_3^*) - \underline{g}_3^{*T}\underline{g}_3^*] \approx \underline{g}_3^{*T}\Delta\underline{G}_3^*
\end{aligned}
\tag{4.382}
$$

where the higher order terms $\Delta\underline{G}_i^{*T}\Delta\underline{G}_j^*$ for $i, j = 1, 2, 3$ are all truncated under the assumption of small strain measures. Referring Eqs. (4.365) and (4.378), the detailed expressions of $\Delta\underline{G}_i^*$ for $i = 1, 2, 3$ are found to be

$$
\begin{aligned}
\Delta\underline{G}_1^* &= R_1^T\underline{X}_{0,1} - R_0^T\underline{x}_{0,1} + (\widetilde{\kappa}_1^1 - \widetilde{\kappa}_1^0)\underline{s}_0 + \widetilde{\kappa}_1^1(\underline{S}_0 - \underline{s}_0) \\
\Delta\underline{G}_2^* &= \underline{S}_{0,2} - \underline{s}_{0,2} \\
\Delta\underline{G}_3^* &= \underline{S}_{0,3} - \underline{s}_{0,3}
\end{aligned}
\tag{4.383}
$$

By the definitions of Lagrangian stretch and curvature measures [31, 32], Eq. (2.152), the above equations can be further simplified to

$$
\begin{aligned}
\Delta\underline{G}_1^* &= \lfloor I,\ \widetilde{s}_0^T \rfloor \underline{\epsilon}_1 + \widetilde{\kappa}_1^1(\underline{S}_0 - \underline{s}_0) \\
\Delta\underline{G}_2^* &= \underline{S}_{0,2} - \underline{s}_{0,2} \\
\Delta\underline{G}_3^* &= \underline{S}_{0,3} - \underline{s}_{0,3}
\end{aligned}
\tag{4.384}
$$

The same relations hold in the case of micro Cosserat continuum, such that

$$
\begin{aligned}
R_1^{\ell T}\underline{S}_{0,2} - R_0^{\ell T}\underline{s}_{0,2} &= \lfloor I,\ \widetilde{s}_0^{\ell T} \rfloor \underline{\epsilon}_2^\ell \\
R_1^{\ell T}\underline{S}_{0,3} - R_0^{\ell T}\underline{s}_{0,3} &= \lfloor I,\ \widetilde{s}_0^{\ell T} \rfloor \underline{\epsilon}_3^\ell
\end{aligned}
\tag{4.385}
$$

where $\underline{\epsilon}_i^\ell$ for $i = 2, 3$ are the Lagrangian stretch and curvature measures, Eq. (2.152), defined in the local micro Cosserat continuum

$$
\underline{\epsilon}_i^\ell = \left| \begin{matrix} R_1^{\ell T}\underline{X}_{0,i}^\ell - R_0^{\ell T}\underline{x}_{0,i}^\ell \\ \underline{\kappa}_i^{1\ell} - \underline{\kappa}_i^{0\ell} \end{matrix} \right|
\tag{4.386}
$$

With the aid of above definitions, the difference of base vectors $\Delta\underline{G}_i^*$ for $i = 1, 2, 3$ are recasted to

$$
\begin{aligned}
\Delta\underline{G}_1^* &= \lfloor I,\ \widetilde{s}_0^T \rfloor \underline{\epsilon}_1 + \widetilde{\kappa}_1^1(\underline{S}_0 - \underline{s}_0) \\
\Delta\underline{G}_2^* &= R_0^\ell \lfloor I,\ \widetilde{s}_0^{\ell T} \rfloor \underline{\epsilon}_2^\ell + (I - R_0^\ell R_1^{\ell T})\underline{S}_{0,2} \\
\Delta\underline{G}_3^* &= R_0^\ell \lfloor I,\ \widetilde{s}_0^{\ell T} \rfloor \underline{\epsilon}_3^\ell + (I - R_0^\ell R_1^{\ell T})\underline{S}_{0,3}
\end{aligned}
\tag{4.387}
$$

Due to the complexity of the problem, the theory of multiscale beam is limited to linear strain description with the assumptions of $\underline{\epsilon}_1 \ll 1$ and $\underline{\epsilon}_i^\ell \ll 1$ for $i = 2, 3$. The assumption of scale separation also shows that the deformations and motions of micro Cosserat continuum are at least one order of magnitude smaller than those of macro beam model. It answers the micro displacements $\underline{u}_s^\ell \ll 1$ and micro rotation are all small quantities, such that $\underline{u}_s^\ell \ll 1$ and $\underline{\phi}^\ell \ll 1$, which leads to

$$\begin{aligned} \Delta\underline{G}_1^* &= \lfloor I,\ \widetilde{s}_0^T \rfloor \underline{\epsilon}_1 + \widetilde{\kappa}_1^0 (\underline{u}_s^\ell + R_0^\ell \widetilde{s}_0^{\ell T} \underline{\phi}^\ell) \\ \Delta\underline{G}_2^* &= R_0^\ell \lfloor I,\ \widetilde{s}_0^{\ell T} \rfloor \underline{\epsilon}_2^\ell + \widetilde{s}_{0,2}^T R_0^\ell \underline{\phi}^\ell \\ \Delta\underline{G}_3^* &= R_0^\ell \lfloor I,\ \widetilde{s}_0^{\ell T} \rfloor \underline{\epsilon}_3^\ell + \widetilde{s}_{0,3}^T R_0^\ell \underline{\phi}^\ell \end{aligned} \tag{4.388}$$

More details of above simplifications can be found in Sect. 4.1.2. After introduced the micro motion measures in its dual bracket

$$\underline{\mathcal{U}}^\ell = \begin{vmatrix} R_0^{\ell T} \underline{u}_s^\ell \\ \underline{\phi}^\ell \end{vmatrix} \tag{4.389}$$

the differences of base vectors can be reformulated to

$$\begin{aligned} \Delta\underline{G}_1^* &= \lfloor I,\ \widetilde{s}_0^T \rfloor \underline{\epsilon}_1 + \widetilde{\kappa}_1^0 R_0^\ell \lfloor I,\ \widetilde{s}_0^{\ell T} \rfloor \underline{\mathcal{U}}^\ell \\ \Delta\underline{G}_2^* &= R_0^\ell \lfloor I,\ \widetilde{s}_0^{\ell T} \rfloor \underline{\epsilon}_2^\ell + \lfloor 0,\ \widetilde{s}_{0,2}^T R_0^\ell \rfloor \underline{\mathcal{U}}^\ell \\ \Delta\underline{G}_3^* &= R_0^\ell \lfloor I,\ \widetilde{s}_0^{\ell T} \rfloor \underline{\epsilon}_3^\ell + \lfloor 0,\ \widetilde{s}_{0,3}^T R_0^\ell \rfloor \underline{\mathcal{U}}^\ell \end{aligned} \tag{4.390}$$

and then are rewritten in the compact form of

$$\begin{aligned} \Delta\underline{G}_1^* &= \lfloor I,\ \widetilde{s}_0^T \rfloor \underline{\epsilon}_1 + C_1^\ell \underline{\mathcal{U}}^\ell \\ \Delta\underline{G}_2^* &= R_0^\ell \lfloor I,\ \widetilde{s}_0^{\ell T} \rfloor \underline{\epsilon}_2^\ell + C_2^\ell \underline{\mathcal{U}}^\ell \\ \Delta\underline{G}_3^* &= R_0^\ell \lfloor I,\ \widetilde{s}_0^{\ell T} \rfloor \underline{\epsilon}_3^\ell + C_3^\ell \underline{\mathcal{U}}^\ell \end{aligned} \tag{4.391}$$

where the following symbols are introduced

$$C_1^\ell = \widetilde{\kappa}_1^0 R_0^\ell \lfloor I,\ \widetilde{s}_0^{\ell T} \rfloor,\ C_i^\ell = \lfloor 0,\ \widetilde{s}_{0,i}^T R_0^\ell \rfloor \tag{4.392}$$

for $i = 2, 3$. In view of above expressions, the vectorial Green-Lagrange strain measures can be readily obtained

$$\begin{aligned} \underline{\varepsilon}_1 &= \mathcal{B}_g^T \lfloor I,\ \widetilde{s}_0^T \rfloor \underline{\epsilon}_1 + \mathcal{B}_g^T C_1^\ell \underline{\mathcal{U}}^\ell \\ \underline{\varepsilon}_2 &= \mathcal{B}_g^T R_0^\ell \lfloor I,\ \widetilde{s}_0^{\ell T} \rfloor \underline{\epsilon}_2^\ell + \mathcal{B}_g^T C_2^\ell \underline{\mathcal{U}}^\ell \\ \underline{\varepsilon}_3 &= \mathcal{B}_g^T R_0^\ell \lfloor I,\ \widetilde{s}_0^{\ell T} \rfloor \underline{\epsilon}_3^\ell + \mathcal{B}_g^T C_3^\ell \underline{\mathcal{U}}^\ell \end{aligned} \tag{4.393}$$

With the definitions of following symbols

$$\mathcal{S}_0^T = \mathcal{B}_g^T \lfloor I,\ \widetilde{s}_0^T \rfloor,\ \mathcal{S}_0^{\ell T} = \mathcal{B}_g^T R_0^\ell \lfloor I,\ \widetilde{s}_0^{\ell T} \rfloor \tag{4.394}$$

and

$$\mathcal{S}_1^{\ell T} = \mathcal{B}_g^T \widetilde{\kappa}_1^0 R_0^\ell \lfloor I,\ \widetilde{s}_0^{\ell T} \rfloor,\ \mathcal{S}_i^{\ell T} = \mathcal{B}_g^T \lfloor 0,\ \widetilde{s}_{0,i}^T R_0^\ell \rfloor \tag{4.395}$$

for $i = 2, 3$, the Green-Lagrange strain tensor can be simplify described as a functions of Lagrangian stretch and curvature measures together with the micro motions like

$$\begin{aligned} \underline{\varepsilon}_1 &= \mathcal{S}_0^T \underline{\epsilon}_1 + \mathcal{S}_1^{\ell T} \underline{\mathcal{U}}^\ell \\ \underline{\varepsilon}_2 &= \mathcal{S}_0^{\ell T} \underline{\epsilon}_2^\ell + \mathcal{S}_2^{\ell T} \underline{\mathcal{U}}^\ell \\ \underline{\varepsilon}_3 &= \mathcal{S}_0^{\ell T} \underline{\epsilon}_3^\ell + \mathcal{S}_3^{\ell T} \underline{\mathcal{U}}^\ell \end{aligned} \tag{4.396}$$

In most of the analysis case, the initial curvature is zero, $\widetilde{\kappa}_1^0 = 0$. Meanwhile, the second term containing the motion of micro continuum can be truncated. The detailed explanations as to this truncation can be found in Sect. 4.1.2. Finally, the above strain formulations are simplified to

$$\begin{aligned} \underline{\varepsilon}_1 &= \mathcal{S}_0^T \underline{\epsilon}_1 \\ \underline{\varepsilon}_2 &= \mathcal{S}_0^{\ell T} \underline{\epsilon}_2^\ell \\ \underline{\varepsilon}_3 &= \mathcal{S}_0^{\ell T} \underline{\epsilon}_3^\ell \end{aligned} \tag{4.397}$$

### 4.5.3 Constitutive Laws for a Special Multiscale Beam

Given the volume with the magnitude of $V_0$ for a beam, the strain energy $\mathcal{U}_e$ can be represented by using the vectorial Green-Lagrange strain measures like

$$\begin{aligned} 2\mathcal{U}_e = \int_{V_0} [&\underline{\varepsilon}_1^T \mathcal{D}_{11} \underline{\varepsilon}_1 + \underline{\varepsilon}_2^T \mathcal{D}_{22} \underline{\varepsilon}_2 + \underline{\varepsilon}_3^T \mathcal{D}_{33} \underline{\varepsilon}_3 \\ &+ 2(\underline{\varepsilon}_2^T \mathcal{D}_{21} \underline{\varepsilon}_1 + \underline{\varepsilon}_3^T \mathcal{D}_{31} \underline{\varepsilon}_1 + \underline{\varepsilon}_3^T \mathcal{D}_{32} \underline{\varepsilon}_2)]\, dV_0 \end{aligned} \tag{4.398}$$

where the constitutive matrices $\mathcal{D}_{ij}$ for $i, j = 1, 2, 3$ are introduced in Sect. 3.1.4. Referring Eq. (4.397), the relations of vectorial Green-Lagrange strains to Lagrangian stretch and curvature measures, the constitutive matrices of Cosserat continuum are readily found

$$\begin{aligned} \mathcal{C}_{11} = \mathcal{S}_0 \mathcal{D}_{11} \mathcal{S}_0^T,\ \mathcal{D}_{\epsilon\alpha}^{11} = \mathcal{S}_0 \mathcal{D}_{11} \mathcal{S}_0^{\ell T} \\ \mathcal{D}_{\alpha\epsilon}^{11} = \mathcal{S}_0^\ell \mathcal{D}_{11} \mathcal{S}_0^T,\ \mathcal{D}_{\alpha\alpha}^{11} = \mathcal{S}_0^\ell \mathcal{D}_{11} \mathcal{S}_0^{\ell T} \end{aligned} \tag{4.399}$$

and only the nonzero constitutive matrices for $i, j = 1, 2, 3$ are presented as below

$$\mathcal{D}^{ij}_{\epsilon\alpha} = \mathcal{S}_0 \mathcal{D}_{ij} \mathcal{S}^{\ell T}_0, \; \mathcal{D}^{ij}_{\alpha\epsilon} = \mathcal{S}^{\ell}_0 \mathcal{D}_{ij} \mathcal{S}^{T}_0, \; \mathcal{D}^{ij}_{\alpha\alpha} = \mathcal{S}^{\ell}_0 \mathcal{D}_{ij} \mathcal{S}^{\ell T}_0 \tag{4.400}$$

Due to the symmetry of constitutive matrices, it can be verified that

$$\mathcal{D}^{ijT}_{\alpha\epsilon} = \mathcal{S}_0 \mathcal{D}^T_{ij} \mathcal{S}^{\ell T}_0 = \mathcal{S}_0 \mathcal{D}_{ji} \mathcal{S}^{\ell T}_0 = \mathcal{D}^{ji}_{\epsilon\alpha} \tag{4.401}$$

In view of above definitions of constitutive matrices, all the terms of strain energy for a beam can be written explicitly one by one like

$$\underline{\varepsilon}^T_1 \mathcal{D}_{11} \underline{\varepsilon}_1 = \underline{\epsilon}^T_1 \mathcal{C}_{11} \underline{\epsilon}_1, \; \underline{\varepsilon}^T_2 \mathcal{D}_{22} \underline{\varepsilon}_2 = \underline{\epsilon}^{\ell T}_2 \mathcal{D}^{22}_{\alpha\alpha} \underline{\epsilon}^{\ell}_2, \; \underline{\varepsilon}^T_3 \mathcal{D}_{33} \underline{\varepsilon}_3 = \underline{\epsilon}^{\ell T}_3 \mathcal{D}^{33}_{\alpha\alpha} \underline{\epsilon}^{\ell}_3 \tag{4.402}$$

and

$$\underline{\varepsilon}^T_2 \mathcal{D}_{21} \underline{\varepsilon}_1 = \underline{\epsilon}^T_1 \mathcal{D}^{12}_{\epsilon\alpha} \underline{\epsilon}^{\ell}_2, \; \underline{\varepsilon}^T_3 \mathcal{D}_{31} \underline{\varepsilon}_1 = \underline{\epsilon}^T_1 \mathcal{D}^{13}_{\epsilon\alpha} \underline{\epsilon}^{\ell}_3, \; \underline{\varepsilon}^T_3 \mathcal{D}_{32} \underline{\varepsilon}_2 = \underline{\epsilon}^{\ell T}_3 \mathcal{D}^{32}_{\alpha\alpha} \underline{\epsilon}^{\ell}_2 \tag{4.403}$$

In summary, the strain energy of multiscale beam can be rewritten as

$$\begin{aligned} \mathcal{U}_e = \int_{V_0} [\; & \frac{1}{2} (\underline{\epsilon}^T_1 \mathcal{C}_{11} \underline{\epsilon}_1 + \underline{\epsilon}^{\ell T}_2 \mathcal{D}^{22}_{\alpha\alpha} \underline{\epsilon}^{\ell}_2 + \underline{\epsilon}^{\ell T}_3 \mathcal{D}^{33}_{\alpha\alpha} \underline{\epsilon}^{\ell}_3) \\ & + \underline{\epsilon}^T_1 \mathcal{D}^{12}_{\epsilon\alpha} \underline{\epsilon}^{\ell}_2 + \underline{\epsilon}^T_1 \mathcal{D}^{13}_{\epsilon\alpha} \underline{\epsilon}^{\ell}_3 + \underline{\epsilon}^{\ell T}_3 \mathcal{D}^{32}_{\alpha\alpha} \underline{\epsilon}^{\ell}_2 \;] \, dV_0 \end{aligned} \tag{4.404}$$

If the characteristic area with the magnitude of $S_c$ for a micro Cosserat continuum was given, the volume magnitude $V_0$ of multiscale beam can be evaluated form $V_0 = S_c \ell_0$. It is possible to rewrite the strain energy of a beam in an alternative formula like

$$\begin{aligned} \mathcal{U}_e = \int_{\ell_0} [\; & \frac{1}{2} (\underline{\epsilon}^T_1 \int_{S_c} \mathcal{C}_{11} dS_c \underline{\epsilon}_1 + \int_{S_c} \underline{\epsilon}^{\ell T}_2 \mathcal{D}^{22}_{\alpha\alpha} \underline{\epsilon}^{\ell}_2 dS_c + \int_{S_c} \underline{\epsilon}^{\ell T}_3 \mathcal{D}^{33}_{\alpha\alpha} \underline{\epsilon}^{\ell}_3 dS_c) \\ & + \underline{\epsilon}^T_1 \int_{S_c} \mathcal{D}^{12}_{\epsilon\alpha} \underline{\epsilon}^{\ell}_2 dS_c + \underline{\epsilon}^T_1 \int_{S_c} \mathcal{D}^{13}_{\epsilon\alpha} \underline{\epsilon}^{\ell}_3 dS_c + \int_{S_c} \underline{\epsilon}^{\ell T}_3 \mathcal{D}^{32}_{\alpha\alpha} \underline{\epsilon}^{\ell}_2 dS_c \;] \, d\alpha_1 \end{aligned} \tag{4.405}$$

Considering the definitions of Lagrangian stretch and curvature measures, Eq. (4.386), for the micro Cosserat continuum, these Lagrangian strains could be explicitly related to the micro motions and its derivatives

$$\underline{\epsilon}^{\ell}_i = \underline{\mathcal{U}}^{\ell}_{,i} + \widetilde{\mathcal{K}}^{0\ell}_i \underline{\mathcal{U}}^{\ell} \tag{4.406}$$

under the assumption of small motions, also with the help of following introduced symbols

$$\underline{\mathcal{U}}^{\ell}_{,i} = \left| \begin{matrix} (R^{\ell T}_0 \underline{u}^{\ell}_s)_{,i} \\ \underline{\phi}^{\ell}_{,i} \end{matrix} \right|, \; \widetilde{\mathcal{K}}^{0\ell}_i = \begin{bmatrix} \widetilde{\kappa}^{0\ell}_i & \widetilde{R^{\ell T}_0 \underline{x}_{0,i}} \\ 0 & \widetilde{\kappa}^{0\ell}_i \end{bmatrix} \tag{4.407}$$

As mentioned before, the classical finite element method can be directly applied to discretize the micro continuum. Giving the shape functions $\mathcal{H}_\epsilon^\ell$, the local motions and the derivatives are interpolated

$$\underline{\mathcal{U}}^\ell = \mathcal{H}_\epsilon^\ell \underline{\mathcal{U}}_n^\ell,\ \underline{\mathcal{U}}_{,i}^\ell = \mathcal{H}_{\epsilon,i}^\ell \underline{\mathcal{U}}_n^\ell \tag{4.408}$$

such that the Lagrangian strain measures can be approximated by

$$\underline{\epsilon}_i^\ell = (\mathcal{H}_{\epsilon,i}^\ell + \widetilde{\mathcal{K}}_i^{0\ell} \mathcal{H}_\epsilon^\ell) \underline{\mathcal{U}}_n^\ell = \mathcal{Q}_i^\ell \underline{\mathcal{U}}_n^\ell \tag{4.409}$$

Under the assumption of scale separation, the global strains $\underline{\epsilon}_1$ are independent of local variables in micro Cosserat continuum. Therefore, the strain energy for a beam can be recasted to

$$\begin{aligned}\mathcal{U}_e = \frac{1}{2}\int_{\ell_0} [\ \underline{\epsilon}_1^T \int_{S_c} \mathcal{C}_{11} dS_c \underline{\epsilon}_1 + 2\underline{\epsilon}_1^T \int_{S_c} (\mathcal{D}_{\epsilon\alpha}^{12} \mathcal{Q}_2^\ell + \mathcal{D}_{\epsilon\alpha}^{13} \mathcal{Q}_3^\ell)\, dS_c \underline{\mathcal{U}}_n^\ell \\ + \underline{\mathcal{U}}_n^{\ell T} \int_{S_c} (\mathcal{Q}_2^{\ell T} \mathcal{D}_{\alpha\alpha}^{22} \mathcal{Q}_2^\ell + \mathcal{Q}_3^{\ell T} \mathcal{D}_{\alpha\alpha}^{33} \mathcal{Q}_3^\ell + 2\mathcal{Q}_3^{\ell T} \mathcal{D}_{\alpha\alpha}^{32} \mathcal{Q}_2^\ell)\, dS_c \underline{\mathcal{U}}_n^\ell\ ]\ d\alpha_1\end{aligned} \tag{4.410}$$

When the equivalent uniform stiffness matrices are computed from the following integrals

$$\mathcal{C}_{\epsilon\epsilon}^{11} = \int_{S_c} \mathcal{C}_{11} dS_c,\ \mathcal{C}_{\epsilon\alpha}^{11} = \int_{S_c} (\mathcal{D}_{\epsilon\alpha}^{12} \mathcal{Q}_2^\ell + \mathcal{D}_{\epsilon\alpha}^{13} \mathcal{Q}_3^\ell) dS_c \tag{4.411}$$

and

$$\mathcal{C}_{\alpha\alpha} = \int_{S_c} (\mathcal{Q}_2^{\ell T} \mathcal{D}_{\alpha\alpha}^{22} \mathcal{Q}_2^\ell + \mathcal{Q}_3^{\ell T} \mathcal{D}_{\alpha\alpha}^{33} \mathcal{Q}_3^\ell) dS_c + 2\int_{S_c} \mathcal{Q}_3^{\ell T} \mathcal{D}_{\alpha\alpha}^{32} \mathcal{Q}_2^\ell dS_c \tag{4.412}$$

the strain energy of multiscale beam are found to be

$$\mathcal{U}_e = \int_{\ell_0} \frac{1}{2} \underline{\epsilon}_1^T \mathcal{C}_{\epsilon\epsilon}^{11} \underline{\epsilon}_1 + \underline{\epsilon}_1^T \mathcal{C}_{\epsilon\alpha}^{11} \underline{\mathcal{U}}_n^\ell + \frac{1}{2} \underline{\mathcal{U}}_n^{\ell T} \mathcal{C}_{\alpha\alpha} \underline{\mathcal{U}}_n^\ell d\alpha_1 \tag{4.413}$$

Comparing the above equations with Eq. (4.330), quickly observations show that two set of equations are identical each other. By taking variation of these equations, the elastic force in the dual bracket is determined, and then the governing equations of multiscale beam can be readily obtained. The governing equations for static and dynamic problems will remain unchanged, as discussed in Sect. 4.4, so there is no necessary to repeat the introductions.

## 4.6 Modal Superelement Based Multiscale Theory

Modal superelement modeling is a very efficient approach relaying on the eigenmode decomposition and truncation of large-scale dynamic systems. The kernel idea of modal superelement based multiscale theory states: by selecting the body attached frame as the floating frame [35–37], the motions of a Cosserat continuum can be broken into two parts: rigid body motions represented by the motions of the floating frame, and superimposed "elastic motions". When the magnitudes of elastic motions are assumed to be infinitesimal, the modal reduction techniques can be applied to describe the small elastic motions in an efficient manner. An apparent advantage of modal reduction arises from the fact that the higher order eigenmodes will be truncated and only some of the lower order eigenmodes are saved, which will lead to the significant reduction in the size of the problem and the resulting computational cost. Based on the Herting reduction technique [38], this section introduces a modal superelement that can be coupled to nonlinear, flexible multibody system models. Since the boundary nodes with physical degrees of freedom are retained, the modal superelement could be readily coupled to the rest of the nonlinear model without requiring special constraint elements. The modal superelement can be treated as a special multiscale element in the sense that the global behavior of Cosserat continuum will be described by the motion of floating frame combined with the retained lower order eigenmodes, and the detailed elastic motions can be recovered by transpose the modal solutions from modal space back to physical space.

### 4.6.1 Herting's Transformation

Consider a Cosserat continuum whose linearized equations of motion are written in the following expressions of motion formalism like

$$\mathcal{M}_n \ddot{\underline{\mathcal{U}}}_n(t) + \mathcal{K}_n \underline{\mathcal{U}}_n(t) = \underline{\mathcal{F}}_n(t) \tag{4.414}$$

where $\underline{\mathcal{U}}_n$ is the array of nodal displacements and rotations in their dual bracket, $\underline{\mathcal{F}}_n(t)$ the array of externally applied nodal forces, $\mathcal{K}_n$ and $\mathcal{M}_n$ the unconstrained stiffness and mass matrices of the Cosserat continuum, respectively. Note that the expressions, Eq. (4.414), form a large set of $N$ algebraic equations, typically obtained from a spatial discretization process such as the finite element method. In current implementation, the structure is assumed to be undamped or lightly damped and hence, damping effects are neglected. Once the mass and stiffness matrices, $\mathcal{M}_n$ and $\mathcal{K}_n$, are given, the eigenvalue decomposition could be performed to evaluate the associated eigenmodes $\Phi$, thus providing a fundamental input for Herting's transformation in addition to $\mathcal{M}_n$ and $\mathcal{K}_n$. The purpose of current implementation is to develop a modal approximation to the Cosserat continuum represented by Eq. (4.414) which will be connected to others components to form the complete system to be analyzed.

The other components could be modeled by a modal representation, or by a multibody formulation. In order to allowing the connection of a specific component to others, the degrees of freedom will be partitioned into boundary and interior degrees of freedom resulting the same partitions of displacement vector and the structural matrices

$$\underline{\mathcal{U}}_n = \begin{vmatrix} \underline{\mathcal{U}}_\ell^{*B} \\ \underline{\mathcal{U}}_\ell^{*I} \end{vmatrix}, \ \mathcal{M}_n = \begin{bmatrix} M^{BB} & M^{BI} \\ M^{IB} & M^{II} \end{bmatrix}, \ \mathcal{K}_n = \begin{bmatrix} K^{BB} & K^{BI} \\ K^{IB} & K^{II} \end{bmatrix} \tag{4.415}$$

where the upper case characters, $(\cdot)^B$ and $(\cdot)^I$, represent the boundary and interior parts of the partition, respectively. Similarly, the truncated eigenmodes are partitioned also

$$\Phi = \begin{bmatrix} \Phi^{BR} & \Phi^{BE} \\ \Phi^{IR} & \Phi^{IE} \end{bmatrix} \tag{4.416}$$

where $(\cdot)^R$ and $(\cdot)^E$ means the rigid and elastic eigenmodes respectively. Then, the Herting transformation introduced a coordinate transformation

$$\underline{\mathcal{U}}_n = \mathcal{H}_H \bar{\underline{\mathcal{U}}}_n \tag{4.417}$$

defined by a set of shape functions

$$\mathcal{H}^H = \begin{bmatrix} I & 0 & 0 \\ G^{IB} & H^{IR} & \Phi^{IE} - G^{IB}\Phi^{BE} \end{bmatrix} = \begin{bmatrix} I & 0 \\ G^{IB} & H^I \end{bmatrix} \tag{4.418}$$

where

$$G^{IB} = -K^{II^{-1}} K^{IB}, \ H^{IR} = -K^{II^{-1}} \bar{M}^{IB} \tag{4.419}$$

and $\bar{M}^{IB}$ is computed from $\bar{M}^{IB} = M^{IB} + M^{II} G^{IB}$. The reduced set of degrees of freedom is expressed to be

$$\bar{\underline{\mathcal{U}}}_n^T = \lfloor \underline{\mathcal{U}}_\ell^{*B^T}, \ \underline{q}^{R^T}, \ \underline{q}^{E^T} \rfloor = \lfloor \underline{\mathcal{U}}_\ell^{*B^T}, \ \underline{q}^T \rfloor \tag{4.420}$$

where the degrees of freedom for boundary points, $\underline{\mathcal{U}}_\ell^{*B}$, keep unchanged as the original system, $\underline{q}^R$ and $\underline{q}^I$ are rigid and elastic eigenmodes associated degrees of freedom. The original linearized systems, Eq. (4.414) reduce to

$$\bar{\mathcal{M}}_n \ddot{\bar{\underline{\mathcal{U}}}}_n(t) + \bar{\mathcal{K}}_n \bar{\underline{\mathcal{U}}}_n(t) = \bar{\underline{\mathcal{F}}}_n(t), \tag{4.421}$$

where the reduced structural matrices and the external forces are introduced as follow

$$\bar{\mathcal{M}}_n = \mathcal{H}^{H^T} \mathcal{M}_n \mathcal{H}^H, \ \bar{\mathcal{K}}_n = \mathcal{H}^{H^T} \mathcal{K}_n \mathcal{H}^H, \ \bar{\underline{\mathcal{F}}}_n = \mathcal{H}^{H^T} \underline{\mathcal{F}}_n \tag{4.422}$$

For the easy understanding the procedure of Herting reduction, the reduced structural matrices, $\bar{\mathcal{M}}_n$, and $\bar{\mathcal{K}}_n$, are given in detail as

$$\begin{aligned} \bar{\mathcal{M}}_n &= \begin{bmatrix} M^{BB} + M^{BI}G^{IB} + G^{IB^T}\bar{M}^{IB} & (M^{BI} + G^{IB^T}M^{II})H^I \\ H^{I^T}(M^{IB} + M^{II}G^{IB}) & H^{I^T}M^{II}H^I \end{bmatrix} \\ \bar{\mathcal{K}}_n &= \begin{bmatrix} K^{BB} + K^{BI}G^{IB} & 0 \\ 0 & H^{I^T}K^{II}H^I \end{bmatrix} \end{aligned} \tag{4.423}$$

The definition of $G^{IB}$, Eq. (4.419), is applied to verify that the reduced stiffness matrix $\bar{\mathcal{K}}_n$ is recasted to the form of diagonal block matrices, and the first diagonal blocks are obtained from a Guyan reduction. Once the reduced linearized equations are solved, the motions $\underline{\mathcal{U}}_\ell^{*B}$ of boundary points can be extracted from the solutions $\bar{\underline{\mathcal{U}}}_n$ of reduced system directly. If the detailed elastic motions $\underline{\mathcal{U}}_n$ of original Cosserat continuum are required, the solution approximations can be obtained with the help of Eq. (4.417).

### 4.6.2 Kinematics of Modal Superelement

Given a Cosserat continuum as shown in Fig. 4.6, the overall rigid body motions are described by a floating frame defined by $\mathcal{F}_0^F = [\mathbf{F}, \mathcal{B}_0^F]$ and $\mathcal{F}_1^F = [\mathbf{F}, \mathcal{B}_1^F]$ in the reference and deformed configurations, respectively. The orthogonal bases $\mathcal{B}_0^F$ and $\mathcal{B}_1^F$ is nothing but the rotation tensors $R_0^F$ and $R_1^F$ describing the orientation of floating frame. The origin of the floating frame labeled by point $\mathbf{F}$ is denoted by $\underline{x}_0^F$ and $\underline{x}_1^F$ in the reference and deformed configurations, respectively. With the aid of all these notations and referring the definitions of motion tensors, Eqs. (2.103) and (2.104), the reference and deformed configurations of the floating frame will be described by using the following motion tensors as

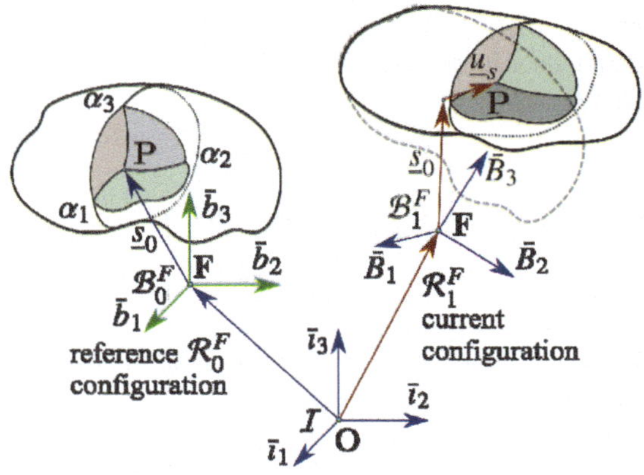


**Fig. 4.6** Kinematics of a material particle in a modal superelement based continuum

$$\mathcal{R}_0^F = \begin{bmatrix} R_0^F & \widetilde{x}_0^F R_0^F \\ 0 & R_0^F \end{bmatrix}, \ \mathcal{R}_1^F = \begin{bmatrix} R_1^F & \widetilde{x}_1^F R_1^F \\ 0 & R_1^F \end{bmatrix} \tag{4.424}$$

Similarly, the configurations of an arbitrary material particle **P** consisting the Cosserat continuum could be defined by two orthogonal motion tensors

$$\mathcal{R}_0^P = \begin{bmatrix} R_0^P & \widetilde{x}_0^P R_0^P \\ 0 & R_0^P \end{bmatrix}, \ \mathcal{R}_1^P = \begin{bmatrix} R_1^P & \widetilde{x}_1^P R_1^P \\ 0 & R_1^P \end{bmatrix} \tag{4.425}$$

where $\underline{x}_0^P$ and $\underline{x}_1^P$ are the position vectors of material particle **P**, meanwhile, $R_0^P$ and $R_1^P$ are the corresponding rotation tensors denoting the orientations of particle **P** in the reference and deformed configurations, respectively. Assuming that the total motion of Cosserat continuum consists of the superposition of a rigid body motion and an elastic motion, the relative motion tensors $\mathcal{R}_0^*$ and $\mathcal{R}_1^*$ are introduced to form the following relations

$$\mathcal{R}_0^P = \mathcal{R}_0^F \mathcal{R}_0^*, \ \mathcal{R}_1^P = \mathcal{R}_1^F \mathcal{R}_1^* \tag{4.426}$$

The rigid body motion, defined by the motion of the floating frame, brings the Cosserat continuum to the "fictitious rigid body configuration" described by the motion tensor $\mathcal{R}_1^F$. Then, the elastic motion moves the position vector $\underline{s}_0$ of point **P** with respected to the floating frame $\mathcal{F}_0^F$ to the new position vector $\underline{s}_1$. Both vectors are measured in the floating frame, such that

$$\underline{s}_0 = R_0^{F^T}(\underline{x}_0^P - \underline{x}_0^F), \ \underline{s}_1 = R_1^{F^T}(\underline{x}_1^P - \underline{x}_1^F) \tag{4.427}$$

The relative motion tensors $\mathcal{R}_0^*$ and $\mathcal{R}_1^*$ measure the configurations of the continuum before and after elastic motion in the floating frame, and the detailed expressions are found to be

$$\mathcal{R}_0^* = \begin{bmatrix} R_0^* & \widetilde{s}_0 R_0^* \\ 0 & R_0^* \end{bmatrix}, \ \mathcal{R}_1^* = \begin{bmatrix} R_1^* & \widetilde{s}_1 R_1^* \\ 0 & R_1^* \end{bmatrix} \tag{4.428}$$

where the relative rotation tensors $R_0^*$ and $R_1^*$ can be computed from $R_0^* = R_0^{FT} R_0^P$ and $R_1^* = R_1^{FT} R_1^P$. According to the definitions of velocity in the skew-symmetric dual tensor form, Eq. (2.123), the local velocity of continuum measured in the floating frame

$$\widetilde{\mathcal{V}}_\ell^{*P} = \mathcal{R}_1^{*T} \dot{\mathcal{R}}_1^* \tag{4.429}$$

can be found to be the functions of

$$\underline{\mathcal{V}}_\ell^{*P} = \mathcal{R}_1^{PT} \underline{\mathcal{V}}^{*F} + \underline{\mathcal{V}}^{*P} \tag{4.430}$$

where $\underline{\mathcal{V}}^{*F}$ and $\underline{\mathcal{V}}^{*P}$ are the velocity of the floating frame and material particle **P** in the dual form, respectively, and

$$\widetilde{\mathcal{V}}^{*F} = \mathcal{R}_1^{FT} \dot{\mathcal{R}}_1^F, \ \widetilde{\mathcal{V}}^{*P} = \mathcal{R}_1^{PT} \dot{\mathcal{R}}_1^P \tag{4.431}$$

Similarly, the virtual elastic motion could be described by using the following relations like

$$\underline{\delta\mathcal{U}_\ell^{*P}} = \mathcal{R}_1^{PT}\underline{\delta\mathcal{U}^{*F}} + \underline{\delta\mathcal{U}^{*P}} \tag{4.432}$$

Under the assumption of infinitesimal elastic motion and according to the definition of axial vector of motion tensor with infinitesimal magnitude, Eq. (1.332), the elastic motion $\underline{\mathcal{U}_\ell^{*P}}$ becomes the axial vector of

$$\underline{\mathcal{U}_\ell^{*P}} = \text{axi}(\mathcal{R}_0^{*T}\mathcal{R}_1^*) \tag{4.433}$$

and then, the explicit formulation of elastic motion $\underline{\mathcal{U}_\ell^{*P}}$ and velocity $\underline{\mathcal{V}_\ell^{*P}}$ is readily obtained to be

$$\underline{\mathcal{U}_\ell^{*P}} = \begin{vmatrix} R_0^{*T}\underline{u_s} \\ \underline{\phi} \end{vmatrix}, \quad \underline{\mathcal{V}_\ell^{*P}} \approx \underline{\dot{\mathcal{U}}_\ell^{*P}} \tag{4.434}$$

where $\underline{u_s}$ defines the displacement of material particle **P** with respected to the floating frame $\mathcal{F}_0^F$ and can be computed from $\underline{u_s} = \underline{s_1} - \underline{s_0}$, $\underline{\phi}$ the rotation vector with small amplitude, such that $R_0^{*T}R_1^* = I + \widetilde{\phi}$. Since Eqs. (4.430) and (4.432) are valid for any point on the Cosserat continuum, they also hold for the boundary points, hence

$$\underline{\mathcal{V}_\ell^{*B}} = \mathcal{R}_1^{BT}\underline{\mathcal{V}^{*F}} + \underline{\mathcal{V}^{*B}}, \ \underline{\mathcal{V}_\ell^{*B}} = \underline{\dot{\mathcal{U}}_\ell^{*B}}, \ \underline{\delta\mathcal{U}_\ell^{*B}} = \mathcal{R}_1^{BT}\underline{\delta\mathcal{U}^{*F}} + \underline{\delta\mathcal{U}^{*B}} \tag{4.435}$$

### 4.6.3 *Linearized Strain Energy of Modal Superelement*

With the help of Herting's transformation, Eq. (4.417), the strain energy of the Cosserat continuum can be written as

$$\mathcal{U}_e = \frac{1}{2}\underline{\bar{\mathcal{U}}_n}^T\bar{\mathcal{K}}_n\underline{\bar{\mathcal{U}}_n} \tag{4.436}$$

With the aid of the expression of virtual displacements $\underline{\delta\mathcal{U}_\ell^{*B}}$, the second equations of Eq. (4.435), taking variations of strain energy $\mathcal{U}_e$ will obtain the elastic forces $\mathcal{F}_e$ to be

$$\delta\mathcal{U}_e = \begin{vmatrix} \underline{\delta\mathcal{U}^{*F}} \\ \underline{\delta\mathcal{U}^{*B}} \\ \underline{\delta q} \end{vmatrix}^T \begin{vmatrix} \mathcal{R}_1^{B\mathcal{L}}\underline{f_e^*} \\ \underline{f_e^*} \\ \underline{f_q^*} \end{vmatrix} = \begin{vmatrix} \underline{\delta\mathcal{U}^{*F}} \\ \underline{\delta\mathcal{U}^{*B}} \\ \underline{\delta q} \end{vmatrix}^T \underline{\mathcal{F}_e} \tag{4.437}$$

where the elastic force components are introduced like

$$\begin{vmatrix} \underline{f_e^*} \\ \underline{f_q^*} \end{vmatrix} = \bar{\mathcal{K}}_n \begin{vmatrix} \underline{\mathcal{U}_\ell^{*B}} \\ \underline{q} \end{vmatrix} \tag{4.438}$$

For simplicity of the exposition, the formulation is presented here for a single boundary point. The case of multiple boundary points is obtained by a straightforward generalization. The evaluation of the strain energy of the modal based element only requires the reduced stiffness matrix $\bar{\mathcal{K}}_n = \mathcal{H}^{H^T} \mathcal{K}_n \mathcal{H}^H$; hence, it is independent of the finite element code used to model the elastic component.

### 4.6.4 Linearized Kinetic Energy of Modal Superelement

The total kinetic energy $\mathcal{K}$ of the Cosserat continuum is given by

$$\mathcal{K} = \frac{1}{2} \underline{\dot{\mathcal{U}}}_n^T \bar{\mathcal{M}}_n \underline{\dot{\mathcal{U}}}_n \tag{4.439}$$

and variations of the kinetic energy becomes

$$\delta\mathcal{K} = \delta\underline{\dot{\mathcal{U}}}_n^T \bar{\mathcal{M}}_n \underline{\dot{\mathcal{U}}}_n = \begin{vmatrix} \delta\underline{\dot{\mathcal{U}}}_\ell^{*B} \\ \delta\underline{\dot{q}} \end{vmatrix}^T \begin{vmatrix} \underline{h}^{*B} \\ \underline{h}^{*q} \end{vmatrix} \tag{4.440}$$

where

$$\begin{vmatrix} \underline{h}^{*B} \\ \underline{h}^{*q} \end{vmatrix} = \bar{\mathcal{M}}_n \underline{\dot{\mathcal{U}}}_n = \bar{\mathcal{M}}_n \begin{vmatrix} \underline{\dot{\mathcal{U}}}_\ell^{*B} \\ \underline{\dot{q}} \end{vmatrix} \tag{4.441}$$

are the momenta components. Then, the inertial forces $\underline{\mathcal{F}}_i$ are readily found from variations of the kinetic energy

$$\delta\mathcal{K} = \begin{vmatrix} \delta\mathcal{U}^{*F} \\ \delta\mathcal{U}^{*B} \\ \delta\underline{q} \end{vmatrix}^T \begin{vmatrix} \mathcal{R}_1^{B\mathcal{L}} \underline{\dot{h}}^{*B} \\ \underline{\dot{h}}^{*B} \\ \underline{\dot{h}}^{*q} \end{vmatrix} = \begin{vmatrix} \delta\mathcal{U}^{*F} \\ \delta\mathcal{U}^{*B} \\ \delta\underline{q} \end{vmatrix}^T \underline{\mathcal{F}}_i \tag{4.442}$$

The evaluation of the kinetic energy of the modal based element only requires the reduced mass matrix $\bar{M}_n = \mathcal{H}^{H^T} \mathcal{M}_n \mathcal{H}^H$; hence, it is independent of the finite element code used to model the elastic component.

As a result, the governing equations of modal superelement can be easily obtained by above formulations of inertial and elastic forces

$$\underline{\mathcal{F}}_i + \underline{\mathcal{F}}_e = \underline{\mathcal{F}}_a \tag{4.443}$$

for a macroscopic model if the external forces $\underline{\mathcal{F}}_a$ were applied also.

## 4.7 Summary of Multiscale Multibody Dynamics

In this chapter, the rigid body assumption of the local region occupied by the director in a Cosserat continuum is abandoned. The local region occupied by the material particles, where the directors attached to, could be a micro-structure composed of micro-shell, micro-beam and micro Cauchy continuum etc. The micro-structures also contribute to the strain energy of the global model consisting of one-dimensional, two-dimensional or three-dimensional Cosserat continua. By considering the deformations of material directors under the assumption of scale separation, this chapter decouples the original Cosserat continuum into a global macroscopic model and a local microscopic model. In current implementations, the microscopic model is valid for static problem, and within the scope of linear analysis, the application of Gaussian elimination makes it possible to solve the macro problem first and then the local micro problem. In detail, the global analysis will predict the global deformation and stress distributions of macro-structures, which can be used to estimate the global mechanical behavior. Through micro-macro mapping, the detailed micro-analysis could be performed with the stress concentration determined in the global analysis as the external input, and then a fine analysis in the level of micro-scale can be accomplished. As the most complicated theoretical part in this book, based on the kinematic description of above macro-micro decoupling, the multiscale theories of Cosserat continuum stemming from the unified formula introduced in previous chapter are then derived. In view of the fact that multiscale modeling affords an efficient approach for detailed analysis of micro-structures at the speed of quickly global analysis, six different multiscale models, covering three-dimensional Cosserat continuum, shell and beam together with modal superelement, are developed in this chapter. Since the Cosserat continua, shells, beams, joints and rigid bodies were composed into multibody systems, the multiscale modeling technologies developed in this chapter are specially used for the numerical simulations of multiscale multibody systems. Hence, the chapter is named as multiscale multibody dynamics.

## References


1. Bauchau, O.A.: Flexible Multibody Dynamics. Springer, Dordrecht (2011)
2. Rubin, M.B.: Cosserat Theories: Shells, Rods and Points. Springer, Dordrecht (2000)
3. Kouznetsova, V.: Computational homogenization for the multi-scale analysis of multi-phase materials. Ph.d. Thesis, Mechanical Engineering, Technische Universiteit Eindhoven (2020). https://doi.org/10.6100/IR560009
4. Kanouté, P., Boso, D.P., Chaboche, J.L., Schrefler, B.A.: Multiscale methods for composites: a review. Arch. Comput. Methods Eng. (2009). https://doi.org/10.1007/s11831-008-9028-8
5. Lloberas-Valls, O., Rixen, D.J., Simone A., Sluys, L.J.: Multiscale domain decomposition analysis of quasi-brittle heterogeneous materials. Int. J. Numer. Meth. Eng. (2012). https://doi.org/10.1002/nme.3286
6. Hassani, B., Hinton, E.: A review of homogenization and topology optimization I-homogenization theory for media with periodic structure. Comput. Struct. **69**, 707–717 (1998)

7. Hassani, B., Hinton, E.: A review of homogenization and topology optimization II-analytical and numerical solution of homogenization equations. Comput. Struct. **69**, 719–738 (1998)
8. Matsui, K., Terada, K., Yuge, K.: Two-scale finite element analysis of heterogeneous solids with periodic microstructures. Comput. Struct. **82**, 593–606 (2004)
9. Yuan, Z., Fish, J.: Toward realization of computational homogenization in practice. Int. J. Numer. Methods Eng. (2008). https://doi.org/10.1002/nme.2074
10. Berdichevskii, V.L.: Variational-asymptotic method of constructing a theory of shells. PMM (1973). https://doi.org/10.1016/0021-8928(79)90157-6
11. Berdichevskii, V.L.: On averaging of periodic systems. PMM (1977). https://doi.org/10.1016/0021-8928(77)90059-4
12. Berdichevsky, V.L.: Variational Principles of Continuum Mechanics I: Fundamentals. Springer, Berlin (2009)
13. Yu, W., Hodges, D.H., Ho, J.C.: Variational asymptotic beam sectional analysis—an updated version. Int. J. Eng. Sci. (2012). https://doi.org/10.1016/j.ijengsci.2012.03.006
14. Zhang, L., Yu, W.: A micromechanics approach to homogenizing elasto-viscoplastic heterogeneous materials. Int. J. Solids Struct. **51**, 3878–3888 (2014)
15. Huet, C.: Application of variational concepts to size effects in elastic heterogeneous bodies. J. Mech. Phys. Solids **38**, 813–841 (1990)
16. Liu, H., Zeng, D., Li, Y., Jiang, L.: Development of RVE-embedded solid elements model for predicting effective elastic constants of discontinuous fiber reinforced composites. Mech. Mater. (2015). https://doi.org/10.1016/j.mechmat.2015.10.011
17. Martin, O.S.: Material spatial randomness: from statistical to representative volume element. Probab. Eng. Mech. **21**, 112–132 (2006)
18. Pecullan, S., Gibiansky, L.V., Torquato, S.: Scale effects on the elastic behavior of periodic and hierarchical two-dimensional composites. J. Mech. Phys. Solids **47**, 1509–1542 (1999)
19. Berger, H., Kari, S., Gabbert, U., Rodriguez-Ramos, R., Bravo-Castillero, J., Guinovart-Diaz, R., Sabina, F., Maugin, G.: Unit cell models of piezoelectric fiber composites for numerical and analytical calculation of effective properties. Smart Mater. Struct. **15**, 451–458 (2006)
20. Liu, H., Yang, D., Wu, J., Zheng, Y., Zhang, H.: An open-source Matlab implementation for elastic analyses of heterogeneous materials using the extended multiscale finite element method. Int. J. Multiscale Comput. Eng. **19**, 19–43 (2021)
21. Hou, T.Y., Wu, X.H.: A multiscale finite element method for elliptic problems in composite materials and porous media. J. Comput. Phys. (1997). https://doi.org/10.1006/jcph.1997.5682
22. Hou, T.Y., Wu, X.H., Cai, Z.: Convergence of a multiscale finite element method for elliptic problems with rapidly oscillating coefficients. Math. Comput. (1999). https://doi.org/10.1090/S0025-5718-99-01077-7
23. Babuška, I., Osborn, J.E.: Generalized finite element methods: their performance and their relation to mixed methods. SIAM J. Numer. Anal. **20**, 510–536 (1983)
24. Zhang, H., Wu, J., Lv, J.: A new multiscale computational method for elasto-plastic analysis of heterogeneous materials. Comput. Mech. **49**, 149–169 (2012)
25. Wu, J., Zhang, H., Zheng, Y.: A concurrent multiscale method for simulation of crack propagation. Acta Mech. Solida Sin. (2015). https://doi.org/10.1016/S0894-9166(15)30011-2
26. Hodges, D.H.: Nonlinear Composite Beam Theory. AIAA, Reston, VA (2006)
27. Bauchau, O.A., Han, S.: Three-dimensional beam theory for flexible multibody dynamics. J. Comput. Nonlinear Dyn. (2014). https://doi.org/10.1115/1.4025820
28. Han, S., Bauchau, O.A.: Nonlinear three-dimensional beam theory for flexible multibody dynamics. Multibody Syst. Dyn. (2014). https://doi.org/10.1007/s11044-014-9433-8
29. Cesnik, C.E.S., Hodges, D.H.: VABS: a new concept for composite rotor blade cross-sectional modeling. J. Am. Helicopter Soc. (1997). https://doi.org/10.4050/JAHS.42.27
30. Yu, W., Hodges, D.H., Volovoi, V.V.: Asymptotic generalization of Reissner-Mindlin theory: accurate three-dimensional recovery for composite shells. Comput. Methods Appl. Mech. Eng. **191**, 5087–5109 (2002)
31. Pietraszkiewicz, W., Eremeyev, V.A.: On vectorially parameterized natural strain measures of the non-linear Cosserat continuum. Int. J. Solids Struct. (2009). https://doi.org/10.1016/j.ijsolstr.2009.01.030

32. Yu, W.: A unified theory for constitutive modeling of composites. J. Mech. Mater. Struct. (2016). https://doi.org/10.2140/jomms.2016.11.379
33. Bathe, K.J.: Finite Element Procedures. Prentice-Hall, New Jersey (1996)
34. Yu, W.: Simplified formulation of mechanics of structure genome. AIAA J. (2019). https://doi.org/10.2514/1.J057500
35. Wasfy, T.M., Noor, A.K.: Computational strategies for flexible multibody systems. Appl. Mech. Rev. **56**, 553–613 (2003)
36. Shabana, A.A.: Flexible multibody dynamics: review of past and recent developments. Multibody Syst. Dyn. **1**, 161–180 (1997)
37. Cardona, A., Géradin, M.: A superelement formulation for mechanism analysis. Comput. Methods Appl. Mech. Eng. **100**, 1–29 (1992)
38. Bauchau, O.A., Rodriguez, J.: Formulation of modal based elements in nonlinear, flexible multibody dynamics. J. Multiscale Comput. Eng. **1**, 161–180 (2003)

# Chapter 5
# Recursive Formulas of Joints

In general, rigid or flexible bodies, the essential parts of a flexible multibody dynamic system, could be connected by using any kind of mechanical joints. It can be said that the relative motions among rigid or flexible bodies are also constrained by the connecting joints. In particular, the lower pair joints [1]: the revolute, prismatic, screw, cylindrical, planar and spherical joints, depicted in Fig. 5.1, present six fundamental types of constrained kinematics. In traditional analysis of multibody systems, the effects of clearance [2–7], friction [8–11], wear [12] and lubrication [13] at the joints are neglected to obtain the ideal kinematic joints, which can be represented by a set of kinematic constraints [14, 15]. As described in reference [16], Lagrange multipliers are generally applied to enforce the kinematic constraints. This enforcement will lead to infinite frequencies because Lagrange multipliers have associated degrees of freedom that are massless. It becomes the main source where the stiff properties of the multibody systems stem from. The book will present the recursive joint models avoiding the definitions of Lagrange multipliers as independent variables with massless degree of freedom.

Another drawback of using the kinematic constraints to model the ideal joints leads to the governing equations of motion for multibody system with joints will be stiff Differential Algebraic Equations (DAE) of index-3. In contrast, recursive theory [17] takes advantage of the system topology and applies the joint coordinates to formulate a nonstiff Ordinary Differential Equations (ODE). Petzold [18] had pointed out that these equations are easier to solve than DAE of index-3. In this chapter, the recursive relations between the bodies connected by six lower pair joints are identified to formulate the recursive models. These new type of recursive models introduce the relative displacements and relative rotations as independent variables to describe the relative motions of joint. If a multibody system contains only the recursive models of joints, its motion will be controlled by a set of nonstiff ODE which can be easily solved.

 
J. Wang, *Multiscale Multibody Dynamics*,
https://doi.org/10.1007/978-981-19-8441-9_5

## 5.1 Motion Formalism of Six Lower Pair Joints

As shown in Fig. 5.2, it is assumed that two bodies, denoted "body $k$" and "body $\ell$," are connected by a revolute joint. In fact, the joint could be anyone of six lower pairs. The bodies $k$ and $\ell$ could be rigid or flexible Cosserat continuum. Quantities pertaining to body $k$ and body $\ell$ will be indicated with superscripts $(\cdot)^k$ and $(\cdot)^\ell$, respectively. The two bodies are labeled by **K** and **L** with the initial position vectors defined as $\underline{x}_0^k$ and $\underline{x}_0^\ell$, where two reference bases $\mathcal{B}_0^k(\underline{b}_i^k)$ and $\mathcal{B}_0^\ell(\underline{b}_i^\ell)$ for $i = 1, 2, 3$ are attached to. Such that two reference frames $\mathcal{F}_0^k = [\mathbf{K}, \mathcal{B}_0^k]$ and $\mathcal{F}_0^\ell = [\mathbf{L}, \mathcal{B}_0^\ell]$ or two motion tensors $\mathcal{R}_0^k$ and $\mathcal{R}_0^\ell$, equivalently, could be constructed to describe the reference configurations of bodies $k$ and $\ell$. The reference configuration of anyone of six lower pair joints can be described by the motion tensor $\mathcal{R}_0^*$ resolved in the body $k$ attached reference frame to get

$$\mathcal{R}_0^\ell = \mathcal{R}_0^k \mathcal{R}_0^* \tag{5.1}$$

where the motion tensor $\mathcal{R}_0^*$ is explicitly expressed as

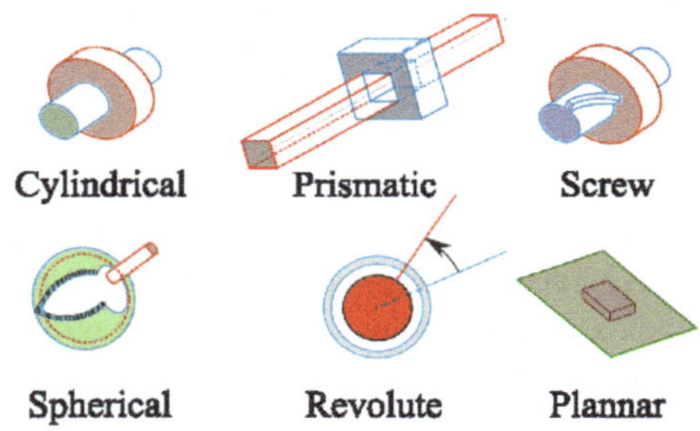


**Fig. 5.1** The six lower pair joints

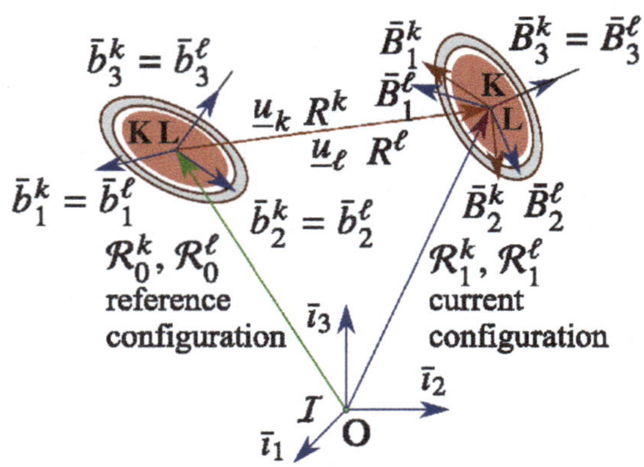


**Fig. 5.2** Motion formalism of six lower pair joints

$$\mathcal{R}_0^* = \begin{bmatrix} R_0^*, & \widetilde{R_0^{kT} \underline{x}_0 R_0^*} \\ 0, & R_0^* \end{bmatrix} \tag{5.2}$$

with the definitions of relative position vector $\underline{x}_0 = \underline{x}_0^\ell - \underline{x}_0^k$ and the relative rotation tensor $R_0^* = R_0^{kT} R_0^\ell$. According to the geometric description of orthogonal basis in Sect. 1.3, the initial rotations tensors $R_0^k$ and $R_0^\ell$ are nothing but the reference bases $\mathcal{B}_0^k$ and $\mathcal{B}_0^\ell$, respectively. In general, for ease of description, the relative position vector $\underline{x}_0$ of any six lower joints is initialized to zero, $\underline{x}_0 = 0$. Here, a nonzero vector is introduced with the purposed of describing the pure rigid connections. In current deformed configurations, the bodies $k$ and $\ell$ will move to the new positions $\underline{x}_1^k$ and $\underline{x}_1^\ell$ under the displacements of $\underline{u}^k$ and $\underline{u}^\ell$ and the rotations of $R^k$ and $R^\ell$, respectively. The reference bases $\mathcal{B}_0^k$ and $\mathcal{B}_0^\ell$ will rotate to the new configurations of bodies $k$ and $\ell$ which can be described by using the rotation tensors $R_1^k$ and $R_1^\ell$ or the deformed bases $\mathcal{B}_1^k$ and $\mathcal{B}_1^\ell$, respectively. Note that the deformed and reference configuarations of bodies $k$ and $\ell$ are related by the motions like

$$\begin{aligned} \underline{x}_1^k &= \underline{x}_0^k + \underline{u}^k, \quad R_1^k = R_0^k R^k \\ \underline{x}_1^\ell &= \underline{x}_0^\ell + \underline{u}^\ell, \quad R_1^\ell = R_0^\ell R^\ell \end{aligned} \tag{5.3}$$

The deformed relative position vector $\underline{x}_1$ of joint can be computed from $\underline{x}_1 = \underline{x}_1^\ell - \underline{x}_1^k$. Meanwhile, the relative rotation tenor $R_1^*$ is readily determined from $R_1^* = R_1^{kT} R_1^\ell$. Alternatively, the deformed configurations of bodies $k$ and $\ell$ can be described by using the deformed frames $\mathcal{F}_1^k = [\mathbf{K}, \mathcal{B}_1^k]$ and $\mathcal{F}_1^\ell = [\mathbf{L}, \mathcal{B}_1^\ell]$ or two motion tensors $\mathcal{R}_1^k$ and $\mathcal{R}_1^\ell$, equivalently. The deformed configuration of joint will be described by using the motion tensor $\mathcal{R}_1^*$ resolved in the body $k$ attached reference frame $\mathcal{B}_1^k$ to find

$$\mathcal{R}_1^\ell = \mathcal{R}_1^k \mathcal{R}_1^* \tag{5.4}$$

where the motion tensor $\mathcal{R}_1^*$ is explicitly expressed as

$$\mathcal{R}_1^* = \begin{bmatrix} R_1^*, & \widetilde{R_1^{kT} \underline{x}_1 R_1^*} \\ 0, & R_1^* \end{bmatrix} \tag{5.5}$$

The virtual motions of bodies $k$ and $\ell$ can be computed from

$$\widetilde{\delta\mathcal{U}}^{*k} = \mathcal{R}_1^{kT} \delta\mathcal{R}_1^k, \quad \widetilde{\delta\mathcal{U}}^{*\ell} = \mathcal{R}_1^{\ell T} \delta\mathcal{R}_1^\ell \tag{5.6}$$

which can be related by

$$\underline{\delta\mathcal{U}}^{*\ell} = \mathcal{R}_1^{*T} \underline{\delta\mathcal{U}}^{*k} + \underline{\delta\mathcal{U}}^* \tag{5.7}$$

where $\underline{\delta\mathcal{U}}^*$ describes the virtual motion of joint and

$$\widetilde{\delta\mathcal{U}}^* = \mathcal{R}_1^{*T} \delta\mathcal{R}_1^* \tag{5.8}$$

The velocities of bodies $k$ and $\ell$ in their dual brackets can be defined in the similar approach like

$$\underline{\widetilde{\mathcal{V}}}^{*k} = \mathcal{R}_1^{kT}\dot{\mathcal{R}}_1^k, \quad \underline{\widetilde{\mathcal{V}}}^{*\ell} = \mathcal{R}_1^{\ell T}\dot{\mathcal{R}}_1^\ell \tag{5.9}$$

which are related by

$$\underline{\mathcal{V}}^{*\ell} = \mathcal{R}_1^{*T}\underline{\mathcal{V}}^{*k} + \underline{\mathcal{V}}^* \tag{5.10}$$

where $\underline{\mathcal{V}}^*$ describes the velocities of joint and is defined as

$$\widetilde{\mathcal{V}}^* = \mathcal{R}_1^{*T}\dot{\mathcal{R}}_1^* \tag{5.11}$$

Taking time derivatives of velocities $\underline{\mathcal{V}}^{*\ell}$ obtains the accelerations of body $\ell$ in their Plücker coordinates like

$$\underline{\dot{\mathcal{V}}}^{*\ell} = \mathcal{R}_1^{*T}\underline{\dot{\mathcal{V}}}^{*k} + \widetilde{\mathcal{V}}^{*T}\mathcal{R}_1^{*T}\underline{\mathcal{V}}^{*k} + \underline{\dot{\mathcal{V}}}^* \tag{5.12}$$

If the relative motions occurred between bodies $k$ and $\ell$, it is necessary to clarify the specific expressions of the relative motions, including the relative displacement $\underline{u}^*$ and the relative rotation $R^*$ of all six lower pair joints. As shown in Table 5.1, the six lower pairs are now formally defined in terms of the relative displacement and/or rotation components that each joint allows, where $u_i$ for $i = 1, 2, 3$ are the components of relative displacements, $R_\phi$ is the special rotation tensor around the axis $\underline{b}_3^k$ of basis $\mathcal{B}_0^k$. For the screw joint, $p$ is the pitch of the screw. Note that if the two bodies are rigidly connected to each other, their relative motion must vanish at the connection point. The special rotation tensor $R_\phi$ describing the rotation with magnitude of $\phi$ around the axis $\underline{b}_3^k$ of basis $\mathcal{B}_0^k$ and measured in the body $k$ attached frame is given as

$$R_\phi = \begin{bmatrix} \cos\phi & -\sin\phi & 0 \\ \sin\phi & \cos\phi & 0 \\ 0 & 0 & 1 \end{bmatrix} \tag{5.13}$$

**Table 5.1** Definition of relative motion of the six lower pair joints

| Joint type | Relative displacements | Relative rotations |
|---|---|---|
| Prismatic | $\underline{u}^* = u_3\bar{\imath}_3$ | $R^* = I$ |
| Screw | $\underline{u}^* = p\phi\bar{\imath}_3$ | $R^* = R_\phi$ |
| Cylindrical | $\underline{u}^* = u_3\bar{\imath}_3$ | $R^* = R_\phi$ |
| Revolute | $\underline{u}^* = \underline{0}$ | $R^* = R_\phi$ |
| Spherical | $\underline{u}^* = \underline{0}$ | $R^* = R_0^{*T}R_1^*$ |
| Planar | $\underline{u}^* = u_1\bar{\imath}_1 + u_2\bar{\imath}_2$ | $R^* = R_\phi$ |

When the motion of body $\ell$ is described independently, the governing equations of motion of body $\ell$ are assumed to be

$$M_e^\ell \underline{\dot{\mathcal{V}}}^{*\ell} = \underline{F}_e^\ell \tag{5.14}$$

where $M_e^\ell$ is the elemental mass matrix of body $\ell$, and $\underline{F}_e^\ell$ the associated elemental right hand side vector. Following the conventional assembling procedure of finite element method, the above elemental equations need to be assembled into the global system. Current implementation will transform elemental equations of body $\ell$ to the sort of recursive form at first, and then it will be assembled. Substituting velocity relations, Eq. (5.12), into the above elemental equation to find

$$\bar{M}_e^\ell \left| \begin{matrix} \underline{\dot{\mathcal{V}}}^{*k} \\ \underline{\dot{\mathcal{V}}}^{*} \end{matrix} \right| = \begin{bmatrix} \mathcal{R}_1^{*\mathcal{L}} \\ \mathcal{I} \end{bmatrix} \left( \underline{F}_e^\ell - M_e^\ell \widetilde{\mathcal{V}}^{*T} \mathcal{R}_1^{*T} \underline{\mathcal{V}}^{*k} \right) \tag{5.15}$$

where $\bar{M}_e^\ell$ is the transformed mass matrix and

$$\bar{M}_e^\ell = \begin{bmatrix} \mathcal{R}_1^{*\mathcal{L}} \\ \mathcal{I} \end{bmatrix} M_e^\ell \left[ \mathcal{R}_1^{*T}, \ \mathcal{I} \right] \tag{5.16}$$

and $\mathcal{I}$ the $6 \times 6$ identity matrix. The symbol $(\cdot)^{\mathcal{L}}$ represents the adjoint representation of *Lie* group $SE(3)$, such that

$$\mathcal{R}_1^{*\mathcal{L}} = \begin{bmatrix} R_1^*, & 0 \\ \widetilde{R_1^{kT} \underline{x}_1} R_1^*, & R_1^* \end{bmatrix} \tag{5.17}$$

Furthermore, similar transformations need to be applied to the elemental stiffness and damping matrices, $K_e^\ell$ and $C_e^\ell$, respectively. With the help of linearization of velocity $\underline{\mathcal{V}}^{*\ell}$ of body $\ell$

$$d\underline{\mathcal{V}}^{*\ell} = \mathcal{R}_1^{*T} d\underline{\mathcal{V}}^{*k} + \widetilde{\mathcal{R}_1^{*T} \underline{\mathcal{V}}^{*k}} d\underline{\mathcal{U}}^* + d\underline{\mathcal{V}}^* \tag{5.18}$$

and linearization of acceleration

$$\begin{aligned} d\underline{\dot{\mathcal{V}}}^{*\ell} &= \mathcal{R}_1^{*T} d\underline{\dot{\mathcal{V}}}^{*k} + \widetilde{\mathcal{V}}^{*T} \mathcal{R}_1^{*T} d\underline{\mathcal{V}}^{*k} + d\underline{\dot{\mathcal{V}}}^* \\ &\quad + \widetilde{\mathcal{R}_1^{*T} \underline{\mathcal{V}}^{*k}} d\underline{\mathcal{V}}^* + \left( \widetilde{\mathcal{R}_1^{*T} \underline{\dot{\mathcal{V}}}^{*k}} + \widetilde{\mathcal{V}}^{*T} \widetilde{\mathcal{R}_1^{*T} \underline{\mathcal{V}}^{*k}} \right) d\underline{\mathcal{U}}^* \end{aligned} \tag{5.19}$$

the stiffness and damping matrices have been transformed to

$$\bar{C}_e^\ell = \begin{bmatrix} \mathcal{R}_1^{*\mathcal{L}} \\ \mathcal{I} \end{bmatrix} \left( C_e^\ell \left[ \mathcal{R}_1^{*T}, \ \mathcal{I} \right] + M_e^\ell \left[ \widetilde{\mathcal{V}}^{*T} \mathcal{R}_1^{*T}, \ \widetilde{\mathcal{R}_1^{*T} \underline{\mathcal{V}}^{*k}} \right] \right) \tag{5.20}$$

and

$$\bar{K}_e^\ell = \begin{bmatrix} \mathcal{R}_1^{*\mathcal{L}} \\ \mathcal{I} \end{bmatrix} \Big( K_e^\ell \, [\, \mathcal{R}_1^{*T}, \, \mathcal{I} \,] + C_e^\ell \big[ 0, \, \widetilde{\mathcal{R}_1^{*T} \underline{\mathcal{V}}^{*k}} \,\big] \Big) + \begin{bmatrix} \mathcal{R}_1^{*\mathcal{L}} \\ \mathcal{I} \end{bmatrix} M_e^\ell \Big[ 0, \, \mathcal{R}_1^{*T} \underline{\dot{\mathcal{V}}}^{*k} + \widetilde{\mathcal{V}}^{*T} \widetilde{\mathcal{R}_1^{*T} \underline{\mathcal{V}}^{*k}} \,\Big] \tag{5.21}$$

It is very necessary to note that the expressions of virtual motion $\delta \underline{\mathcal{U}}^*$, the relative velocity $\underline{\mathcal{V}}^*$ and acceleration $\underline{\dot{\mathcal{V}}}^*$, together with the linearized quantities need to be determined for each of six lower pair joints. The specific recursive formulas will be given in the next sub-sections one by one for six lower pair joints.

## 5.2 Motion Formalism of Prismatic Joint

The prismatic joint only allow the relative translation along the axis $\bar{b}_3^k$ of basis $\mathcal{B}_1^k$ attached to the body $k$. The relative rotations are constrainted to be zero, such that $R_0^* = R_1^* = I$. The relative translation for a prismatic joint is given in the Table 5.1 as

$$\underline{u}^* = u_3 \bar{\imath}_3 \tag{5.22}$$

Together with the clarified initial configurations $\underline{x}_0 = \underline{x}_0^\ell - \underline{x}_0^k = 0$, and $R_0^k = R_0^\ell$, the motion tensors $\mathcal{R}_1^*$ and $\mathcal{R}_1^{*\mathcal{L}}$ are simplified to be

$$\mathcal{R}_1^* = \begin{bmatrix} I & \widetilde{u}^* \\ 0 & I \end{bmatrix}, \quad \mathcal{R}_1^{*\mathcal{L}} = \begin{bmatrix} I & 0 \\ \widetilde{u}^* & I \end{bmatrix} \tag{5.23}$$

According to the defintion, Eq. (5.8), the virtual relative motion of prismatic joint is readily found to be

$$\underline{\delta \mathcal{U}^*} = \begin{vmatrix} \bar{\imath}_3 \\ \underline{0} \end{vmatrix} \delta u_3 = \bar{\mathcal{I}}_3 \delta u_3 \tag{5.24}$$

The velocity $\underline{\mathcal{V}}^*$ and acceleration $\underline{\dot{\mathcal{V}}}^*$ of prismatic joint becomes

$$\underline{\mathcal{V}}^* = \begin{vmatrix} \bar{\imath}_3 \\ \underline{0} \end{vmatrix} \dot{u}_3 = \bar{\mathcal{I}}_3 \dot{u}_3, \quad \underline{\dot{\mathcal{V}}}^* = \begin{vmatrix} \bar{\imath}_3 \\ \underline{0} \end{vmatrix} \ddot{u}_3 = \bar{\mathcal{I}}_3 \ddot{u}_3 \tag{5.25}$$

When the motion of body $\ell$ is described independently, the governing equations of motion of body $\ell$ are assumed to be

$$M_e^\ell \underline{\dot{\mathcal{V}}}^{*\ell} = \underline{F}_e^\ell \tag{5.26}$$

where $M_e^\ell$ is the elemental mass matrix of body $\ell$, and $\underline{F}_e^\ell$ the associated elemental right hand side vector. The elemental equations of body $\ell$ will be transformed to the sort of recursive form at first like

$$\bar{M}_e^\ell \begin{vmatrix} \dot{\underline{\mathcal{V}}}^{*k} \\ \ddot{u}_3 \end{vmatrix} = \begin{bmatrix} \mathcal{R}_1^{*\mathcal{L}} \\ \bar{\mathcal{I}}_3^T \end{bmatrix} \left(\underline{F}_e^\ell - M_e^\ell \widetilde{\mathcal{V}}^{*T} \mathcal{R}_1^{*T} \underline{\mathcal{V}}^{*k}\right) \tag{5.27}$$

where $\bar{M}_e^\ell$ is the transformed mass matrix and in detail

$$\bar{M}_e^\ell = \begin{bmatrix} \mathcal{R}_1^{*\mathcal{L}} \\ \bar{\mathcal{I}}_3^T \end{bmatrix} M_e^\ell \left[\mathcal{R}_1^{*T},\ \bar{\mathcal{I}}_3\right] \tag{5.28}$$

Similar transformation needs to be applied to the elemental stiffness and damping matrices, $K_e^\ell$ and $C_e^\ell$, respectively. With the help of linearization of velocity $\underline{\mathcal{V}}^{*\ell}$ of body $\ell$

$$d\underline{\mathcal{V}}^{*\ell} = \mathcal{R}_1^{*T} d\underline{\mathcal{V}}^{*k} + \widetilde{\mathcal{R}_1^{*T} \underline{\mathcal{V}}^{*k}} \bar{\mathcal{I}}_3 du_3 + \bar{\mathcal{I}}_3 d\dot{u}_3 \tag{5.29}$$

and linearization of acceleration

$$\begin{aligned} d\dot{\underline{\mathcal{V}}}^{*\ell} &= \mathcal{R}_1^{*T} d\dot{\underline{\mathcal{V}}}^{*k} + \widetilde{\mathcal{V}}^{*T} \mathcal{R}_1^{*T} d\underline{\mathcal{V}}^{*k} + \bar{\mathcal{I}}_3 d\ddot{u}_3 \\ &+ \widetilde{\mathcal{R}_1^{*T} \underline{\mathcal{V}}^{*k}} \bar{\mathcal{I}}_3 d\dot{u}_3 + \left(\widetilde{\mathcal{R}_1^{*T} \dot{\underline{\mathcal{V}}}^{*k}} + \widetilde{\mathcal{V}}^{*T} \widetilde{\mathcal{R}_1^{*T} \underline{\mathcal{V}}^{*k}}\right) \bar{\mathcal{I}}_3 du_3 \end{aligned} \tag{5.30}$$

the stiffness and damping matrices have been transformed to

$$\bar{C}_e^\ell = \begin{bmatrix} \mathcal{R}_1^{*\mathcal{L}} \\ \bar{\mathcal{I}}_3^T \end{bmatrix} \left(C_e^\ell \left[\mathcal{R}_1^{*T},\ \bar{\mathcal{I}}_3\right] + M_e^\ell \left[\widetilde{\mathcal{V}}^{*T} \mathcal{R}_1^{*T},\ \widetilde{\mathcal{R}_1^{*T} \underline{\mathcal{V}}^{*k}} \bar{\mathcal{I}}_3\right]\right) \tag{5.31}$$

and

$$\begin{aligned} \bar{K}_e^\ell &= \begin{bmatrix} \mathcal{R}_1^{*\mathcal{L}} \\ \bar{\mathcal{I}}_3^T \end{bmatrix} \left(K_e^\ell \left[\mathcal{R}_1^{*T},\ \bar{\mathcal{I}}_3\right] + C_e^\ell \left[0,\ \widetilde{\mathcal{R}_1^{*T} \underline{\mathcal{V}}^{*k}} \bar{\mathcal{I}}_3\right]\right) \\ &+ \begin{bmatrix} \mathcal{R}_1^{*\mathcal{L}} \\ \bar{\mathcal{I}}_3^T \end{bmatrix} M_e^\ell \left[0,\ \left(\widetilde{\mathcal{R}_1^{*T} \dot{\underline{\mathcal{V}}}^{*k}} + \widetilde{\mathcal{V}}^{*T} \widetilde{\mathcal{R}_1^{*T} \underline{\mathcal{V}}^{*k}}\right) \bar{\mathcal{I}}_3\right] \end{aligned} \tag{5.32}$$

Then the mass, stiffness and damping matrices, $\bar{M}_e^\ell$, $\bar{K}_e^\ell$ and $\bar{C}_e^\ell$, together with transformed right hand side vector, need to be assembled into the global governing equations of motion. Observations of recursive model, Eq. (5.27), show that the governing equations are exactly ODE and relative translation $u_3$ associated degree of freedom are not massless. The recursive model has avoided formulating the stiff DAE of index-3. When linear spring and dash-pot damper are considered, the forces of spring and damper are easily obtained as $k_s u_3$ and $c_s \dot{u}_3$, respectively.

## 5.3 Motion Formalism of Screw Joint

The screw joint allows the relative translation along the axis $\bar{b}_3^k$ of basis $\mathcal{B}_1^k$ attached to the body $k$ together with the relative rotation with the magnitude of $\phi$ around the same axis of basis $\mathcal{B}_1^k$. The relative translation and rotation, constrainted by using the pitch $p$ of the screw, for a screw joint is given in the Table 5.1 as

$$\underline{u}^* = p\phi\bar{\imath}_3, \quad R_\phi = \begin{bmatrix} \cos\phi & -\sin\phi & 0 \\ \sin\phi & \cos\phi & 0 \\ 0 & 0 & 1 \end{bmatrix} \tag{5.33}$$

Given the initial configurations $\underline{x}_0 = \underline{x}_0^\ell - \underline{x}_0^k = 0$, and $R_0^k = R_0^\ell$, the motion tensors $\mathcal{R}_1^*$ and $\mathcal{R}_1^{*\mathcal{L}}$ are simplified to be

$$\mathcal{R}_1^* = \begin{bmatrix} R_\phi & \widetilde{u}^* R_\phi \\ 0 & R_\phi \end{bmatrix}, \quad \mathcal{R}_1^{*\mathcal{L}} = \begin{bmatrix} R_\phi & 0 \\ \widetilde{u}^* R_\phi & R_\phi \end{bmatrix} \tag{5.34}$$

According to the defintion, Eq. (5.8), the virtual relative motion of screw joint is readily found to be

$$\underline{\delta\mathcal{U}}^* = \begin{vmatrix} pR_\phi^T\bar{\imath}_3 \\ \bar{\imath}_3 \end{vmatrix} \delta\phi = \underline{\mathcal{I}}_p \delta\phi \tag{5.35}$$

The velocity $\underline{\mathcal{V}}^*$ and acceleration $\dot{\underline{\mathcal{V}}}^*$ of screw joint becomes

$$\underline{\mathcal{V}}^* = \begin{vmatrix} pR_\phi^T\bar{\imath}_3 \\ \bar{\imath}_3 \end{vmatrix} \dot{\phi} = \underline{\mathcal{I}}_p \dot{\phi}, \quad \dot{\underline{\mathcal{V}}}^* = \begin{vmatrix} pR_\phi^T\bar{\imath}_3 \\ \bar{\imath}_3 \end{vmatrix} \ddot{\phi} = \underline{\mathcal{I}}_p \ddot{\phi} \tag{5.36}$$

When the motion of body $\ell$ is described independently, the governing equations of motion of body $\ell$ are assumed to be

$$M_e^\ell \dot{\underline{\mathcal{V}}}^{*\ell} = \underline{F}_e^\ell \tag{5.37}$$

where $M_e^\ell$ is the elemental mass matrix of body $\ell$, and $\underline{F}_e^\ell$ the associated elemental right hand side vector. The elemental equations of body $\ell$ will be transformed to the sort of recursive form at first like

$$\bar{M}_e^\ell \begin{vmatrix} \dot{\underline{\mathcal{V}}}^{*k} \\ \ddot{\phi} \end{vmatrix} = \begin{bmatrix} \mathcal{R}_1^{*\mathcal{L}} \\ \underline{\mathcal{I}}_p^T \end{bmatrix} \left( \underline{F}_e^\ell - M_e^\ell \widetilde{\mathcal{V}}^{*T} \mathcal{R}_1^{*T} \underline{\mathcal{V}}^{*k} \right) \tag{5.38}$$

where $\bar{M}_e^\ell$ is the transformed mass matrix and in detail

$$\bar{M}_e^\ell = \begin{bmatrix} \mathcal{R}_1^{*\mathcal{L}} \\ \underline{\mathcal{I}}_p^T \end{bmatrix} M_e^\ell \left[ \mathcal{R}_1^{*T}, \ \underline{\mathcal{I}}_p \right] \tag{5.39}$$

Similar transformation needs to be applied to the elemental stiffness and damping matrices, $K_e^\ell$ and $C_e^\ell$, respectively. With the help of linearization of velocity $\underline{\mathcal{V}}^{*\ell}$ of body $\ell$

$$d\underline{\mathcal{V}}^{*\ell} = \mathcal{R}_1^{*T} d\underline{\mathcal{V}}^{*k} + \widetilde{\mathcal{R}_1^{*T}\underline{\mathcal{V}}^{*k}}\underline{\mathcal{I}}_p d\phi + \underline{\mathcal{I}}_p d\dot{\phi} \tag{5.40}$$

and linearization of acceleration

$$\begin{aligned} d\dot{\underline{\mathcal{V}}}^{*\ell} &= \mathcal{R}_1^{*T} d\dot{\underline{\mathcal{V}}}^{*k} + \widetilde{\mathcal{V}}^{*T}\mathcal{R}_1^{*T} d\underline{\mathcal{V}}^{*k} + \underline{\mathcal{I}}_p d\ddot{\phi} \\ &+ \widetilde{\mathcal{R}_1^{*T}\underline{\mathcal{V}}^{*k}}\underline{\mathcal{I}}_p d\dot{\phi} + \left(\widetilde{\mathcal{R}_1^{*T}\dot{\underline{\mathcal{V}}}^{*k}} + \widetilde{\mathcal{V}}^{*T}\widetilde{\mathcal{R}_1^{*T}\underline{\mathcal{V}}^{*k}}\right)\underline{\mathcal{I}}_p d\phi \end{aligned} \tag{5.41}$$

the stiffness and damping matrices have been transformed to

$$\bar{C}_e^\ell = \begin{bmatrix} \mathcal{R}_1^{*\mathcal{L}} \\ \underline{\mathcal{I}}_p^T \end{bmatrix} \left( C_e^\ell \left[ \mathcal{R}_1^{*T},\ \underline{\mathcal{I}}_p \right] + M_e^\ell \left[ \widetilde{\mathcal{V}}^{*T}\mathcal{R}_1^{*T},\ \widetilde{\mathcal{R}_1^{*T}\underline{\mathcal{V}}^{*k}}\underline{\mathcal{I}}_p \right] \right) \tag{5.42}$$

and

$$\begin{aligned} \bar{K}_e^\ell &= \begin{bmatrix} \mathcal{R}_1^{*\mathcal{L}} \\ \underline{\mathcal{I}}_p^T \end{bmatrix} \left( K_e^\ell \left[ \mathcal{R}_1^{*T},\ \underline{\mathcal{I}}_p \right] + C_e^\ell \left[ 0,\ \widetilde{\mathcal{R}_1^{*T}\underline{\mathcal{V}}^{*k}}\underline{\mathcal{I}}_p \right] \right) \\ &+ \begin{bmatrix} \mathcal{R}_1^{*\mathcal{L}} \\ \underline{\mathcal{I}}_p^T \end{bmatrix} M_e^\ell \left[ 0,\ \left( \widetilde{\mathcal{R}_1^{*T}\dot{\underline{\mathcal{V}}}^{*k}} + \widetilde{\mathcal{V}}^{*T}\widetilde{\mathcal{R}_1^{*T}\underline{\mathcal{V}}^{*k}} \right) \underline{\mathcal{I}}_p \right] \end{aligned} \tag{5.43}$$

Then the mass, stiffness and damping matrices, $\bar{M}_e^\ell$, $\bar{K}_e^\ell$ and $\bar{C}_e^\ell$, together with transformed right hand side vector, need to be assembled into the global governing equations of motion. The recursive model guarantees the relative rotation $\phi$ associated degree of freedom are not massless. When torsional spring and dash-pot damper are considered, the forces of spring and damper are easily obtained as $k_s\phi$ and $c_s\dot{\phi}$, respectively.

## 5.4 Motion Formalism of Cylindrical Joint

The cylindrical joint allows the relative translation along the axis $\bar{b}_3^k$ of basis $\mathcal{B}_1^k$ attached to the body $k$ together with the relative rotation with the magnitude of $\phi$ around the same axis of basis $\mathcal{B}_1^k$. The relative translation and rotation for a cylindrical joint is given in the Table 5.1 as

$$\underline{u}^* = u_3\bar{\imath}_3, \quad R_\phi = \begin{bmatrix} \cos\phi & -\sin\phi & 0 \\ \sin\phi & \cos\phi & 0 \\ 0 & 0 & 1 \end{bmatrix} \tag{5.44}$$

Given the initial configurations $\underline{x}_0 = \underline{x}_0^\ell - \underline{x}_0^k = 0$, and $R_0^k = R_0^\ell$, the motion tensors $\mathcal{R}_1^*$ and $\mathcal{R}_1^{*\mathcal{L}}$ are simplified to be

$$\mathcal{R}_1^* = \begin{bmatrix} R_\phi & \widetilde{u}^* R_\phi \\ 0 & R_\phi \end{bmatrix}, \quad \mathcal{R}_1^{*\mathcal{L}} = \begin{bmatrix} R_\phi & 0 \\ \widetilde{u}^* R_\phi & R_\phi \end{bmatrix} \tag{5.45}$$

According to the defintion, Eq. (5.8), the virtual relative motion of cylindrical joint is readily found to be

$$\underline{\delta \mathcal{U}^*} = \begin{bmatrix} R_\phi^T \bar{\imath}_3 & 0 \\ 0 & \bar{\imath}_3 \end{bmatrix} \begin{vmatrix} \delta u_3 \\ \delta \phi \end{vmatrix} = \mathcal{I}_\phi \delta \underline{u}_\phi \tag{5.46}$$

where the following symbols are introduced as

$$\mathcal{I}_\phi = \begin{bmatrix} R_\phi^T \bar{\imath}_3 & 0 \\ 0 & \bar{\imath}_3 \end{bmatrix}, \quad \underline{u}_\phi = \begin{vmatrix} u_3 \\ \phi \end{vmatrix} \tag{5.47}$$

The velocity $\underline{\mathcal{V}}^*$ and acceleration $\underline{\dot{\mathcal{V}}}^*$ of cylindrical joint becomes

$$\underline{\mathcal{V}}^* = \mathcal{I}_\phi \underline{\dot{u}}_\phi, \quad \underline{\dot{\mathcal{V}}}^* = \mathcal{I}_\phi \underline{\ddot{u}}_\phi \tag{5.48}$$

When the motion of body $\ell$ is described independently, the governing equations of motion of body $\ell$ are assumed to be

$$M_e^\ell \underline{\dot{\mathcal{V}}}^{*\ell} = \underline{F}_e^\ell \tag{5.49}$$

where $M_e^\ell$ is the elemental mass matrix of body $\ell$, and $\underline{F}_e^\ell$ the associated elemental right hand side vector. The elemental equations of body $\ell$ will be transformed to the sort of recursive form at first like

$$\bar{M}_e^\ell \begin{vmatrix} \underline{\dot{\mathcal{V}}}^{*k} \\ \ddot{\phi} \end{vmatrix} = \begin{bmatrix} \mathcal{R}_1^{*\mathcal{L}} \\ \mathcal{I}_\phi^T \end{bmatrix} \left( \underline{F}_e^\ell - M_e^\ell \widetilde{\mathcal{V}}^{*\mathcal{T}} \mathcal{R}_1^{*\mathcal{T}} \underline{\mathcal{V}}^{*k} \right) \tag{5.50}$$

where $\bar{M}_e^\ell$ is the transformed mass matrix and in detail

$$\bar{M}_e^\ell = \begin{bmatrix} \mathcal{R}_1^{*\mathcal{L}} \\ \mathcal{I}_\phi^T \end{bmatrix} M_e^\ell \left[ \mathcal{R}_1^{*\mathcal{T}}, \mathcal{I}_\phi \right] \tag{5.51}$$

Similar transformation needs to be applied to the elemental stiffness and damping matrices, $K_e^\ell$ and $C_e^\ell$, respectively. With the help of linearization of velocity $\underline{\mathcal{V}}^{*\ell}$ of body $\ell$

$$d\underline{\mathcal{V}}^{*\ell} = \mathcal{R}_1^{*\mathcal{T}} d\underline{\mathcal{V}}^{*k} + \widetilde{\mathcal{R}_1^{*\mathcal{T}} \underline{\mathcal{V}}^{*k}} \mathcal{I}_\phi d\underline{u}_\phi + \mathcal{I}_\phi d\underline{\dot{u}}_\phi \tag{5.52}$$

and linearization of acceleration

$$
\begin{aligned}
d\dot{\underline{\mathcal{V}}}^{*\ell} &= \mathcal{R}_1^{*T} d\dot{\underline{\mathcal{V}}}^{*k} + \widetilde{\mathcal{V}}^{*T}\mathcal{R}_1^{*T} d\underline{\mathcal{V}}^{*k} + \mathcal{I}_\phi d\ddot{\underline{u}}_\phi \\
&+ \widetilde{\mathcal{R}_1^{*T}\underline{\mathcal{V}}^{*k}}\mathcal{I}_\phi d\dot{\underline{u}}_\phi + \left(\widetilde{\mathcal{R}_1^{*T}\dot{\underline{\mathcal{V}}}^{*k}} + \widetilde{\mathcal{V}}^{*T}\widetilde{\mathcal{R}_1^{*T}\underline{\mathcal{V}}^{*k}}\right)\mathcal{I}_\phi d\underline{u}_\phi
\end{aligned} \tag{5.53}
$$

the stiffness and damping matrices have been transformed to

$$
\bar{C}_e^\ell = \begin{bmatrix} \mathcal{R}_1^{*\mathcal{L}} \\ \mathcal{I}_\phi^T \end{bmatrix} \left( C_e^\ell \left[ \mathcal{R}_1^{*T},\ \mathcal{I}_\phi \right] + M_e^\ell \left[ \widetilde{\mathcal{V}}^{*T}\mathcal{R}_1^{*T},\ \widetilde{\mathcal{R}_1^{*T}\underline{\mathcal{V}}^{*k}}\mathcal{I}_\phi \right] \right) \tag{5.54}
$$

and

$$
\begin{aligned}
\bar{K}_e^\ell &= \begin{bmatrix} \mathcal{R}_1^{*\mathcal{L}} \\ \mathcal{I}_\phi^T \end{bmatrix} \left( K_e^\ell \left[ \mathcal{R}_1^{*T},\ \mathcal{I}_\phi \right] + C_e^\ell \left[ 0,\ \widetilde{\mathcal{R}_1^{*T}\underline{\mathcal{V}}^{*k}}\mathcal{I}_\phi \right] \right) \\
&+ \begin{bmatrix} \mathcal{R}_1^{*\mathcal{L}} \\ \mathcal{I}_\phi^T \end{bmatrix} M_e^\ell \left[ 0,\ \left( \widetilde{\mathcal{R}_1^{*T}\dot{\underline{\mathcal{V}}}^{*k}} + \widetilde{\mathcal{V}}^{*T}\widetilde{\mathcal{R}_1^{*T}\underline{\mathcal{V}}^{*k}} \right)\mathcal{I}_\phi \right]
\end{aligned} \tag{5.55}
$$

Then the mass, stiffness and damping matrices, $\bar{M}_e^\ell$, $\bar{K}_e^\ell$ and $\bar{C}_e^\ell$, together with transformed right hand side vector, need to be assembled into the global governing equations of motion. The recursive model guarantees the relative translation $u_3$ and relative rotation $\phi$ associated degree of freedom are not massless. When spring and dash-pot damper are considered, the forces of spring and damper are easily obtained as $k_s \underline{u}_\phi$ and $c_s \dot{\underline{u}}_\phi$, respectively.

## 5.5 Motion Formalism of Revolute Joint

The revolute joint only allows the relative rotation with the magnitude of $\phi$ around the axis $\bar{b}_3^k$ of basis $\mathcal{B}_1^k$ attached to the body $k$. The relative rotation for a revolute joint is given in the Table. 5.1 as

$$
\underline{u}^* = \underline{0}, \quad R_\phi = \begin{bmatrix} \cos\phi & -\sin\phi & 0 \\ \sin\phi & \cos\phi & 0 \\ 0 & 0 & 1 \end{bmatrix} \tag{5.56}
$$

Given the initial configurations $\underline{x}_0 = \underline{x}_0^\ell - \underline{x}_0^k = 0$, and $R_0^k = R_0^\ell$, the motion tensors $\mathcal{R}_1^*$ and $\mathcal{R}_1^{*\mathcal{L}}$ are simplified to be

$$
\mathcal{R}_1^* = \mathcal{R}_1^{*\mathcal{L}} = \begin{bmatrix} R_\phi & 0 \\ 0 & R_\phi \end{bmatrix} \tag{5.57}
$$

According to the defintion, Eq. (5.8), the virtual relative motion of revolute joint is readily found to be

$$\underline{\delta \mathcal{U}^*} = \begin{vmatrix} 0 \\ \bar{\imath}_3 \end{vmatrix} \delta\phi = \bar{\mathcal{I}}_3^0 \delta\phi \tag{5.58}$$

The velocity $\underline{\mathcal{V}}^*$ and acceleration $\underline{\dot{\mathcal{V}}}^*$ of revolute joint becomes

$$\underline{\mathcal{V}}^* = \begin{vmatrix} 0 \\ \bar{\imath}_3 \end{vmatrix} \dot{\phi} = \bar{\mathcal{I}}_3^0 \dot{\phi}, \quad \underline{\dot{\mathcal{V}}}^* = \begin{vmatrix} 0 \\ \bar{\imath}_3 \end{vmatrix} \ddot{\phi} = \bar{\mathcal{I}}_3^0 \ddot{\phi} \tag{5.59}$$

When the motion of body $\ell$ is described independently, the governing equations of motion of body $\ell$ are assumed to be

$$M_e^\ell \underline{\dot{\mathcal{V}}}^{*\ell} = \underline{F}_e^\ell \tag{5.60}$$

where $M_e^\ell$ is the elemental mass matrix of body $\ell$, and $\underline{F}_e^\ell$ the associated elemental right hand side vector. The elemental equations of body $\ell$ will be transformed to the sort of recursive form at first like

$$\bar{M}_e^\ell \begin{vmatrix} \underline{\dot{\mathcal{V}}}^{*k} \\ \ddot{\phi} \end{vmatrix} = \begin{bmatrix} \mathcal{R}_1^{*\mathcal{L}} \\ \bar{\mathcal{I}}_3^{0\,T} \end{bmatrix} \left( \underline{F}_e^\ell - M_e^\ell \widetilde{\mathcal{V}}^{*T} \mathcal{R}_1^{*T} \underline{\mathcal{V}}^{*k} \right) \tag{5.61}$$

where $\bar{M}_e^\ell$ is the transformed mass matrix and in detail

$$\bar{M}_e^\ell = \begin{bmatrix} \mathcal{R}_1^{*\mathcal{L}} \\ \bar{\mathcal{I}}_3^{0\,T} \end{bmatrix} M_e^\ell \left[ \mathcal{R}_1^{*T}, \ \bar{\mathcal{I}}_3^0 \right] \tag{5.62}$$

Similar transformation needs to be applied to the elemental stiffness and damping matrices, $K_e^\ell$ and $C_e^\ell$, respectively. With the help of linearization of velocity $\underline{\mathcal{V}}^{*\ell}$ of body $\ell$

$$d\underline{\mathcal{V}}^{*\ell} = \mathcal{R}_1^{*T} d\underline{\mathcal{V}}^{*k} + \widetilde{\mathcal{R}_1^{*T} \underline{\mathcal{V}}^{*k}} \bar{\mathcal{I}}_3^0 d\phi + \bar{\mathcal{I}}_3^0 d\dot{\phi} \tag{5.63}$$

and linearization of acceleration

$$\begin{aligned} d\underline{\dot{\mathcal{V}}}^{*\ell} &= \mathcal{R}_1^{*T} d\underline{\dot{\mathcal{V}}}^{*k} + \widetilde{\mathcal{V}}^{*T} \mathcal{R}_1^{*T} d\underline{\mathcal{V}}^{*k} + \bar{\mathcal{I}}_3^0 d\ddot{\phi} \\ &+ \widetilde{\mathcal{R}_1^{*T} \underline{\mathcal{V}}^{*k}} \bar{\mathcal{I}}_3^0 d\dot{\phi} + \left( \widetilde{\mathcal{R}_1^{*T} \underline{\dot{\mathcal{V}}}^{*k} + \widetilde{\mathcal{V}}^{*T} \widetilde{\mathcal{R}_1^{*T} \underline{\mathcal{V}}^{*k}}} \right) \bar{\mathcal{I}}_3^0 d\phi \end{aligned} \tag{5.64}$$

the stiffness and damping matrices have been transformed to

$$\bar{C}_e^\ell = \begin{bmatrix} \mathcal{R}_1^{*\mathcal{L}} \\ \bar{\mathcal{I}}_3^{0\,T} \end{bmatrix} \left( C_e^\ell \left[ \mathcal{R}_1^{*T}, \ \bar{\mathcal{I}}_3^0 \right] + M_e^\ell \left[ \widetilde{\mathcal{V}}^{*T} \mathcal{R}_1^{*T}, \ \widetilde{\mathcal{R}_1^{*T} \underline{\mathcal{V}}^{*k}} \bar{\mathcal{I}}_3^0 \right] \right) \tag{5.65}$$

and

$$\bar{K}_e^\ell = \begin{bmatrix} \mathcal{R}_1^{*\mathcal{L}} \\ \bar{\mathcal{I}}_3^{0\,T} \end{bmatrix} \left( K_e^\ell \left[ \mathcal{R}_1^{*T},\ \bar{\mathcal{I}}_3^0 \right] + C_e^\ell \left[ 0,\ \widetilde{\mathcal{R}_1^{*T} \underline{\mathcal{V}}^{*k} \bar{\mathcal{I}}_3^0} \right] \right) + \begin{bmatrix} \mathcal{R}_1^{*\mathcal{L}} \\ \bar{\mathcal{I}}_3^{0\,T} \end{bmatrix} M_e^\ell \left[ 0,\ \left( \mathcal{R}_1^{*T} \dot{\underline{\mathcal{V}}}^{*k} + \widetilde{\mathcal{V}}^{*T} \widetilde{\mathcal{R}_1^{*T} \underline{\mathcal{V}}^{*k}} \right) \bar{\mathcal{I}}_3^0 \right] \tag{5.66}$$

Then the mass, stiffness and damping matrices, $\bar{M}_e^\ell$, $\bar{K}_e^\ell$ and $\bar{C}_e^\ell$, together with transformed right hand side vector, need to be assembled into the global governing equations of motion. The recursive model guarantees the relative rotation $\phi$ associated degree of freedom are not massless. When torsional spring and dash-pot damper are considered, the forces of spring and damper are easily obtained as $k_s\phi$ and $c_s\dot{\phi}$, respectively.

## 5.6 Motion Formalism of Spherical Joint

The spherical joint allows the arbitrary relative rotation, given in the Table 5.1. If the reference configurations are initialised to $\underline{x}_0 = \underline{x}_0^\ell - \underline{x}_0^k = 0$, and $R_0^k = R_0^\ell$, the motion tensors $\mathcal{R}_1^*$ and $\mathcal{R}_1^{*\mathcal{L}}$ are then simplified to

$$\mathcal{R}_1^* = \mathcal{R}_1^{*\mathcal{L}} = \begin{bmatrix} R_1^* & 0 \\ 0 & R_1^* \end{bmatrix} \tag{5.67}$$

According to the defintion, Eq. (5.8), the virtual relative motion of spherical joint is readily found to be

$$\underline{\delta\mathcal{U}^*} = \begin{bmatrix} 0 \\ I \end{bmatrix} \underline{\delta\psi}^* = \mathcal{I}_3^0 \underline{\delta\psi}^* \tag{5.68}$$

The velocity $\underline{\mathcal{V}}^*$ and acceleration $\dot{\underline{\mathcal{V}}}^*$ of spherical joint becomes

$$\underline{\mathcal{V}}^* = \begin{bmatrix} 0 \\ I \end{bmatrix} \underline{\omega}^* = \mathcal{I}_3^0 \underline{\omega}^*, \quad \dot{\underline{\mathcal{V}}}^* = \begin{bmatrix} 0 \\ I \end{bmatrix} \dot{\underline{\omega}}^* = \mathcal{I}_3^0 \dot{\underline{\omega}}^* \tag{5.69}$$

When the motion of body $\ell$ is described independently, the governing equations of motion of body $\ell$ are assumed to be

$$M_e^\ell \dot{\underline{\mathcal{V}}}^{*\ell} = \underline{F}_e^\ell \tag{5.70}$$

where $M_e^\ell$ is the elemental mass matrix of body $\ell$, and $\underline{F}_e^\ell$ the associated elemental right hand side vector. The elemental equations of body $\ell$ will be transformed to the sort of recursive form at first like

$$\bar{M}_e^\ell \left| \begin{matrix} \dot{\underline{\mathcal{V}}}^{*k} \\ \dot{\underline{\omega}}^* \end{matrix} \right| = \begin{bmatrix} \mathcal{R}_1^{*\mathcal{L}} \\ \mathcal{I}_3^{0\,T} \end{bmatrix} \left( \underline{F}_e^\ell - M_e^\ell \widetilde{\mathcal{V}}^{*T} \mathcal{R}_1^{*T} \underline{\mathcal{V}}^{*k} \right) \tag{5.71}$$

where $\bar{M}_e^\ell$ is the transformed mass matrix and in detail

$$\bar{M}_e^\ell = \begin{bmatrix} \mathcal{R}_1^{*\mathcal{L}} \\ \mathcal{I}_3^{0\,T} \end{bmatrix} M_e^\ell \left[ \mathcal{R}_1^{*T},\ \mathcal{I}_3^0 \right] \tag{5.72}$$

Similar transformation needs to be applied to the elemental stiffness and damping matrices, $K_e^\ell$ and $C_e^\ell$, respectively. With the help of linearization of velocity $\underline{\mathcal{V}}^{*\ell}$ of body $\ell$

$$d\underline{\mathcal{V}}^{*\ell} = \mathcal{R}_1^{*T} d\underline{\mathcal{V}}^{*k} + \widetilde{\mathcal{R}_1^{*T}\underline{\mathcal{V}}^{*k}} \mathcal{I}_3^0 d\underline{\psi}^* + \mathcal{I}_3^0 d\underline{\omega}^* \tag{5.73}$$

and linearization of acceleration

$$\begin{aligned} d\dot{\underline{\mathcal{V}}}^{*\ell} &= \mathcal{R}_1^{*T} d\dot{\underline{\mathcal{V}}}^{*k} + \tilde{\mathcal{V}}^{*T} \mathcal{R}_1^{*T} d\underline{\mathcal{V}}^{*k} + \mathcal{I}_3^0 d\dot{\underline{\omega}}^* \\ &+ \widetilde{\mathcal{R}_1^{*T}\underline{\mathcal{V}}^{*k}} \mathcal{I}_3^0 d\underline{\omega} + \left( \widetilde{\mathcal{R}_1^{*T}\dot{\underline{\mathcal{V}}}^{*k}} + \tilde{\mathcal{V}}^{*T} \widetilde{\mathcal{R}_1^{*T}\underline{\mathcal{V}}^{*k}} \right) \mathcal{I}_3^0 d\underline{\psi}^* \end{aligned} \tag{5.74}$$

the stiffness and damping matrices have been transformed to

$$\bar{C}_e^\ell = \begin{bmatrix} \mathcal{R}_1^{*\mathcal{L}} \\ \mathcal{I}_3^{0\,T} \end{bmatrix} \left( C_e^\ell \left[ \mathcal{R}_1^{*T},\ \mathcal{I}_3^0 \right] + M_e^\ell \left[ \tilde{\mathcal{V}}^{*T} \mathcal{R}_1^{*T},\ \widetilde{\mathcal{R}_1^{*T}\underline{\mathcal{V}}^{*k}} \mathcal{I}_3^0 \right] \right) \tag{5.75}$$

and

$$\begin{aligned} \bar{K}_e^\ell &= \begin{bmatrix} \mathcal{R}_1^{*\mathcal{L}} \\ \mathcal{I}_3^{0\,T} \end{bmatrix} \left( K_e^\ell \left[ \mathcal{R}_1^{*T},\ \mathcal{I}_3^0 \right] + C_e^\ell \left[ 0,\ \widetilde{\mathcal{R}_1^{*T}\underline{\mathcal{V}}^{*k}} \mathcal{I}_3^0 \right] \right) \\ &+ \begin{bmatrix} \mathcal{R}_1^{*\mathcal{L}} \\ \mathcal{I}_3^{0\,T} \end{bmatrix} M_e^\ell \left[ 0,\ \left( \widetilde{\mathcal{R}_1^{*T}\dot{\underline{\mathcal{V}}}^{*k}} + \tilde{\mathcal{V}}^{*T} \widetilde{\mathcal{R}_1^{*T}\underline{\mathcal{V}}^{*k}} \right) \mathcal{I}_3^0 \right] \end{aligned} \tag{5.76}$$

Then the mass, stiffness and damping matrices, $\bar{M}_e^\ell$, $\bar{K}_e^\ell$ and $\bar{C}_e^\ell$, together with transformed right hand side vector, need to be assembled into the global governing equations of motion. The recursive model guarantees the relative rotation $R^*$ associated degree of freedom are not massless.

## 5.7 Motion Formalism of Planar Joint

The planar joint allows the relative translation along the axis $\bar{b}_3^k$ of basis $\mathcal{B}_1^k$ attached to the body $k$ together with the relative rotation with the magnitude of $\phi$ around the same axis of basis $\mathcal{B}_1^k$. The relative translation and rotation for a planar joint is given in Table 5.1 as

$$\underline{u}^* = u_1 \bar{\imath}_1 + u_2 \bar{\imath}_2, \quad R_\phi = \begin{bmatrix} \cos\phi & -\sin\phi & 0 \\ \sin\phi & \cos\phi & 0 \\ 0 & 0 & 1 \end{bmatrix} \tag{5.77}$$

Given the initial configurations $\underline{x}_0 = \underline{x}_0^\ell - \underline{x}_0^k = 0$, and $R_0^k = R_0^\ell$, the motion tensors $\mathcal{R}_1^*$ and $\mathcal{R}_1^{*\mathcal{L}}$ are simplified to be

$$\mathcal{R}_1^* = \begin{bmatrix} R_\phi & \widetilde{u}^* R_\phi \\ 0 & R_\phi \end{bmatrix}, \quad \mathcal{R}_1^{*\mathcal{L}} = \begin{bmatrix} R_\phi & 0 \\ \widetilde{u}^* R_\phi & R_\phi \end{bmatrix} \tag{5.78}$$

According to the defintion, Eq. (5.8), the virtual relative motion of planar joint is readily found to be

$$\underline{\delta \mathcal{U}^*} = \begin{bmatrix} R_\phi^T \bar{\imath}_1 & R_\phi^T \bar{\imath}_2 & 0 \\ 0 & 0 & \bar{\imath}_3 \end{bmatrix} \begin{vmatrix} \delta u_1 \\ \delta u_2 \\ \delta \phi \end{vmatrix} = \mathcal{I}_\phi^3 \delta \underline{u}_\phi^3 \tag{5.79}$$

where the following symbols are introduced

$$\mathcal{I}_\phi^3 = \begin{bmatrix} R_\phi^T \bar{\imath}_1 & R_\phi^T \bar{\imath}_2 & 0 \\ 0 & 0 & \bar{\imath}_3 \end{bmatrix}, \quad \underline{u}_\phi^3 = \begin{vmatrix} u_1 \\ u_2 \\ \phi \end{vmatrix} \tag{5.80}$$

The velocity $\underline{\mathcal{V}}^*$ and acceleration $\dot{\underline{\mathcal{V}}}^*$ of planar joint becomes

$$\underline{\mathcal{V}}^* = \mathcal{I}_\phi^3 \dot{\underline{u}}_\phi^3, \quad \dot{\underline{\mathcal{V}}}^* = \mathcal{I}_\phi^3 \ddot{\underline{u}}_\phi^3 + \widetilde{\mathcal{V}}^* \bar{\mathcal{I}}_3^0 \dot{\phi} \tag{5.81}$$

After introducing new symbol

$$\mathcal{I}_0 = \begin{bmatrix} \underline{0}, & \underline{0}, & \underline{0} \\ \underline{0}, & \underline{0}, & \bar{\imath}_3 \end{bmatrix} \tag{5.82}$$

The acceleration $\dot{\underline{\mathcal{V}}}^*$ of planar joint can be rewritten as

$$\dot{\underline{\mathcal{V}}}^* = \mathcal{I}_\phi^3 \ddot{\underline{u}}_\phi^3 + \widetilde{\mathcal{V}}^* \mathcal{I}_0 \dot{\underline{u}}_\phi^3 \tag{5.83}$$

When the motion of body $\ell$ is described independently, the governing equations of motion of body $\ell$ are assumed to be

$$M_e^\ell \dot{\underline{\mathcal{V}}}^{*\ell} = \underline{F}_e^\ell \tag{5.84}$$

where $M_e^\ell$ is the elemental mass matrix of body $\ell$, and $\underline{F}_e^\ell$ the associated elemental right hand side vector. The elemental equations of body $\ell$ will be transformed to the sort of recursive form at first like

$$\bar{M}_e^\ell \begin{vmatrix} \dot{\underline{\mathcal{V}}}^{*k} \\ \ddot{\underline{u}}_\phi^3 \end{vmatrix} = \begin{bmatrix} \mathcal{R}_1^{*\mathcal{L}} \\ \mathcal{I}_\phi^{3\,T} \end{bmatrix} [\underline{F}_e^\ell - M_e^\ell (\widetilde{\mathcal{V}}^{*T} \mathcal{R}_1^{*T} \underline{\mathcal{V}}^{*k} + \widetilde{\mathcal{V}}^* \mathcal{I}_0 \dot{\underline{u}}_\phi^3)] \tag{5.85}$$

where $\bar{M}_e^\ell$ is the transformed mass matrix and in detail

$$\bar{M}_e^\ell = \begin{bmatrix} \mathcal{R}_1^{*\mathcal{L}} \\ \mathcal{I}_\phi^{3T} \end{bmatrix} M_e^\ell \left[ \mathcal{R}_1^{*T}, \ \mathcal{I}_\phi^3 \right] \tag{5.86}$$

With the help of linearization of velocity $\underline{\mathcal{V}}^{*\ell}$ of body $\ell$

$$d\underline{\mathcal{V}}^{*\ell} = \mathcal{R}_1^{*T} d\underline{\mathcal{V}}^{*k} + \left( \widetilde{\mathcal{R}_1^{*T} \underline{\mathcal{V}}^{*k}} \mathcal{I}_\phi^3 + \widetilde{\mathcal{V}}^* \mathcal{I}_0 \right) d\underline{u}_\phi^3 + \mathcal{I}_\phi^3 d\underline{\dot{u}}_\phi^3 \tag{5.87}$$

and linearization of acceleration

$$\begin{aligned} d\underline{\dot{\mathcal{V}}}^{*\ell} &= \mathcal{R}_1^{*T} d\underline{\dot{\mathcal{V}}}^{*k} + \widetilde{\mathcal{V}}^{*T} \mathcal{R}_1^{*T} d\underline{\mathcal{V}}^{*k} + \mathcal{I}_\phi^3 d\underline{\ddot{u}}_\phi^3 \\ &+ \left( \widetilde{\mathcal{R}_1^{*T} \underline{\mathcal{V}}^{*k}} \mathcal{I}_\phi^3 + \widetilde{\mathcal{V}}^* \mathcal{I}_0 + \mathcal{I}_\phi^3 \right) d\underline{\dot{u}}_\phi^3 + \left( \widetilde{\mathcal{R}_1^{*T} \underline{\dot{\mathcal{V}}}^{*k}} + \widetilde{\mathcal{V}}^{*T} \widetilde{\mathcal{R}_1^{*T} \underline{\mathcal{V}}^{*k}} \right) \mathcal{I}_\phi^3 d\underline{u}_\phi^3 \\ &+ \left( \widetilde{\mathcal{R}_1^{*T} \underline{\mathcal{V}}^{*k}} \widetilde{\mathcal{V}}^* + \widetilde{\mathcal{I}_\phi^3 \underline{\ddot{u}}_\phi^3} + \widetilde{\mathcal{I}_0 \underline{\dot{u}}_\phi^3}^T \widetilde{\mathcal{V}}^* \right) \mathcal{I}_0 d\underline{u}_\phi^3 \end{aligned} \tag{5.88}$$

similar transformation could be applied to the elemental stiffness and damping matrices, $K_e^\ell$ and $C_e^\ell$, respectively. Then the transformed mass, stiffness and damping matrices, $\bar{M}_e^\ell$, $\bar{K}_e^\ell$ and $\bar{C}_e^\ell$, together with transformed right hand side vector, need to be assembled into the global governing equations of motion. The recursive model guarantees the relative displacement $\underline{u}_\phi^3$ associated degree of freedom are not massless.

## 5.8 Motion Formalism of Rigid Connection

As shown in Fig. 5.2, two bodies, denoted "body $k$" and "body $\ell$," are rigidly connected. The reference configurations of bodies $k$ and $\ell$ could be described by using the motion tensors $\mathcal{R}_0^k$ and $\mathcal{R}_0^\ell$, respectively. Given the relative motion tensor $\mathcal{R}_0^*$ of rigid connection like

$$\mathcal{R}_0^* = \begin{bmatrix} R_0^*, & \widetilde{R_0^{kT} \underline{x}_0} R_0^* \\ 0, & R_0^* \end{bmatrix} \tag{5.89}$$

the reference motion tensor of body $\ell$ can be determined to be the composition of

$$\mathcal{R}_0^\ell = \mathcal{R}_0^k \mathcal{R}_0^* \tag{5.90}$$

For deformed configurations, the similar relations hold

$$\mathcal{R}_1^\ell = \mathcal{R}_1^k \mathcal{R}_0^* \tag{5.91}$$

where the relative position vector defined as $\underline{x}_0 = \underline{x}_0^\ell - \underline{x}_0^k$ and the relative rotation tensor $R_0^* = R_0^{kT} R_0^\ell$ will keeps unchanged during the motion. The virtual motions of bodies $k$ and $\ell$ computed from

$$\widetilde{\delta\mathcal{U}}^{*k} = \mathcal{R}_1^{kT} \delta\mathcal{R}_1^k, \quad \widetilde{\delta\mathcal{U}}^{*\ell} = \mathcal{R}_1^{\ell T} \delta\mathcal{R}_1^\ell \tag{5.92}$$

will be related by

$$\underline{\delta\mathcal{U}}^{*\ell} = \mathcal{R}_0^{*T} \underline{\delta\mathcal{U}}^{*k} \tag{5.93}$$

The velocities of bodies $k$ and $\ell$ in the dual vector forms defined in the similar approach like

$$\widetilde{\mathcal{V}}^{*k} = \mathcal{R}_1^{kT} \dot{\mathcal{R}}_1^k, \quad \widetilde{\mathcal{V}}^{*\ell} = \mathcal{R}_1^{\ell T} \dot{\mathcal{R}}_1^\ell \tag{5.94}$$

are readily verified to be constrainted as

$$\underline{\mathcal{V}}^{*\ell} = \mathcal{R}_0^{*T} \underline{\mathcal{V}}^{*k} \tag{5.95}$$

Taking time derivatives of velocities $\underline{\mathcal{V}}^{*\ell}$ will obtain the accelerations of body $\ell$ in their dual brackets like

$$\underline{\dot{\mathcal{V}}}^{*\ell} = \mathcal{R}_0^{*T} \underline{\dot{\mathcal{V}}}^{*k} \tag{5.96}$$

When the motion of body $\ell$ is described independently, the governing equations of motion of body $\ell$ are assumed to be

$$M_e^\ell \underline{\dot{\mathcal{V}}}^{*\ell} = \underline{F}_e^\ell \tag{5.97}$$

where $M_e^\ell$ is the elemental mass matrix of body $\ell$, and $\underline{F}_e^\ell$ the associated elemental right hand side vector. Current implementation will transform elemental equations of body $\ell$ to the sort of recursive form at first, and then it will be assembled following the conventional assembling procedure of finite element method. Substituting velocity relations, Eq. (5.96), into the above elemental equation to find

$$\bar{M}_e^\ell \underline{\dot{\mathcal{V}}}^{*k} = \mathcal{R}_0^{*\mathcal{L}} \underline{F}_e^\ell \tag{5.98}$$

where $\bar{M}_e^\ell$ is the transformed mass matrix and

$$\bar{M}_e^\ell = \mathcal{R}_0^{*\mathcal{L}} M_e^\ell \mathcal{R}_0^{*T} \tag{5.99}$$

The symbol $(\cdot)^{\mathcal{L}}$ represents the adjoint representation of *Lie* group *SE*(3), such that

$$\mathcal{R}_0^{*\mathcal{L}} = \begin{bmatrix} R_0^*, & 0 \\ \widetilde{R_0^{kT} \underline{x}_0} R_0^*, & R_0^* \end{bmatrix} \tag{5.100}$$

Furthermore, similar transformations need to be applied to the elemental stiffness and damping matrices, $K_e^\ell$ and $C_e^\ell$, respectively. With the help of linearization of velocity $\underline{\mathcal{V}}^{*\ell}$ of body $\ell$

$$d\underline{\mathcal{V}}^{*\ell} = \mathcal{R}_0^{*\mathcal{T}} d\underline{\mathcal{V}}^{*k} \tag{5.101}$$

and linearization of acceleration

$$d\underline{\dot{\mathcal{V}}}^{*\ell} = \mathcal{R}_0^{*\mathcal{T}} d\underline{\dot{\mathcal{V}}}^{*k} \tag{5.102}$$

the stiffness and damping matrices have been transformed to

$$\bar{C}_e^\ell = \mathcal{R}_0^{*\mathcal{L}} C_e^\ell \mathcal{R}_0^{*\mathcal{T}} \tag{5.103}$$

and

$$\bar{K}_e^\ell = \mathcal{R}_0^{*\mathcal{L}} K_e^\ell \mathcal{R}_0^{*\mathcal{T}} \tag{5.104}$$

Then the mass, stiffness and damping matrices, $\bar{M}_e^\ell$, $\bar{K}_e^\ell$ and $\bar{C}_e^\ell$, together with transformed right hand side vector, need to be assembled into the global ones. Observations of recursive model, Eq. (5.98), show that the governing equations are exactly ODE. The recursive model has avoided formulating the stiff DAE of index-3.

## 5.9 Motion Formalism of Rigid Rotation

Rigid rotation model is only applicable to the static analysis that sets the velocities of all types of finite elements as zero, except for elements with rigid rotation. The rigid rotation could be considered as a special type of revolute joint with a given rotation speed $\omega_r$. It means the relative rotation angle $\phi$ associated degrees of freedom will disappear from the governing equations of the multibody system, and $\phi$ can be computed from $\phi = \omega_r t$. The available elements featuring rigid rotation could be any type of Cosserat continuum, such as three-dimensional Cosserat continuum, plate/shell, beam and rigid body etc. Given a revolute joint with rigid rotation, its relative motion tensors $\mathcal{R}_0^*$ for reference configuration and $\mathcal{R}_1^*$ for deformed configuration could be identified to be

$$\mathcal{R}_0^* = \begin{bmatrix} R_\phi(0), & 0 \\ 0, & R_\phi(0) \end{bmatrix}, \quad \mathcal{R}_1^* = \begin{bmatrix} R_\phi(t), & 0 \\ 0, & R_\phi(t) \end{bmatrix} \tag{5.105}$$

The motion tensor of body $\ell$ can be determined to be the composition of

$$\mathcal{R}_0^\ell = \mathcal{R}_0^k \mathcal{R}_0^* \tag{5.106}$$

and

$$\mathcal{R}_1^\ell = \mathcal{R}_1^k \mathcal{R}_1^* \tag{5.107}$$

for reference and deformed configurations, respectively. The virtual motions of bodies $k$ and $\ell$ will be related by

$$\underline{\delta\mathcal{U}}^{*\ell} = \mathcal{R}_1^{*T}\,\underline{\delta\mathcal{U}}^{*k} \tag{5.108}$$

Similarly, according to the definition of velocity in the dual motion form,

$$\widetilde{\mathcal{V}}^{*\ell} = \mathcal{R}^{\ell T}\dot{\mathcal{R}}^{\ell} = \mathcal{R}_1^{*T}\dot{\mathcal{R}}_1^{*} \tag{5.109}$$

it is not difficult to assemble the velocity of body $\ell$ into the dual bracket

$$\underline{\mathcal{V}}^{*\ell} = \left|\frac{0}{\bar{\imath}_3\omega_r}\right| \tag{5.110}$$

Taking time derivatives of virtual motion of body $\ell$ leads to

$$\underline{\delta\dot{\mathcal{U}}}^{*\ell} = \widetilde{\mathcal{V}}^{*\ell T}\,\underline{\delta\mathcal{U}}^{*\ell} = \widetilde{\mathcal{V}}^{*\ell T}\mathcal{R}_1^{*T}\,\underline{\delta\mathcal{U}}^{*k} \tag{5.111}$$

For any type of Cosserat continuum, the motion tensor of an arbitrary particle **P** needs to be defined in the body attached local frame $\mathcal{B}_1$ with the help of the position vector $\underline{p}_1$ and total rotation tensor $R_1$ like

$$\mathcal{R}_1 = \begin{bmatrix} R_1 & \widetilde{p}_1 R_1 \\ 0 & R_1 \end{bmatrix} \tag{5.112}$$

in deformed configuration. When Cosserat continuum is connected to a revolut joint with described rotation speed $\omega_r$, the postion vector $\underline{p}_1$ and rotation tensor $R_1$ of an arbitrary particle **P** in the Cosserat continuum can be represented by

$$\underline{p}_1 = \underline{p}^{\ell} + \underline{p}_1^{b}, \quad R_1 = R^{\ell} R_1^{*b} \tag{5.113}$$

where $\underline{p}_{\ell}$ and $R_{\ell}$ are the position vector and rotation tensor of body $\ell$, the rotating part of a revolute joint, $\underline{p}_1^{b}$ and $R_1^{*b}$ are the relative position vector and relative rotation tensor of the particle **P** respected to the revolute joint, respectively. Denoting the relative motion tensor and its transpose as

$$\mathcal{R}_1^{*b} = \begin{bmatrix} R_1^{*b} & \widetilde{R^{\ell T}\underline{p}_b} R_1^{*b} \\ 0 & R_1^{*b} \end{bmatrix}, \quad \mathcal{R}_1^{*bT} = \begin{bmatrix} R_1^{*bT} & R_1^{*bT}\widetilde{R^{\ell T}\underline{p}_b}^{T} \\ 0 & R_1^{*bT} \end{bmatrix} \tag{5.114}$$

the motion tensor $\mathcal{R}_1$ will be related to the motion tensor $\mathcal{R}_1^{\ell}$ of body $\ell$ like

$$\mathcal{R}_1 = \mathcal{R}_1^{\ell}\mathcal{R}_1^{*b} \tag{5.115}$$

According to the above expressions, the velocity of particle **P** can be determined from its definition alternatively as

$$\widetilde{\mathcal{V}}_1^* = \mathcal{R}_1^T \dot{\mathcal{R}}_1 = \mathcal{R}_1^{*bT} \widetilde{\mathcal{V}}^{*\ell} \mathcal{R}_1^{*b} \tag{5.116}$$

and in detail, the velocity in the dual vector form can be found to be

$$\underline{\mathcal{V}}_1^* = \mathcal{R}_1^{*bT} \underline{\mathcal{V}}^{*\ell} \tag{5.117}$$

The definitions of virtual motions are similar to those of velocities of particle **P** like

$$\widetilde{\delta\mathcal{U}}_1^* = \mathcal{R}_1^T \delta\mathcal{R}_1 \tag{5.118}$$

Referring composition of motion tensors, Eq. (5.115), it is readily to find the dual vector of virtual motions become

$$\underline{\delta\mathcal{U}}_1^* = \mathcal{R}_1^{*bT} \underline{\delta\mathcal{U}}^{*\ell} + \underline{\delta\mathcal{U}}_1^{*b} \tag{5.119}$$

where $\underline{\delta\mathcal{U}}_1^{*b}$ represents the relative virtual motions which can be obtained from its definitions as

$$\widetilde{\delta\mathcal{U}}_1^{*b} = \mathcal{R}_1^{*bT} \delta\mathcal{R}_1^{*b} \tag{5.120}$$

By taking the time derivative of virtual motion vector, it will leads to

$$\underline{\delta\dot{\mathcal{U}}}_1^* = \mathcal{R}_1^{*bT} \widetilde{\mathcal{V}}^{*\ell T} \underline{\delta\mathcal{U}}^{*\ell} \tag{5.121}$$

Referring variation relations, Eq. (2.160), the variation of the dual velocity vector can be derived from the following relations

$$\delta\underline{\mathcal{V}}_1^* = \underline{\delta\dot{\mathcal{U}}}_1^* + \widetilde{\mathcal{V}}_1^* \underline{\delta\mathcal{U}}_1^* \tag{5.122}$$

With the aid of the time derivatives of virtual motion, the above expressions are recasted to

$$\delta\underline{\mathcal{V}}_1^* = \mathcal{R}_1^{*bT} \widetilde{\mathcal{V}}^{*\ell T} \underline{\delta\mathcal{U}}^{*\ell} + \widetilde{\mathcal{V}}_1^* \underline{\delta\mathcal{U}}_1^* \tag{5.123}$$

The contribution of rigid rotation to the kinetic energy can be computed from following formulations for a three-dimensional Cosserat continuum like

$$K = \frac{1}{2} \int_{V_0} \underline{\mathcal{V}}_1^{*T} \mathcal{M}^* \underline{\mathcal{V}}_1^* dV_0 \tag{5.124}$$

where $\mathcal{M}^*$ is the mass matrix. The variation of the kinetic energy is expressed as

$$\delta K = \int_{V_0} \delta\underline{\mathcal{V}}_1^{*T} \mathcal{M}^* \underline{\mathcal{V}}_1^* dV_0 \tag{5.125}$$

Substituting Eq. (5.123) into above expressions then leads to

$$\delta K = \int_{V_0} \left(\underline{\delta \mathcal{U}}^{*\ell T} \widetilde{\mathcal{V}}^{*\ell \mathcal{L}} \mathcal{R}_1^{*b\mathcal{L}} + \underline{\delta \mathcal{U}}_1^{*T} \widetilde{\mathcal{V}}_1^{*T}\right) \mathcal{M}^* \underline{\mathcal{V}}_1^* d V_0 \tag{5.126}$$

From which, the detailed expressions of the centrifugal force, $\underline{\mathcal{F}}_{cc}$, applied to the three-dimensional Cosserat continuum

$$\underline{\mathcal{F}}_{cc} = -\int_{V_0} \mathcal{H}^T \widetilde{\mathcal{V}}_1^{*T} \mathcal{M}^* \underline{\mathcal{V}}_1^* d V_0 \tag{5.127}$$

and the centrifugal force, $\underline{\mathcal{F}}_{c\ell}$, applied to the body $\ell$ of revolute joint with rigid rotation

$$\underline{\mathcal{F}}_{c\ell} = -\int_{V_0} \widetilde{\mathcal{V}}^{*\ell \mathcal{L}} \mathcal{R}_1^{*b\mathcal{L}} \mathcal{M}^* \underline{\mathcal{V}}_1^* d V_0 \tag{5.128}$$

are obtained with the aid of finite element interpolations.

## References


1. Bauchau, O.A., Bottasso, C.L.: Contact conditions for cylindrical, prismatic, and screw joints in flexible multibody systems. Multibody Syst. Dynam. (2001). https://doi.org/10.1023/A:1009710818135
2. Funabashi, H., Ogawa, K., Horie, M.: A dynamic analysis of mechanisms with clearances. Bullet. JSME. (1978). https://doi.org/10.1007/s11044-016-9562-3
3. Flores, P., Ambrósio, J.: Revolute Joints with Clearance in Multibody Systems. Comput. Struct. (2004). https://doi.org/10.1016/j.compstruc.2004.03.031
4. Flores, P., Ambrósio, J., Claro, J.C.P., Lankarani, H.M.: Spatial revolute joints with clearances for dynamic analysis of multi-body systems. In: Proceedings of the Institution of Mechanical Engineers, Part K: Journal of Multi-body Dynamics. (2006). https://doi.org/10.1023/A:1009710818135
5. Flores, P., Leine, R., Glocker, C.: Modeling and analysis of rigid multibody systems with translational clearance joints based on the nonsmooth dynamics approach. Multi. Syst. Dynam. (2010). https://doi.org/10.1007/s11044-009-9178-y
6. Marques, F., Isaac, F., Dourado, N., Flores, P.: An enhanced formulation to model spatial revolute joints with radial and axial clearances. Mech. Mach. Theo. (2017). https://doi.org/10.1016/j.mechmachtheory.2017.05.020
7. Akhadkar, N., Acary, V., Brogliato, B.: Multibody systems with 3D revolute joints with clearances: an industrial case study with an experimental validation. Mult. Syst. Dynam. (2017). https://doi.org/10.1007/s11044-017-9584-5
8. Marques, F., Flores, P., Lankarani, H.M.: On the frictional contacts in multibody system dynamics. In: Font-Llagunes, J. (ed.) Multibody Dynamics, vol. 42, pp. 67–91. Springer, Cham. (2016)
9. Canudas de Wit, C., Olsson, H., Astrom, K.J., Lischinsky, P.: A new model for control of systems with friction. IEEE Trans. Autom. Cont. **40**(3), 419–425, (1995)
10. Oden, J.C., Martins, J.A.C.: Models and computational methods for dynamic friction phenomena. Comput. Meth. Appl. Mech. Eng. **52**(1–3), 527–634 (1985)
11. Pennestrí, E., Rossi, V., Salvini, P., Valentini, P.P.: Review and comparison of dry friction force models. Nonlinear Dynam. (2016). https://doi.org/10.1023/A:1009710818135

12. Askari, E., Flores, P., Dabirrahmani, D., Appleyard, R.: Dynamic modeling and analysis of wear in spatial hard-on-hard couple hip replacements using multibody systems methodologies. Nonlinear Dynam. (2015). https://doi.org/10.1007/s11071-015-2216-9
13. Tian, Q., Sun, Y., Liu, C., Hu, H., Flores, P.: ElastoHydroDynamic lubricated cylindrical joints for rigid-flexible multibody dynamics. Comput. Struct. (2013). https://doi.org/10.1016/j.compstruc.2012.10.019
14. Shigley, J.E., Uicker, J.J.: Theory of Machines and Mechanisms. McGraw-Hill Book Company, New York (1980)
15. Shigley, J.E., Mischke, C.R.: Mechanical Engineering Design. McGraw-Hill Book Company, New York (1989)
16. Cardona, A., Gerardin, M.: Time integration of the equations of motion in mechanism analysis. Comput. Struct. **33**(3), 801–820 (1989)
17. Bae, D.S., Haug, E.J.: A recursive formulation for constrained mechanical system dynamics: part II. closed loop systems. Mech. Struct. Mach. (1987). https://doi.org/10.1080/08905458708905130
18. Petzold, L.: Differential/algebraic equations are not ODEs. Siam J. Sci. Stat. Comput. **3**(3), 367–384 (1982)

# Part IV
# Implicit Solver Based on Radau IIIA Algorithms

# Chapter 6
# Implicit Stiff Solvers with Post-error Estimation

Considering the efficient and elegant manner by which dual vector and motion tensor deal with motion and deformation, the unified theory and multiscale modeling of Cosserat continuum, together with the recursive formulas of kinematic constraints have been presented successively in Chaps. 3, 4 and 5. While, the motion formalism implementations of flexible multibody dynamic systems arise the new challenge for the stiff integrators that aim at solving the large-scale discrete system. Even though, Arnold and Brüls [1, 2] had developed the global parameterization-free generalized-$\alpha$ scheme with second-order accuracy to integrate the equations of motion in time domain. The high-order computing algorithms for the numerical simulation of constrained flexible multibody dynamics are still required with the purpose of affording versatile stiff solvers in a more general sense for the finite element based flexible multibody codes. The investigation [3] shows that the Radau IIA algorithms, belonging to the family of implicit integrators, feature excellent accuracy properties. Based on 2- and 3-stage Radau IIA algorithms [4], and tied up with post-error estimation, this chapter presents the new efficient solvers of third and fifth order, respectively.

As mentioned in Chap. 4, the multibody system is composed of rigid and flexible bodies in arbitrary motion as to each other, in which the magnitude of the motion could be large or arbitrary. This chapter pays a special attention to the prediction of arbitrary motion of multibody systems with the kinematic constraints, also called holonomic constraints, among various bodies. When modeled the multibody system by finite element method, the corresponding equations of motion present distinguishing features: stiff, nonlinear and Differential Algebraic Equations (DAE) of index-3. The enforcement of kinematic constraints by means of Lagrange multipliers [5] leads to infinite frequencies since the multiplier associated degrees of freedom are massless. This becomes the main source where the stiffness of the multibody system stems from. The designing of new stiff solvers has been closely related to the way how the constraints are enforced. Nevertheless, the recursive formulas of six low-pair of joints have been given in Chap. 5 to avoid the enforcing of holonomic constrains to the multibody systems. This chapter still concentrates on the developing of stiff

 
J. Wang, *Multiscale Multibody Dynamics*,
https://doi.org/10.1007/978-981-19-8441-9_6

solvers for DAE of index-3 in a more general sense, because the direct integration of Ordinary Differential Equations (ODE) can be readily accomplished as a special case of DAE of index-3 without kinematic constraints. Bauchau and Laulusa [6] reviewed the approaches for the constraint enforcements. Among of them, Baumgarte's stabilization method [7], an index reduction technique, is a common approach. However, it is not easy to determine the proper Baumgarte's parameters, and difficult to transform DAE of index-3 into the Ordinary Differential Equations (ODE) in most of the cases. The current implementation enforces the holonomic constraints by the preferable $h$-scaling technique [8], which makes it possible to integrate the DAE of index-3 directly.

Since the early 1950s, many numerical methods have been proposed for the solution of stiff problems satisfying both stability and accuracy requirements. The associated theoretical framework can be found in the book of Hairer and Wanner [9]. When solving stiff problems, the implicit algorithms are the first choice because a larger step size is permitted during the simulation compared to the explicit algorithms. The well-known single-step implicit solvers belong to the Runge-Kutta family, such as the Radau IIA algorithms. The commonly used multistep implicit solvers are the Backward Differentiation Formula (BDF) and its extension [10]. Currently, several mathematical codes based on the implicit methods are available. For example, RADAU5 [11] solves the stiff problems by using the 3-stage Radau IIA algorithm; DASSL and DASPK [12] are the public BDF codes developed for the solutions of DAE. However, RADAU5 and public BDF codes mainly focused on the mathematical implementation of the implicit algorithm and cannot directly solve the dynamic problems of multibody systems featuring arbitrary motions. The reason is due to the fact the motion, containing translation and rotation to span a six-dimensional space, can be described strictly by using the second order tensors, and the classical algebraic operations don't work for. Nevertheless, RADAU5 and public BDF codes allow users to define the right hand side functions $\underline{f}(t, \underline{x})$ of a DAE in its general form $\dot{\underline{x}} = \underline{f}(t, \underline{x})$, and assemble all the unknowns into a vector $\underline{x}$. Only the algebraic operations of vectors are permissible, such as vector addition $\underline{x}_1 = \underline{x}_1^0 + d\underline{x}_1$, which limits its application to the motion tensor. Furthermore, the skyline storage and the infrastructure of finite element code directly hinder the application of RADAU5 and BDF codes to solve the discrete problem of multibody system. Taking into account all these reasons, it is necessary to design new Radau IIA algorithms for the simulation of constrained flexible multibody dynamics in their motion formalism.

When designing new numerical algorithms for flexible multibody system, a key step is the vectorial parameterization of motion tensor before integrating the governing equations of motion. Typically, as described in Chap. 2, the Wiener-Milenković parameters [13] are the minimal set representation of motion containing only six independent variables. Furthermore, the range of validity of Wiener-Milenković parameters can be extended by using a rescaling operation, which means these preferred parameters are able to handle motions of arbitrary magnitudes. When motions expressed in terms of the Wiener-Milenković parameters, the derivatives of these parameters turn out to be very complicated and the linearization of these quantities become cumbersome especially for the second-order derivatives, which prevents its widely application. In current implementation, the total velocities and the incremental motions are carefully selected as the unknowns, which lead to a fairly new formula of governing equations of motion named as mixed formula in the book. For this mixed formula, the linearization of acceleration is not required. Furthermore, with the application of simplified Newton iterations, the cumbersome linearization associated with Wiener-Milenković is dramatically simplified. The new 2- and 3-stage Radau IIA algorithms have to be re-designed to fit into the mixed formula. Briefly speaking, solving the mix formula of multibody systems can be treated as the function extension of ODE solvers developed by Wang et al. [4]. Different from the previous work, the new algorithms enforce the kinematic constraints through the augmented functional of Hamilton's principle [14], and stabilizes the new DAE solvers by using the preferable $h$-scaling technique. All these improvements enhance the stability and efficiency of the new solvers. Instead of using the classical embedded formulation [9], a new approach based on post-error estimations is developed to predict the local truncation error from the associated deferred correction equations. The drawbacks of embedded formulation, overestimating the truncation error leads to a useless prediction of step size for stiff problem, are successfully overcome.

In summary, the features of current implementation include: (1) The new 2- and 3-stage stiff solvers are developed for the integration of DAE of index-3. (2) The mixed formula of governing equations of motion is presented and a vectorial parameterization of motion, Wiener-Milenković parameters, is implemented in a simplified approach. (3) The new solvers are stabilized by using the preferable $h$-scaling technique. (4) A new scheme based on post-error estimation is designed to compute the local truncation error without overestimation. All these features of the new 2- and 3-stage Radau IIA algorithm will be addressed clearly in this chapter.

## 6.1 Linearized Governing Equations of Motion for a Beam

For the easy description of new Radau IIA algorithms, the governing equations of motion and its linearization for a beam, also a one-dimensional Cosserat continuum equivalently, are addressed in detail. The associated algorithms can be readily extended to two- and three-dimensional Cosserat continua. The semi-discretized governing equations of motion for one-dimensional Cosserat continuum are verified to be

$$\mathcal{M}_n \underline{\dot{\mathcal{V}}}_n + \underline{\mathcal{F}}_g + \underline{\mathcal{F}}_e = \underline{\mathcal{F}}_a \tag{6.1}$$

The detailed derivations of above equations can be found in Sect. 3.4. Note that the above governing equations are discretized by using the Galerkin method. The choice of Galerkin test function $\mathcal{W}$ utilizes the shape functions $\mathcal{H}_\epsilon$ of finite element method and

$$\underline{\mathcal{W}} = \mathcal{H}_\epsilon \delta \underline{\mathcal{U}}_n \tag{6.2}$$

where $\underline{\delta \mathcal{U}}_n$ are the virtual nodal motions. It is necessary to point out $\mathcal{H}_\epsilon$ is the interpolation matrix containing the shape functions, $h_i$, at all $n$ nodes of a beam element. Since the velocities spanned a linear six-dimensional space, it can be interpolated directly by using the same interpolation matrix

$$\underline{\mathcal{V}}_1 = \mathcal{H}_\epsilon \underline{\mathcal{V}}_n \tag{6.3}$$

where $\underline{\mathcal{V}}_n$ are the nodal velocities resolved in the body attached frame $\mathcal{F}_1$. The detailed expressions of interpolated matrix $\mathcal{H}_\epsilon$, and nodal velocities $\underline{\mathcal{V}}_n$ are given as below

$$\mathcal{H}_\epsilon^T = \begin{bmatrix} h_1 I_{3\times3}, & 0 \\ 0, & h_1 I_{3\times3} \\ h_2 I_{3\times3}, & 0 \\ 0, & h_2 I_{3\times3} \\ \vdots, & \vdots \\ h_n I_{3\times3}, & 0 \\ 0, & h_n I_{3\times3} \end{bmatrix}_{6n\times6}, \quad \underline{\mathcal{V}}_n = \begin{vmatrix} \underline{\mathcal{V}}_1 \\ \underline{\mathcal{V}}_2 \\ \vdots \\ \underline{\mathcal{V}}_n \end{vmatrix}_{6n\times1} \tag{6.4}$$

With the aid of finite element method, the detailed expressions of discrete mass matrix $\mathcal{M}_n$, gyroscopic force $\underline{\mathcal{F}}_g$, elastic force $\underline{\mathcal{F}}_e$, and external force $\underline{\mathcal{F}}_a$ are summarized

$$\begin{aligned} \mathcal{M}_n &= \int_{\ell_0} \mathcal{H}_\epsilon^T \mathcal{M}_{\epsilon\epsilon} \mathcal{H}_\epsilon d\alpha_1 \\ \underline{\mathcal{F}}_g &= \int_{\ell_0} \mathcal{H}_\epsilon^T \widetilde{\mathcal{V}}_1^T \mathcal{M}_{\epsilon\epsilon} \underline{\mathcal{V}}_1 d\alpha_1 \\ \underline{\mathcal{F}}_e &= \int_{\ell_0} \left(\mathcal{H}_{\epsilon,1}^T + \mathcal{H}_\epsilon^T \widetilde{\mathcal{K}}_1^{1\,T}\right) \underline{F}_{\epsilon\epsilon}^{11} d\alpha_1 \\ \underline{\mathcal{F}}_a &= \int_{\ell_0} \mathcal{H}_\epsilon^T \underline{F}_{\epsilon\alpha} d\alpha_1 \end{aligned} \tag{6.5}$$

The linearization of gyroscopic force $\underline{\mathcal{F}}_g$ yields the damping matrix $\mathcal{G}_n$ and in detail

$$d\underline{\mathcal{F}}_g = \int_{\ell_0} \mathcal{H}_\epsilon^T (\widetilde{\mathcal{P}}_1^{\mathcal{L}^o} + \widetilde{\mathcal{V}}_1^T \mathcal{M}_{\epsilon\epsilon}) \mathcal{H}_\epsilon d\alpha_1 d\underline{\mathcal{V}}_n = \mathcal{G}_n d\underline{\mathcal{V}}_n \tag{6.6}$$

where the momentum vector with the definition of $\underline{\mathcal{P}}_1 = \mathcal{M}_{\epsilon\epsilon}\underline{\mathcal{V}}_1$ is introduced. The linearization procedure is then repeatedly applied to the elastic forces and external forces. With the aid of compatibility attribute, Eq. (3.330), the following stiffness matrices

$$\begin{aligned} \mathcal{K}_{sym} &= \int_{\ell_0} (\mathcal{H}_{\epsilon,1}^T + \mathcal{H}_\epsilon^T \widetilde{\mathcal{K}}_1^{1\,T}) \mathcal{C}_{11} (\mathcal{H}_{\epsilon,1} + \widetilde{\mathcal{K}}_1^1 \mathcal{H}_\epsilon) d\alpha_1 \\ \mathcal{K}_{non} &= \int_{\ell_0} \mathcal{H}_\epsilon^T \underline{\widetilde{F}}_{\epsilon\epsilon}^{11\mathcal{L}^o} (\mathcal{H}_{\epsilon,1} + \widetilde{\mathcal{K}}_1^1 \mathcal{H}_\epsilon) d\alpha_1 \end{aligned} \tag{6.7}$$

are identified after linearized the elastic forces that can be represented by

$$d\underline{\mathcal{F}}_e = (\mathcal{K}_{sym} + \mathcal{K}_{non}) \underline{d\mathcal{U}}_n \tag{6.8}$$

In the process of linearization, the motion increments are defined in a way similar to virtual motion, Eq. (2.154), like

$$\widetilde{d\mathcal{U}}_1 = \mathcal{R}_1^T d\mathcal{R}_1 \tag{6.9}$$

and discretized into

$$\underline{d\mathcal{U}}_1 = \mathcal{H}_\epsilon \underline{d\mathcal{U}}_n \tag{6.10}$$

Since the external force $\underline{F}$ and torque $\underline{T}$, defined by Eq. (3.324), are measured in the body attached frame $\mathcal{F}_1$, the linearization of external forces also contributes to the stiffness matrices as

$$d\underline{\mathcal{F}}_a = \int_{\ell_0} \mathcal{H}_\epsilon^T \begin{bmatrix} 0 & \widetilde{F} \\ 0 & \widetilde{T} \end{bmatrix} \mathcal{H}_\epsilon d\alpha_1 \underline{d\mathcal{U}}_n = \mathcal{K}_{ext} \underline{d\mathcal{U}}_n \tag{6.11}$$

Finally, the linearized governing equations of motion for a beam element are found to be

$$\mathcal{M}_n d\dot{\underline{\mathcal{V}}}_n + \mathcal{G}_n d\underline{\mathcal{V}}_n + \mathcal{K}_n \underline{d\mathcal{U}}_n = d\underline{\mathcal{F}}_n \tag{6.12}$$

where $d\underline{\mathcal{F}}_n$ is the out-of-balance forces, and $\mathcal{K}_n$ represents the elemental stiffness matrix, also the summation of all the stiffness contributions $\mathcal{K}_n = \mathcal{K}_{sym} + \mathcal{K}_{non} - \mathcal{K}_{ext}$.

## 6.2 Mixed Formula of Governing Equations of Motion for a Beam

The discrete governing equations of motion and the linearizations, Eqs. (6.1) and (6.12), have been presented in Sect. 6.1. It is observed that only the virtual motion $\underline{\delta\mathcal{U}}_1$, velocity $\underline{\mathcal{V}}_1$, and incremental motion $\underline{d\mathcal{U}}_1$ are interpolated by using the shape functions of finite element method. Since these variables span the six-dimensional linear spaces, so their finite element interpolations are exactly defined. However, motion tensor, $\mathcal{R}_1$, form a nonlinear space, cannot approximate directly by using the classical finite element interpolations. We even cannot obtain the explicit formula of motion vector $\underline{\mathcal{U}}_1$ from motion tensor $\mathcal{R}_1$ when the motion features large magnitude. This is always a challenge associated with the parameterization and interpolation of motion tensor. Current implementation avoids the interpolation of motion tensor, $\mathcal{R}_1$, by solving the governing equations of motion for incremental variables instead of for total variables. The trick arises from the following observations. When 2- and 3-stage Radau IIA algorithms [4, 15], the implicit solvers of third and fifth order, are applied to integrate the governing equations of motion, the integral has to be advanced step by step in time domain. For the guarantee of accuracy and stability, the time steps are constrained to be small. As a result, the magnitudes of motion increments inside each time step keep small. This phenomenon inspired the idea of incremental formula.

The motion, including translation and rotation, could be described by three different configurations, the reference, the current and the next configurations, respectively. Reference and current configurations are known configurations at times $t_0$ and $t_n$. It is assumed the integral has been successed to time $t_n$. Next configuration is the unknown deformed configuration needs to be determined at time $t_{n+1}$. The time step, $h = t_{n+1} - t_n$, is small. If denoted the motion tensors for three different configurations like, $\mathcal{R}_0$, $\mathcal{R}_n$ and $\mathcal{R}_1$ measured in inertial frame $\mathcal{I}$ and also the incremental motion from current to deformed configuration as $\mathcal{R}$ in body attached frame $\mathcal{B}_n$, the following composition of motion tensors holds

$$\mathcal{R}_1 = \mathcal{R}_n \mathcal{R} \tag{6.13}$$

and $\mathcal{R}_n = \mathcal{R}_0$ when $t_n = t_0$. Note the incremental motion $\mathcal{R}$ is small, and this is not an introduced assumption but the real situation of direct integration. Current and next motions $\mathcal{R}_n$ and $\mathcal{R}_1$ could be motions of arbitrary magnitudes. It is concluded the next motion of arbitrary magnitude can be described by current motion of arbitrary magnitude and incremental motion of small magnitude. For Eq. (6.13), there is no assumption of small motion introduced but a tricky of motion decomposition applied. The incremental motion tensor could be determined from $\mathcal{R} = \mathcal{R}_n^T \mathcal{R}_1$ and in detail

$$\mathcal{R} = \begin{bmatrix} R & \widetilde{R_n^T \underline{u}} R \\ 0 & R \end{bmatrix} \tag{6.14}$$

where $R_n$ and $R$ are the current and incremental rotation tensors, and $\underline{u}$ the incremental displacement of reference particle **P** for a beam element. The incremental motion tensor offers an alternative approach to evaluate the dual velocity, $\widetilde{\mathcal{V}}_1 = \mathcal{R}^T \dot{\mathcal{R}}$, and the virtual motion $\widetilde{\delta \mathcal{U}}_1 = \mathcal{R}^T \delta \mathcal{R}$, respectively.

Current implementation carefully selects the Wiener-Milenković parameterization [16] to vectorial parameterize the motion tensors, $\mathcal{R}_0$, $\mathcal{R}_n$ and $\mathcal{R}$, all of them. Referring Eq. (2.196), given the generating function $\underline{\mathcal{P}}$, the incremental motion tensor $\mathcal{R}$ can be determined from

$$\mathcal{R}(\underline{\mathcal{P}}) = \mathcal{I} + \frac{\wp^2}{2} \wp_0 \widetilde{\mathcal{P}} + \frac{\wp^2}{2} \widetilde{\mathcal{P}} \widetilde{\mathcal{P}} \tag{6.15}$$

Meanwhile, the dual velocity $\underline{\mathcal{V}}_1$ can be computed from these Wiener-Milenković parameterizations like

$$\underline{\mathcal{V}}_1 = \mathcal{H}^T(\underline{\mathcal{P}}) \dot{\underline{\mathcal{P}}} \tag{6.16}$$

where the tangent operator $\mathcal{H}(\underline{\mathcal{P}})$ is defined by Eq. (2.202), and given as follow

$$\mathcal{H}(\underline{\mathcal{P}}) = \wp (\mathcal{I} + \frac{\wp}{2} \widetilde{\mathcal{P}} + \frac{\wp}{8} \widetilde{\mathcal{P}} \widetilde{\mathcal{P}}) \tag{6.17}$$

for the completeness of the description. Since all these parameterizations are still valid for each of the beam nodes, the discrete vector of nodal motions for a beam element can be replaced by the following column array

$$\underline{\mathcal{U}}_n^T = \lfloor \underline{\mathcal{P}}_1^T,\ \underline{\mathcal{P}}_2^T,\ \ldots,\ \underline{\mathcal{P}}_n^T \rfloor \tag{6.18}$$

where $\underline{\mathcal{P}}_i$ for $i = 1, 2, \ldots, n$ are the generating functions of Wiener-Milenković parameterization for beam nodes. The nodal velocities can then be strictly computed from

$$\underline{\mathcal{V}}_n = \mathcal{H}_n^T \dot{\underline{\mathcal{U}}}_n \tag{6.19}$$

where the tangent operators $\mathcal{H}_i(\underline{\mathcal{P}}_i)$ for $i = 1, 2, \ldots, n$ have been assembled into the tangent matrix $\mathcal{H}_n$ in the diagonal blocks form of

$$\mathcal{H}_n^T = \begin{bmatrix} \mathcal{H}_1^T(\underline{\mathcal{P}}_1) & & \\ & \mathcal{H}_2^T(\underline{\mathcal{P}}_2) & \\ & & \ddots \end{bmatrix} \tag{6.20}$$

Even though $\underline{\mathcal{V}}_n$ are the total velocity of beam nodes, they are evaluated by using the incremental parameters $\underline{\mathcal{U}}_n$. Hence, Eq. (6.19) can be viewed as incremental formula of total velocity. With the aid of these incremental formulas of discrete velocities, the governing equations of motion, Eq. (6.1), becomes solvable and can be recasted to the closed form, called *fixed formula* in state space, like

$$\begin{aligned} \mathcal{H}_n \dot{\underline{\mathcal{U}}}_n &= \underline{\mathcal{V}}_n \\ \mathcal{M}_n \dot{\underline{\mathcal{V}}}_n &= \underline{\mathcal{F}}_n \end{aligned} \tag{6.21}$$

where the discrete force vector $\underline{\mathcal{F}}_n$ is introduced and $\underline{\mathcal{F}}_n = \underline{\mathcal{F}}_a - \underline{\mathcal{F}}_g - \underline{\mathcal{F}}_e$.

## 6.3 Dynamics of Constrained Flexible Multibody System

For flexible multibody system with holonomic constraints, $\underline{c}_H = 0$, the governing equations of motion can be obtained from the augmented functional of Hamilton's principle [14]. The contributed items of holonomic constraints can be added to the discrete governing equations of motion for a beam element, Eq. (6.21), like

$$\begin{aligned} \mathcal{H}_n \dot{\underline{\mathcal{U}}}_n &= \underline{\mathcal{V}}_n \\ \mathcal{M}_n \dot{\underline{\mathcal{V}}}_n &= \underline{\mathcal{F}}_n - B^T(k\underline{\lambda} + 2p\underline{c}_H) \\ 0 &= k\underline{c}_H(\mathcal{R}) \end{aligned} \tag{6.22}$$

where the matrix $B$ is the space derivatives of holonomic constraints, $B = \partial \underline{c}_H / \partial \underline{\mathcal{U}}_n$, and $\underline{\lambda}$ the Lagrange multipliers, $k$ and $p$ the scaling and penalty factors, respectively.

For the determination of arbitrary motion, the unknowns, incremental motion tensor $\mathcal{R}$, total velocity $\underline{\mathcal{V}}_n$ together with Lagrange multipliers $\underline{\lambda}$, need to be solved from Eq. (6.22), the DAE of index-3. For the easy description of the 2- and 3-stage Radau IIA algorithms in the next section, the compact formulas of these DAEs are introduced

$$\begin{aligned} M(\underline{y})\, \dot{\underline{y}} &= \underline{g}(t, \underline{y}, \underline{\lambda}) \\ 0 &= k\underline{c}_H \end{aligned} \tag{6.23}$$

where the following symbols are defined

$$\underline{y} = \begin{vmatrix} \underline{\mathcal{U}}_n \\ \underline{\mathcal{V}}_n \end{vmatrix}, \ \underline{g} = \begin{vmatrix} \underline{\mathcal{V}}_n \\ \underline{\mathcal{F}}_n - B^T(k\underline{\lambda} + 2p\underline{c}_H) \end{vmatrix}, \ M = \begin{bmatrix} \mathcal{H}_n & \\ & \mathcal{M}_n \end{bmatrix} \tag{6.24}$$

## 6.4 Implementation of 2-Stage Radau IIA Algorithm

In this section, a new algorithm, based on incremental formula of 2-stage Radau IIA [4], is designed to directly solve the dynamic problem of multibody systems with holonomic constraints.

### 6.4.1 *Solving Nonlinear Algebraic Equations*

When a new 2-stage Radau IIA algorithm is designed to solve the DAE of index-3, Eq. (6.23), the following Nonlinear Algebraic Equations (NAE) need to be solved at each time step, $t \in [t_n, t_{n+1}]$, alternatively

$$\underline{G}(\underline{z}) = \begin{vmatrix} \underline{g}(t_n + c_1 h,\ \underline{y}_n + \underline{z}_1,\ \underline{\lambda}_1) \\ -k\underline{c}_H(\underline{y}_n + \underline{z}_1) \\ ------------ \\ \underline{g}(t_n + c_2 h,\ \underline{y}_n + \underline{z}_2,\ \underline{\lambda}_2) \\ -k\underline{c}_H(\underline{y}_n + \underline{z}_2) \end{vmatrix} \tag{6.25}$$
$$- \frac{1}{h} \begin{bmatrix} M_1 & & & \\ & 0 & & \\ & & M_2 & \\ & & & 0 \end{bmatrix} \begin{bmatrix} a_{11}I & a_{12}I \\ a_{21}I & a_{22}I \end{bmatrix}^{-1} \begin{vmatrix} \underline{z}_1 \\ \underline{\lambda}_1 \\ -- \\ \underline{z}_2 \\ \underline{\lambda}_2 \end{vmatrix} = 0$$

where the notation $(\cdot)_i$ represents the quantities measured at stage $i$ for $i = 1, 2$. The definitions of symbols, $h, \underline{z}_i, a_{ij}$ and $c_i$ for $i, j = 1, 2$ can be found in Ref [4]. The new algorithm solves Eq. (6.25) by using the so called inner-outer iteration scheme [4]: the simplified Newton iterations are performed as an outer iteration, and then the resulting linearized block triangular system is solved by an inner iteration scheme. For inner iteration, the block diagonal terms of triangular system are identical, and the equations of first block, needs to be solved by LU decomposition for $j^{\text{th}}$ inner iteration and $k^{\text{th}}$ outer iteration, is given by

$$A\left(\frac{\sqrt{6}}{h}\right)\begin{vmatrix}\Delta\underline{w}_{11}^{k(j)}\\ \Delta\underline{w}_{12}^{k(j)}\\ \Delta\underline{w}_{1\lambda}^{k(j)}\end{vmatrix}=\begin{vmatrix}\underline{b}_1\\ \underline{b}_2\\ \underline{b}_\lambda\end{vmatrix} \tag{6.26}$$

where $\Delta\underline{w}_{1i}^{k(j)}$, $i = 1, 2, \lambda$, is the transformed incremental variables, $\underline{b}_1$, $\underline{b}_2$ and $\underline{b}_\lambda$ the components of residual vector. $A$ the diagonal block matrix and in detail

$$A\left(\frac{\sqrt{6}}{h}\right)=\begin{bmatrix}\frac{\sqrt{6}}{h}\mathcal{H}_n & -\mathcal{I} & 0\\ \mathcal{K}_n & \frac{\sqrt{6}}{h}\mathcal{M}_n+\mathcal{G}_n & kB^T\\ kB & 0 & 0\end{bmatrix} \tag{6.27}$$

where $\mathcal{M}_n$, $\mathcal{K}_n$ and $\mathcal{G}_n$ are the mass, stiffness and damping matrices of the flexible multibody system, as defined in the linearized systems, Eq. (6.12), respectively. $\mathcal{H}_n$ is the diagonal block matrix containing the tangent operators $\mathcal{H}_i$ of beam nodes. It is worthy to note that only the linearizations of motion and velocity, $d\underline{\mathcal{U}}_n$ and $d\underline{\mathcal{V}}_n$ are required and the linearizations of accelerations, $d\dot{\underline{\mathcal{V}}}_n$, very cumbersome works, are avoid by selecting the incremental motions and total velocities as unknowns.

The numerical practices show that LU decomposition of matrix $A$ is the most time consuming computation when solving Eq. (6.25) by simplified Newton method. In current implementation, solving the fixed formulas of governing equations enhances the computing efficiency because matrix $A$ will be further simplified due to the following fact: at reference time $t_n$, the incremental motion is zero so that $\mathcal{R}=\mathcal{I}$, which leads to $\mathcal{H}=\mathcal{I}$ and $\underline{\mathcal{P}}=0$. Furthermore, when advancing the integration of multibody dynamics, only a small step size $h$ is used to guarantee the accuracy of the solution. In such a situation, $\sqrt{6}/h\mathcal{M}_n$ has a dominant contribution to the coefficient matrix $A$ and the contribution of $\mathcal{H}_n$ becomes negligible. So neglecting the matrix $\mathcal{H}_n$ and with the given incremental quantities at reference time $t_n$, matrix $A$ is simplified to

$$A\left(\frac{\sqrt{6}}{h}\right)=\begin{bmatrix}\frac{\sqrt{6}}{h}\mathcal{I} & -\mathcal{I} & 0\\ \mathcal{K}_n & \frac{\sqrt{6}}{h}\mathcal{M}_n+\mathcal{C}_n & kB^T\\ kB & 0 & 0\end{bmatrix} \tag{6.28}$$

### 6.4.2 *Solving Block Triangular Equations*

Due to the special data structure of Eq. (6.28), the associated linearized equations of the first block can be split up into two parts so that the cost of LU decomposition is further reduced, and only the following split system needs to be solved

$$J \begin{vmatrix} \Delta \underline{w}_{12}^{k(j)} \\ \Delta \underline{w}_{1\lambda}^{k(j)} \end{vmatrix} = \begin{vmatrix} \underline{b}_2 \\ \underline{b}_\lambda \end{vmatrix} - \frac{h}{\sqrt{6}} \begin{bmatrix} \mathcal{K}_n \\ kB \end{bmatrix} \underline{b}_1 \tag{6.29}$$

where $J$ is the Jacobian matrix

$$J = \begin{bmatrix} \frac{h}{\sqrt{6}}\mathcal{K}_n + \frac{\sqrt{6}}{h}\mathcal{M}_n + \mathcal{G}_n & kB^T \\ \frac{h}{\sqrt{6}}kB & 0 \end{bmatrix} \tag{6.30}$$

and the rest of the solution computes from

$$\Delta \underline{w}_{11}^{k(j)} = \frac{h}{\sqrt{6}}(\underline{b}_1 + \Delta \underline{w}_{12}^{k(j)}) \tag{6.31}$$

It was pointed out [17] the condition number of Jacobian matrix, Eq. (6.30), is $O(h^{-3})$, which means these equations are severely ill conditioned for small $h$. In most of the case, implicit solvers, such as RADAU5 and public BDF codes, become unstable and fail to converge when the Jacobian matrix is ill conditioned. Current implementation stabilizes the new algorithm by using the preferable $h$-scaling technique to improve the condition number to $O(h^{-1})$. With $h$-scaling, the linearized equations are modified to

$$\hat{J} \begin{vmatrix} \Delta \underline{w}_{12}^{k(j)} \\ \frac{h}{\sqrt{6}} \Delta \underline{w}_{1\lambda}^{k(j)} \end{vmatrix} = \left( \begin{vmatrix} \underline{b}_2 \\ \frac{6}{h^2}\underline{b}_\lambda \end{vmatrix} - \begin{bmatrix} \frac{h}{\sqrt{6}}\mathcal{K}_n \\ \frac{\sqrt{6}}{h}B \end{bmatrix} \underline{b}_1 \right) \tag{6.32}$$

where

$$\hat{J} = \begin{bmatrix} \frac{h}{\sqrt{6}}\mathcal{K}_n + \frac{\sqrt{6}}{h}\mathcal{M}_n + \mathcal{G}_n & \frac{\sqrt{6}}{h}B^T \\ \frac{\sqrt{6}}{h}B & 0 \end{bmatrix} \tag{6.33}$$

and the scaling factor $k$ set as $k = 1$.

Once the solutions for $\Delta \underline{w}_1^{k(j)}$ are obtained, the same processing can be performed for $\Delta \underline{w}_2^{k(j)}$. When solving for $\Delta \underline{w}_2^{k(j)}$, there is no additional LU factorizations required. This leads to a significant reduction in the computational cost.

## 6.5 Implementation of 3-Stage Radau IIA Algorithm

Similar to the 2-stage Radau IIA algorithm, the incremental formula of 3-stage Radau IIA algorithm [4] is used in designing a new algorithm.

### 6.5.1 Solving Nonlinear Algebraic Equations

The new 3-stage Radau IIA algorithm in current implementation converts the DAE of index-3, Eq. (6.23), into the following nonlinear algebraic equations

$$\left|\begin{array}{c} \underline{g}(t_n + c_1 h,\ \underline{y}_n + \underline{z}_1) \\ -k\underline{c}_H(\underline{y}_n + \underline{z}_1) \\ \hline \underline{g}(t_n + c_2 h,\ \underline{y}_n + \underline{z}_2) \\ -k\underline{c}_H(\underline{y}_n + \underline{z}_2) \\ \hline \underline{g}(t_n + c_3 h,\ \underline{y}_n + \underline{z}_3) \\ -k\underline{c}_H(\underline{y}_n + \underline{z}_3) \end{array}\right| - \frac{1}{h}\begin{bmatrix} M_1 & & & & & \\ & 0 & & & & \\ & & M_2 & & & \\ & & & 0 & & \\ & & & & M_3 & \\ & & & & & 0 \end{bmatrix} \begin{bmatrix} a_{11}I & a_{12}I & a_{13}I \\ a_{21}I & a_{22}I & a_{23}I \\ a_{31}I & a_{32}I & a_{33}I \end{bmatrix}^{-1} \left|\begin{array}{c} \underline{z}_1 \\ \underline{\lambda}_1 \\ \hline \underline{z}_2 \\ \underline{\lambda}_2 \\ \hline \underline{z}_3 \\ \underline{\lambda}_3 \end{array}\right| = 0 \tag{6.34}$$

where $(\cdot)_3$ represents the quantities measured at the third stage, and the definitions of symbols, $\underline{z}_i$, $a_{ij}$ and $c_i$ for $i, j = 1, 2, 3$ can be found in Ref [4]. Here again, the simplified Newton iterations are performed to release the heavy computational burden. The NAE, Eq. (6.34), are linearized at first, and then the linearized equations are transformed into a decoupled system with the help of orthogonal decomposition [4]. The decoupled system is composed of two independent subsystems and written as

$$\begin{bmatrix} A(\frac{\gamma}{h}) & & \\ & A(\frac{\alpha}{h}) & -\frac{\beta}{h}M \\ & \frac{\beta}{h}M & A(\frac{\alpha}{h}) \end{bmatrix} \left|\begin{array}{c} \Delta\underline{w}_1^k \\ \Delta\underline{w}_2^k \\ \Delta\underline{w}_3^k \end{array}\right| = \left|\begin{array}{c} \underline{G}_1^* \\ \underline{G}_2^* \\ \underline{G}_3^* \end{array}\right| \tag{6.35}$$

where the definitions of scalars $\alpha$, $\beta$ and $\gamma$ can be found in Ref [4]. it is amazing note that the first subsystem is consistent with Eq. (6.26). Thus, the procedure for solving $\Delta\underline{w}_1^{k(j)}$ described in Sect. 6.4.2 can be applied again to solve $\Delta\underline{w}_1^k$ efficiently. For the second subsystem, the dimension of equations is readily reduced when the subsystem transforms into a complex one

$$(J_r + \imath J_m)\left(\left|\begin{array}{c} \Delta\underline{w}_{22}^k \\ \Delta\underline{w}_{2\lambda}^k \end{array}\right| + \imath \left|\begin{array}{c} \Delta\underline{w}_{32}^k \\ \Delta\underline{w}_{3\lambda}^k \end{array}\right|\right) = \left|\begin{array}{c} \underline{b}_r^2 \\ \underline{b}_r^\lambda \end{array}\right| + \imath \left|\begin{array}{c} \underline{b}_m^2 \\ \underline{b}_m^\lambda \end{array}\right| \tag{6.36}$$

where $\imath = \sqrt{-1}$, $J_r$ and $J_m$ are the real and imagine parts of Jacobian matrix, respectively

$$J_r = \begin{bmatrix} \frac{h\alpha}{\rho^2}\mathcal{K}_n + \frac{\alpha}{h}\mathcal{M}_n + \mathcal{G}_n & kB^T \\ \frac{h\alpha}{\rho^2}kB & 0 \end{bmatrix}$$
$$J_m = \begin{bmatrix} -\frac{h\beta}{\rho^2}\mathcal{K}_n + \frac{\beta}{h}\mathcal{M}_n & 0 \\ -\frac{h\beta}{\rho^2}kB & 0 \end{bmatrix} \tag{6.37}$$

where $\rho^2 = \alpha^2 + \beta^2$. The complex right hand side vector writes in detail

$$\begin{vmatrix} \underline{b}_r^2 \\ \underline{b}_r^\lambda \end{vmatrix} = \begin{vmatrix} \underline{G}_{22}^* \\ \underline{G}_{2\lambda}^* \end{vmatrix} - \begin{bmatrix} \mathcal{K}_n \\ kB \end{bmatrix} \left( \frac{h\beta}{\rho^2}\underline{G}_{31}^* + \frac{h\alpha}{\rho^2}\underline{G}_{21}^* \right)$$
$$\begin{vmatrix} \underline{b}_m^2 \\ \underline{b}_m^\lambda \end{vmatrix} = \begin{vmatrix} \underline{G}_{32}^* \\ \underline{G}_{3\lambda}^* \end{vmatrix} - \begin{bmatrix} \mathcal{K}_n \\ kB \end{bmatrix} \left( \frac{h\alpha}{\rho^2}\underline{G}_{31}^* - \frac{h\beta}{\rho^2}\underline{G}_{21}^* \right) \tag{6.38}$$

The rest of the solution, $\Delta\underline{w}_{21}^k$ and $\Delta\underline{w}_{31}^k$, computes from

$$\Delta\underline{w}_{21}^k = \frac{h\alpha}{\rho^2}\underline{G}_{21}^* + \frac{h\beta}{\rho^2}\underline{G}_{31}^* + \frac{h\alpha}{\rho^2}\Delta\underline{w}_{22}^k + \frac{h\beta}{\rho^2}\Delta\underline{w}_{32}^k$$
$$\Delta\underline{w}_{31}^k = -\frac{h\beta}{\rho^2}\underline{G}_{21}^* + \frac{h\alpha}{\rho^2}\underline{G}_{31}^* - \frac{h\beta}{\rho^2}\Delta\underline{w}_{22}^k + \frac{h\alpha}{\rho^2}\Delta\underline{w}_{32}^k \tag{6.39}$$

As mentioned above, the new algorithm will be stabilized by using the preferable $h$-scaling technique. With the application of this technique, the complex subsystem is modified to

$$\left(\hat{J}_r + \imath\hat{J}_m\right)\left( \begin{vmatrix} \Delta\underline{w}_{22}^k \\ s_1\Delta\underline{w}_{2\lambda}^k \end{vmatrix} + \imath \begin{vmatrix} \Delta\underline{w}_{32}^k \\ s_1\Delta\underline{w}_{3\lambda}^k \end{vmatrix} \right)$$
$$= \begin{vmatrix} \underline{b}_r^2 \\ s_2\underline{b}_r^\lambda \end{vmatrix} + \imath \begin{vmatrix} \underline{b}_m^2 \\ s_2\underline{b}_m^\lambda \end{vmatrix} \tag{6.40}$$

where $s_1$ and $s_2$ are scalars, $s_1 = h/\alpha$ and $s_2 = \rho^2/h^2$, respectively. The condition number of Jacobian matrix has been improved to $O(h^{-1})$, and the real and imagine parts of the Jacobian become

$$\hat{J}_r = \begin{bmatrix} \frac{h\alpha}{\rho^2}\mathcal{K}_n + \frac{\alpha}{h}\mathcal{M}_n + \mathcal{G}_n & \frac{\alpha}{h}B^T \\ \frac{\alpha}{h}B & 0 \end{bmatrix}$$
$$\hat{J}_m = \begin{bmatrix} -\frac{h\beta}{\rho^2}\mathcal{K}_n + \frac{\beta}{h}\mathcal{M}_n & 0 \\ -\frac{\beta}{h}B & 0 \end{bmatrix} \tag{6.41}$$

The Gaussian elimination is applied to solve the complex system, Eq. (6.40), and the cost is dramatically reduced because the dimension of Eq. (6.40) has recovered to the original size of Eq. (6.22).

## 6.6 Step Size Selection

The step size adaptive scheme is closely related to the accuracy control of the computational process. For the proposed approaches, the local truncation error is evaluated at each step to test the accuracy of integration and predict the new step size. The most popular error estimation techniques involve the embedding of two schemes within a single integral step and named as embedded formulation as implemented in RADUA5. The details of this technique for 2-stage and 3-stage Radau IIA algorithms can be found in Ref [4]. However, numerical practices show this technique always overestimates the error especially for stiff problem. The overestimation will mislead the step adaptive scheme to predict a very small step size, which makes no sense. In current framework, a new scheme based on post-error estimation is designed with the purpose of improving the accuracy of error estimation. Denoting $\underline{y}$ and $\underline{\lambda}$ as the real solutions of Eq. (6.23), $\hat{\underline{y}}$ and $\hat{\underline{\lambda}}$ the solution approximations, the associated truncation errors are defined as $\underline{\epsilon} = \underline{y} - \hat{\underline{y}}$ and $\underline{\epsilon}_\lambda = \underline{\lambda} - \hat{\underline{\lambda}}$, respectively. With these notations, the deferred correction equations are formulated by

$$\begin{aligned} M(\hat{\underline{y}} + \underline{\epsilon})\,\dot{\underline{\epsilon}} &= \underline{g}(t, \hat{\underline{y}} + \underline{\epsilon}, \hat{\underline{\lambda}} + \underline{\epsilon}_\lambda) - \underline{g}(t, \hat{\underline{y}}, \hat{\underline{\lambda}}) - \delta_r \\ 0 &= k\underline{c}_H(t, \hat{\underline{y}}_q + \underline{\epsilon}_q) \end{aligned} \tag{6.42}$$

where $\hat{\underline{y}}_q$ is the motion components of solution approximations $\hat{\underline{y}}$, and $\underline{\epsilon}_q$ the motion components of $\underline{\epsilon}$. $\delta_r$ denotes the residual of differential equations

$$\delta_r = M(\hat{\underline{y}})\,\dot{\hat{\underline{y}}} - \underline{g}(t, \hat{\underline{y}}, \hat{\underline{\lambda}}) \tag{6.43}$$

The truncation errors $\underline{\epsilon}$ and $\underline{\epsilon}_\lambda$ can be solved from the correction equations by using the proposed algorithms. The book has no means to improve the accuracy of integration by repeatedly applying the same algorithm inside a single step. It also pointed out [18] applying the same algorithm inside a single step iteratively cannot improve the accuracy of solution. However, it is a valuable practice to predict the truncation error by using the same algorithm, Eq. (6.42). After estimated the truncation error, the integration accuracy is evaluated by computing the weighted root-mean-square norm

$$\|\underline{\epsilon}_{n+1}\| = \sqrt{1/(2N)\sum_{i=1}^{2N}(\epsilon_i/\tau_i)^2} \tag{6.44}$$

where $N$ is the size of discrete system, $\epsilon_i$ the $i$th component of $\underline{\epsilon}_{n+1}$ and $\tau_i = \tau_a + \tau_r|y_i|$, $\tau_a$ and $\tau_r$ are user defined absolute and relative tolerances, respectively. There is no specific criteria for determining the tolerances, the quantitative values can vary from $1 \times 10^{-6}$ to $1 \times 10^{-3}$ in accordance with actual condition. $y_i$ is the $i$th component of computed solution $\underline{y}_{n+1}$. If the quantity is small enough, $\|\underline{\epsilon}_{n+1}\| < 1$, the iteration results $\underline{y}_{n+1}$ and $\underline{\lambda}_{n+1}$ are accepted as the solution of differential algebraic equations.

Another parameter needs to be determined at each integral step is the step size. In current implementation, the step adaptive scheme [4] is applied to predict the new step size automatically. User needs to input the initial size, varied from $1 \times 10^{-5}$ to $1 \times 10^{-3}$ second, or constraint the algorithms running with a constant size.

## 6.7 Validation of 2- and 3-Stage Radau IIA Algorithms

Two examples are presented in this section to validate the 2- and 3-stage Radau IIA algorithms developed in this chapter. Then, two Radau IIA algorithms will be used in the rest part of this chapter to predict the response of Cosserat continua under external excitation, and solve the engineering problems.

### *6.7.1 Beam Actuated by a Tip Crank*

The first example checks the accuracy and convergence properties of developed stiff solvers. As depicted in Fig. 6.1, the flexible multibody system is consisted of a cantilevered beam actuated by a crank mechanism. The beam with a length of 1 m has a rectangular cross-section of depth 10 mm and width 5 mm. The beam is made

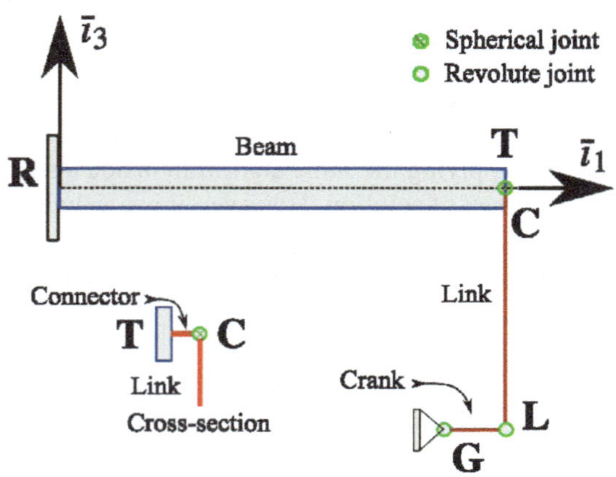


**Fig. 6.1** Beam actuated by a tip crank

of aluminum with Young's modulus 73 GPa and Poisson's ratio 0.3. It connects to a spherical joint at point **C** by a short connector of length 5 mm featuring physical properties identical with these of beam. The spherical joint is connected to a flexible link made of steel with Young's modulus 206 GPa and Poisson's ratio 0.28. The link with a length of 0.5 m has a hollow circular cross-section of outer radius 15 mm and thickness 8 mm. The link connects to a steel crank of length 30 mm through a revolute joint at point **L**; the crank cross-section is the same as that of the link. The crank is attached to the ground with a revolute joint at point **G**.

For numerical simulation, the beam-crank mechanism is discretized by cubic beam elements. In details, the beam is discretized into 8 elements, the connecter 2, the link 6 and the crank 2 elements. The spherical and revolute joints are modeled as holonomic constraints, the formula of all six low-pair joints can be found in Book [13]. The rotation of revolute joint at point **G** is prescribed as $\phi = 1.6(1 - \cos 2\pi t/T)$ rad, where $T = 1.6$ s. Under the framework of finite element method, the beam, connecter, link and crank have been discretized by geometrically exact beam elements. Each beam element and the holonomic constraints of spherical and revolute joints, all of them will be assembled into the global ones. The governing equations of motion of discrete beam-crank mechanism are exactly DAE of index-3. The new algorithms developed in the book are suitable for solving these equations.

At first, the simulation runs over 3.2 s by using new 2-stage Radau IIA algorithm with constant step $h = 1.0$ ms and user defined tolerance $\tau_a = \tau_r = 0.001$. The time histories of beam's tip displacements are shown in Fig. 6.2.

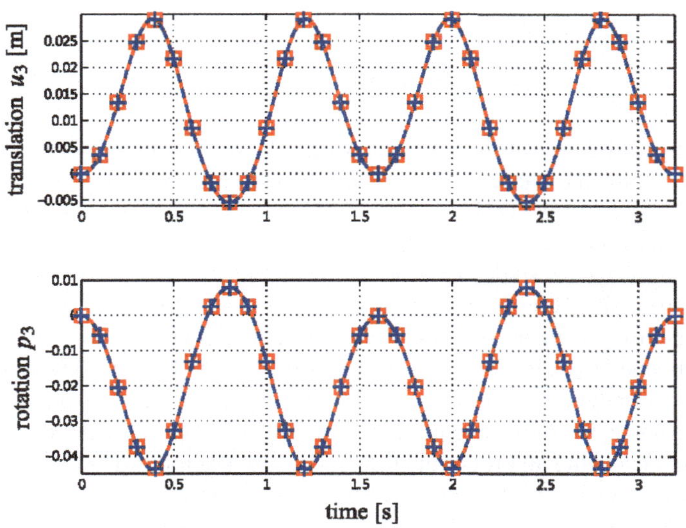


**Fig. 6.2** Time history of beam's tip displacements. Results of 2-stage scheme: solid red line with (□), results of DYMORE: dashed blue line with (+)

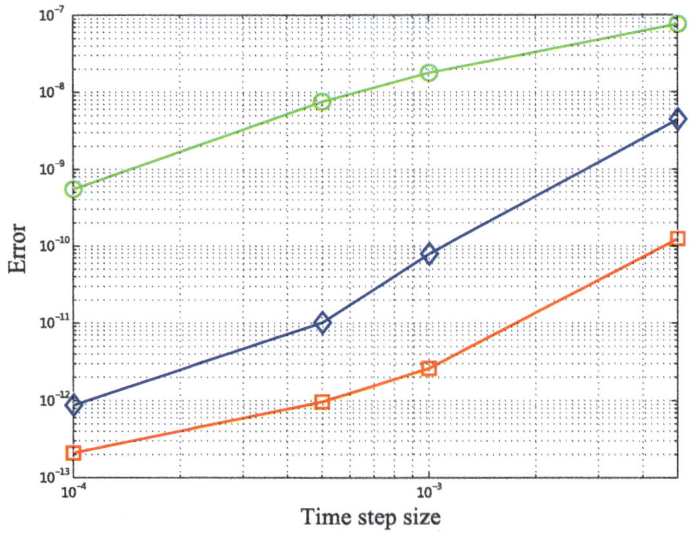


**Fig. 6.3** Error versus time step size computed by energy decaying algorithm (○), 2-stage Radau IIA (◇), 3-stage Radau IIA (□)

The result of new algorithm, presented by solid red line, has been compared to those of DYMORE [19], dashed blue line. The relative error

$$\epsilon_r = \frac{1}{N_s} \sum_{i=1}^{N_s} \frac{|u_{3i}^f - u_{3i}^d|}{|u_{3i}^d|} \tag{6.45}$$

measures the difference of these results, where $N_s$ is the total number of sampling points, $u_{3i}^f$ and $u_{3i}^d$ the transverse displacements predicted by 2-stage Radau IIA and DYMORE, respectively. This quantity is calculated as, $\epsilon_r = 0.0003$, which means the user desired accuracy is obtained.

Next, a convergence study is performed by running the same example at constant step size repeatedly. Here, a set of steps is used, $h = 5.0, 1.0, 0.5, 0.1$ ms. The error in transverse displacement of beam's tip is calculated with respect to a reference solution obtained by using 3-stage Radau IIA method with a step size $h = 0.01$ ms because of no analytical solution is available. The error as a function of step size plots in Fig. 6.3, in which the convergence characteristics of proposed algorithms compare to the energy decaying algorithm in DYMORE. Observation shows the slope of convergence curve for 2-stage Radau IIA algorithm is estimated to 3, and that for energy decaying algorithm is 2, which proves the 2-stage Radau IIA algorithm features third order accuracy and energy decaying algorithm second order accuracy. However, the slope of convergence curve for 3-stage Radau IIA algorithm is approximated to 3, which means the order degradation happened for this stiff problem.

**Fig. 6.4** Windmill resonance problem

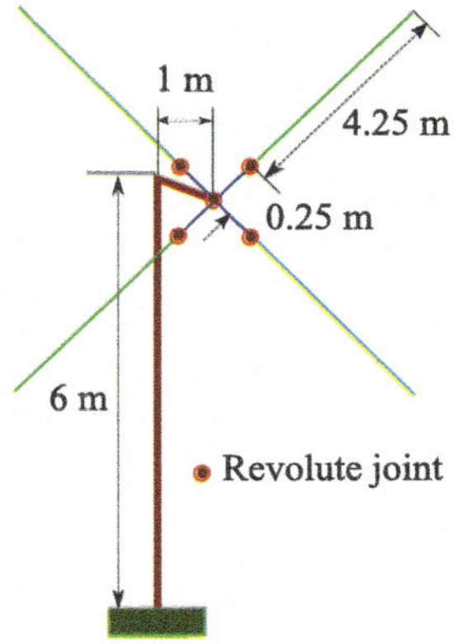


### 6.7.2 *The Windmill Resonance Problem*

The second example solves a problem of windmill resonance [19]. The efficiency of new Radau IIA algorithms is compared with that of the H.H.T. algorithm [20], a second order algorithm with high efficiency. As described in Fig. 6.4, a four-bladed rotor mounted on an elastic tower.

The tower of height 6.0 m features the following properties: bending stiffnesses $I_{22} = I_{33} = 3.87 \times 10^6$ Nm$^2$, torsional stiffness $GJ = 2.97$ Nm$^2$, and a mass per unit span 12.73 kg/m. A concentrated mass of 30 kg is located at the top of the tower, and a nacelle is connected to the tip of the tower and projects 1 m forward. The properties of the nacelle are the same as those of the tower. The rotor hub models as a concentrated mass of 20 kg and finds at the tip of the nacelle. Each uniform blade of length 4.25 m has a mass 12.75 kg, out-of-plane bending stiffness $I_{22} = 0.547 \times 10^3$ Nm$^2$ and in-plane bending stiffness $I_{33} = 4.71 \times 10^3$ Nm$^2$. The blade root hinge is located at distance 0.25 m from the bub. For the finite element modeling, the tower and nacelle are modeled by 2 and 1 cubic beam elements, respectively. Each blade is discretized with 3 cubic beam elements. A revolute joint located at the tip of the nacelle prescribes rotation of the rotor, and 4 revolute joints model 4 lead-lag hinges of the blades. Here again, revolute joints are modeled as kinematic constraints, so that the governing equations of motion are formulated as DAE of index-3.

At first, the resonance of windmill is studied with the initial rotor speed 30.06 rad/s, which corresponds to unstable rotor speed as pointed in Ref [19]. Figure 6.5 shows the time history of the lag angles, $\phi_1$ and $\phi_3$, predicted by new 3-stage Radau IIA and DYMORE [19], respectively. The relative error of $\phi_1$ between prediction of DYMORE and that of 3-stage algorithm is evaluated as $\epsilon_r = 2 \times 10^{-5}$ from Eq. (6.45). The trajectory of rotor center is presented in Fig. 6.6, in which the predictions of new Radau IIA algorithms are compared with those of H.H.T. [20] and the perfect curve fitting is obtained.

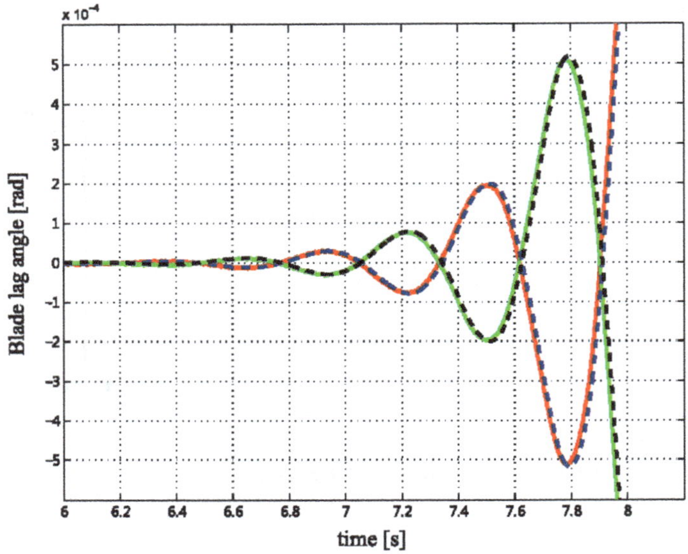


**Fig. 6.5** Time history of lag angles. Results of 3-stage Radau IIA: $\phi_1$ solid red line, $\phi_3$ solid green line; results of DYMORE: $\phi_1$ dashed blue line, $\phi_3$ dashed black line

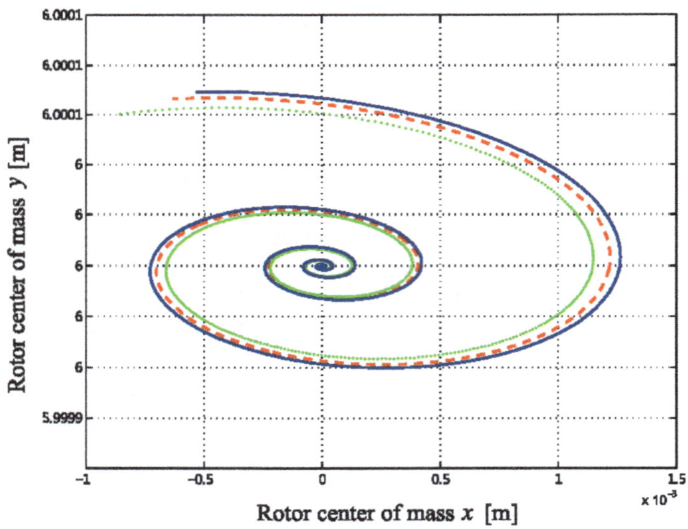


**Fig. 6.6** Trajectory of rotor center of mass predicted by H.H.T.(dashed red line), new 2-stage Radau IIA(solid blue line), and new 3-stage Radau IIA (dotted green line)

**Table 6.1** Numerical statistics with time adaptive scheme

| Scheme | H.H.T. algorithm | 2-stage scheme | 3-stage scheme |
|---|---|---|---|
| nbJacs | 12663 | 8001 | 12616 |
| nbFacs | 24488 | 8003 | 46496 |
| nbItes | 207333 | 45021 | 65746 |
| nbStps | 12666 | 8000 | 13951 |
| cpuTim | 106.0 | 73.0 | 102.0 |

Next, the efficiencies of new Radau IIA are compared with that of the H.H.T. algorithm. Table 6.1 summarizes the statistical information of numerical simulations, including number of Jacobian evaluations (nbJacs), number of LU factorizations (nbFacs), number of Newton iterations (nbItes), number of time steps (nbStps) and the CUP time in seconds (cpuTim).

All simulations run on a desktop computer with processor Intel(R) Core(TM) i3 CPU 3.20 GHz and memory(RAM) 4.0 GB. Theoretically, the H.H.T. algorithm is the most efficient solver for the same number of operations. However, the new algorithms enhance the efficiency due to the improvements of current implementation. It can be observed from Table 6.1, the new algorithms finish the integration faster than H.H.T. algorithm. The 2 and 3-stage schemes spent 73 and 102 s respectively and H.H.T. algorithm spent 106 s.

Finally, the behavior of windmill is studied for the stable initial rotor speed 14.025 rad/s. As shown in Fig. 6.7, the convergence characteristics of new algorithms are double-checked following the same procedure described in Sect. 6.7.1. The out-of-plane displacement of the first blade tip is predicted by new algorithms and H.H.T. algorithm, respectively. The associated truncation errors are estimated with respect to a reference solution obtained by using the new 3-stage algorithm with a constant time step size $h = 0.01$ ms.

The slopes of convergence curves have been evaluated as 2.0 for H.H.T. algorithm, 2.6 for 2-stage Radau IIA algorithm and 3.5 for 3-stage Radau IIA algorithm.

## 6.8 Validation of Finite Element Models for Cosserat Continuum

The analytical models of Cosserat continua have been addressed in detail in Chap. 3. The finite elements with the type of beam, shell and rigid body, etc. are also accomplished. Since their validations totally relay on the stiff solvers developed in this chapter. Therefore, the numerical models are presented in this section to verify the ability of these elements dealing with the large deformations and arbitrary motions.

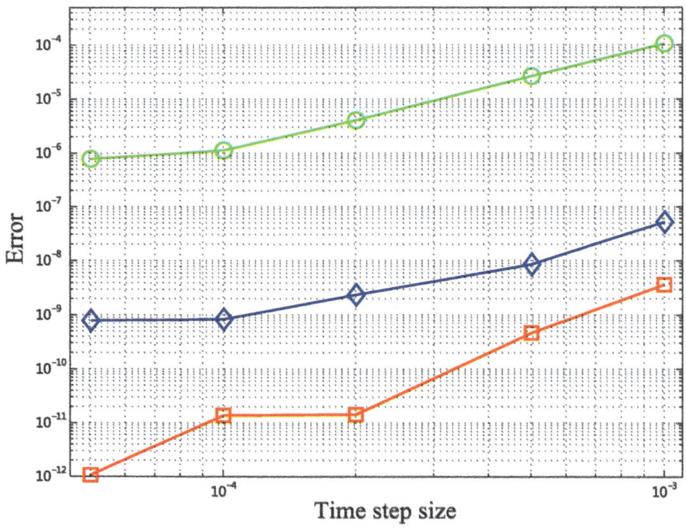


**Fig. 6.7** Error versus time step size computed by H.H.T. (○), 2-stage Radau IIA (◇), 3-stage Radau IIA (□)

## *6.8.1 Pure Bending of a Cantilevered Beam*

This example deals with the problem of a cantilevered beam subjected to a concentrated end moments. It has been widely used as a test case [21, 22] for large deformation analysis.

The beam has a span of $L = 10.0$ m, and a rectangular cross-section of size 0.5 m $\times$ 0.01 m. Due to the particularity of these geometric sizes, the cantilevered beam can be considered to be a thin-walled strip structure. Apparently, the characteristic length of the cross-section is much smaller than the span, one-dimensional Cosserat continuum is very suitable for the finite element modeling of the cantilevered beam. Meanwhile, the thickness of 0.01 m is much smaller than the width of 0.5 m and the length of 1 m for this strip structure. The shell/plate elements are also applicable to model the cantilevered beam. When neglecting the small scale deformations, one-dimensional Cosserat continuum shows the powerful ability to predict large deformations.

The material properties are given as follow: young's modulus $E = 1.0 \times 10^8$ Pa, poison's ratio $\nu = 0$. Based on the classical beam theory, it can be concluded the component of curvature, $\kappa_3$, becomes uniform under the pure end moment $M$, and $EI_{33}\kappa_3 = M$. When setting the moment with the amplitude $M_{cr} = 2\pi EI_{33}/L$, the deformed configuration of beam forms a curve of pure circle, and the component of curvature will be $\kappa_{3cr} = 2\pi/L$. Current simulation applies the pure moment $M$ to the end of the beam, in which the amplitude of moment increases linearly from zero to $M_{cr}$ in 1 s. The beam has been discretized into $10 \times 1$ shell elements, each

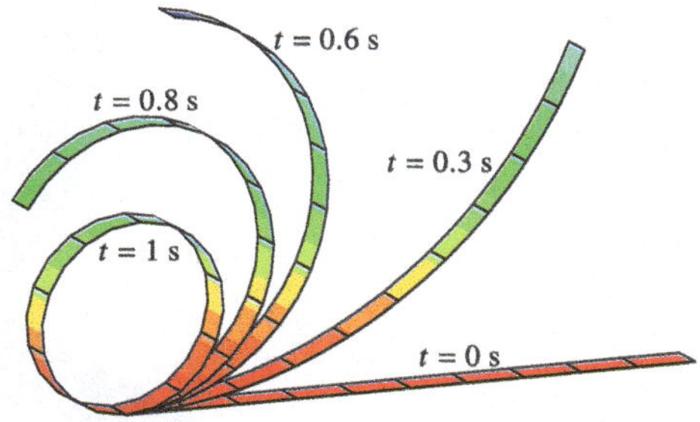


**Fig. 6.8** Deformed configurations meshed by shell elements for a cantilevered beam under pure moment

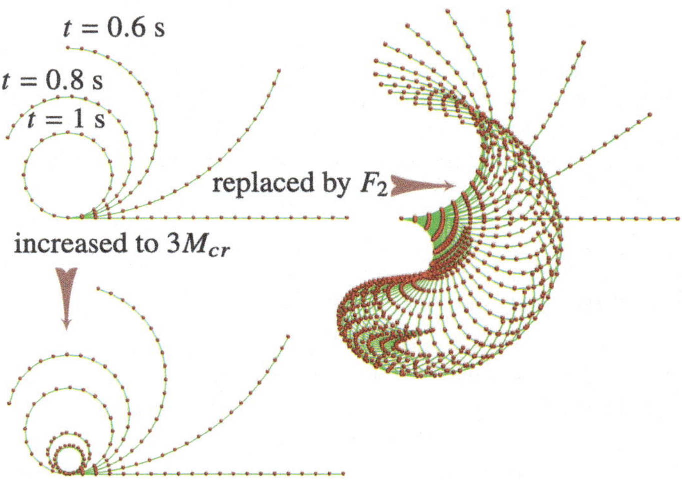


**Fig. 6.9** Deformed configurations meshed by beam elements for a cantilevered beam under pure moment or transverse force

element contains 9 nodes, then the Newton-Raphson iteration is applied to predict the deformations. As depicted in Fig. 6.8, the deformed configuration at time $t = 1$ s is predicted to be a curve of pure circle. Meanwhile, the configurations of the beam under different time steps are presented also. Obviously, the shell element can predict the large deformation exactly.

Next, the same problem is solved by using the finite elements of one-dimensional Cosserat continuum. After discretized the cantilevered beam into 10 beam elements of second order, the numerical simulations are performed. As depicted in Fig. 6.9, it is observed again that the cantilevered beam deforms into a pure circle under the same critical end moment that is linearly loaded to $M_{cr}$ in 1 s. If continuously loaded the moment triple times to $3M_{cr}$, the cantilevered beam continues to deform until the other two circles are formed, and the curvature increases its magnitude from $2\kappa_{3cr}$ to $3\kappa_{3cr}$. These results are described in Fig. 6.9. At last, a crude test is performed

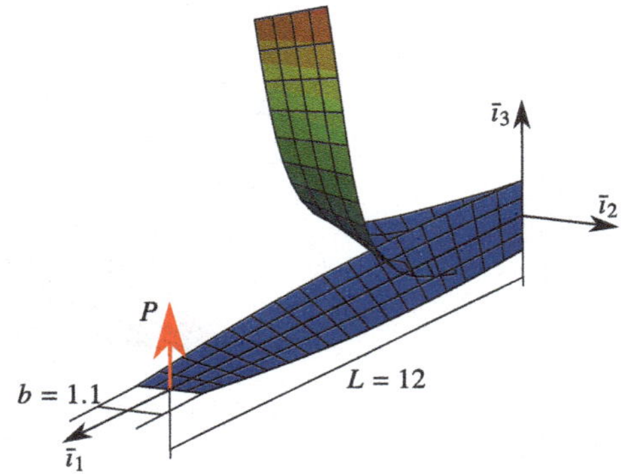


**Fig. 6.10** Reference and deformed configurations of twisted plate

by replacing the end moment $M$ with the follower force $F_2$ applied at the tip of the beam. When the follower force is set to $F_2 = 10M_{cr}/L$, the deformation of the beam apparently exceeds the normal range of large deformation [23]. As depicted in Fig. 6.9, the deformation behavior of the beam is similar to that of hyper elastic materials.

### *6.8.2 Response of a Twisted Plate*

The second example predicts the dynamic response of a cantilever plate of rectangular cross-section. The plate twisted 90° over its length and subjected to the load of magnitude $P = 5 \times 10^3$ lb at its free end.

As depicted in Fig. 6.10, the size of twisted plate is set as: span $L = 12$ in, width $b = 1.1$ in and thickness $h = 0.32$ in. The material properties are: Young's modulus $E = 29 \times 10^6$ Psi and Poisson's ratio $\nu = 0.22$. The plate is meshed by a $2 \times 8$ grid (2 element in width and 8 elements along span) using shell elements. For comparison, a beam model with 8 cubic elements is defined as reference. The dynamic simulation runs over 5 s by using 2-stage Radau IIA algorithm. The deformation and motion of the plate has been predicted. As depicted in Fig. 6.10, The final configuration shows that the large deformation is occurred.

The time history of dynamic response of the centroid point at the free end of the plate is presented in Fig. 6.11. From these results, it can be concluded that the shell elements can predict the response of this type structure with good accuracy even a crude mesh ($2 \times 8$) is used.

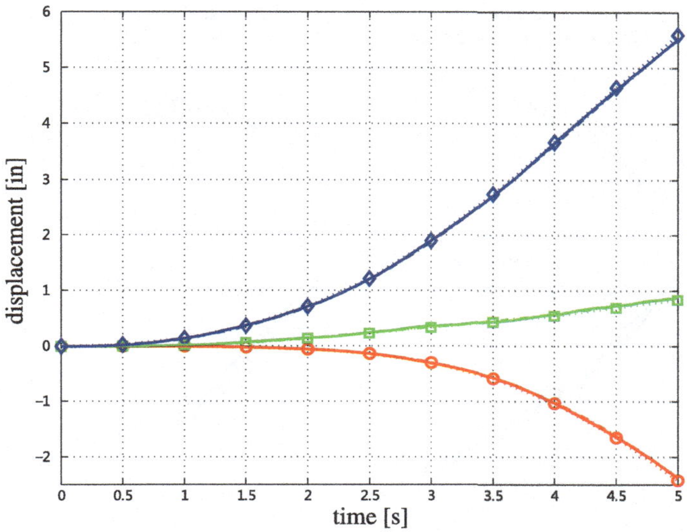


**Fig. 6.11** Time history of displacement response of the centroid point at the free end. Components of displacement, $u_1$ : (○), $u_2$ : (□), $u_1$ : (◇). Result of beam model: dotted line; results of shell model: solid line

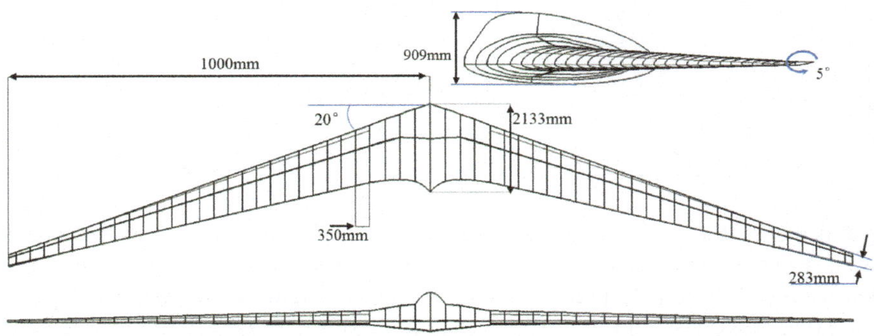


**Fig. 6.12** Geometry of the Horten IV flying wing

## 6.8.3 *Eigenmode Analysis of Horten IV Flying Wing*

The third example performs eigenmode analysis of a flying wing with the geometry resembling Horten IV [24, 25] as depicted in Fig. 6.12.

The fly wing is consisted of two tapered wing swept 20°, 61 equally spaced ribs and the skins with the thickness of 1 mm. Its geometry is shown in Fig. 6.12, in which the wing tip is twisted counter-clockwise by 5° relative to the wing root. During the numerical modeling, the spar with the thickness of 4 mm is made of aluminum alloy, the ribs with the thickness of 4 mm are made of wood, and the

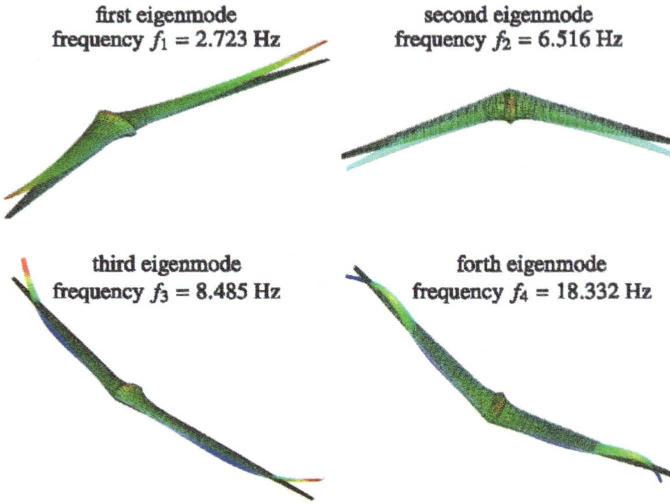


**Fig. 6.13** First four eigenmodes of Horten IV flying wing

**Table 6.2** The frequencies of first four eigenmodes (Hz)

| No | Shell | S8R5 | Relative error (%) |
|---|---|---|---|
| 1 | 2.7230 | 2.7271 | 0.15 |
| 2 | 6.5155 | 6.4140 | 1.58 |
| 3 | 8.4846 | 8.4985 | −0.16 |
| 4 | 18.332 | 18.279 | 0.29 |

skin is made of a composite laminate. The following material properties have been used: Young's modulus $E = 7.24 \times 10^4$ MPa, Poisson's ratio $\nu = 0.33$, material density $\rho = 2.78 \times 10^{-9}$ tonne/mm$^3$ for aluminum, Young's modulus $E = 1.50 \times 10^4$ MPa, Poisson's ratio $\nu = 0.35$, material density $\rho = 7.50 \times 10^{-9}$ tonne/mm$^3$ for wood, and Young's modulus $E = 3.50 \times 10^4$ MPa, Poisson's ratio $\nu = 0.30$, material density $\rho = 2.10 \times 10^{-9}$ tonne/mm$^3$ for composite laminate.

Due to symmetry, only half of the fly wing is discretized and the clamped boundary conditions are applied to the middle part of the aircraft. Two different simulation cases are performed for the eigenmode analysis of flying wing. The first case utilizes the geometrically exact shell element of 9 nodes to discretize the wing. The discrete model contains 3613 shell elements with the total degrees of freedom 67,781, and then the Arnoldi algorithm is applied to perform the eigenmode analysis. For the first four mode shapes, the prediction results are depicted in Fig. 6.13. The second case applies the commercial software ABAQUS to perform the eigenmode analysis. The numerical model is consisted of 4786 S8R5 elements. The predicted natural frequencies are summarized in Table 6.2, and compared with the results of shell model.

The low level of relative error shows that the new shell element has the same accuracy as the S8R5 element of ABAQUS.

### 6.8.4 Bio-inspired Flight Flapping Wing

The last example deals with the kinetic problem of a stick model of bio-inspired flying robot, which is designed to simulate the flapping motion of birds. The internal structure of wing has been simplified to the four-bar linkages. Due to symmetry, only a half of the basic geometric configuration is depicted in Fig. 6.14. Note that 16 points are numbered in the figure to determine the geometric topology of the four-bar linage, where the left linkage will connect to the right one through points 13 and 16. The whole structure is clamped at point 16. The coordinates of all the points are given in Table 6.3 with the unit of length cm.

In current implementation, the numerical model of four-bar linkage is consisted of beam elements and revolute joints. A user-defined function, $\phi(t) = \omega t$ with constant angular speed $\omega = 12.56$ rad/s, prescribes rotation of crank shaft. The beam elements share the same material properties as following: mass per unit length, $\rho = 0.1032$ kg/m, bending stiffness, $I_{22} = I_{33} = 29.856$ Nm$^2$, torsional stiffness, $J = 23.303$ Nm$^2$. Two different test cases have been performed. For case $\mathcal{I}$, all the revolute joints have been modeled by using classical kinematic constraints [26]. The governing equations of motion for discrete model will be DAE of index-3 with degrees of freedom, $N = 1368$, containing 42 Lagrange multipliers. The 2-stage Radau IIA algorithm is applied to integrate the governing equations and the predicted trajectory of four-bar linkage is shown in Fig. 6.15.

The cyclic flapping of four-bar linkage apparently mimics the flapping motion of birds. As depicted in Fig. 6.16, the time histories of relative rotations for revolute joint located at points 8 and 9 are sensed and compared with the results of DYMORE [19], in which all the revolute joints are modeled by using the classical formula. The relative error between these two predictions for $\phi_8$ is computed as $\epsilon_8 = 5.04 \times 10^{-6}$.

For case $\mathcal{II}$, the kinematic constraints at point 2 and point 4 for both left and right linkages are replaced by four recursive models, with the rest remaining unchanged. The relative rotation of recursive model located at point 4 is described in Fig. 6.17.

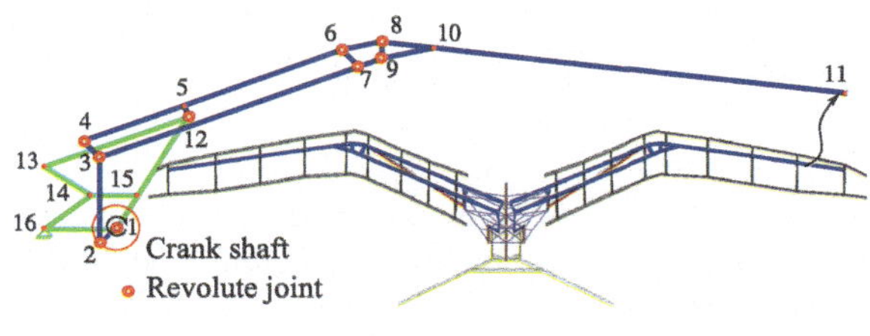


**Fig. 6.14** Configuration of four-bar linkage

**Table 6.3** Coordinates of the four-bar linages control points

| No. | Left wing | No. | Right wing |
|---|---|---|---|
| L1 | (−4.91071, 0.00000, 0) | R1 | (4.91071, 0.00000, 0) |
| L2 | (−2.25663, 2.38975, 0) | R2 | (2.25663, 2.38975, 0) |
| L3 | (−2.22156, 9.08609, 0) | R3 | (2.22156, 9.08609, 0) |
| L4 | (−0.76893, 12.3488, 0) | R4 | (0.76893, 12.3488, 0) |
| L5 | (−11.4767, 12.7227, 0) | R5 | (11.4767, 12.7227, 0) |
| L6 | (−44.0461, 13.8600, 0) | R6 | (44.0461, 13.8600, 0) |
| L7 | (−45.4988, 10.5974, 0) | R7 | (45.4988, 10.5974, 0) |
| L8 | (−47.5828, 13.3630, 0) | R8 | (47.5828, 13.3630, 0) |
| L9 | (−48.0479, 10.7251, 0) | R9 | (48.0479, 10.7251, 0) |
| L10 | (−52.9394, 13.4378, 0) | R10 | (52.9394, 13.4378, 0) |
| L11 | (−100.256, 14.0985, 0) | R11 | (100.256, 14.0985, 0) |
| L12 | (−11.5390, 10.9380, 0) | R12 | (11.5390, 10.9380, 0) |
| L13 | ( 0.00000, 8.75726, 0) | R13 | (0.00000, 8.75726, 0) |
| L14 | (−4.91071, 4.39570, 0) | R14 | (4.91071, 4.39570, 0) |
| L15 | (−7.36607, 4.39570, 0) | R15 | (7.36607, 4.39570, 0) |
| L16 | ( 0.00000, 0.00000, 0) | R16 | (0.00000, 0.00000, 0) |

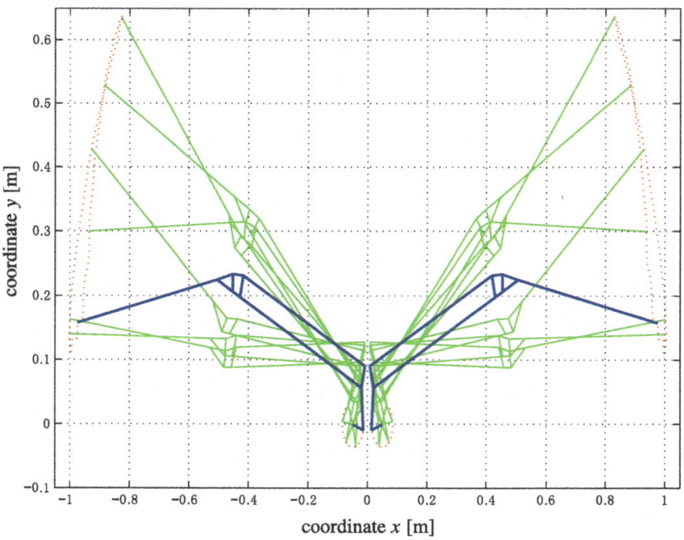


**Fig. 6.15** Trajectory of four-bar linkage

The time history has been compared with this of case $\mathcal{I}$. The relative error between the responses of $\phi_4$ for these two different models, $\epsilon_4 = 1.44 \times 10^{-8}$, shows that the recursive model has the same accuracy as the kinematic constraint model.

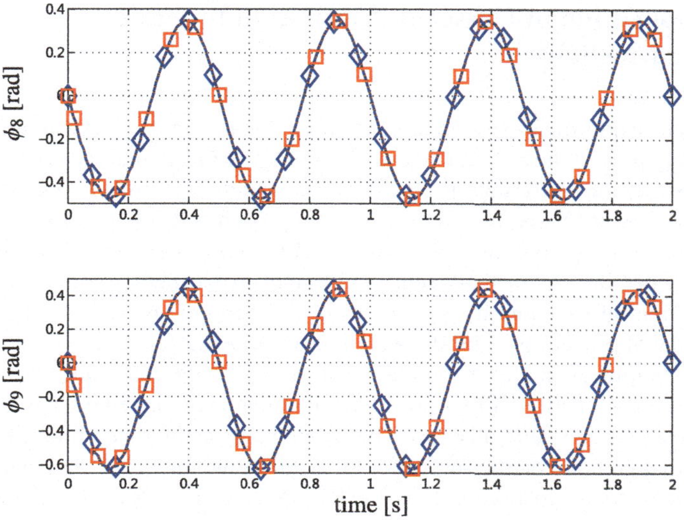


**Fig. 6.16** Relative rotations of revolute joints at points 8 and 9, current implementation: ◇, results of DYMORE: □

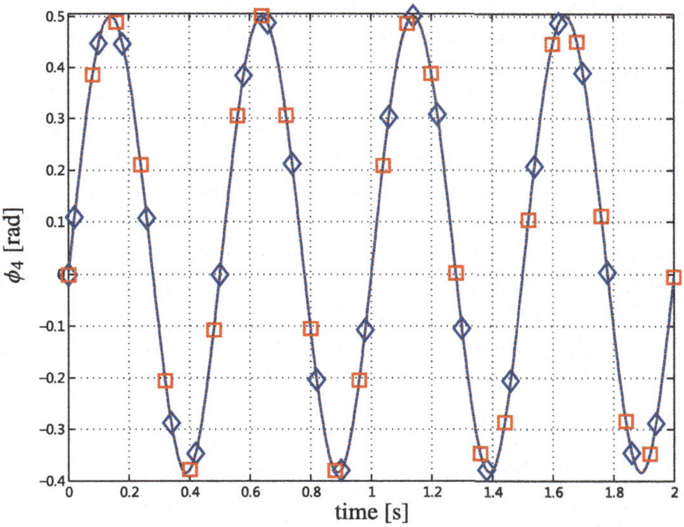


**Fig. 6.17** Relative rotations of revolute joints at point 4, recursive model: ◇, kinematic constraints: □

## 6.9 Application of Cosserat Continuum to Static Aeroelasticity

For transonic experiments of civil aircraft, it is becoming increasingly important [27] to accurately predict the aeroelastic deformations of wind-tunnel model, because the effect of structural deformation on the aerodynamic performance becomes more serious. In view of this point, a high fidelity approach is developed for static aeroelastic simulations of civil aircraft wind-tunnel model in transonic flow, and aimed to improve the correlation between wind-tunnel experimental and computational data by taking into account the real aeroelastic deformations. Instead of applying a linear modal approach, the real deformation of the wind-tunnel model is predicted by a fully nonlinear model of one-dimensional Cosserat continuum developed in Chap. 4, the combination of two-dimensional planar elements of cross-sections and one-dimensional geometrically exact beam element of reference line. Meanwhile, the aerodynamic force on the surface of the wind-tunnel model is predicted by solving the fundamental equations of flow, the Reynolds Average Navier-Stokes equation (RANS), which introduces relatively few assumptions.

In general, aeroelastic problems of civil aircraft due to the interactions between elastic and aerodynamic forces are determined by flow nonlinearities and at times by the deformations of sub-structures, such as wings, tails and control surfaces. Therefore, coupled approaches are necessary to solve such problems [28], accurately. These approaches are usually categorized in two types, loosely-coupled or fully-coupled. The loosely-coupled methods iteratively solve the fluid and structure equations by introducing the coupling scheme without altering the source code of either fluid or structure analysis tool. Because it allows the use of a variety of existing fluid/structure codes, the loosely-coupled methods have developed quickly. Fully-coupled methods [29, 30] require the solution of the fluid and structure equations simultaneously, which necessitates the reformulation of the governing equations of each discipline [31]. The stiffness matrices associated with the structures are orders of magnitude stiffer than those associated with fluids. Thus, it is numerically inefficient to solve both systems by using a monolithic numerical algorithm [28, 32, 33]. Till now, the fully-coupled methods are still the research hot topics. In view of this situation, a special attentions paid to the loosely-coupled approaches, and explored the available loosely-coupled tools [34–36]. For example, the structured NASA code CFL3D v6.44 [37]developed as a means of aeroelastic simulation, and the unstructured NASA code FUN3D v5.6 [38] was given aeroelastic capability as well. These NASA CFD codes simulated the aeroelastic phenomena by using a nonlinear flow field solver and a modal representation of the structural equations. The fluids and structures were modeled independently and exchanged boundary information to obtain aeroelastic solutions in an efficient manner. In Refs [39, 40], a loosely-coupled approach, based on high-fidelity numerical fluid dynamics and structural analysis methods, was designed to performed aeroelastic analyses. An in-house simulation procedure was built around DLR's flow solver TAU and the commercial finite-element analysis code NASTRAN, and the numerical results were validated against experimental data. Similarly, a static fluid-structure simulation [41] on a complete aircraft configuration has been performed using the commercial CFD++ Navier-Stokes code and the NASTRAN structural analysis code. Soon after, a numerical static aeroelastic analysis procedure [42] is presented by applying a modal approach in coupling the fluid dynamic and structural solutions. This approach constructed a modal model for aircraft wind-tunnel model through a preliminary structural modal analysis from which a number of natural modes are selected. Recently, the fluid structure interaction methodology [43] couples high-fidelity computational fluid dynamics to a simplified beam representation of the finite element model. The correlation between the numerical simulations and wind tunnel data for varying angles of attack is analyzed and the results revealed the importance of considering structural model deformations in the aerodynamic simulations.

The above investigations show that CFD solvers with high resolution was commonly applied in a loosely-coupled approach, that are necessary to capture the nonlinear behavior of the aerodynamics in the transonic regime, such as shocks, vortices and flow separation. Whereas, the modal representation based on linear structural assumption or simplified beam representation were popular selections describing the structural deformations that had been limited to deal with deformations with small amplitudes. In this chapter, a new static aeroelastic analysis method with high fidelity is proposed that takes into accounts both the nonlinearity of transonic flow and geometric nonlinearity of structural deformation.

This new loosely-coupled approach has been implemented in the computational code AeLas, and in detail the structural deformation of wind-tunnel model is described by the fully nonlinear multiscale beam model, while the aerodynamic force applied to the wind-tunnel model is obtained by solving the RANS based fundamental equations of flow. Aeroelastic coupling procedure is built around this fully nonlinear beam element and the in-house RANS solver with structured and unstructured CFD grids.

The new algorithms of fluid-structural interpolation are designed and compared with the classical interpolation scheme, the Radial Basis Function (RBF) interpolation [44–46]. This new interpolation operates on sets of totally arbitrary point clouds, and allows multiple intersecting surfaces and components without any modification. According to the principle of virtual work, the aerodynamic forces are strictly transferred to the beam nodes, and the structural surface deformations of wind-tunnel model are evaluated by the finite element interpolation, such that the accuracy of force/displacement mapping is improved.

### 6.9.1 Aeroelastic Coupling Procedure

A general loosely-coupled procedure is designed by which static aeroelastic solutions of aircraft with full configuration can be obtained with high fidelity. The in-house computational fluid dynamics (CFD) code in conjunction with a newly developed multiscale beam models captures the full nonlinearity of both flow and structural

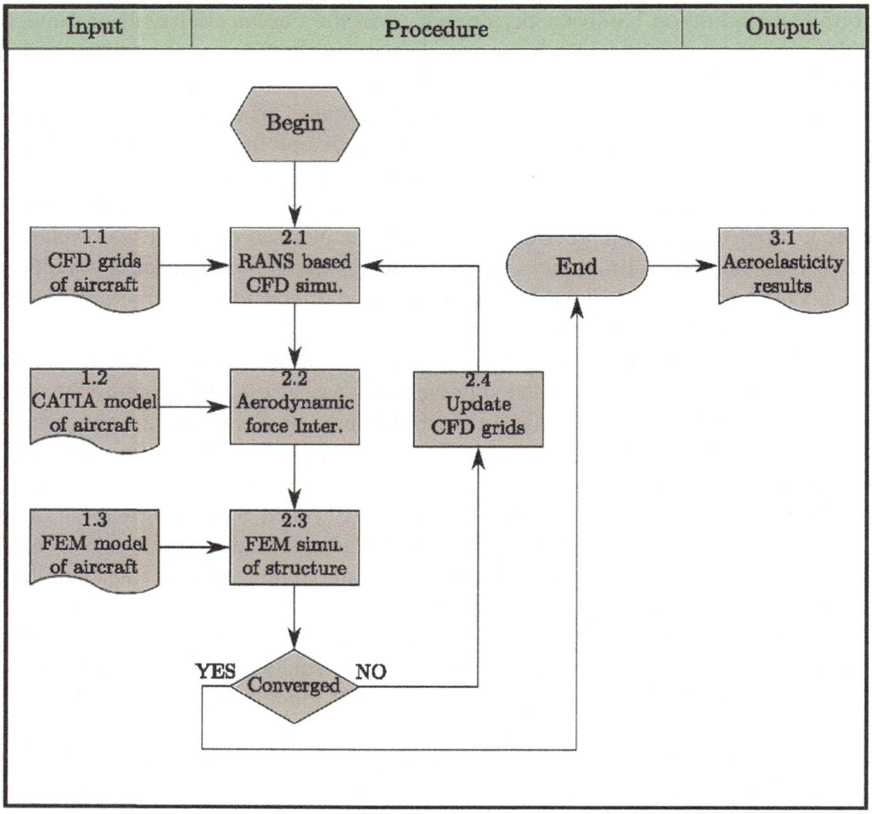


**Fig. 6.18** Aeroelastic coupled procedure

deformation. As depicted in Fig. 6.18, the aeroelastic coupled procedure includes following key steps:

1. RANS based CFD simulation: solving RANS equations with $k - \omega$ SST model to obtain an intermediate or rigid steady state CFD solution
2. Aerodynamic force Interpolation: computing the pressures at the CFD grid points on the aerodynamic surface, and then transferring pressures at the CFD grid points to the structural finite element nodes
3. Finite Element Method (FEM) simulation of structure: predicting the structural deformations of the wind-tunnel model by solving the nonlinear governing equations of deformation
4. Update CFD grids: deforming the aerodynamic surface according to the structural deformation of the wind-tunnel model, and then deforming the entire three-dimensional CFD grids
5. Repeat steps 1–4 until preselected convergence criteria is satisfied.

The above steps are repeated in an iterative manner until a converged solution is obtained. The inputs of coupled procedure contain the structured or unstructured CFD grids of full configuration for an aircraft, the CATIA model of the same aircraft used for multiscale beam modeling, or a linearized FEM model created by using the commercial software such as NASTRAN. The outputs of coupled procedure are mainly the aerodynamic parameters taking into account the influence of structural deformations. Note that aeroelasticity simulations are automatically invoked by the iterative algorithm running in the main process. For the computation of each key step, the main process will create a child process to complete the associated simulations. Especially for the key step 1, the CFD simulation, the efficiency of computation is greatly improved by executing large-scale parallel computations in the computer groups.

### *6.9.2 Solving Reynolds Averaged Navier-Stokes Equations*

Current implementation utilizes the in-house CFD code to solve the Reynolds averaged Navier-Stokes equations with $k - \omega$ SST turbulence model. The numerical solver is designed with the aid of a three-factor, implicit, finite-volume algorithm. The convective and pressure terms are differenced using the upwind flux-difference splitting technique of Roe [47]. Both structured and unstructured CFD grids are available for the in-house CFD code by invoking different CFD solvers. As depicted in Figs. 6.19, and 6.20, the structured and unstructured three-dimensional CFD grids are created by using commercial software Pointwise and ICEM CFD, respectively, in our routinely CFD simulations. Since the multiple structured grids require less memory than unstructured girds but are more accurate when simulating the boundary layers, the procedure of static aeroelastic analysis is introduced by using the structured grids of flow field. Meanwhile, the parallel computing based on MPI mode is implemented to enhance the computing speed.

**Fig. 6.19** Structured CFD grid for the full configuration of civil aircraft

**Fig. 6.20** Unstructured CFD grid for the full configuration of civil aircraft

### *6.9.3 Interface Mappings*

For a loosely-coupled and modular approach, the boundary information between the in-house CFD code and structural analysis code must be exchanged at iteration through the data interface. First, the mapping of one element to multiple grids is automatically established to distinguish which aerodynamic grids belong to a specific structure element. The identifying process is performed by AeLas for each aerodynamic grid until all the grids are processed. As shown in Fig. 6.21, different colors label the set of aerodynamic grids contained by different structural elements. Then, the aerodynamic forces $\underline{F}_a$ at the center of mesh quadrilateral on aerodynamic surface are evaluated from the pressures results of the CFD code. Based on the principle of virtual work, the aerodynamic forces are translated to the beam element where the center of mesh quadrilateral located in. According to the mapping relations between aerodynamic grids and beam elements, the position vector of center of mesh quadrilateral on aerodynamic surface can be written as

$$\underline{p}_a = \underline{p}_s + \underline{d} \tag{6.46}$$

where $\underline{p}_s$ is the position vector of the projected point of grids center inside the beam element, and $\underline{d}$ denotes the perpendicular distance between aerodynamic grid center and mapped beam element. Then the weighted interpolation of the aerodynamic forces is evaluated, and transmitted to the nodes of this beam element.

$$\underline{F}_s = \begin{vmatrix} \mathcal{H}_\epsilon^T \underline{F}_a \\ \mathcal{H}_\epsilon^T \tilde{d} \underline{F}_a \end{vmatrix} \tag{6.47}$$

Note that $\underline{F}_s$ are the concentrated loads applied to the structural nodes. On the contrary, the aerodynamic surface is required to deform once the structural deformations of the wind-tunnel model are predicted. Here, a displacement mapping algorithm is

**Fig. 6.21** Mapping of structural elements to aerodynamic grids for the full configuration of civil aircraft

designed to interpolate the CFD surface grids based on the updated positions of structural nodes

$$\underline{p}_a = \mathcal{H}_\epsilon \underline{p}_n + \underline{d} \tag{6.48}$$

After updated the surface grids as discussed above, the in-house CFD code will automatically deform the volume mesh before simulating the flow field with these updated CFD grids.

### *6.9.4 Static Aeroelastic Analysis of Aircraft Wind-Tunnel Model*

The wing-body model, as depicted in Fig. 6.22, is the CATIA model of a real wind-tunnel model designed for wind-tunnel experiments. This model is made of 30CRMnSi steel. The Young's modulus of the steel material is $2 \times 10^{11}$ Pa, and the Poisson's ratio 0.226.

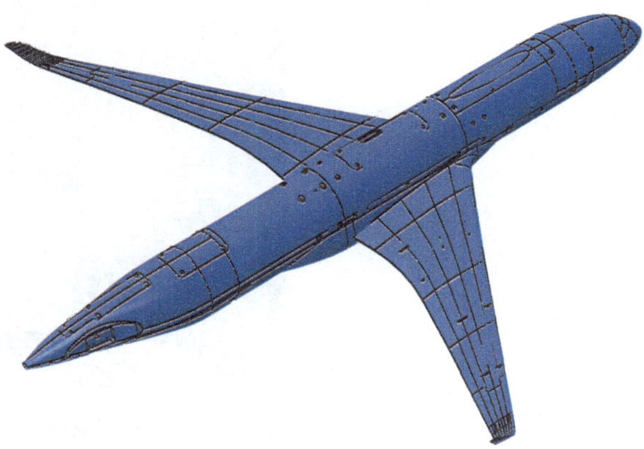

**Fig. 6.22** Aircraft wing model for wind-tunnel experiments

The works required for static aeroelastic analysis include: First, the CFD model with a fine mesh is constructed to perform the CFD simulation of the rigid aerodynamic surface. Next, the associated structural model is constructed by using the fully nonlinear multiscale beam elements. The virtual/actual loading experiments are performed, and the numerical predictions of the structural deformations are compared with the actual experimental measurements. The stiffness of the structural numerical model is calibrated to ensure that the numerical model can predict the real deformation of the wind-tunnel model under aerodynamic pressures. Finally, the static aeroelastic analysis is performed in a loosely-coupled manner to predict the difference of CFD simulation on rigid and flexible aerodynamic surface.

As depicted in Fig. 6.23, the static loading experiment measures the wing deformation under a set of concentrated loads, where these loads are equivalent to the aerostatic pressure measured in the wind tunnel experiments. The purpose of this experiment is to provide sufficient information for the stiffness calibrations of the structural numerical model. In current implementation, the stiffness properties, the cross-sectional $6 \times 6$ stiffness matrices are automatically computed by using the multiscale technology address in Chap. 4. The predictions of stiffness matrices will be slightly higher than the real values due to the numerical discretization, which leads to the deformation evaluations are smaller than the actual deformations measured in the static loading experiments. The stiffness has to be calibrated according to the experimental measurements to ensure the numerical model has sufficient accuracy for subsequent aeroelasticity analysis. The calibration procedure includes the finite element modeling of structural model, virtual/actual static loading experiments, and stiffness

**Fig. 6.23** Static loading experiments of wing model designed for wind tunnel experiments

**Fig. 6.24** Multiscale beam element model for the wing of wind-tunnel model

calibrating. For finite element modeling, the real wing model has very complicated inner structures to install the pressure measuring sets, which makes it difficult to discretize the wing model into three-dimensional solid elements. The multiscale beam model affords a powerful tool to discretize the wing model with high-aspect-ratio into two-dimensional cross-sectional planar elements and one-dimensional reference line beam elements, as described in Fig. 6.24. Along the wing span from the wing root to the tip, the wing model is discretized into 41 fully nonlinear beam elements with 42 cross-sections, each section is modeled by using planar elements of 4 nodes. The winglet and the body are simplified to the rigid bodies and rigidly connected to the beam element. The approach proposed in Chap. 4 can automatically evaluate the sectional stiffness parameters for all the 42 cross-sections. Due to the openings and slots along the wingspan, the stiffness of the wing changes significantly in local, as depicted in Fig. 6.25. For virtual static loading experiments, the flexible beams, rigid bodies and revolute joints are utilized to model the loading devices in the experiment, a cable-cantilever system. As described in Fig. 6.26, the load distribution over the wing surface is approximated through the virtual loading devices, by which the total transvers loads $F_1$ and $F_2$ are applied, respectively.

The deformations of wing model are predicted under a set of vertical loads and the predictions are compared with the experimental measurements. For stiffness calibrating, the maximum relative error is used to modify the beam stiffness, which makes the beam more flexible and then evaluates the deformations again. Finally, the maximum relative error is evaluated to less than 3.75%, Fig. 6.25 gives the stiffness curves after calibration, and Fig. 6.27 shows the total applied loads as a function of the deformation of a measure point at wingspan 0.923. A good curve fitting indicates the numerical simulations can accurately predict the actual bending deformations of the wing.

Following the loosely-coupled approach, as depicted in Fig. 6.18, the static aeroelastic analysis is performed in an efficient manner. At each iteration, the pure CFD simulation will give an intermediate or rigid steady state CFD solution of the wing wind-tunnel model. For this steady CFD simulation, the external flow field around

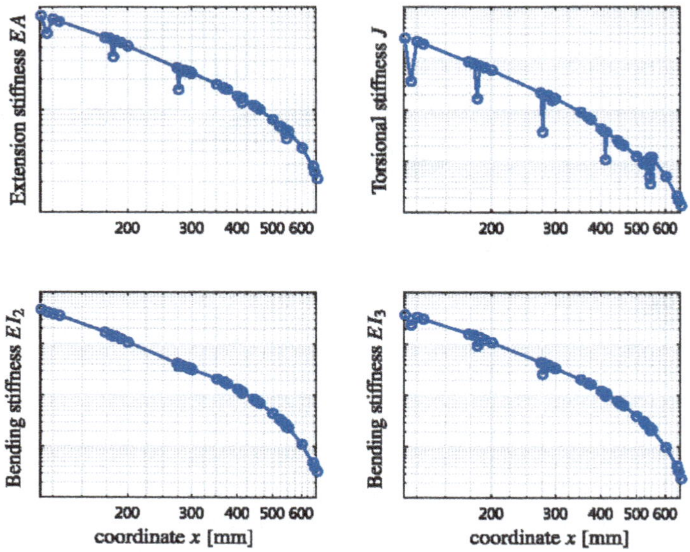


**Fig. 6.25** Stiffness curve of the wing model

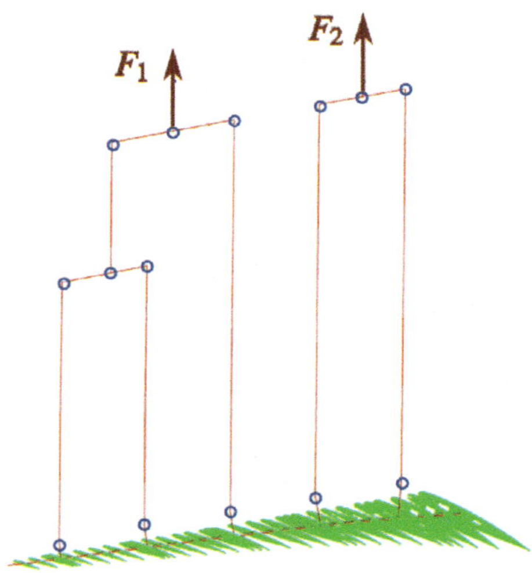


**Fig. 6.26** Loading device simulation for virtual experiment, (○): revolute joints

the wing-body model has been meshed into a 3D block with structured grids of total number $2.3 \times 10^7$, and especially the wing surface mesh grids with the number of $5.5 \times 10^4$. The initial flow conditions are: Mach number $M_a = 0.85$, Reynolds number $R_e = 2.12 \times 10^7$, angle of attack, $\alpha = 3.0°$. Next, the state equations of perfect gas is used to predict the pressure

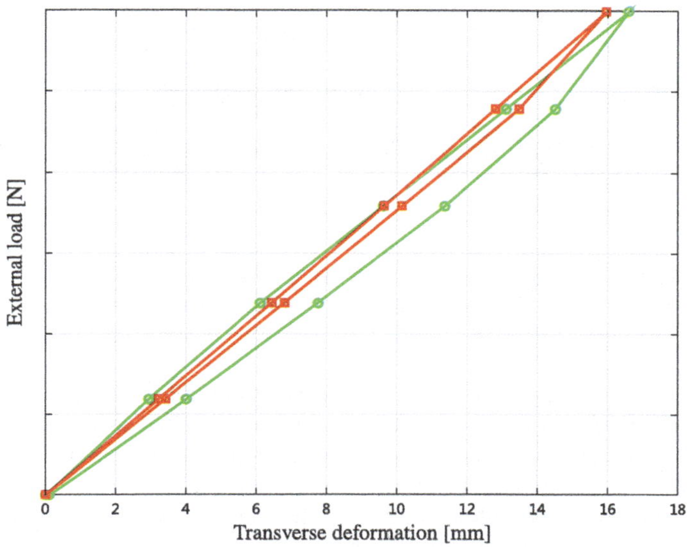


**Fig. 6.27** Load-displacement curve at the wingspan 0.923 of wing model. *Case 1*: solid line (○), experimental results; *case 2*: solid line (□), results of current approach

$$p = (\gamma - 1)[E - \frac{\rho}{2}(u^2 + v^2 + w^2)] \tag{6.49}$$

where $u$, $v$ and $w$ are three components of flow velocity, and the variables $\rho$ and $E$, are the flow density, total energy per unit volume, respectively. As shown in Fig. 6.28, the pressures $p_i$ at the grid points on the aerodynamic surface is evaluated once the numerical results of conservative variables $\rho$, $\rho v$, $\rho w$ and $E$ are obtained. Third, the aerodynamic load at the center of each mesh quadrilateral on the aerodynamic surface is evaluated from

$$\underline{F}_a = -\frac{1}{4}(\sum_{i=1}^{4} p_i) ds\bar{n} \tag{6.50}$$

where $ds$ is the area of the mesh quadrilateral, $\bar{n}$ the associated outer normal vector of the quadrilateral. Then, the evaluated loads $\underline{F}_a$ is translated to the structural nodes. Under these mapped aerodynamic loads, the nonlinear bending deformation and torsion of the wing wind-tunnel model is predicted by using the multiscale beam model discussed before. Once obtaining the deformations of the master nodes, the new position of the CFD grid points on the aerodynamic surface can be interpolated by using the finite element method, as depicted in Fig. 6.28. It shows that the CFD surface mesh grid points have significant deformations. Finally, the entire CFD volume grids are deformed by in-house CFD code automatically. Typically, 5–6 iterations are sufficient for this coupling computation converged to the equilibrium configuration. The cross-sectional pressure coefficients of the initial configuration

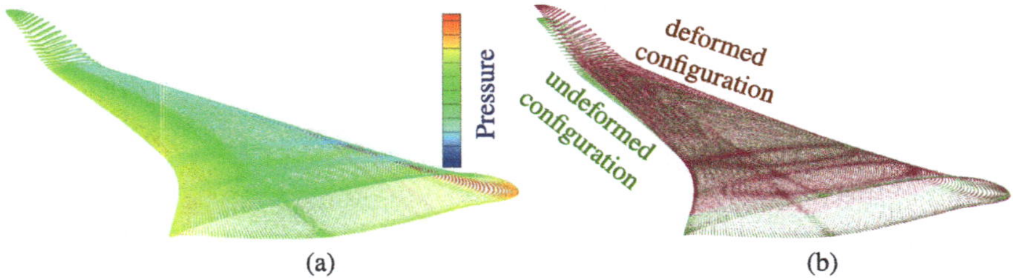


**Fig. 6.28** First iteration data in a loosely coupling. **a** Pressures at the grid points on the aerodynamic surface. **b** Undeformed/Deformed meshes grids on the aerodynamic surface

and equilibrium configuration along the wingspans 0.259, 0.392, 0.583, 0.773 are shown in Fig. 6.29 and also compared with the wind-tunnel experimental results. The static aeroelastic simulations present much better agreement with experiment than the RANS simulations. The effect of deformation is to decrease the suction on the upper surface and provide a more accurate shock location, particularly toward the tip of the wing, as depicted in Fig. 6.29. When the loosely-coupled iterations are converged after 6 iterations, the bending and torsional deformations along the wingspan are predicted, as described in Fig. 6.30. The maximal bending deformation is measured to be 1.53% of half wing span, and the maximal twist angle along the $z$ axis of the wind frame is 1.20°, which closes to the wind tunnel experiment measurements. After validated the results of aeroelastic simulation for a given angle of attack, $\alpha = 3.0$, a large number of numerical simulations were carried out for the angle of attack in the range of $\alpha \in [-2, 4]$ degree. The aerodynamic parameters as a function of angle of attack $\alpha$ can be extracted from the rigid and flexible steady state CFD solutions, respectively. In Fig. 6.31, the curves of lift coefficient $C_L$ with angle of attack are presented. The predictions of numerical simulation have been compared with the experimental results. From Fig. 6.31, it can be observed the slope of $C_L$-$\alpha$ curve plotted by using the aeroelastic predictions is much more close to that of wind-tunnel experiments, which indicates apparently that the influence of structural deformations on the aerodynamic parameters cannot be ignored.

## 6.10 Application of Cosserat Continuum to Buffeting

Buffeting [48] is an important aerodynamic phenomenon that can cause severe vibrations of aircraft structures. The vibrations may further lead to structural fatigue or structural damage [49], and even result in aircraft stalling and flight accidents. Therefore, the term of China Civil Aviation Regulations (CCAR) Part 25, CCAR-25.305(e), clearly specifies that aircraft must be designed to withstand any buffeting that might occur under any possible operating conditions up to $V_D/M_D$, where $V_D$ is the design dive speed and $M_D$ the design Mach number. This must be verified through analysis,

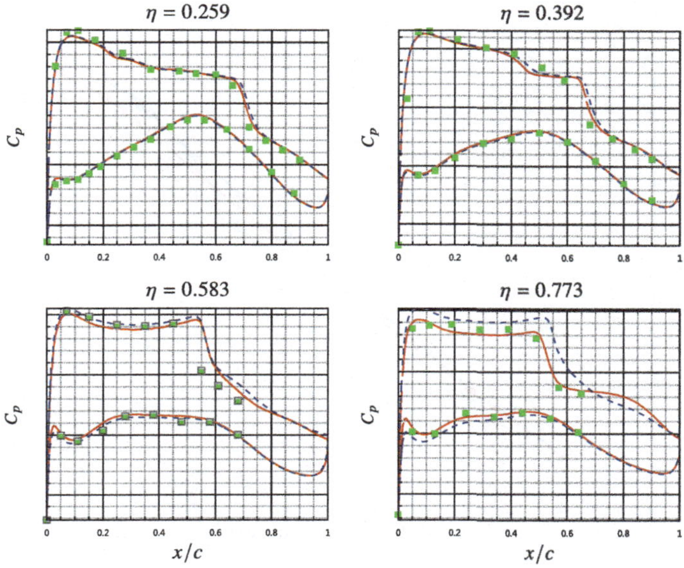


**Fig. 6.29** Chordwise pressure distribution at angle of attack $\alpha = 3.0°$. Solid line, aeroelastic analysis result; Dashed line, pure CFD simulation result; □, Wind-tunnel experiment results

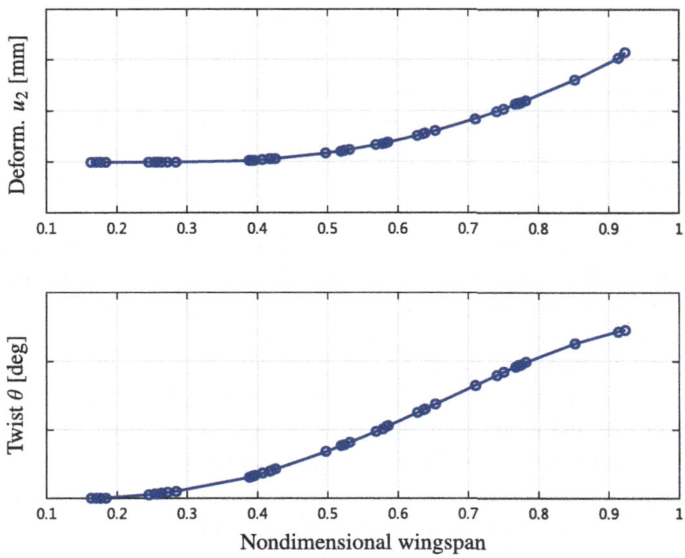


**Fig. 6.30** Bending and torsional deformations along wingspan

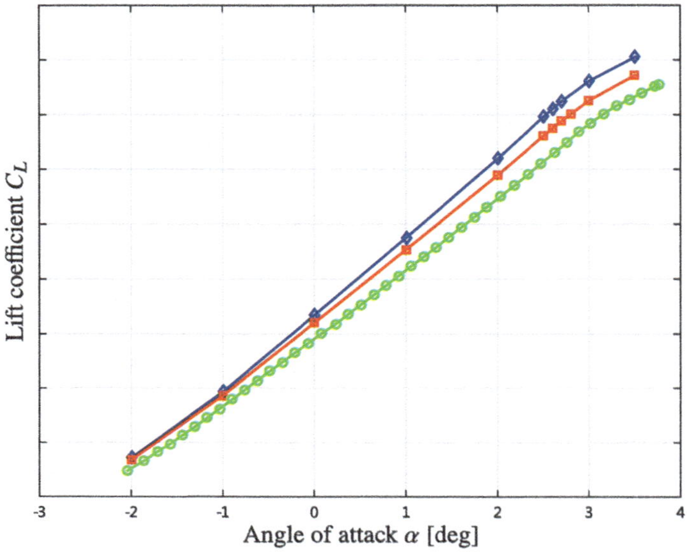


**Fig. 6.31** The curve of lift coefficient with the angle of attack. Solid line with □: aeroelastic predictions; Solid line with ○: wind-tunnel experimental results; Solid line with ◇: RANS predictions

flight tests, or other tests found necessary by the Agency. Till now, buffet analysis has become an important part of the civil aircraft design. For modern supercritical wing design with thick profiles, the shock-induced buffeting is particularly severe. A special attention has been paid to the shock buffeting, which is a buffeting phenomenon of lift-type commonly occurring due to the shock movement and flow separation in the transonic regime. When shock buffeting occurs, instabilities in the interaction between the shock and the separation bubble will cause self-sustaining periodic oscillations in the shock position [50], which cause large fluctuations in pressure with a frequency on the order 10 Hz [51]. Consequently, unsteady aerodynamic loads of a stochastic nature are generated that buffets the wing. Studies have shown that the transonic shock buffeting is closely related to the strong nonlinearity of shock movement, shock/boundary-layer interaction and three-dimensional (3D) effect of flow separation. Hence, the numerical simulation of shock buffeting is difficult. So far, there are still no effective approaches to accurately predict the unsteady response of aircraft wing due to shock buffeting. The innovations of this work lies on solving the unsteady response problem of shock buffeting efficiently, and develops a new type of approach to predict the buffet response of aircraft wing in time domain. The analysis procedure includes buffet envelope identification, buffet loads computation and buffet response prediction.

Since 1970s, buffeting phenomenon has been investigated through wind tunnel experiments and numerical analysis. Under transonic flow conditions, McDevitt and Levy [52] observed the airfoil surface-pressure fluctuations, oscillatory regions of trailing-edge and shock-induced separation through wind tunnel experiments. Afterwards, turbulent flows with separation caused by a shock [53] were also qualitatively reproduced by means of numerical simulations. Lee [54, 55] conducted the experimental studies on the physical mechanisms of the periodic shock motion over the Bauer-Garabedian-Korn(BGK) No.1 supercritical airfoil at transonic speed. The buffet envelop was determined from the divergence of the unsteady balance normal force. Steady and unsteady pressure measurements were obtained for shock-induced separation with reattachment as well as fully separated flows. Harmann et al. [56] determined the sound wave propagation in the flow field outside the separation of a transonic buffet flow over a DRA 2303 supercritical airfoil using high-speed particle-image velocimetry. They studied the mechanism of transonic buffeting by introducing artificial noise at the trailing edge of the airfoil to support Lee's buffet mechanism model [55]. Farhangnia et al. [57] predicted the transonic-buffet response of a supercritical wing using the Navier-Stokes (N-S) equations for fluid flow and the modal equations for the structures. The coupling and interaction of the shock oscillations and the modal response were studied. The Power Spectral Density (PSD) is used to identify the modal frequencies and those due to shock oscillations. They found that shock oscillations are independent of the structure response but trigger the structural response. These results keep consistent with the mechanisms of shock buffeting, which states that shock buffet is primarily an aerodynamic phenomenon because the frequencies of the shock-induced vibrations are at least one order of magnitude greater than the natural frequencies of the wing's primary elastic modes, so no aeroelastic computations are required to predict it. Lee [58] presented a method for predicting the response of a wing to buffet loads, and it was shown that unsteady aerodynamic forces have a significant effect on the response predictions. On the basis of the research work of Farhangnia and others, many researches have been carried out on the transonic shock buffeting problems. Bartels [59] studied the interactive boundary layer and the thin-layer N-S equations to solve the problem of shock buffeting onset regions for conventional and supercritical airfoils. The computational analysis compares well with experiment. Chung et al. [60] developed a steady approach to predict the transonic buffet onset for an airfoil. Steady solutions have been computed for the transonic flow over NACA0012 airfoil based on steady N-S solver with Baldwin-Lomax(B-L) turbulence model. The distinct change in the variations of aerodynamic parameters, such as lift, trailing edge pressure deviation, reversal in the shock movement, pitching moment and center of pressure, indicates the onset of transonic buffet. It concluded that the variation in the center of pressure provides a clearest indication of buffet onset. Gillan [61] solved the full mass-averaged N-S equations with the zero equation algebraic B-L turbulence models to accurately predict the shock-induced oscillation boundaries, the reduction frequency, and the hysteresis regime of a circular-arc airfoil. Goncalves and Houdeville [62] performed unsteady two-dimensional computations of the transonic buffet over a supercritical airfoil, in which the uncoupled RANS/turbulent system is solved. Various popular two-equation turbulence models were used, and the Shear Stress Transport (SST) turbulence model was found to be more accurate.

Recently, numerous researchers have modeled transonic buffet for airfoils with unsteady CFD using large-eddy simulations [63, 64], detached eddy simulations [65], and unsteady Reynolds-Averaged Navier-Stokes equations(RANS) [66, 67]. Ref [68] has revealed that unsteady RANS simulations can successfully reproduce the buffet unsteadiness using various turbulence models [69–72]. They provided a complete description of the low-frequency shock motions, and found that the shock movements strongly depend on the shock-induced separated zone. Consequently, this work also applies the unsteady RANS-based solver together with the $k - \omega$ SST turbulence model to identify the onset of shock buffeting and to predict the buffeting load for transonic flow. Extensive investigations have found that many works have focused on the mechanism study of the transonic buffeting and developed new approaches for the determination of transonic buffeting onset. However, few works have been done to predict the transonic buffeting response. The main contribution of this work is developing a new type of approach to predict the buffeting response of aircraft wings in transonic regime. The innovations include: (1) A new prediction procedure of buffeting response has been designed based on an open-loop, including the works of identifying the transonic buffeting onset, computation of the associated buffeting loads and the prediction of the buffet response of aircraft wings. (2) A new type of nonlinear shell element is developed to accomplish the finite element modeling of wings accurately. (3) The interface for data exchange between the CFD and CSD has been designed according to CVT method. 4) Dynamic response of aircraft wing can be predicted efficiently under the buffeting loads, and the detailed strength analyses are then performed.

## 6.10.1 Prediction Procedure of Buffeting Response

As mentioned above, shock buffeting is primarily an aerodynamic phenomenon because the frequencies of the shock-induced vibrations are at least one order of magnitude greater than the natural frequencies of the wing's primary elastic modes, so no aeroelastic computations are required to predict it. In the proposed approach, the buffeting loads will be evaluated either using an inner CFD code with unsteady RANS solver or the commercial code CFX, but the wing is assumed to remain a rigid body, i.e., all elastic deformations are ignored. This assumption simplifies the unsteady CFD analysis greatly: a known, stationary grid is used to perform the computation. Unsteady aerodynamic pressures are then computed over the entire surface of the wing, i.e. the data of pressure value at each point over the wing outer surface results from the CFD computation. The next step of the procedure is to predict the wing dynamic response to the stochastic excitation generated by the aerodynamic pressure distribution. A finite element model of the wing structure based on the Reissner-Mindlin shear deformable shell theory will be available for this study. Meanwhile, the finite element models of finite element code ABAQUS or NASTRAN

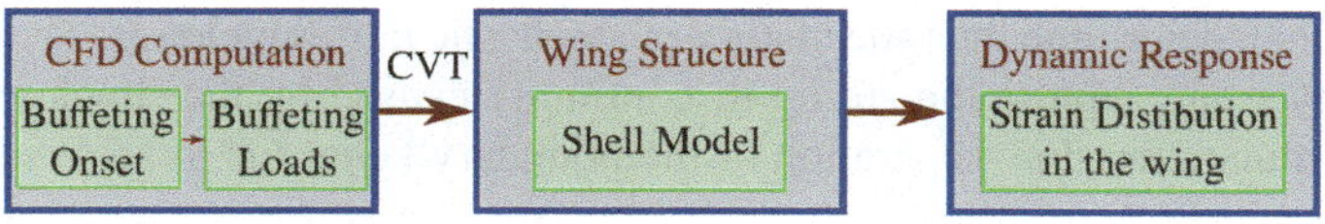


**Fig. 6.32** Schematic of the proposed approach

can be available also. Applying the aerodynamic pressure to the finite element model then yields the desired stress and strain components at any point within the wing structure. The overall analysis procedure is depicted in Fig. 6.32.

Because the wing is assumed to be a rigid body, the proposed approach is not an aeroelastic analysis: as shown in Fig. 6.32, the aerodynamic results are fed to the structural analysis program, but the structural deformations do not affect the aerodynamic analysis. The proposed analysis methodology is an "open loop procedure." Typically, buffeting loads will create minor overall deflections of the wing, although resulting stress levels are a concern for strength, and specially fatigue life of the wing. This observation justifies the assumption of a rigid wing.

Indeed, the aerodynamic pressures are available at the nodes of the CFD grid used for the fluid dynamics computation. On the other hand, the magnitudes of the vibration mode shapes are defined at the nodes of the finite element model of the wing. In general, these two computational grids do not match, and hence, an interpolation technique is required for the computation of the generalized forces. The Constant Volume Tetrahedron (CVT) Method [73], a simple geometry based method, is implemented in current work. This algorithm guarantees the proper transfer of information between the fluid and structure computational domains, thereby ensuring consistency and stability of the computation.

### 6.10.2 *Buffeting Onset Identification and Buffeting Loads Prediction*

The inner CFD code has been used to identify the buffeting onset and to predict the buffeting loads in transonic regime. Based on upwind-biased spatial differencing, in-house code solves the Unsteady Reynolds Averaged Navier-Stokes equations (URANS) by using a three-factor, implicit, finite-volume algorithm. Two turbulence models are available: the equilibrium, algebraic, eddy-viscosity B-L model, and the $k - \omega$ SST model. This inner CFD code has been verified by a large number of numerical tests with good accuracy. The verifications will not repeat here. For the identification of shock buffeting onset, the steady computations are performed at first to solve the RANS equations with $k - \omega$ SST turbulence model. The distinct change in the variations of lift will indicate the onset of transonic buffet. For the critical points where the variation of lift has no significant change, the unsteady RANS computations will capture the reversal in the shock movement to identify the buffeting onset and then determine the time history of stochastic buffeting loads. The unsteady RANS solver can capture the strong discontinuity caused by the shock movement and can simulate the flow separation of the boundary layer due to adverse pressure gradient.

The convective and pressure terms are differenced using the upwind flux-difference splitting technique of Roe [47]. Unsteady computation of RANS equations is advanced in time with an LU-SGS implicit integrator. The multiple structured grids and parallel computing based on MPI mode are implemented to enhance the computing speed.

### 6.10.3 CFD/CSD Data Interface

The Constant Volume Tetrahedron (CVT) Method, a simple geometry based method, is implemented in current work to transfer data between CFD and CSD, such as force and displacement information of coupling boundary. The basic idea behind CVT method is the building of tetrahedron. For example, each aerodynamic grid point $\underline{x}_a$ together with the closest three structural nodes $\underline{x}_{si}$ for $i = 1, 2, 3$ assembles a tetrahedron. The relative vectors $\underline{a} = \underline{x}_{s3} - \underline{x}_{s1}$, $\underline{b} = \underline{x}_{s2} - \underline{x}_{s1}$ and the normal vector $\underline{c} = \tilde{\underline{a}}\underline{b}$ are selected to build a local coordinate system. The aerodynamic grid point can project into this local frame

$$\underline{c} = \underline{x}_a - \underline{x}_{s1} = \alpha\underline{a} + \beta\underline{b} + \gamma\underline{d} \tag{6.51}$$

where $\alpha$, $\beta$ and $\gamma$ are the scalars. The projection of grid point $\underline{x}_a$ onto the structural surface writes

$$\underline{x}_p = \underline{x}_{s1} + \alpha\underline{a} + \beta\underline{b} = \sum_{i=1}^{3} \phi_i \underline{x}_{si} \tag{6.52}$$

where $\phi_1 = 1 - \alpha - \beta, \phi_2 = \alpha, \phi_3 = \beta$. Assuming the interpolation factors $\phi_i$ become constant, the displacement of projected point $\underline{x}_p$ on the surface of structure can be interpolated from

$$\underline{u}_p = \sum_{i=1}^{3} \phi_i \underline{u}_{si} \tag{6.53}$$

The constant quantities $\alpha$ and $\beta$ are determined from

$$\alpha = \frac{\underline{b}^T\underline{b}\,\underline{a}^T\underline{c} - \underline{a}^T\underline{b}\,\underline{b}^T\underline{c}}{\underline{a}^T\underline{a}\,\underline{b}^T\underline{b} - \underline{a}^T\underline{b}\,\underline{a}^T\underline{b}}, \quad \beta = \frac{\underline{a}^T\underline{a}\,\underline{b}^T\underline{c} - \underline{a}^T\underline{b}\,\underline{a}^T\underline{c}}{\underline{a}^T\underline{a}\,\underline{b}^T\underline{b} - \underline{a}^T\underline{b}\,\underline{a}^T\underline{b}} \tag{6.54}$$

The interpolation of grid point $\underline{x}_a$ needs to be determined by the distance $\gamma\underline{d}$ between $\underline{x}_a$ and its projection $\underline{x}_p$. The initial value of parameter $\gamma$ can be written as

$$\gamma_0 = \frac{\underline{c}_0^T \underline{d}_0}{\underline{d}_0^T \underline{d}_0} \tag{6.55}$$

For deformed configuration, it was advised [73] to compute $\gamma$ using the following formulation to guarantee the volume of the tetrahedron assembled by vectors $\underline{a}$, $\underline{b}$ and $\underline{c}$ keeps unchanged

$$\gamma = \frac{\underline{d}_0^T \underline{d}_0}{\underline{d}^T \underline{d}} \gamma_0 \tag{6.56}$$

### *6.10.4 Buffeting Response of M6 Wing*

The turbulent flow over the ONERA M6 wing [74] is a classic test case, both experimental and numerous computational results [75] are available. The dynamic response of ONEAR M6 wing under the buffeting loads is predicted to validate the proposed prediction procedure as depicted by Fig. 6.32. Following the proposed procedure, the buffeting onset for M6 wing is identified by the distinct change in the variations of lift at first, then unsteady aerodynamic pressures are integrated over the entire surface of the wing after solving the unsteady RANS equations with $k - \omega$ SST turbulent model. The next step is the prediction of the wing dynamic response to the stochastic excitation generated by the aerodynamic pressure distribution. For structural modeling, two different approaches are available. One approach is using the shear deformable shell elements developed in Chap. 3, the other is the shell element S4R in ABAQUS. Finally, applying the aerodynamic pressure to the finite element model yields the desired stress and strain components at any point within the M6 wing structure.

For the steady/unsteady RANS simulation, the external flow field around the ONERA M6 wing has been meshed into a 3D block with structured grids of total number $289 \times 65 \times 49$, and especially the mesh size on the M6 wing surface sets as $6 \times 10^{-6}$. The spatial grids, the meshing conditions near the wing, and also the wall grids are depicted in Fig. 6.33. A bunch of steady computations with different initial flow conditions have been performed for the identification of buffeting onset. As summarized in Table 6.4, a set of angles of attack, $\alpha$, are selected for each Mach number to predict the curves of lift and moment coefficients. Finally, a large amount of RANS simulations have been completed with a total of 152 different initial conditions. As depicted in the Fig. 6.34, the curves of lift and moment coefficients for Mach number, $Ma = 0.85$, have been predicted through the above steady RANS simulations. Figure 6.34 clearly shows that the lift coefficient, $C_L$, keep increasing as the angle of attack increases initially. When the angle of attack reaches 3.75°, the slopes of $C_L$ and $C_M$ curves feature distinct change. This distinct change indicates the buffeting onset of ONERA M6 wing at Mach number, $Ma = 0.85$. The same works are repeated to identify the buffeting onset for all the Mach numbers in Table 6.4. All

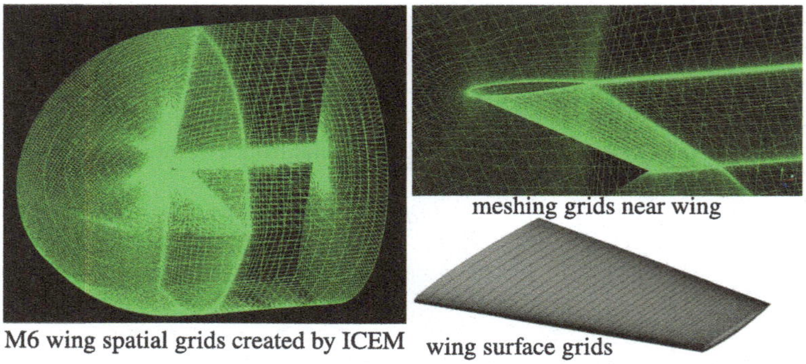


**Fig. 6.33** M6 wing spatial grids created by ICEM for CFD simulation, the meshing near the wing and wing surface meshing are presented

**Table 6.4** Input conditions for steady RANS simulations

| No. | Mach number | Angle of attack |
|---|---|---|
| 1 | 0.50 | 1.0,1.5,2.0,2.5,3.0,3.5,4.0,4.5,5.0,5.5,6.0,6.5,7.0,7.5,8.0,8.5,9.0,9.5,10.0 |
| 2 | 0.55 | 1.0,1.5,2.0,2.5,3.0,3.5,4.0,4.5,5.0,5.5,6.0,6.5,7.0,7.5,8.0,8.5,9.0 |
| 3 | 0.60 | 1.0,1.5,2.0,2.5,3.0,3.5,4.0,4.5,5.0,5.5,6.0,6.5,7.0,7.5,8.0,8.5,9.0 |
| 4 | 0.65 | 1.0,1.5,2.0,2.5,3.0,3.5,4.0,4.5,5.0,5.5,6.0,6.5,7.0,7.5,8.0,8.5 |
| 5 | 0.70 | 1.0,1.5,2.0,2.5,3.0,3.5,4.0,4.5,5.0,5.5,6.0,6.5,7.0,7.5,8.0 |
| 6 | 0.75 | 1.0,1.5,2.0,2.5,3.0,3.5,4.0,4.5,5.0,5.5,6.0,6.5,7.0,7.5 |
| 7 | 0.80 | 1.0,1.5,2.0,2.5,3.0,3.5,4.0,4.5,5.0,5.5,6.0,6.5,7.0,7.5 |
| 8 | 0.84 | 1.0,1.5,2.0,2.5,3.0,3.5,4.0,4.5,5.0,5.5,6.0,6.5,7.0,7.5 |
| 9 | 0.85 | 1.0,1.5,2.0,2.5,3.0,3.5,4.0,4.5,5.0,5.5,6.0,6.5,7.0,7.5 |
| 10 | 0.9 | 1.0,1.5,2.0,2.5,3.0,3.5,4.0,4.5,5.0,5.5,6.0,6.5,7.0 |

the identified values of angles of attack are then fitted to obtain the curve of buffeting onset as shown in Fig. 6.35.

Once the buffeting onset has been identified, the unsteady RANS simulation is performed to obtain the unsteady aerodynamic pressures over the entire surface of the ONERA M6 wing. A bunch of buffeting loads required to be determined along the buffeting onset. In current implementation, only the buffeting load is predicted, as described in Fig. 6.36, for a specific case to validate the designed procedure. Figure 6.36 takes snapshots of unsteady pressure distributions at an angle of attack of 5.5° with a Mach number of $M_a = 0.80$ in 9 s. Apparent shock movement and flow separation of the boundary layer have been observed.

For structural modeling, the first approach discretize the ONERA M6 wing into $6 \times 12$ shell elements, as depicted in Fig. 6.37. The second approach discretizes the wing into $12 \times 24$ S4R shell elements by using ABAQUS to validate the shell element developed in Chap. 3. The swept wing has a semi-span of 1000 mm with

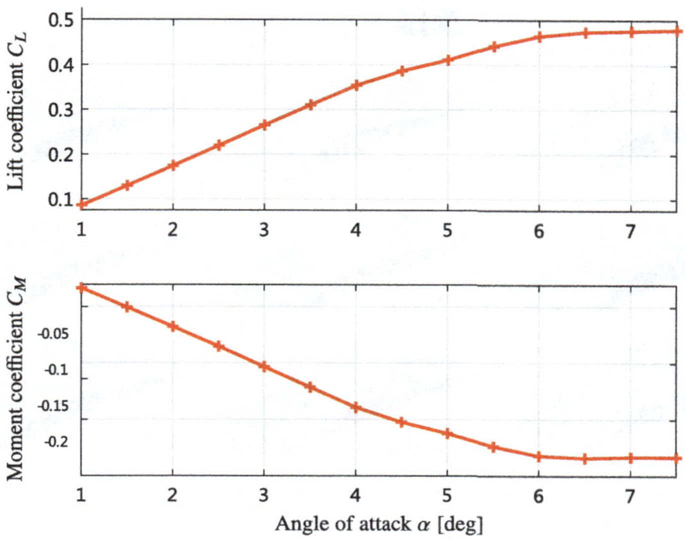


**Fig. 6.34** ONERA M6 wing lift and moment coefficient curves for Mach number $Ma = 0.85$

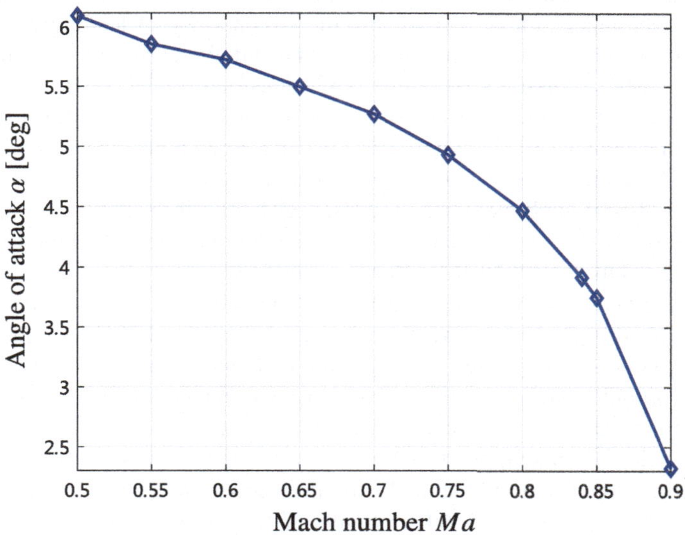


**Fig. 6.35** Buffeting boundary for M6 wing identified by using the lift coefficient method

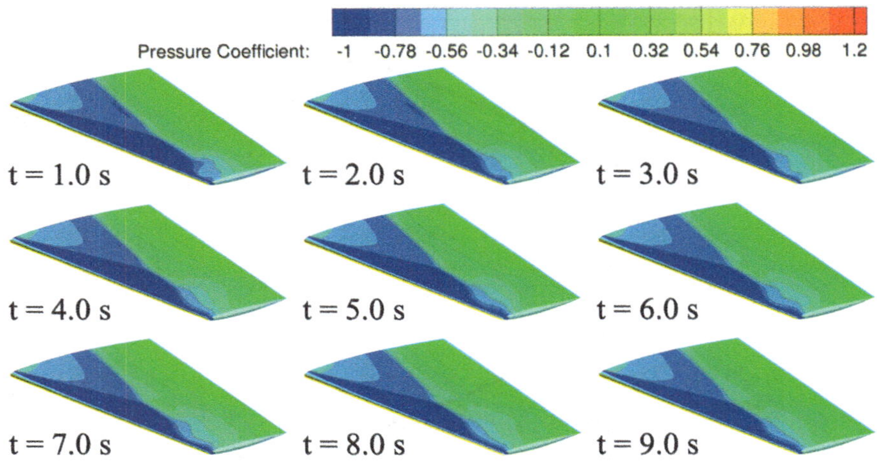


**Fig. 6.36** Unsteady pressure on the surface of M6 wing for Mach number $M_a = 0.80$ and AOA = 5.5°

no twist, and a symmetric airfoil using the ONERA D section. The aspect ratio is $A = 3.8$, the taper ratio $\lambda = 0.56$, and the sweep angle of leading edge is $\Lambda = 30°$. The following structural properties have been defined: Young's modulus, $E = 7.102 \times 10^4$ MPa, Poisson's ratio $\nu = 0.32$, density $\rho = 2.7 \times 10^{-9}$ tonne/mm$^3$. The eigenvalue problem has been solved by using Arnoldi algorithm to compute the natural frequencies of the wing. The first four eigenmodes have been shown in Fig. 6.38. These eigenmodes have been compared with the results of ABAQUS by using 12 × 24 S4R shell elements. Table 6.5 gives the frequencies results of both structural models. The small relative errors indicate both numerical models predict the dynamic behavior of ONERA M6 wing correctly.

The surface nodes of CFD grids and the finite element nodes of the Onera M6 wing are plotted in Fig. 6.39. The figure apparently shows that CFD grids used for the fluid dynamics computation and the finite element grids used for the structural dynamics computation do not match. An interpolation technique based on CVT has been implemented to convert the aerodynamic pressure over the surface of the wing to the concentrated force applied to the structural finite element nodes, as shown in Fig. 6.39. The unsteady aerodynamic loads applied to finite element nodes 7 and 153 with the positions (781.5, 958.3, 0) mm and (727.3, 500, 0) mm, respectively, are presented in Fig. 6.40. For buffeting response prediction, the first approach applies 2-stage Radau IIA algorithm [15] to integrate the dynamic problem of shell element model. The second approach uses H.H.T. algorithm to solve the dynamic problem of S4R model in ABAQUS. The time histories of transverse displacement and velocity of the node 7 with position (781.5, 958.3, 0) mm, by using these two approaches are described in Figs. 6.41 and 6.42. The perfect curve fitting has been observed.

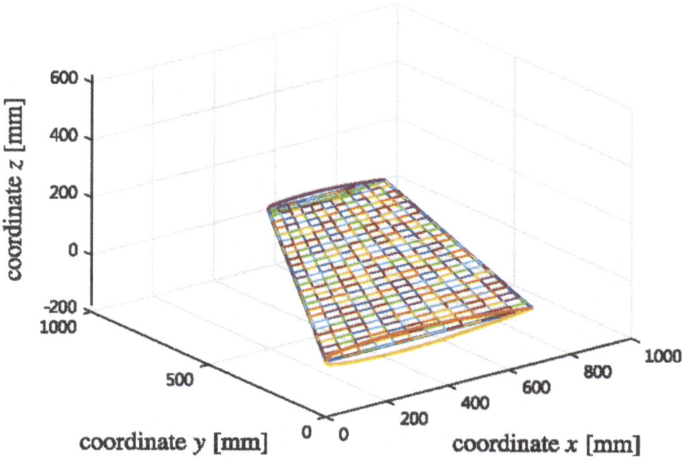


**Fig. 6.37** Structural finite element discretization of Onera M6 wing

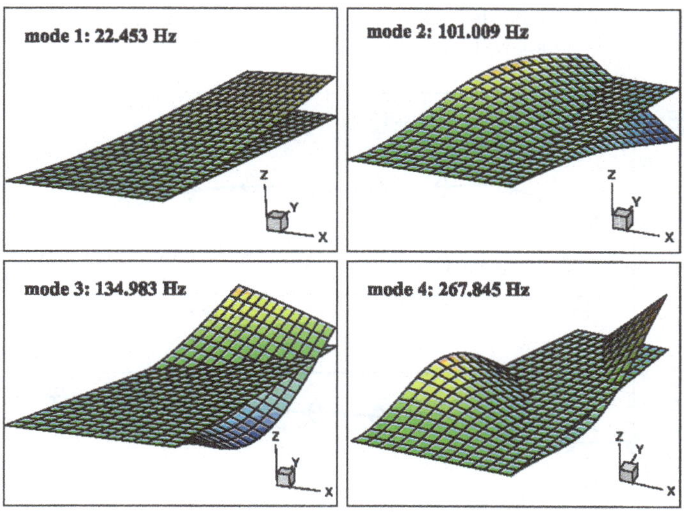


**Fig. 6.38** The first four eigenmodes of structural finite element model for Onera M6 wing

**Table 6.5** The frequencies (Hz) for ONERA M6 wing

| No | shell | S4R | Relative error (%) |
|---|---|---|---|
| 1 | 22.453 | 22.435 | 0.08 |
| 2 | 101.009 | 100.95 | 0.06 |
| 3 | 134.983 | 135.20 | −0.16 |
| 4 | 267.845 | 268.87 | −0.38 |

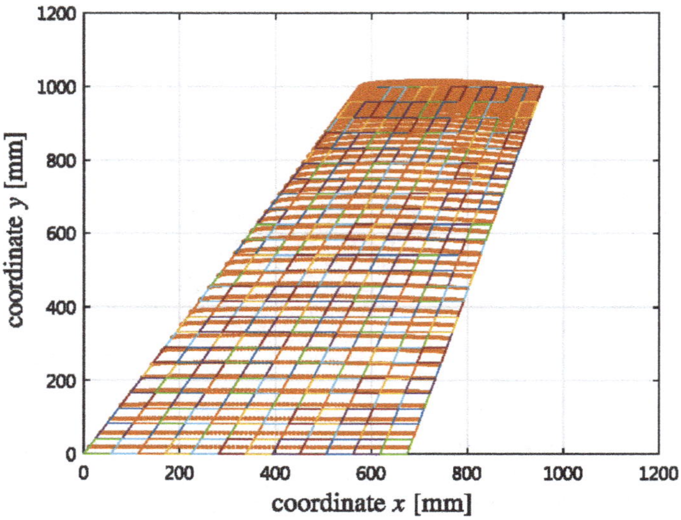


**Fig. 6.39** Surface grids for the fluid dynamics computation and structural finite element grids of Onera M6 wing. CFD surface grids: dots, Finite element grids: solid line

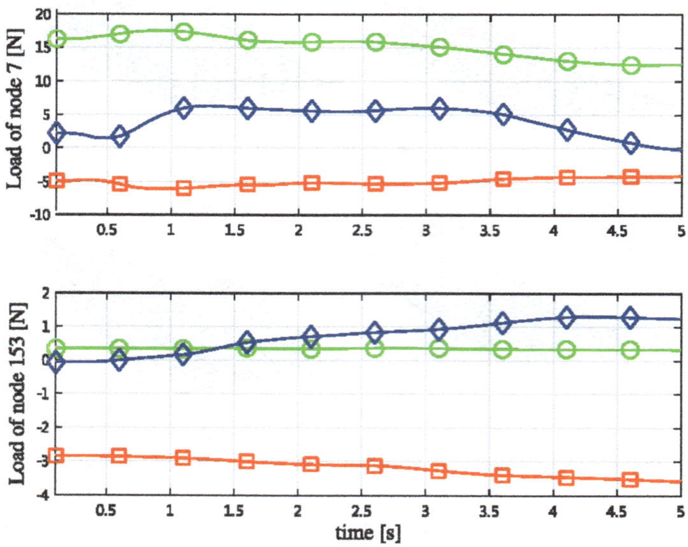


**Fig. 6.40** The unsteady aerodynamic loads applied to finite element node 7 and node 153 with the positions (781.5, 958.3, 0) mm and (727.3, 500, 0) mm, respectively. $F_x$: □, $F_y$: ∘, $F_z$: ◊

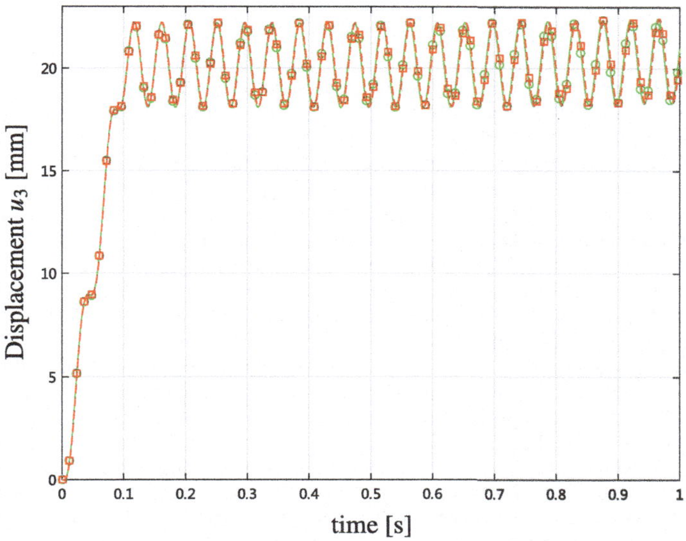


**Fig. 6.41** The time history of transvers displacement of node 7 with position (781.5, 958.3, 0) mm. INSH result: solid line with ∘, S4R results: dashed line with □

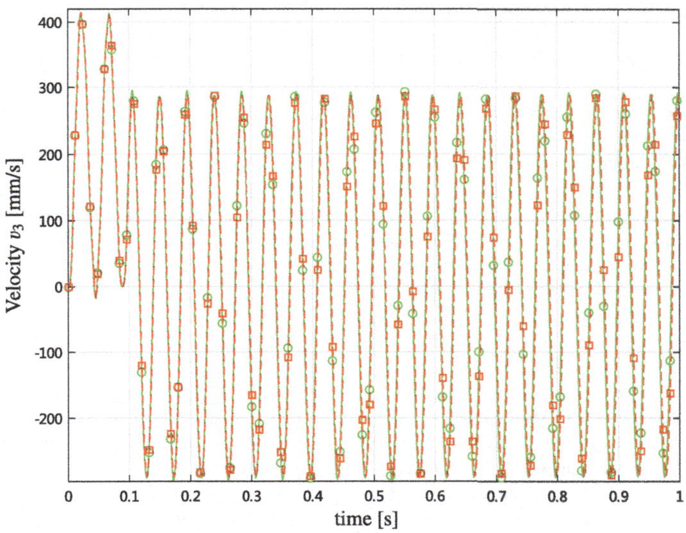


**Fig. 6.42** The time history of transvers velocity of node 7 with position (781.5, 958.3, 0) mm. INSH result: solid line with ∘, S4R results: dashed line with □

## References

1. Arnodl, M., Brüls, O.: Convergence of the generalized-$\alpha$ scheme for constrained mechanical systems. Multibody Syst. Dyn. (2007). https://doi.org/10.1007/s11044-007-9084-0
2. Brüls, O., Cardona, A.: On the use of Lie group time integrators in multibody dynamics. J. Comput. Nonlinear Dyn. (2010). https://doi.org/10.1115/1.4001370
3. Cash, J.: Review paper. Efficient numerical methods for the solution of stiff initial-value problems and differential algebraic equations. In: Proceedings: Mathematical, Physical and Engineering Sciences (2003). https://doi.org/10.1098/rspa.2003.1130
4. Wang, J., Rodriguez, H., Keribar, R.: Integration of flexible multibody systems using Radau IIA algorithms. J. Comput. Nonlinear Dyn. (2010). https://doi.org/10.1115/1.4001907
5. Cardona, A., Gerardin, M.: Time integration of the equations of motion in mechanism analysis. Comput. Struct. **33**, 801–820 (1989)
6. Bauchau, O., Laulusa, A.: Review of contemporary approaches for constraint enforcement in multibody systems. J. Comput. Nonlinear Dyn. (2008). https://doi.org/10.1115/1.4001907
7. Baumgarte, J.: Stabilization of constraints and integrals of motion in dynamic systems. Comput. Methods Appl. Mech. Eng. **1**, 1–16 (1972)
8. Schaub, M., Simeon, B.: Automatic $h$-scaling for the efficient time integration of stiff mechanical systems. Multibody Syst. Dyn. (2002). https://doi.org/10.1023/A:1020973630828
9. Hairer, E., Wanner, G.: Solving Ordinary Differential Equations II: Stiff and Differential-Algebraic Problems. Springer, New York (1996)
10. Shampine, L., Reichelt, M.: The MATLAB ODE suite. SIAM J. Sci. Comput. (1997). https://doi.org/10.1115/1.4001907
11. Hairer, E., Wanner, G.: Stiff differential equations solved by Radau methods. J. Comput. Appl. Math. (1999). https://doi.org/10.1016/S0377-0427(99)00134-X
12. Brown, P.N., Hinmarsh, A.C., Petzold, L.R.: Using Krylov methods in the solution of large-scale differential-algebraic systems. SIAM J. Sci. Comput. (1994). https://doi.org/10.1137/0915088
13. Bauchau, O.: Flexible Multibody Dynamics. Springer, Dordrecht, Heidelberg, London, New York (2011)
14. Brüls, O.: Integrated simulation and reduced-order modeling of controlled flexible multibody systems. PhD Thesis, Université de Liége, Liége, Belgium (2005)
15. Wang, J.: Application of Radau IIA algorithms to flexible multibody system with holonomic constraints. Nonlinear Dyn. (2017). https://doi.org/10.1007/s11071-017-3384-6
16. Han, S., Bauchau, O.A.: Manipulation of motion via dual entities. Nonlinear Dyn. (2016) https://doi.org/10.1007/s11071-016-2703-7
17. Epple, A.: Methods for increased computational efficiency of multibody simulations. PhD Thesis, Georgia Institute of Technology, School of Aerospace Engineering, Georiga, US (2008). See also http://www.gatech.edu
18. Shampine, L.F.: Error estimation and control for ODEs. J. Sci. Comput. (2005). https://doi.org/10.1007/s10915-004-4629-3
19. Bauchau, O.: Computational schemes for flexible, nonlinear multi-body systems. Multibody Syst. Dyn. (1998). https://doi.org/10.1023/A:1009710818135
20. Wang, J., Li, Z.: Implementation of HHT algorithm for numerical integration of multibody dynamics with holonomic constraints. J. Nonlinear Dyn. (2015). https://doi.org/10.1007/s11071-015-1908-5
21. Dufva, K.E., Sopanen, J.T., Mikkola, A.M.: A two-dimensional shear deformable beam element based on the absolute nodal coordinate formulation. J. Sound Vib. **280**, 719–738 (2005)
22. Simo, J.C., Quoc, L.V.: A three-dimensional finite-strain rod model. part II: computational aspects. Comput Methods Appl. Mech. Eng. **58**, 79–116 (1986)
23. Timoshenko, S., Young, D.H., Rao, J.V., Pati, S.: Engineering Mechanics. McGraw Hill Education (2020)
24. Myhra, D.: The Horten Brothers and Their All-Wing Aircraft. Schiffer Publ, PA, Sept, Schiffer Military/Aviation History (1998)

25. Gyorgy-Falvy, D.: Performance Analysis of the "Horten IV" Flying Wing. Mississippi State University State College, Department of Aerophysics, AD-480 202/1, Starkville, MS (1960)
26. Wang, J.: Modified models for revolute joints coupling flexibility of links in multibody systems. Multibody Syst. Dyn. (2019). https://doi.org/10.1007/s11044-018-9616-9
27. Hooker, J.R., Burner, A.W., Valla, R.: Static Aeroelastic Analysis of Transonic Wind Tunnel Models Using Finite Element Methods, AIAA Paper, 97-2243 (1997)
28. Guruswamy, G.P.: Coupled Finite-Difference/Finite-Element Approach for Wing-Body Aeroelasticity, AIAA Paper, 92-4680 (1992)
29. Bendiksen, O.O.: A New Approach to Computational Aeroelasticity, AIAA Paper, 91-0939 (1991)
30. Felker, F.F.: A New Method for Transonic Static Aeroelastic Problems, AIAA Paper 92-2123 (1992)
31. Bauchau, O.A., Ahmad, J.U.: Advanced CFD and CSD Methods for Multidisciplinary Applications in Rotorcraft Problems, AIAA Paper, 96-4151 (1996)
32. Dunne, T., Rannacher, R., Richter, T.: Numerical simulation of fluid-structure interaction based on monolithic variational formulations. In: Galdi, G.P., Rannacher, R. (eds.) Fundamental Trends in Fluid-Structure Interaction, Contemporary Challenges in Mathematical Fluid Dynamics and its Applications, pp. 1–75. World Scientific, Singapore (2010)
33. Hoffmana, J., Janssona, J., Janssona, N.: Simulation of 3D Unsteady Incompressible Flow Past a NACA 0012 Wing Section, KTH-CTL-4023 (2012)
34. Guruswamy, G.P.: Vortical flow computations on Swept flexible wings using Navier-Stokes equations. AIAA J. (1990). https://doi.org/10.2514/3.10524
35. Bond, V.L., Canfield, R.A., Suleman, A., Blair, M.: Aeroelastic scaling of a joined wing for nonlinear geometric stiffness. AIAA J. (2012). https://doi.org/10.2514/1.41139
36. Prananta, B.B., Meijer, J.J., Muijden, J.V.: Static Aeroelastic Simulation Using CFD, Comparison with Linear Method, NLR-TP-2003-530 (2003)
37. Bartels, R.E.: Development of Advanced Computational Aeroelasticity Tools at NASA Langley Research Center, AVT-154-003 (2008)
38. Silva, W.A., Vatsa, V.N., Biedron, R.T.: Development of Unsteady Aerodynamic and Aeroelastic Reduced-Order Models Using the FUN3D Code, IFASD 2009-030 (2009)
39. Keye, S., Rudnik, R.: Aeroelastic effects in maximum lift prediction of a transport aircraft and comparison to flight data. In: 32nd AIAA Applied Aerodynamics Conference, pp. 1–11. AIAA, Washington, DC (2014)
40. Keye, S., Rudnik, R.: Validation of wing deformation simulations for the NASA CRM model using fluid-structure interaction computations. In: 53rd AIAA Aerospace Sciences Meeting, pp. 1–13. AIAA, Washington, DC (2015)
41. Cella, U., Biancolini, M.E.: Aeroelastic analysis of aircraft wind-tunnel model coupling structural and fluid dynamic codes. J. Aircr. (2012). https://doi.org/10.2514/1.C031293
42. Biancolini, M.E., Evangelos, M., Cella, U., Groth, C., Genta, M.: Static aeroelastic analysis of an aircraft wind-tunnel model by means of modal RBF mesh updating. J. Aerosp. Eng. (2016). https://doi.org/10.1061/(ASCE)AS.1943-5525.0000627
43. Bdeiwi, H., Ciarella, A., Peace, A., Hahn, M.: Model structure effect on static aeroelastic deformation of the NASA CRM. Int. J. Numer. Methods Heat Fluid Flow (2019). https://doi.org/10.1108/HFF-07-2018-0352
44. Rendall, T.C.S., Allen, C.B.: Unified fluid-structure interpolation and mesh motion using radial basis functions. Int. J. Numer. Methods Eng. (2008). https://doi.org/10.1002/nme.2219
45. Rendall, T.C.S., Allen, C.B.: Efficient mesh motion using radial basis functions with data reduction algorithms. J. Comput. Phys. (2009). https://doi.org/10.1016/j.jcp.2009.05.013
46. Rendall, T.C.S., Allen, C.B.: Reduced surface point selection options for efficient mesh deformation using radial basis functions. J. Comput. Phys. (2010). https://doi.org/10.1016/j.jcp.2009.12.006
47. Roe, P.L.: Approximate Riemann solvers, parameter vectors, difference schemes. J. Comput. Phys. (1981). https://doi.org/10.1016/0021-9991(81)90128-5

48. Iovnovich, M., Raveh, D.E.: Reynolds-averaged Navier-Stokes study of the shock-Buffet instability mechanism. AIAA J. (1986). https://doi.org/10.2514/1.J051329
49. Timme, S., Sartor, F.: Passive control of transonic Buffet onset on a half wing-body configuration. In: Proceedings of the 16th International Forum on Aeroelasticity and Structural Dynamics, IFASD-2015-088, vol. 1, pp. 1–15, St Petersburg, Russia (2015)
50. Kenway, G.K., Martins, J.R.: Buffet-onset constraint formulation for aerodynamic shape optimization. AIAA J. (2007). https://doi.org/10.2514/1.J055172
51. Vos, R., Farokhi, S.: Introduction to Transonic Aerodynamics, 2nd edn. Springer, New York (2015)
52. McDevitt, J.B., Levy, L.L.J., Deiwert, G.S.: Transonic flow about a thick circular-arc airfoil. AIAA J. (1976). https://doi.org/10.2514/3.61402
53. Levy, L.L.J.: Experimental and computational steady and unsteady transonic flows about a thick airfoil. AIAA J. (1978). https://doi.org/10.2514/3.60935
54. Lee, B.H.K.: Transonic buffet on a supercritical aerofoil. Aeronaut. J. (1990). https://doi.org/10.1017/S0001924000022752
55. Lee, B.H.K.: Self-sustained shock oscillations on airfoils at transonic speeds. Progr. Aerosp. Sci. (2001). https://doi.org/10.1016/S0376-0421(01)00003-3
56. Hartmann, A., Feldhusen, A., Schröder, W.: On the interaction of shock waves and sound waves in transonic buffet flow. Phys. Fluids (2001). https://doi.org/10.1063/1.4791603
57. Farhangnia, M., Guruswamy, G., Biringgen, S.: On the interaction of shock waves and sound waves in transonic buffet flow. In: 34th Aerospace Sciences Meeting and Exhibit, Aerospace Sciences Meetings, pp. 1–8. AIAA, Reno, NV (1996)
58. Lee, B.H.K.: A method for predicting wing response to buffet loads. J. Aircr. (1984). https://doi.org/10.2514/3.48227
59. Bartels, R.E.: Flow and Turbulence Modeling and Computation of Shock Buffet Onset for Conventional and Supercritical Airfoils, NASA/TP-1998-206908 (1998)
60. Chung, I., Lee, D., Reu, T.: Prediction of transonic Buffet onset for an airfoil with shock induced separation bubble using steady Navier-Stokes solver. In: 20th AIAA Applied Aerodynamics Conference, pp. 1–8. AIAA, St. Louis, Missouri (2002)
61. Gillan, M.A.: Navier-Stokes simulation of self-excited shock induced oscillations. In: 13th Applied Aerodynamics Conference, pp. 1–8. AIAA, San Diego, CA (1995)
62. Goncalves, E., Houdeville, R.: Turbulence model and numerical scheme assessment for buffet computations. Int. J. Numer. Methods Fluids (2004). https://doi.org/10.1002/fld.777
63. Garnier, E., Deck, S.: Large-Eddy simulation of transonic Buffet over a supercritical airfoil. Direct and Large-Eddy Simulation VII. In: Proceedings of the Seventh International ERCOFTAC Workshop on Direct and Large-Eddy Simulation, pp. 549–554. Springer, Dordrecht, The Netherlands (2010)
64. Fukushima, Y., Kawai, S.: Wall-modeled large-eddy simulation of transonic buffet over a supercritical airfoil at high Reynolds number. Direct and Large-Eddy Simulation VII. In: Proceedings of the Seventh International ERCOFTAC Workshop on Direct and Large-Eddy Simulation, pp. 1–15. AIAA, Grapevine, Texas (2017)
65. Deck, S.: Numerical simulation of transonic Buffet over a supercritical airfoil. AIAA J. (2005). https://doi.org/10.2514/1.9885
66. Raveh, D., Dowell, E.: Frequency lock-in phenomenon for oscillating airfoils in buffeting flows. J. Fluids Struct. (2011). https://doi.org/10.1016/j.jfluidstructs.2010.10.001
67. Giannelis, N.F., Vio, G.A., Dimitriadis, G.: Dynamic interactions of a supercritical aerofoil in the presence of transonic shock buffet. In: Proceedings of the 27th International Conference on Noise and Vibration Engineering, Leuven, Belgium, pp. 1–15 (2016)
68. Sartor, F., Mettot, C., Sipp, D.: Stability, receptivity, sensitivity analyses of buffeting transonic flow over a profile. AIAA J. (2015). https://doi.org/10.2514/1.J053588
69. Barakos, G., Drikakis, D.: Numerical simulation of transonic Buffet flows using various turbulence closures. Int. J. Heat Fluid Flow (2000). https://doi.org/10.1016/S0142-727X(00)00053-9

70. Brunet, V.: Computational study of Buffet phenomenon with unsteady RANS equations. In: 21st AIAA Applied Aerodynamics Conference, pp. 1–8. AIAA, Orlando, Florida (2003)
71. Deck, S.: Numerical simulation of transonic Buffet over the OAT15A airfoil. AIAA J. (2005). https://doi.org/10.2514/1.9885
72. Thiery, M., Coustols, E.: Numerical prediction of shock induced oscillations over a 2-D airfoil: influence of turbulence modelling and test section walls. Int. J. Heat Fluid Flow (2006). https://doi.org/10.1016/j.ijheatfluidflow.2006.02.013
73. Goura, G.S.L., Badcock, K.J., Woodgate, M.A., Richards, B.E.: Transformation methods for the time marching analysis of flutter. AIAA J. (2001). https://doi.org/10.1016/j.ijheatfluidflow.2006.02.013
74. ONERA-M6 Wing, Star of CFD. https://www.onera.fr/en/news/onera-m6-wing-star-of-cfd (2013)
75. Mani, M., Ladd, J.A., Cain, A.B., Bush, R.H.: An assessment of one- and two-equation turbulence models for internal and external flows. In: 28th Fluid Dynamics Conference, pp. 1–8. AIAA, Snowmass Village, CO (1997)

# Correction to: Vector and Tensor

## Correction to:
## Chapter 1 in: J. Wang, *Multiscale Multibody Dynamics*, https://doi.org/10.1007/978-981-19-8441-9_1

In the original version of the book, the following belated correction has been incorporated in Chapter 1: Figure 1.10 has been replaced with revised figure. The correction chapter and the book has been updated with the changes.

---

The updated version of this chapter can be found at
https://doi.org/10.1007/978-981-19-8441-9_1



J. Wang, *Multiscale Multibody Dynamics*,
https://doi.org/10.1007/978-981-19-8441-9_7

# Index




J. Wang, *Multiscale Multibody Dynamics*,
https://doi.org/10.1007/978-981-19-8441-9_1

The manufacturer's authorised representative in the EU is Springer Nature Customer Service Centre GmbH, Europaplatz 3, 69115 Heidelberg, Germany. If you have any concerns regarding our products, please contact ProductSafety@springernature.com

Printed and bound by CPI Group (UK) Ltd, Croydon, CR0 4YY
30/09/2026
02225242-0015